The Theory of Chemical Reaction Dynamics

NATO ASI Series

Advanced Science Institutes Series

A series presenting the results of activities sponsored by the NATO Science Committee, which aims at the dissemination of advanced scientific and technological knowledge, with a view to strengthening links between scientific communities.

The series is published by an international board of publishers in conjunction with the NATO Scientific Affairs Division

A Life Sciences B Physics	Plenum Publishing Corporation London and New York
C Mathematical and Physical Sciences	D. Reidel Publishing Company Dordrecht, Boston, Lancaster and Tokyo
D Behavioural and Social Sciences E Engineering and Materials Sciences	Martinus Nijhoff Publishers The Hague, Boston and Lancaster
F Computer and Systems Sciences G Ecological Sciences	Springer-Verlag Berlin, Heidelberg, New York and Tokyo

Series C: Mathematical and Physical Sciences Vol. 170

The Theory of Chemical Reaction Dynamics

edited by

D. C. Clary

University Chemical Laboratory,
University of Cambridge, U.K.

D. Reidel Publishing Company

Dordrecht / Boston / Lancaster / Tokyo

Published in cooperation with NATO Scientific Affairs Division

Proceedings of the NATO Advanced Research Workshop on
The Theory of Chemical Reaction Dynamics
Orsay, France
June 17-30, 1985

Library of Congress Cataloging in Publication Data

NATO Advanced Research Workshop on the Theory of Chemical Reaction Dynamics (1985:
Cambridge, Cambridgeshire)
The theory of chemical reaction dynamics.

(NATO ASI series. Series C, Mathematical and physical sciences; vol. 170)
"Sponsored by the NATO Science Committee"
"Published in cooperation with NATO Scientific Affairs Division."
Includes index.
1. Chemical reaction, Conditions and laws of—Congresses. I. Clary, D. C.
II. NATO Science Committee. III. Title. IV. Series: NATO ASI series. Series C,
Mathematical and physical sciences; vol. 170.
QD501.N355 1985 541.3'9 86–451
ISBN 90–277–2202–1

Published by D. Reidel Publishing Company
P.O. Box 17, 3300 AA Dordrecht, Holland

Sold and distributed in the U.S.A. and Canada
by Kluwer Academic Publishers,
190 Old Derby Street, Hingham, MA 02043, U.S.A.

In all other countries, sold and distributed
by Kluwer Academic Publishers Group,
P.O. Box 322, 3300 AH Dordrecht, Holland

D. Reidel Publishing Company is a member of the Kluwer Academic Publishers Group

Printed in The Netherlands

CONTENTS

PREFACE

The calculation of cross sections and rate constants for chemical reactions in the gas phase has long been a major problem in theoretical chemistry. The need for reliable and applicable theories in this field is evident when one considers the significant recent advances that have been made in developing experimental techniques, such as lasers and molecular beams, to probe the microscopic details of chemical reactions. For example, it is now becoming possible to measure cross sections for chemical reactions state selected in the vibrational-rotational states of both reactants and products. Furthermore, in areas such as atmospheric, combustion and interstellar chemistry, there is an urgent need for reliable reaction rate constant data over a range of temperatures, and this information is often difficult to obtain in experiments. The classical trajectory method can be applied routinely to simple reactions, but this approach neglects important quantum mechanical effects such as tunnelling and resonances. For all these reasons, the quantum theory of reactive scattering is an area that has received considerable attention recently.

This book describes the proceedings of a NATO Advanced Research Workshop held at CECAM, Orsay, France in June, 1985. The Workshop concentrated on a critical examination and discussion of the recent developments in the theory of chemical reaction dynamics, with particular emphasis on quantum theories. Several papers focus on exact theories for reactions. Exact calculations on three-dimensional reactions are very hard to perform, but the results are valuable in testing the accuracy of approximate theories which can be applied, with less expense, to a wider variety of reactions. Indeed, critical discussions of the merits and defects of approximate theories, such as sudden, distorted-wave, reduced dimensionality and transition-state methods, form a major part of the book. The theories developed for chemical reactions have found useful extensions into other areas of chemistry and physics. This is illustrated by papers describing topics such as photodissociation, electron-scattering, molecular vibrations and collision-induced dissociation. Furthermore, the important topic of how to treat potential energy surfaces in reaction dynamics calculations is also discussed.

The articles demonstrate that substantial progress has been made in chemical reaction theory in recent years, and they also show that the field is likely to develop at an even faster pace in the future as we get closer to the goal of making accurate predictions from first principles for a wide range of chemical reactions.

We would like to acknowledge financial support for the Workshop from the NATO Scientific Affairs Division (Programme Director Dr Mario di Lullo) and also from Carl Moser at the CECAM Organization, where the Workshop was held.

D C Clary,
University Chemical Laboratory,
University of Cambridge.

RECENT QUANTUM SCATTERING CALCULATIONS ON THE H + H_2 REACTION AND ITS ISOTOPIC COUNTERPARTS

George C. Schatz*
Northwestern University
Department of Chemistry
Evanston, IL 60201 USA

This paper reviews recent developments in the theoretical description of the H + H_2 reaction and its isotopic counterparts. Both methods and applications are considered, with an emphasis on quantum dynamics applications in three dimensional reactive collisions. Among the methods discussed are coupled channel (CC) and coupled states (CS) reactive scattering methods, reduced dimensionality exact quantum (RDEQ) methods, infinite order sudden (IOS) methods, coupled channel distorted wave (CCDW) methods and approximations thereto, and one dimensional reaction path methods used in variational transition state theory (VTST). Applications discussed center on four topics: (1) integral cross sections and rate constants for H + H_2, D + H_2, H + HD/DH and Mu (muonium) + H_2; (2) vibrationally excited cross sections and rate constants for H + H_2 and D + H_2; (3) product state distributions in H + D_2, and (4) reactive scattering resonances in H + H_2. These applications generally indicate a high level of agreement between accurate theory and experiment, with several of the more approximate methods exhibiting nearly quantitative accuracy compared to CC or CS, and at a fraction of the computational effort. Some disagreement between theory and experiment still remains, however, with the rate constant for D + H_2 (ν=1) being the worst example in this regard. One of the most exciting new results in our review is that H + H_2 possesses an apparently rich spectrum of resonance states associated with excitation of the bend mode simultaneously with the symmetric stretch.

I. Introduction

The H + H_2 reaction has long served as a focal point for theoretical studies of gas phase chemical reaction dynamics. As the simplest of chemical reactions from the point of view of electronic structure, it has been the subject of numerous potential surface calculations (as reviewed in Ref. 1), and it is still the only reaction for which the potential surface is known to within a few tenths of a kcal/mol. From

*Camille and Henry Dreyfus Teacher-Scholar.

D. C. Clary (ed.), The Theory of Chemical Reaction Dynamics, 1–26.

a dynamics point of view it has provided an often used benchmark for testing exact and approximate theories. It can, in fact, claim to be the first reaction studied using collinear[2] and three dimensional[3,4] quasiclassical methods, the first studied by collinear[5] and three dimensional[6,7,8] accurate quantum scattering methods, the first reaction for which the thermal rate constant was calculated reliably from first principles to within better than a factor of two[9], and the first reaction for which resonances were discovered in collinear[10,11] and three dimensional[12] scattering calculations.

In view of all these "firsts", one might well question what remains to be done in research on H + H_2. The answer to this question, if based on the amount of activity recently devoted to this reaction, is evidently "plenty"! It is the purpose of this paper to review this recent activity in dynamical studies of H + H_2 and its isotopic counterparts, with an emphasis on topics where significant progress has been made or where important challenges remain to be answered. Although the primary emphasis will be on theoretical dynamical applications, computational methods (especially new ones) will be discussed, and the results of recent experiments will be highlighted.

There have been many reviews of H + H_2 and related topics, and it is useful to summarize them here so that the stage can be set for the more recent results that are of interest to this paper. The most extensive reviews devoted specifically to H + H_2 are two by Truhlar and Wyatt[1,13]. These reviews cover all aspects of the H_3 system up through 1976, including the potential surface, theoretical dynamical studies of the reactive, nonreactive and dissociative collision dynamics, and experimental studies. More recent reviews which consider H + H_2 as a special topic include reviews of reactive scattering by Walker and Light[14], by Schatz[15], and by Connor[16]; of variational transition state theory (VTST) by Truhlar and Garrett[17] (see Ref. 17 also for references to earlier VTST reviews); of reduced dimensionality exact quantum (RDEQ) dynamical approximations by Bowman[18]; and of infinite order sudden approximations (IOSA) by Baer[19]. Also, a monograph on resonances[20] has recently appeared, and there have been two monographs[21,22] which cover very wide areas in reactive collision processes.

The exclusive focus of this paper will be on <u>reactive</u> collisions involving H + H_2 and its isotopic counterparts in the gas phase. Within this area, we will first discuss the methods that are used to describe the reactive quantum dynamics, and then four specific applications: (1) calculations of ground state integral cross sections, thermal rate constants and isotope effects, (2) vibrationally excited cross sections and rate constants, (3) translationally hot (fast hydrogen atom) reactions, and (4) resonances. Space limitations preclude us from considering a few other topics where recent progress has been made, including studies of product angular distributions[23] and rate processes involving tritium atoms[24].

II. Theoretical Dynamics Methods

With the advent of accurate quantum methods[6-8] for describing the H + H_2 dynamics, the emphasis in methodology development has changed somewhat. Existing quantum reactive scattering methods are now being applied to isotopes of H + H_2 (H + D_2, D + H_2, Mu + H_2) and to other reactions (O + H_2, F + H_2), while new methods development is increasingly emphasizing either very efficient exact algorithms or very accurate approximate theories that can be generalized easily to much more complex reactions. Semiclassical methods are being used only in evaluating one dimensional tunnelling expressions, and the quasiclassical trajectory method is being used mainly for describing high energy collision processes including dissociation. In the following paragraphs the current state of the art in quantum dynamics methods is surveyed from the point of view of applications to 3D H + H_2.

At the top of the hierarchy of quantum dynamical methods are the coupled channel (CC) methods that involve solving the Schrodinger equation exactly by expanding the wavefunction in a basis set, propagating the resulting coupled equations for the expansion coefficients, matching solutions generated in different arrangement channels on appropriate matching surfaces, and applying scattering boundary conditions. The technology for doing these calculations is described in several places[14,25,26] and recent improvements[27] in this technology have extended the range of systems that can be studied to essentially all the isotopic variants of H_3. However, because of the rapid growth in the number of projection quantum numbers needed in the rotational basis as the energy is increased, it is unlikely that CC methods will be extensively applied to any system other than H + H_2.

The best approximation to CC is called coupled states or CS[28-30], and involves neglecting the kinematic couplings between the different projection quantum states. This approximation greatly reduces computer time so that reactions involving all the H_3 isotopes as well as a few other atom-diatom reactions are feasible. Tests of accuracy of the most recent versions of CS[30] indicate that errors of 25% or smaller in cross sections or rate constants compared to CC are to be expected for H + H_2 close to threshold. More accurate CS approximations have also been studied[29], but computational effort has generally made such methods prohibitively expensive to use. Recently, an approach which combines simplicity with physical accuracy has been proposed[31] but not yet tested.

Further simplifications upon CS involve approximations to the rotational motions. Here there are two schools of thought as to how best to do this. One school argues that since the rotational periods are usually slow compared to vibrational periods, it makes sense to use the rotational sudden approximation wherein the atom-diatom orientation angle is fixed for motion in the reagent and product arrangement channels[32,33]. The other school argues that since rotational motion correlates into bend motion along the reaction path, and the bend is only weakly coupled by curvature to reaction path motions, while at the same time the bend frequency is comparable to the other perpendicular modes near the reaction bottleneck, it is more appropriate to treat the bend as adiabatic, with correlation to the asymptotic rotational levels described statistically[18]. The first approach is generally called the "infinite order sudden approximation" (IOSA)[19], while the second has been labelled "reduced dimensionality exact quantum" (RDEQ) dynamics[18]. Both methods can be applied in a number of different ways, with the IOSA applications being distinguished by the mechanism for interrelating the reagent and product channel rotor orientation angles, and RDEQ being distinguished by precisely how the bend is taken to be adiabatic, and whether additional approximations in the treatment of the centrifugal potentials are included. For $H + H_2$ where the reaction path is associated with collinear H_3 geometries, the different applications of RDEQ considered to date have been termed CEQ[34], CEQB[35] and BCRLM[36]. In CEQ, a Collinear Exact Quantum calculation is done to determine a reduced dimensional transition probability. This probability is then shifted by the saddlepoint bending zero point energy to define an approximate three dimensional cumulative probability from which cross sections and rate constants can be calculated using a canonical statistical approximation in the treatment of centrifugal forces. CEQB incorporates the local bending energy into the collinear potential surface so that the calculated transition probability is directly a three dimensional cumulative probability. BCRLM (bending corrected rotating linear model) is (essentially[18]) a CEQB calculation with the centrifugal potential explicitly included.

The relative accuracy of the IOSA and RDEQ treatments of the reactions is still under active investigation[37]. Some of the applications we consider later will touch upon this, as will other chapters in this book. From these applications, we will see that the comparison is a rather strong function of what types of dynamical information are being considered, with the most noticeable differences arising in the study of product rotational distributions and reaction cross section threshold energies. For rotational distributions, the IOSA treats rotation explicitly while RDEQ assumes that they are microcanonical, while for energy thresholds, IOSA omits explicit incorporation of bending zero point energy contributions while RDEQ includes them directly.

Another approach for developing approximations to CC and CS reactive scattering calculations is to use distorted wave theory. In this approach, one considers that reaction is only a small perturbation on the nonreactive collision dynamics. As a result, the reactive scattering matrix can be approximated by the matrix element of a perturbative Hamiltonian operator using reagent and product nonreactive wavefunctions. Variations on this idea can be developed by using different approximations to the nonreactive wavefunctions. At the top of the hierachy of these methods is the coupled channel distorted wave (CCDW) method, followed by coupled states distorted wave (CSDW). Below this are a variety of reduced dimensionality approaches, including adiabatic distorted wave theory (treating either vibration (VADW), rotation (RADW) or both (VRADW), adiabatically) and at the bottom of the hierarchy is the frozen molecule distorted wave (FMDW) approximation. In recent years, applications of all of these distorted wave approaches have been made to H + H_2, with only CCDW[38], CSDW[38] and to a lesser extent VRADW[39] capable of providing quantitative cross section evaluations. VADW[40], RADW[41] and FMDW[42] give inaccurate cross section magnitudes, although the relative distributions of product states are often quite good. An improvement upon even CCDW can be obtained by determining the Green's function associated with the expression for the transition amplitude. A recent application to collinear H + H_2 of this approach was quite successful[43].

An important advantage of distorted wave methods over CC, CS, IOSA or RDEQ is the absence of coordinate system related kinematics problems in the determination of the reagent and product wavefunctions and their overlap. The reagent to product coordinate transformation process is confined to the S-matrix evaluation where there is often much flexibility since the integrand is in part a numerically determined function. This contrasts with matching based methods (in either full or reduced dimensions) where coordinates must be chosen to interrelate reagent and products smoothly. Of course, a big disadvantage of DWBA is its perturbative underpinnings. Generally it is found that CCDW and CSDW results are inaccurate if the total reaction probability exceeds 0.1. This restricts reliable use of these methods to low energies or temperatures. Recently it has been shown that for H + D_2 VADW can produce qualitatively accurate product rotational distributions even at high energies[44,45] where reaction is not a perturbation.

Another class of methods which has been used in studies of H + H_2 is based on the use of hyperspherical coordinates[46]. Use of these coordinates for collinear reactive scattering is described in several articles[47-51] as well as elsewhere in this volume. When applied in the CC sense, this approach provides an alternative to the reaction path based methods described above with the advantage of a simplified procedure for interrelating reagent and product coordinates. Although applications of this approach to 3D H + H_2 have not yet been published, it appears that results will be soon forthcoming[52] and when they do, it is likely that CC capabilities for studying systems other than H + H_2 will be substantially enhanced. CS, IOSA, RDEQ and DWBA versions of the hyperspherical approach can all be imagined but have yet to be implemented.

The last class of methods which shall be discussed is based on one dimensional "reaction path Hamiltonian"[53] methods. At its most fundamental level the reaction path concept simply involves the introduction of a certain coordinate system for representing the three body dynamics, and as such one can imagine applications using any of the methods (CS, CC, IOSA, RDEQ, DWBA) discussed above. A few applications of this sort have been considered[54] but by far the most extensive use of the reaction path concept has been for the purpose of calculating one dimensional transmission coefficients for variational transition state theory (VTST) applications. The concept of using a one dimensional barrier has, of course, been in widespread use for decades[1], but earlier theories in which curvature along the reaction path was neglected were found to be inaccurate for most reactions[55]. Recently, however, several methods have been developed for directly incorporating curvature into a one dimension description of the dynamics. Perhaps the first was the path of Marcus and Coltrin[56] (MCP) which follows the locus of outer vibrational turning points. This works well for $H + H_2$[56] and can be approximately justified based on more rigorous theory[57]. It does however run into artifacts with other reactions[58], but Truhlar, Garrett and coworkers have provided improved methods for generalizing the one dimensional path idea to a wide spectrum of reactions[58-61]. For reactions involving small curvature (i.e., light + heavy-light, L + HL), the small curvature tunnelling (SCT) approximation[58] using either semiclassical or quantum vibrationally adiabatic ground state (SAG or VAG) potential curves has proved to be as good as or better than MCP in many applications. For large curvature reactions (i.e., H + LH), the use of corner cutting paths is essential, as embodied in the large curvature ground state (LCG) method[59]. A method which searches families of paths for the one having the least action along the ground state adiabatic potential curves (LAG) has been developed recently[60] which smoothly interpolates between the SCT and LCG limits. When applied using the WKB approximation to generate the adiabatic curves, and combined with canonical or improved canonical variational transition state theory (CVT or ICVT, respectively)[61], this method can be used to generate VTST rate constants which are in excellent agreement with accurate quantal values[61]. The next Section will provide a number of applications and tests of the ICVT-WKB/LAG approximation.

III. Applications

A. Ground State Integral Cross Sections, Rate Constants and Isotope Effects.

In this Section we will attempt to answer the two important questions: "How well do we really know the integral reactive cross sections and rate constants for $H + H_2$ and its isotopic counterparts"; and "How accurate are the approximate dynamical methods in determining cross sections and rate constants". The question of potential surfaces will be addressed only in the sense of the accuracy of comparisons with experiment. For the most part, our discussion will emphasize results obtained using the so-called LSTH surface of Liu and Siegbahn[62], Truhlar and Horowitz[63]. Recent *ab initio* and quantum monte carlo studies[64-66] have confirmed the accuracy of this surface to within a few tenths of a kcal/mol. Since a large number of theoretical studies using the PK2 surface of Porter and Karplus[67] have been made, we will discuss results on it where comparable LSTH results are not available. Several theoretical studies[9] have demonstrated that the PK2 barrier (0.396 eV) is too low (the LSTH barrier is 0.425 eV), leading to rate constants that are noticeably too high.

$H + H_2$: The most accurate quantum dynamical studies done on $H + H_2$ on the LSTH surface have been CC calculations due to Walker *et al*[8] and CS calculations by Colton and Schatz[68]. Very few results from the Walker *et al* calculation have been published, but the comparison of integral cross sections with those from Colton and Schatz is quite good. If we define $Q_{\nu j}$ as the distinguishable atom cross section (summed over product arrangement channels) from initial vibration/rotation state (νj), then Q_{00} = 2.8 x 10^{-4} and 5.3 x 10^{-2} a_0^2 at total energies (relative to H + H_2 at equilibrium) E = 0.5 and 0.6 eV, respectively, from the Walker et al calculation, and Q_{00} = 2.4 x 10^{-4} and 4.2 x 10^{-2} a_0^2 at the same energies from Colton and Schatz. The observed agreement to within 20% is typical of previous comparisons between CC and CS calculations[27-29].

The CS cross sections from Colton and Schatz are plotted in Fig. 1

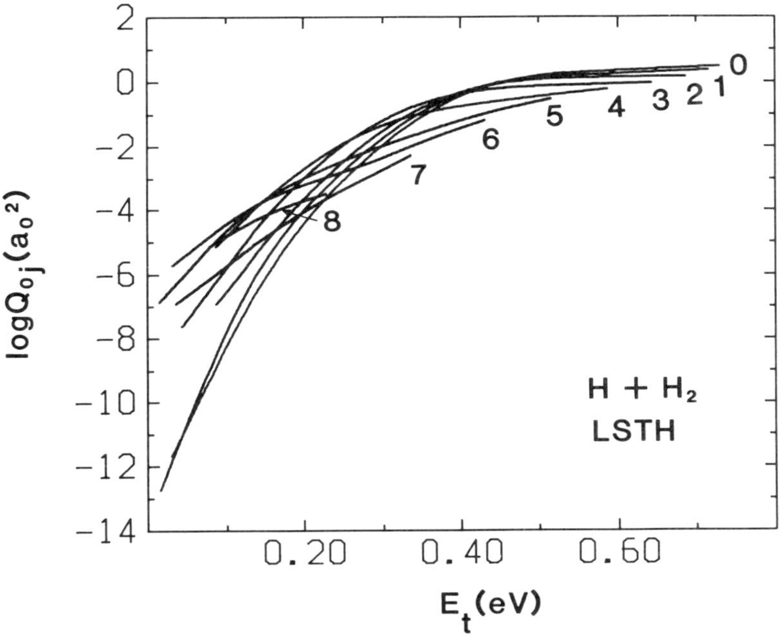

Figure 1. Logarithm of reactive cross section Q_{0j} versus translational energy E_t for H + H_2 on the LSTH potential surface. These results are from Ω = 0 CS calculations by Colton and Schatz[68].

(as a function of translational energy E_t) for (νj) = (00) to (08). A curious feature of the results is the strong dependence on initial rotational state, with rotational enhancement seen at low energies and low rotational quantum numbers, and rotational suppression at high energies or for any energy at high rotational quantum numbers. These CS results refer to the Ω = 0 body fixed projection quantum number. At higher energy (above E_t = 0.3 eV), the contributions from higher Ω's become important, and need to be included in order to compare with quasiclassical trajectory (QCT) results. Since recent trajectory studies of H + H_2 on the LSTH surface show rotational suppression of the cross section at low j followed by enhancement at high j[69], it is apparent that the comparison of Ω summed CS cross sections with quasiclassical cross sections will provide an interesting test of the ability of trajectory simulations to predict reagent state dependent effects.

Fig. 2 presents the comparison of CS and QCT cross sections for (νj) = (00). This comparison would not be changed by adding higher

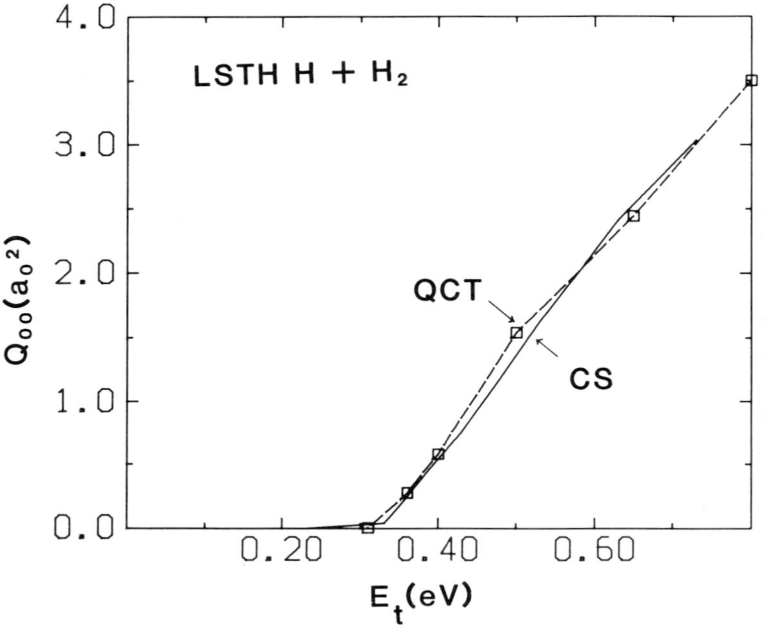

Figure 2. Comparison of the CS cross section Q_{00} from Fig. 1 with the corresponding QCT cross section from Ref. 69 as a function of E_t.

Ω's since these do not contribute to the j = 0 cross section. Evidently, the agreement in Fig. 2 is quite good, comparable in fact to analogous comparisons between CC and QCT cross sections that were done for the PK2 surface long ago[25]. This sort of comparison provides evidence that QCT calculations can be accurate for integral cross sections at energies that are not too close to threshold. Unfortunately, we will see that this comparison does not hold for other isotopes of H + H_2.

There are actually very few additional comparisons possible between the CS results in Fig. 1 and those from other dynamical theories on the LSTH surface. Comparisons of IOSA[32,33], CEQ[34] and CEQB[34] results with CC have, however, been made on the PK2 surface, and the results are summarized in Table 1. This table indicates that of the three approximate methods only CEQB describes both the tunnelling region and post threshold region accurately. IOSA tends to be too

Table 1. Cross Sections (a_0^2) for H + H_2 on the PK2 Surface

(A) From initial state νj = (00)

E(eV)	CC[a]	IOSA[b]	QCT[c]	CSDW[d]
0.50	0.50(-2)	0.10	--	0.57(-2)
0.55	0.057	0.25	--	--
0.60	0.35	0.57	0.35	0.42
0.65	0.93	0.80	0.90	--
0.70	1.52	1.25	1.40	--

(B) Rotationally averaged

E(eV)	CC[a]	CEQ[e]	CEQB[e]
0.50	1.1(-3)	4.3(-5)	1.1(-3)
0.55	0.02	0.004	0.02
0.60	0.064	0.06	0.09
0.65	0.18	0.23	0.27
0.70	0.28	0.43	0.51

a. Ref. 25 c. Ref. 4 e. Ref. 34
b. Ref. 32 d. Ref. 38

high at low energy and too low at high energy, while CEQ is too low at low energy. These comparisons suggest that the threshold constraints embodied in the CEQB method are more realistic than IOSA or CEQ. This also justifies recent proposals[37] that IOS cross sections should be shifted by bending zero point energies before comparing with CC.

Other recent comparisons between approximate and exact results on the PK2 surface include several variants of distorted wave theories[38,39]. One interesting result of one of these comparisons[38] is the perfect agreement between CC and CCDW reaction probabilities for the J = 0 partial wave over a wide range of translational energies ranging from deep tunnelling (pre-threshold) to just above threshold. The corresponding agreement of the CSDW cross sections at two energies is indicated in Table 1, and while not as good as for the J = 0 probabilities, it is still very good.

The thermal rate constants for H + H_2 that have been obtained from a variety of methods on the LSTH and PK2 surfaces are summarized in Table 2. This table indicates that on PK2, QCT[4] underestimates the rate constant compared to CC by a factor of 3.1 at 300 K. The VTST based methods (CVT[9] and ICVT[61]) do much better, with the most recent ICVTWKB/LAG results high by 21% at 300 K and 29% at 600 K. The TST-CEQB results[35] are in comparably good agreement with CC. On the LSTH surface, CS or CC rate constants are not yet available, but the comparison with experiment[70] indicates that QCT[71] again is substantially low at 300 K, while the VTST based methods[9,61] are quite close to experiment. Based on the PK2 comparisons, it seems unlikely that the CC rate constant on LSTH at 300 K will be above 2.3×10^{-16}, which

Table 2. Summary of Distinguishable Atom Rate Constants (cm^3/molec/sec) for $H + H_2$ ($\nu = 0$)

(A) PK2 Surface

T(K)	QCT[a]	TST-CEQB[b]	CVT/MCPVAG	ICVT-WKB/LAG	CC[e]
200	8.3(-19)	1.0(-17)	8.0(-18)[c]	1.6(-17)[h]	1.6(-17)
300	3.1(-16)	1.0(-15)	8.0(-16)[c]	1.2(-15)[d]	9.9(-16)
400	5.9(-15)	1.6(-14)	1.1(-14)[c]	1.2(-14)[h]	1.1(-14)
500	3.7(-14)	6.0(-14)	5.5(-14)[c]	5.8(-14)[h]	5.0(-14)
600	1.3(-13)	1.8(-13)	1.8(-13)[c]	1.8(-13)[d]	1.4(-13)
1000	1.9(-12)	2.0(-12)	2.3(-12)[h]	2.2(-12)[h]	

(B) LSTH Surface

T(K)	QCT[f]	CVT/MCPVAG	ICVT-WKB/LAG	Exp
200	2.4(-19)	1.2(-18)[h]	1.8(-18)[h]	
300	1.2(-16)	2.0(-16)[c]	2.3(-16)[h]	2.6(-16)[g]
400	3.1(-15)	3.8(-16)[c]	3.9(-15)[h]	
500	2.2(-14)	2.5(-14)[c]	2.5(-14)[h]	
600	8.4(-14)	9.5(-14)[c]	9.3(-14)[h]	
1000	---	1.8(-12)[c]	1.7(-12)[h]	

a. Ref. 4 d. Ref. 61a g. Ref. 70
b. Ref. 35 e. Ref. 25 h. Ref. 61b
c. Ref. 9 f. Ref. 71

means that the 27% discrepancy between the LAG rate constant and experiment must be due to either error in the experiment or to defects in the potential surface. Recent estimates of the $H + H_2$ barrier[64,66] which lower this from the LSTH value (0.425 eV) to about 0.418 eV will in fact remove most of the discrepancy.

<u>$D + H_2$, $H + HD$</u>: No CC or CS studies of $D + H_2$ on the LSTH surface have been made yet, but numerous studies using more approximate quantum methods have been done. Table 3 summarizes the current status of the $D + H_2$ rate constants as obtained from QCT[71], variational transition state theory[72], CEQB[73], IOSA[74], VADW[75,76] and experiment[77,78]. Here we see that QCT is slightly above experiment at all but the lowest temperature listed. In view of the above comments about the error in the LSTH barrier, it seems likely that the CC rate constant will be at least 30% below experiment at 300 K, i.e., below 2.2×10^{-16}. This would make it slightly less than the CVT/MCPSAG and CEQB results, the latter two being very close. IOSA is, on the other hand, above even the QCT rate constant, and the latter is well above experiment at 300 K, in contrast to the $H + H_2$ comparison in Table 2. Again we can infer from the IOSA result that bending zero point effects are important at low temperature. Note that VRADW is very close to CEQB

Table 3. Summary of Rate Constants for D + H_2 (ν = 0) (cm^3/molec sec) Using LSTH Potential Surface

T(K)	QCT[a]	CVT/MCPSAG[b]	TST-CEQB[c]	IOSA[d]	VRADW[e]	Exp
200	1.4(-18)	1.2(-18)	1.3(-18)	3.6(-18)	4.0(-18)	1.5(-18)[f]
300	3.6(-16)	2.2(-16)	2.3(-16)	5.6(-16)	2.9(-16)	3.1(-16)[f]
400	6.2(-15)	4.1(-15)	4.6(-15)	8.1(-15)	4.5(-14)	---
500	3.7(-14)	---	3.0(-14)	4.4(-14)	2.9(-14)	---
600	1.3(-13)	9.8(-14)	1.1(-13)	1.4(-13)	1.0(-13)	1.2(-13)[g]
750	---	3.9(-13)	4.2(-13)	---	---	4.4(-13)[g]
1000	---	1.7(-12)	1.8(-12)	1.9(-12)	---	---

a. Ref. 71
b. Ref. 72
c. Ref. 73
d. Ref. 74
e. Ref. 75,76
f. Ref. 77
g. Ref. 78

except at low temperature. This represents a substantial improvement over previous implementations of distorted wave theory where excessively low rate constants were obtained[75].

On the PK2 surface there have been several studies of D + H_2, the most accurate of which is probably the CS calculation of Schatz and Kuppermann[79]. Fig. 3 presents the cross sections Q_{00} from that calculation as a function of translational energy. Included are the cross sections for D + H_2, H + HD $\rightarrow$ H_2 + D, H + DH $\rightarrow$ HD + H and H + H_2. The first three of these all come from the same calculation, corresponding to different parts of the same scattering matrix. Fig. 3 indicates that the D + H_2 cross sections are shifted down in energy by 0.01 eV relative to the H + H_2 cross sections. This shift is also seen in collinear exact quantum calculations[79] and can be rationalized in terms of symmetric stretch zero point energy differences. There is also a 0.04 eV shift upward in energy in going from D + H_2 to H + HD which is mostly related to diatomic zero point energy differences and microscopic reversibility. The comparison between H + HD and H + DH is more complicated because saddle point zero point differences favor H + HD, but reagent and product zero point differences favor the thermoneutral H + DH over the endoergic H + HD. Fig. 3 indicates that H + DH has the larger cross section at low energy where tunnelling dominates while H + HD is larger at higher energies.

The results in Fig. 3 also have their counterparts for thermal isotope ratios, as is indicated in Table 4. Here are listed the D + H_2, H + HD, H + DH and H + H_2 rate constants and isotope ratios for temperatures in the range 100-300 K. This range is all that could be considered by Schatz and Kuppermann because of convergence problems in their cross section calculations at high energies.

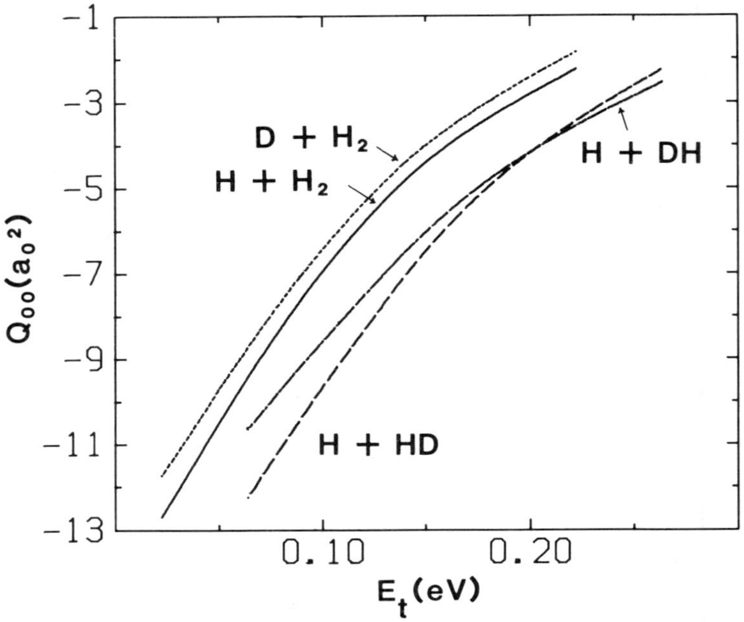

Figure 3. Logarithm of CS reactive cross section Q_{00} from Ref. 79 versus E_t for $D + H_2$, $H + H_2$, H + DH and H + HD.

Table 4. Thermal Rate Constant Isotope Ratios ($H + H_2 = 1$) from CS Calculations[a] (PK2 Surface)

Reaction	T = 100	150	200	250	300K	‡[b](300K)	ICVT/LAG[c](300K)
$D + H_2$	2.2	1.7	1.6	1.5	1.4	2.0	1.1
H + DH	0.083	0.067	0.074	0.084	0.10	0.21	0.12
H + HD	0.029	0.083	0.15	0.19	0.25	0.28	0.17
$H + H_2$[c]	1.0	1.0	1.0	1.0	1.0	1.0	1.0
absolute value (cm^3/molec sec)	9.6(-21)	6.5(-19)	1.6(-17)	1.7(-16)	9.8(-16)	4.3(-17)	9.1(-16)

a. Ref. 79
b. Ref. 80
c. Value given is the distinguishable atom rate constant consistent with what is used for $D + H_2$.

As expected from Fig. 3, we find that the $D + H_2$ rate constant is larger than $H + H_2$ while H + HD and H + DH are considerably smaller, with a crossing between 100 and 200 K. The results at 300 K are in very good agreement with results from variational transition state theory (ICVT/LAG)[80]. Conventional transition state theory (labelled ‡) also shows the same qualitative trends as CS, although the D + H_2 isotope ratio is too high and H + DH too low. The absolute magnitude of the ‡ result is too low by a factor of 23 at 300 K. Overall these results illustrate the importance of tunnelling and variational effects in determining accurate isotope ratios.

Mu + H_2: Muonium (Mu) is an isotope of the hydrogen atom with a positive muon as the nucleus. Since it is 8.8 times lighter than H, tunnelling and zero point effects are enormous, and in fact the Mu + H_2 dynamics is quite different from H + H_2. Since thermal rate constants have recently been measured for Mu + H_2[81], it is an important subject for theoretical study. The first predictions of the Mu + H_2 rate constants (based on VTST) were made well before the experiment was done[82], and subsequent experimental results have been in good agreement with these predictions[83]. More recently, CS calculations[84] have confirmed that the VTST estimates for the LSTH surface are in fact quite accurate.

Table 5. Thermal Rate Constants (cm^3/molec sec) for Mu + H_2

T(K)	CS	Experiment[a]	ICVT-WKB/LAG[b]	QCTF[c]	QCTR[c]
200	4.9(-24)				
250	1.3(-21)				
300	7.2(-20)		1.92(-20)		
350	1.4(-18)				
400	1.4(-17)		5.54(-18)		
444	6.8(-17)			1.2(-14)	
608	3.6(-15)	1.3±1.8(-15)	2.49(-15)		7.0(-15)
624	4.8(-15)	4.7±1.3(-15)			
669	9.7(-15)	7.5±2.0(-15)			
708	1.7(-14)	2.5±0.9(-14)			
745	2.7(-14)	2.2±0.4(-14)			
802	5.0(-14)	5.4±1.3(-14)			
845	7.7(-14)	6.1±1.0(-14)	7.60(-14)		
875	1.0(-13)		1.03(-13)	1.1(-12)	1.6(-13)
1000	2.6(-13)		3.06(-13)		

a. Ref. 81. Note that there are temperature uncertainties of about ±20 K associated with the temperature at which each measurement is reported.
b. Ref. 61a
c. R3f. 83

Table 5 summarizes the results of the experimental and theoretical results to date on Mu + H_2 while Fig. 4 plots the CS, ICVT-WKB/LAG[61a] and experimental results over the range of temperatures where the experiments have been done. The Figure and Table indicate that CS,

ICVT-WKB/LAG and experimental results are all in excellent agreement, generally to within the experimental error bars. At low temperatures, the LAG results dip well below CS, with the difference being a factor of 6 at 300 K. This indicates that the LAG tunnelling calculation is inaccurate in the deep tunnelling regime.

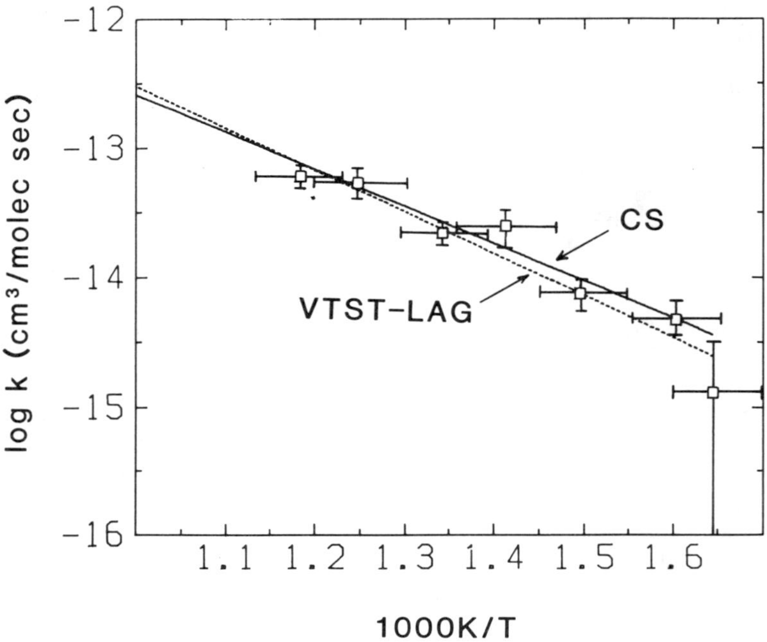

Figure 4. Arrhenius plot of thermal rate constant k versus 1/T for Mu + H_2, comparing the experimentally measured results[82] (with error bars) with the results of ICVT/LAG[61] and CS[84] calculations.

Table 5 also includes results from QCT calculations, run in both the forward (QCTF) and reverse (QCTR) sense. The QCTF results are in very poor agreement with CS, being high by one or more orders of magnitude. QCTR is much closer, being high by a factor of two. This comparison illustrates an extreme case where classical nonadiabatic "leak" leads to a gross overestimation of the rate constant by classical mechanics. The fact that QCTF is the poorer of the two methods simply reflects the fact that the reaction bottleneck is in the product channel region where adiabatic zero point shifts are smaller coming from the products than from the reagents.

B. Vibrationally Excited Cross Sections and Rate Constants

The reactions H + H_2 (ν=1) and D + H_2 (ν=1) have been a source of controversy between experimentalists and theoreticians ever since Gordon et al[85] and Kneba et al[86] published values for the rate constants for these reactions which were higher than theoretical estimates by factors of 10-80[15]. The topic was reviewed in 1981[15], but since then there has been much activity. Table 6 summarizes the current state of

Table 6. Rate Constants (cm^3/molec sec $\times 10^{13}$) for $H + H_2$ ($\nu = 1$) and $D + H_2$ ($\nu = 1$) at 300 K

	$H + H_2$			$D + H_2$
	1 → 0 reactive	1 → 0 total	1 → all reactive	1 → all reactive
(A) PK2 Surface				
QCT	1.3[a], 3.0[b]	1.8[a], 3.0[b]	4.0[a], 6.1[c], 8.4[b]	
IOSA	1.6[d]		5.6[d], 4.0[i]	
CEQB	0.77[e]		2.6[e]	
SCAD			1.8[f]	
CS	0.36[g]	0.62[g]	1.7[g]	
ICVT/LA			3.3[v]	2.9[v]
(B) LSTH Surface				
QCT	0.55[h]	0.73[h]	1.3[h]	1.7[j]
IOSA				2.9[k]
RAIOS				10.[l]

	$H + H_2$			$D + H_2$
	1 → 0 reactive	1 → 0 total	1 → all reactive	1 → all reactive
(B) LSTH Surface				
VADW	0.54[m]			5.0[m]
BCRLM			4.8[n]	2.8[n]
SCAD			2.0[f]	2.1[f]
ICVT/LA			2.0[v]	2.1[v]
CEQB				0.86[o]
(C) Experiment		3.0[p]	52.[q]	120.[r]
				9.8[s]
				16.[t]
				8.3[u]

a. Ref. 91 g. Ref. 27 m. Ref. 98 s. Ref. 87
b. Ref. 93 h. Ref. 71 n. Ref. 36 t. Ref. 88
c. Ref. 92 i. Ref. 94 o. Ref. 97 u. Ref. 89
d. Ref. 95 j. Ref. 71 p. Ref. 90 v. Ref. 61c
e. Ref. 35 k. Ref. 74 q. Ref. 85
f. Ref. 99 l. Ref. 100 r. Ref. 86

affairs, comparing the (1 → all) reactive, (1 → 0) reactive and (1 → 0) total rate constants for $H + H_2$ and (1 → all) total for $D + H_2$ on the PK2 and LSTH potential surfaces with corresponding experimental values[85-90] at 300 K.

The largest change which has taken place in the last few years has been in the $D + H_2$ experimental value. Recent measurements by Glass and Chaturvedi[87], Rozenshteyn et al[88] and Wellhausen and Wolfrum[98] have dropped the estimated value by over an order of magnitude to around 10^{-12} cm^3/molec sec.

A number of new theoretical estimates of the ν = 1 rate constants have also appeared recently. On the PK2 surface, Schatz[27] has done a CS calculation which indicates that if anything the earlier theoretical estimates based on trajectories[71,91-93] and IOSA[74,94,95] are too high rather than too low. Fig. 5 indicates the reason for this. In

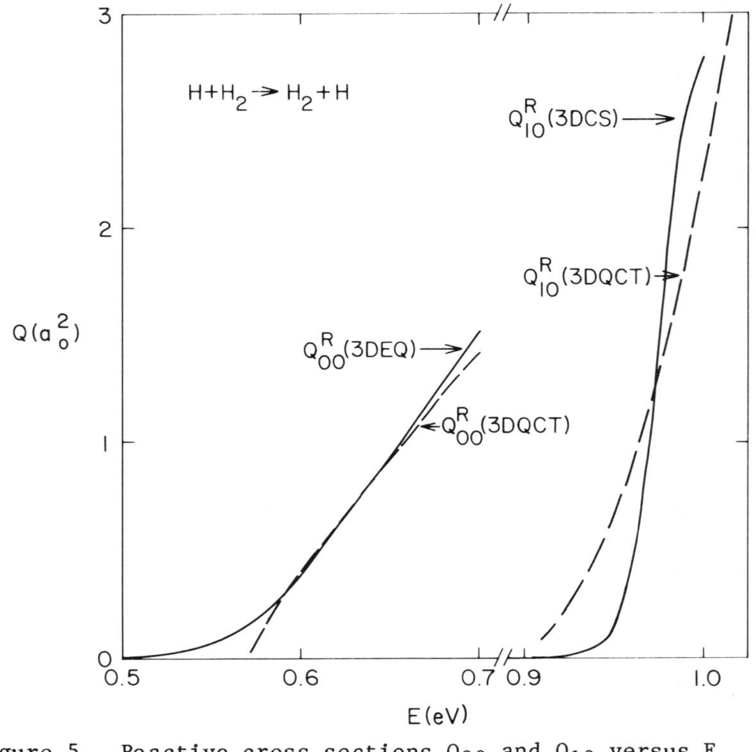

Figure 5. Reactive cross sections Q_{00} and Q_{10} versus E, comparing the results of CC[25] with QCT[5] for ν = 0 and CS[27] with QCT[91] for ν = 1.

this figure we have compared the cross sections Q_{00} and Q_{10} for $H + H_2$ on PK2 as obtained from either CC[25] or CS[27] calculations with those from QCT[5,91]. For ν = 0 the agreement is comparable to what was seen in Fig. 2, but for ν = 1, the QCT cross section threshold energy is too low and the value of the QCT cross section near threshold too high compared to CS. The reason for this can be traced[96] to bending energy zero point constraints which are present in the quantum dynamics but absent in QCT. This leads to a QCT rate constant in Table 6 on PK2 that is a factor of about 3 higher than CS. The results of methods that incorporate bending zero point effects explicitly, such as CEQB[35] are seen to be much closer to CS.

Other recent calculations presented in Table 6 include IOSA[74], BCRLM[36] and CEQB[97] calculations on LSTH for D + H_2, VADW[98] for H + H_2, a harmonic model semiclassical adiabatic (SCAD) calculation[99] for H + H_2 and D + H_2, a rotationally averaged IOS (RAIOS) calculation[100] for D + H_2, and ICVT calculations using a least action (LA) tunnelling coefficient for H + H_2 and D + H_2[61c]. The comparison of the D + H_2 results with experiment shows some variation, but except for the RAIOS calculation, none of the other results are within a factor of 3 of 1 x 10^{-12}. Moreover, based on the comparisons in Tables 2 and 3, it might be expected that the CEQB rate constant is the most accurate of those given for D + H_2, and this is over a factor of 10 below the average of the new experiments. Thus the controversy between theory and experiment continues in this area.

C. Fast Hydrogen Atom Chemistry

A major experimental development in the past two years has been in the application of laser photolysis techniques for producing translationally hot hydrogen atoms to the H + D_2 → HD + D reaction[101-106]. This has enabled the direct observation of product vibration/rotation distributions in this reaction for the first time. As might be expected, these experiments have prompted a flurry of theoretical activity[44,45,107-112], and by and large the comparison between theory and experiment has been very good.

Fig. 6 shows the comparison of experimental HD vibration/rotation distributions from Gerrity and Valentini[102], and Marinero et al[106] with QCT results due to Blais and Truhlar[107]. The translational energy is a mixture of 0.55 eV and 1.3 eV, with 1.3 eV contributing most to the measured distributions. The Marinero et al results are only available for HD(ν = 1,2), and for these states it is clear that the Marinero results are rotationally colder than those of Gerrity and Valentini. Possible reasons for this have been discussed at length in Ref. 102, but it is still not clear which set of experimental results is more accurate. The comparison of experiment with QCT cross sections is closest for the Gerrity and Valentini results, with a general trend for QCT distributions to be rotationally hotter than experiment. In lower energy experiments[103], the agreement with QCT is as good as that seen in Fig. 6.

Given that the H + D_2 experiments have all considered translational energies above 0.98 eV[103], it seems likely that trajectory methods should be able to describe the experiments quite well, and indeed, the only significant difference between Gerrity and Valentini's results and trajectories noted so far[104] has been in the

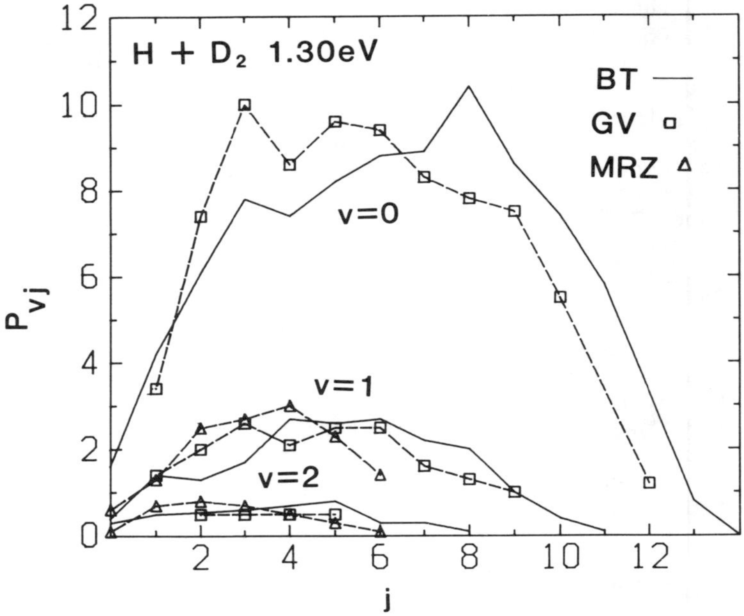

Figure 6. Relative cross section for $H + D_2 \rightarrow HD(\nu j) + D$ at 1.3 eV translational energy (see text) as a function of j for $\nu = 0,1,2$ comparing the experimental results of Ref. 106 and 102 with QCT calculations from Ref. 107.

branching ratio between reactive and nonreactive collisions which produce $\nu = 1$ product. Comparison between trajectory and CS results has also been good[109], although the energy used for the comparison (0.55 eV) is well below that used in all of the experiments. Fig. 7 shows this comparison for HD ($\nu = 0$) using QCT cross sections from Blais and Truhlar and CS from Schatz[111]. In the CS calculations it was found that contributions to the cross sections from $|\Omega| > 0$ are moderately important, so Fig. 7 includes both $\Omega = 0$ and the sum over $\Omega = 0,1$. The comparison with QCT on both a relative and absolute basis is quite good, and in fact the only significant discrepancy between CS and QCT is in the magnitude of the very tiny HD($\nu = 1$) cross sections[111]. The question of reactive versus nonreactive branching has not yet been explored.

D. Resonances

Although there is as yet no experimental confirmation that resonances exist in the $H + H_2$ reaction, the study of resonances in this reaction has been of great interest to theoreticians. Just what is meant by the term "resonance" is perhaps best illustrated in Fig. 8.

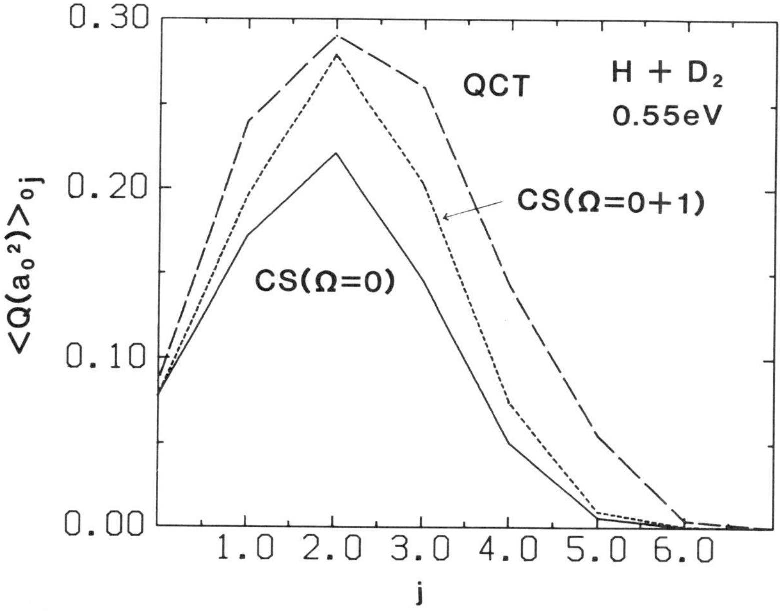

Figure 7. Reactive cross section for H + D_2 → HD (νj) + D at 0.55 eV translational energy as a function of j for ν = 0. Included in the comparison are QCT results from Ref. 107, and CS Ω = 0 and Ω = 0 + 1 results from Ref. 111.

There for the PK2 surface we plot the collinear transition probabilities for the $\nu = 0 \rightarrow \nu' = 1$, $1 \rightarrow 0$ (which is identical to $0 \rightarrow 1$) and $1 \rightarrow 1$ transitions, along with their (J = 0) three dimensional counterparts 00 → 1 (summed over product rotation), 10 → 0 and 10 → 1[27]. The collinear results indicate that over a fairly small energy range (0.05 eV), the 0 → 1 probability goes through a peak. At the same time, other transition probabilities such as 0 → 0 exhibit a sharp dip. Physically this behavior occurs when the total energy equals that of a metastable excited state of the three atom complex, as in this limit a substantial amount of reactive flux is diverted temporarily to this excited state and thence into excited states of the product diatomic. In the case of collinear H + H_2 we find that the 0 → 1/1 → 0 transition probability is strongly modified by the resonance while 1 → 1 is not. The corresponding three dimensional probabilities in Fig. 8 also show a resonance, although the resonance energy in this case is shifted upward in energy relative to collinear by about 0.1 eV. Such shifts have been noted before[12] and are due to bending zero point effects which are present in 3D but not 1D.

Although resonances are at the heart of chemical processes such as atom-molecule recombination and Van der Waals molecule predissociation[113], their influence on quantities that can be measured in the H +

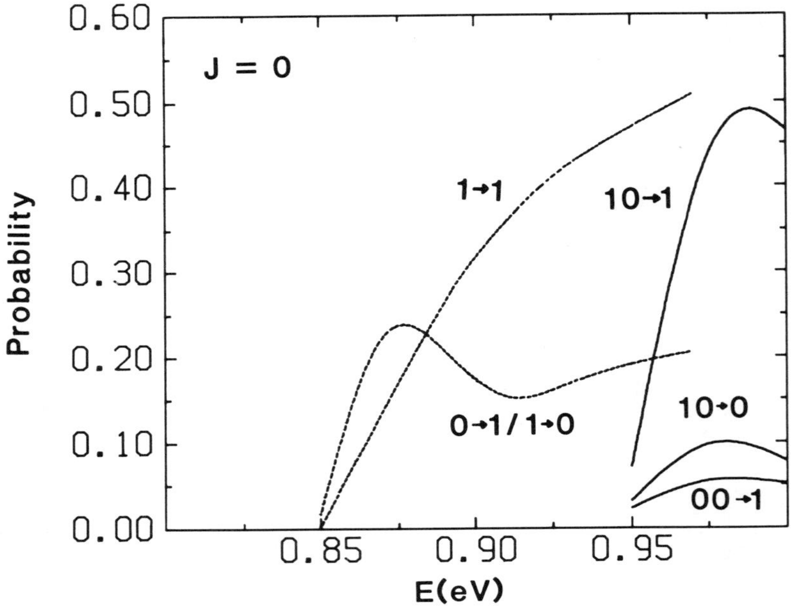

Figure 8. Collinear exact quantum and 3D exact quantum (J = 0) transition probabilities versus total energy for H + H_2 on the PK2 surface. The collinear probabilities plotted are from Ref. 57, and include the transitions 0 → 1/1 → 0 and 1 → 1. The 3D probabilities are from Ref. 114 and are for the states 00 → 1, 10 → 0 and 10 → 1.

H_2 reaction is actually quite subtle. The details of exactly what does happen were recently studied by Schatz[114] on the PK2 surface, and the results are shown in Figs. 9 and 10. Fig. 9 plots the 3D reaction probability for the 00 → 1 transition as a function of J. At E = 0.95 eV, this probability is seen to decay smoothly to zero, typical of nonreactive scattering. At 0.975 eV, the peak near J = 0 broadens substantially, while at 0.99 eV a secondary peak at J ≅ 3 has appeared. This peak shifts to higher J at 1.00 eV and eventually disappears at still higher energy. The standard interpretation of this behavior[115,116] is that the resonance which appears for J = 0 at about E = 0.975 eV (Fig. 8), moves to higher J as E increases just as the energy levels of a rigid rotor increase as J increases. Associated with this is a shift of the differential cross section from backwards to sidewards peaking, as is indicated in Fig. 10. There we see that at 0.95 eV, both the 10 → 0 and 10 → 1 cross sections are backward peaked. At 0.99 eV, however, the (12^00) and (12^20) resonances

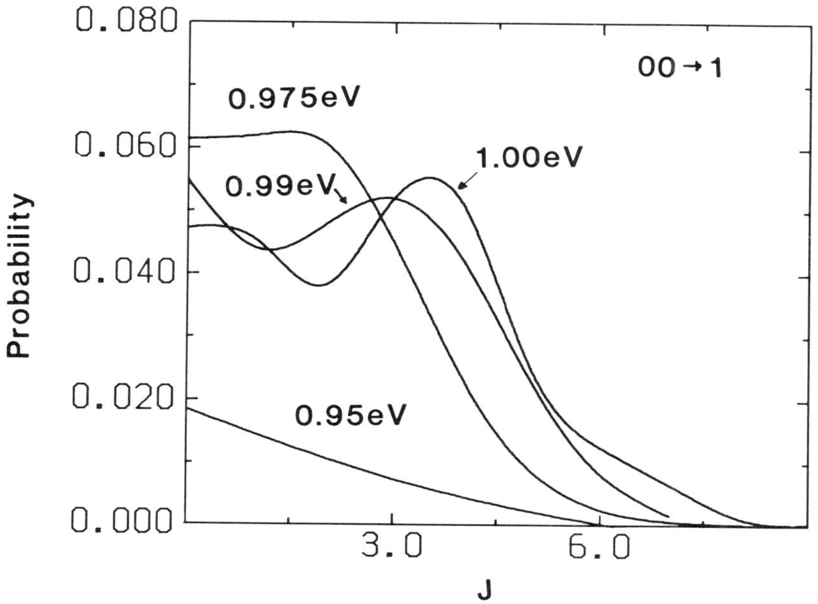

Figure 9. CS reaction probbility for the 00 → 1 transition versus J for H + H_2 on the PK2 surface at E = 0.95, 0.975, 0.99 and 1.00 eV.

are also in good agreement with CEQB. Since SCSA and CEQB indicate that other higher energy resonances should exist (Table 7), it appears

Table 7. J = |Ω| Resonance Energies (eV) for H + H_2 in 3D

Assignment	SCSA[a]	RPO[b]	CEQB[c]	Accurate
(A) PK2 Surface				
$(10^0 0)$	0.983	0.954	0.965	0.975[d]
$(11^1 0)$	1.107	N.R.[e]	--	--
(B) LSTH Surface				
$(10^0 0)$	0.979		0.973	0.984[f]
$(11^1 0)$	1.092		1.09	1.10[g]
$(12^0 0)$	N.P.[h]		1.20	1.20[g]
$(12^2 0)$	N.P.		1.21	1.22-1.24[g]
$(20^0 0)$	1.242		1.35	--
$(20^0 1)$	1.382		--	--

a. Ref. 116
b. Ref. 118
c. Ref. 113 and J.M. Bowman, private communication.
d. Ref. 12
e. No resonance is predicted to exist.
f. Ref. 8
g. Ref. 120
h. No prediction of the energy of this resonance was made.

that the excited state vibrational spectrum of H_3 is both rich and interesting.

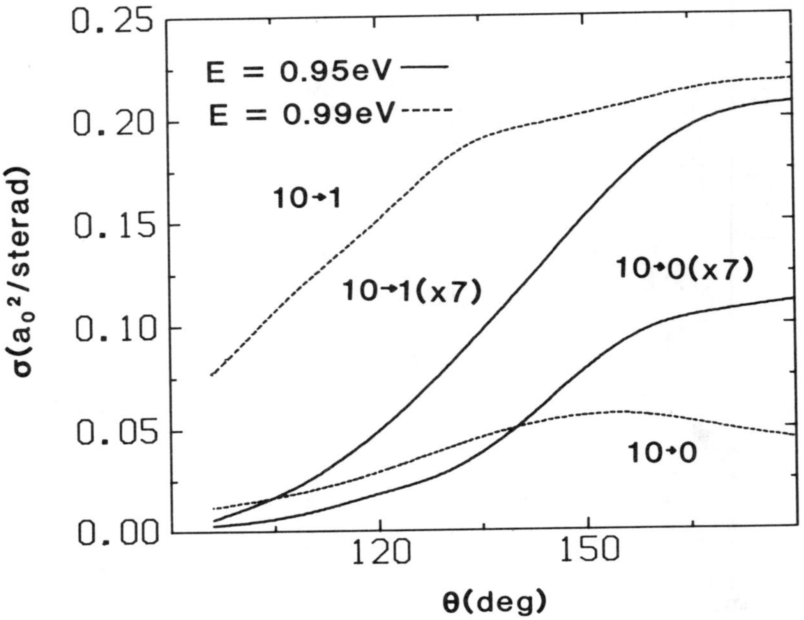

Figure 10. Differential cross section σ versus reactive scattering angle θ for the 10 → 1 and 10 → 0 transitions in H + H_2 from CS calculations on the PK2 surface at E = 0.95 and 0.99 eV.

IV. Acknowledgement

It is a pleasure to acknowledge helpful discussions concerning this manuscript with J.M. Bowman, M.C. Colton, A. Kuppermann, D.G. Truhlar and A.F. Wagner. This research was supported by a grant from the National Science Foundation (CHE-84 16026).

References

1. D.G. Truhlar and R.E. Wyatt, Ann. Rev. Phys. Chem. 27, 1, 1976.
2. (a) J.O. Hirschfelder, H. Eyring and B. Topley, J. Chem. Phys. 4, 170 (1936).
   (b) F.T. Wall, L.A. Hiller, and J. Mazur, J. Chem. Phys. 29, 255 (1958).

3. F.T. Wall, L.A. Hiller, and J. Mazur, J. Chem. Phys. 35, 1284 (1961).
4. M. Karplus, R.N. Porter and R.D. Sharma, J. Chem. Phys. 43, 3259 (1965).
5. (a) E.M. Mortensen and K.S. Pitzer, Chem. Soc. Spec. Publ. 16, 57 (1962).
(b) E.M. Mortenson, J. Chem. Phys. 48, 4029 (1968).
6. G.C. Schatz and A. Kuppermann, J. Chem. Phys. 62, 2502 (1975).
7. A.B. Elkowitz, and R.E. Wyatt, J. Chem. Phys. 62, 2504 (1975).
8. R.B. Walker, E. Stechel and J.C. Light, J. Chem. Phys. 69, 2922 (1978).
9. B.C. Garrett and D.G. Truhlar, Proc. Natl. Acad. Sci. USA 76, 4755 (1979).
10. S. Wu and R.D. Levine, Mol. Phys. 22, 881 (1971).
11. G.C. Schatz and A. Kuppermann, J. Chem. Phys. 59, 964 (1973).
12. G.C. Schatz and A. Kuppermann, Phys. Rev. Lett. 35, 1266 (1975).
13. D.G. Truhlar and R.E. Wyatt, Adv. Chem. Phys. 36, 141 (1977).
14. R.B. Walker and J.C. Light, Ann. Rev. Phys. Chem. 31, 401 (1980).
15. G.C. Schatz, in Potential Energy Surfaces and Reaction Dynamics Calculations (edited by D.G. Truhlar), Plenum, New York, 1981, p. 287.
16. J.N.L. Connor, Computer Physics Comm. 17, 117 (1979).
17. D.G. Truhlar and B.C. Garrett, Ann. Rev. Phys. Chem. 35, 159 (1984).
18. J.M. Bowman, Adv. Chem. Phys. (1985), in press.
19. M. Baer, Adv. Chem. Phys. 49, 191 (1982).
20. Resonances (edited by D.G. Truhlar), ACS Symp. Series Vol. 263 (1984).
21. R.B. Bernstein, ed. Atom-Molecule Collision Theory: A Guide for the Experimentalist (Plenum, New York, 1979).
22. M. Baer, ed., Theory of Chemical Reaction Dynamics CRC Press, Boca Raton, 1985, in press.
23. R. Gotting, H.R. Mayne and J.P. Toennies, J. Chem. Phys. 80, 2230 (1984).
24. B.C. Garrett, D.G. Truhlar, R.S. Grev and R.B. Walker, J. Chem. Phys. 73, 235 (1980).
25. G.C. Schatz and A. Kuppermann, J. Chem. Phys. 65, 4642 (1976).
26. A.B. Elkowitz and R.E. Wyatt, J. Chem. Phys. 63, 702 (1975).
27. G.C. Schatz, Chem. Phys. Lett. 108, 532 (1984).
28. A.B. Elkowitz and R.E. Wyatt, Mol. Phys. 31, 189 (1976).
29. A. Kuppermann, G.C. Schatz and J. Dwyer, Chem. Phys. Lett. 45, 71 (1977).
30. G.C. Schatz, Chem. Phys. Lett. 94, 183 (1983).
31. R.T Pack, Chem. Phys. Lett. 108, 333 (1984).
32. J.M. Bowman and K.T. Lee, J. Chem. Phys. 72, 5071 (1980).
33. D.J. Kouri, V. Khare and M. Baer, J. Chem. Phys. 75, 1179 (1981);
M. Baer, V. Khare and D.J. Kouri, Chem. Phys. Lett. 68, 378 (1979).

34. J.M. Bowman, G-Z. Ju, and K-T. Lee, J. Chem. Phys. 75, 5199 (1981).
35. J.M. Bowman, G-Z. Ju and K-T. Lee, J. Phys. Chem. 86, 2232 (1982).
36. R.B. Walker and E.F. Hayes, J. Phys. Chem. 87, 1255 (1983).
37. E. Pollak, J. Chem. Phys. 82, 106 (1985); N. Abu-Salbi, D.J. Kouri, M. Baer and E. Pollak, J. Chem. Phys. 82, 4500 (1985).
38. G.C. Schatz, L.M. Hubbard, P.S. Dardi and W.H. Miller, J. Chem. Phys. 81, 231 (1984).
39. J.C. Sun, B.H. Choi, R.T. Poe and K.T. Tang, J. Chem. Phys. 78, 4523 (1983); 79, 5376 (1983).
40. B.H. Choi and K.T. Tang, J. Chem. Phys. 61, 5147 (1974), and references therein.
41. D.C. Clary and J.N.L. Connor, Mol. Phys. 43, 621 (1981) and references therein.
42. S.H. Suck, Phys. Rev. A27, 187 (1982) and references therein.
43. P.S. Dardi, S.H. Shi and W.H. Miller, to be published.
44. J.N.L. Connor and W.J.E. Southall, Chem. Phys. Lett. 108, 527 (1984).
45. S.H. Suck Salk, C.R. Klein and C.K. Lutrus, Chem. Phys. Lett. 110, 112 (1984); M.S. Bowers, B.H. Choi, R.T. Poe and K.T. Tang, Chem. Phys. Lett. 116, 239 (1985).
46. L.M. Delves, Nucl. Phys. 9, 391 (1959); 20, 275 (1960).
47. A. Kuppermann, J.A. Kaye and J.P. Dwyer, Chem. Phys. Lett. 74, 257 (1980); J.A. Kaye and A. Kuppermann, Chem. Phys. Lett. 77, 573 (1981).
48. A. Kuppermann, Theor. Chem.: Advances and Perspectives, 6A, 79 (1981).
49. G. Hauke, J. Manz and J. Romelt, J. Chem. Phys. 73, 5040 (1980); J. Romelt, Chem. Phys. Lett. 74, 263 (1980); 87, 259 (1982).
50. D.K. Bondi and J.N.L. Connor, Chem. Phys. Lett. 92, 570 (1982).
51. J.M. Launay and M. LeDourneuf, J. Phys. B15, 455 (1982).
52. A. Kuppermann, private communication.
53. W.H. Miller, N.C. Handy and J.E. Adams, J. Chem. Phys. 72, 99 (1980).
54. S.D. Schwartz and W.H. Miller, J. Chem. Phys. 77, 2378 (1982); 79, 3759 (1983).
55. D.G. Truhlar and A. Kuppermann, J. Chem. Phys. 56, 2232 (1972).
56. R.A. Marcus and M.E. Coltrin, J. Chem. Phys. 67, 2609 (1977).
57. R.I. Altkorn and G.C. Schatz, J. Chem. Phys. 72, 3337 (1980).
58. R.T. Skodje, D.G. Truhlar and B.C. Garrett, J. Chem. Phys. 77, 5955 (1982).
59. D.K. Bondi, J.N.L. Connor, B.C. Garrett and D.G. Truhlar, J. Chem. Phys. 78, 5981 (1983).
60. B.C. Garrett and D.G. Truhlar, J. Chem. Phys. 79, 4931 (1983).
61. (a) B.C. Garrett and D.G. Truhlar, J. Chem. Phys. 81, 309 (1984).
(b) B.C. Garrett and D.G. Truhlar, private communication.
(c) B.C. Garrett and D.G. Truhlar, J. Phys. Chem. 89, 2204 (1985).

62. B. Liu, J. Chem. Phys. 58, 1925 (1973); P. Siegbahn and B. Liu, J. Chem. Phys. 68, 2457 (1978).
63. D.G. Truhlar and C.J. Horowitz, J. Chem. Phys. 68, 2468 (1978); E71, 1514 (1979).
64. B. Liu, J. Chem. Phys. 80, 581 (1984).
65. M.R.A. Blomberg and B. Liu, J. Chem. Phys. 82, 1050 (1985).
66. R.N. Barnett, P.J. Reynolds and W.A. Lester, J. Chem. Phys. 82, 2700 (1965); D.M. Ceperley and B.J. Alder, J. Chem. Phys. 81, 5833 (1984).
67. R.N. Porter and M. Karplus, J. Chem. Phys. 40, 1105 (1964).
68. M.C. Colton and G.C. Schatz, to be published.
69. H.R. Mayne and C.A. Boonenberg, Chem. Phys. Lett. 108, 67 (1984).
70. W.R. Schultz and D.J. LeRoy, J. Chem. Phys. 42, 3869 (1965).
71. H.R. Mayne and J.P. Toennies, J. Chem. Phys. 75, 1794 (1981).
72. B.C. Garrett and D.G. Truhlar, J. Chem. Phys. 72, 3460 (1980).
73. J.M. Bowman, K.T. Lee and R.B. Walker, J. Chem. Phys. 79, 3742 (1983).
74. N. Abu Salbi, D.J. Kouri, Y. Shima and M. Baer, J. Chem. Phys. 82, 2650 (1985).
75. Y.Y. Yung, B.H. Choi and K.T. Tang, J. Chem. Phys. 72, 621 (1980).
76. J.C. Sun, B.H. Choi, R.T. Poe and K.T. Tang, J. Chem. Phys. 79, 5376 (1983).
77. D.N. Mitchell and D.J. LeRoy, J. Chem. Phys. 58, 3449 (1973).
78. A.A. Westenberg and N. deHaas, J. Chem. Phys. 58, 1393 (1967).
79. G.C. Schatz and A. Kuppermann, to be published.
80. B.C. Garrett, D.G. Truhlar and G.C. Schatz, to be published.
81. D.M. Garner, D.G. Fleming and R.J. Mikula, Chem. Phys. Lett., submitted.
82. D.K. Bondi, D.C. Clary, J.N.L. Connor, B.C. Garrett and D.G. Truhlar, J. Chem. Phys. 76, 4986 (1982).
83. N.C. Blais, D.G. Truhlar and B.C. Garrett, J. Chem. Phys. 78, 2363 (1983).
84. G.C. Schatz, J. Chem. Phys., submitted.
85. E.B. Gordon, B.I. Ivanov, A.P. Perminov, V.E. Balalev, A.V. Ponomarev, V.V. Filatov, Chem. Phys. Lett. 58, 425 (1978).
86. M. Kneba, U. Wellhausen, and J. Wolfrum, Ber. Bunsenges Phys. Chem. 83, 940 (1979).
87. G.P. Glass and B.K. Chaturverdi, J. Chem. Phys. 77, 3478 (1982).
88. V.B. Rozenshteyn, Yu.M. Gershenzon, A.V. Ivanov and S.I. Kucheryavii, Chem. Phys. Lett. 105, 423 (1984).
89. U. Wellhausen and J. Wolfrum, Phys. Rev. Lett., unpublished, cited in Ref. 100.
90. R.F. Heidner and J.V.V. Kasper, Chem. Phys. Lett. 15, 179 (1972).
91. H.R. Mayne, Chem. Phys. Lett. 66, 487 (1979).
92. M. Karplus and I. Wang, unpublished cited in Ref. 85.
93. V.I. Osherov, V.G. Ushakov, L.A. Lomakin, Chem. Phys. Lett. 55, 513 (1978).

94. V. Khare, D.J. Kouri, J. Jellinek, and M. Baer, Potential Energy Surfaces and Reaction Dynamics Calculations, D.G. Truhlar, ed., Plenum, New York, 1981 p. 475.
95. J.M. Bowman, K.T. Lee, Chem. Phys. Lett. 64, 291 (1979).
96. G.C. Schatz, J. Chem. Phys. 79, 1808 (1983).
97. J.M. Bowman, K.T. Lee and R.B. Walker, J. Chem. Phys. 79, 3742 (1983).
98. J.C. Sun, B.H. Choi, R.T. Poe, K.T. Tang, Phys. Rev. Lett. 44, 1211 (1980).
99. E. Pollak, J. Chem. Phys. 82, 106 (1985).
100. E. Pollak, N. Abu-Salbi and D.J. Kouri, preprint.
101. D.P. Gerrity and J.J. Valentini, J. Chem. Phys. 79, 5202 (1983).
102. D.P. Gerrity and J.J. Valentini, J. Chem. Phys. 81, 1298 (1984).
103. D.P. Gerrity and J.J. Valentini, J. Chem. Phys. 82, 1323 (1985).
104. D.P. Gerrity and J.J. Valentini, J. Chem. Phys., submitted.
105. C.T. Rettner, E.E. Marinero, and R.N. Zare, in Physics of Electronic and Atomic Collisions : Invited Papers from the XIIIth ICPEAC, Berlin, 1983, edited by J. Eichler, I.V. Hertel and N. Stolterfoht (North Holland, Amsterdam).
106. E.E. Marinero, C.T. Rettner and R.N. Zare, J. Chem. Phys. 80, 4142 (1984).
107. N.C. Blais and D.G. Truhlar, Chem. Phys. Lett. 102, 120 (1983).
108. E. Zamir, R.D. Levine and R.B. Bernstein, Chem. Phys. Lett. 107, 217 (1984).
109. R.D. Levine and R.B. Bernstein, Chem. Phys. Lett. 105, 467 (1984).
110. N.C. Blais, R.B. Bernstein and R.D. Levine, J. Phys. Chem. 89, 10 (1985).
111. G.C. Schatz, Chem. Phys. Lett., 108, 532 (1984).
112. N.C. Blais, D.G. Truhlar and B.C. Garrett, J. Chem. Phys. 82, 2300 (1985).
113. J.M. Bowman, K.T. Lee, H. Romanowski and L.B. Harding, in Resonances, ACS Symp. Series 263, 43 (1984).
114. G.C. Schatz, unpublished.
115. E. Pollak and R.E. Wyatt, J. Chem. Phys. 77, 7689 (1982); 78, 4464 (1983).
116. B.C. Garrett, D.W. Schwenke, R.T. Skodje, D. Thirumalai, T.C. Thompson and D.G. Truhlar, in Resonances, ACS Symp. Series 263, 375 (1984).
117. T.C. Thompson and D.G. Truhlar, J. Chem. Phys. 76, 1790 (1982); E77, 3777 (1982).
118. J. Romelt and E. Pollak, in Resonances, ACS Symp. Series 263, 353 (1984).
119. A. Kuppermann, in Potential Energy Surfaces and Reaction Dynamics Calculations, edited by D.G. Truhlar, Plenum, New York, 1981, p.375.
120. M.C. Colton and G.C. Schatz, to be published.

REACTION PATH MODELS FOR POLYATOMIC REACTION DYNAMICS--FROM TRANSITION STATE THEORY TO PATH INTEGRALS

William H. Miller
Department of Chemistry, University of California, and Materials and Molecular Research Division of the Lawrence Berkeley Laboratory, Berkeley, California 94720 U.S.A.

ABSTRACT. The reaction path Hamiltonian model for the dynamics of general polyatomic systems is reviewed. Various dynamical treatments based on it are discussed, from the simplest statistical approximations (e.g., transition state theory, RRKM, etc.) to rigorous path integral computational approaches that can be applied to chemical reactions in polyatomic systems. Examples are presented which illustrate this "menu" of dynamical possibilities.

1. Introduction.

This paper reviews recent (and current) work in my research group which is aimed at developing practical methods for describing reaction dynamics in polyatomic systems in as *ab initio* a framework as possible. To overcome the "dimensionality dilemma" of polyatomic systems--i.e., the fact that the potential energy surface depends on 3N-6 internal coordinates for an N atom system--we have developed dynamical models based on the "intrinsic reaction path", i.e., the steepest descent path which connects reactants and products through the transition state (i.e., saddle point) on the potential energy surface.[1,2]

The determination of such reaction paths using *ab initio* quantum chemistry is feasible nowadays because computational quantum chemists have developed accurate and efficient ways for computing the gradient of the potential energy surface (i.e., the derivative of the Born-Oppenheimer electronic energy with respect to nuclear coordinates).[3] The procedure is that one starts at the transition state on the potential energy surface and then follows the (mass-weighted) gradient vector forward to products and backward to reactants. This is an inherently one-dimensional procedure even though it is taking place in a 3N-6 dimensional space. The internal coordinates of the system are then chosen to be the reaction coordinate, the distance along this path, plus (3N-7) local normal mode coordinates which describe motion orthogonal to the reaction path. Three Euler angles are introduced in

D. C. Clary (ed.), The Theory of Chemical Reaction Dynamics, 27–45.

the usual manner to describe rotation of the N-atom system in 3-dimensional space.

Once a Hamiltonian is constructed in terms of these coordinates and their conjugate momenta--the reaction path Hamiltonian[4]--one needs dynamical theories to describe the reaction dynamics. Section II first discusses the form of the reaction path Hamiltonian, and then Section III describes the variety of dynamical models that have been based on it. These range from the simplest, statistical models (i.e., transition state theory) all the way to rigorous path integral methods that are essentially exact. Various applications are discussed to illustrate the variety of dynamical treatments.

2. The Reaction Path Hamiltonian.

The classical form of the reaction path Hamiltonian is given here; the quantum mechanical Hamiltonian operator corresponding to it can be obtained in standard ways.[4] In the following equations (s,P_s) denote the reaction coordinate and its conjugate momentum; (Q_k,P_k), $k=1,\ldots,F-1$ the local normal coordinates and momenta for vibration orthogonal to the reaction path ($F = 3N-6$); and (K,q_K) the projection of total angular momentum along a body-fixed axis and its conjugate angle variable. It is useful to write the reaction path Hamiltonian in the following form

$$H_J(P_s,s,\{P_k,Q_k\},K,q_K) = H_0^J + H_1^J + H_2^J + \ldots \tag{2.1}$$

where the zeroth order term is

$$H_0^J = \frac{1}{2} P_s^2 + V_0(s) + \sum_{k=1}^{F-1} \left(\frac{1}{2} P_k^2 + \frac{1}{2} \omega_k(s)^2 Q_k^2\right) + \varepsilon_{rot}^J(K,q_K;s) \tag{2.2a}$$

with

$$\begin{aligned}\varepsilon_{rot}^J &= A(s)(J^2-K^2)\cos^2 q_K \\ &+ B(s)(J^2-K^2)\sin^2 q_K \\ &+ C(s)K^2 \quad .\end{aligned} \tag{2.2b}$$

$\{\omega_k(s)\}$ are the frequencies for harmonic motion orthogonal to the reaction path, and $A(s)$, $B(s)$, $C(s)$ the three rotation constants, all as functions of the distance s along the reaction path. $V_0(s)$ is the potential energy along the reaction path. The first order coupling interaction is

$$H_1^J = -P_s^2 \sum_{k=1}^{F-1} Q_k\, B_{k,3N}(s)$$

$$- 2\, P_s \sum_{k=1}^{F-1} Q_k\, \underset{\sim}{B}_{k,3N}(s)\cdot \underset{\approx}{I}_0^{-1}(s)\cdot \underset{\sim}{J}$$

$$- \sum_{k=1}^{F-1} Q_k\, \underset{\sim}{J}\cdot \underset{\approx}{I}_0^{-1}(s)\cdot \underset{\approx}{b}_k \cdot \underset{\approx}{I}_0^{-1}(s)\cdot \underset{\sim}{J} \quad , \tag{2.3}$$

where the coupling functions are defined by

$$B_{k,3N}(s) = \sum_{i=1}^{N} \frac{\partial \underset{\sim}{L}_{i,k}(s)}{\partial s} \cdot \underset{\sim}{L}_{i,3N}(s) \tag{2.4a}$$

$$\underset{\sim}{B}_{k,3N}(s) = \sum_{i=1}^{N} \underset{\sim}{L}_{i,k}(s) \times \underset{\sim}{L}_{i,3N}(s) \tag{2.4b}$$

$$\underset{\approx}{b}_k(s) = \sum_{i=1}^{N} (\underset{\sim}{L}_{i,k}\cdot \underset{\sim}{a}_i - \cdot \underset{\sim}{L}_{i,k}\underset{\sim}{a}_i \cdot) \quad . \tag{2.4c}$$

In Eq. (2.4) $\underset{\approx}{I}_0(s)$ is the 3x3 inertia tensor, $\underset{\sim}{a}_i(s)$ is the Cartesian coordinate vector of atom i, $\underset{\sim}{L}_{i,k}(s)$ is the component of the kth normal mode eigenvector on atom i, all evaluated on the reaction path at distance s along it. The higher order couplings, i.e., H_2, etc., can be obtained from the general expression for the reaction path Hamiltonian.[4]

The three terms in H_1 have the interpretation as curvature couplings, coriolis coupling, and centrifugal distortion, respectively. The latter two obviously involve rotation-vibration coupling--i.e., they are zero if J=0. The first (curvature coupling) term describes energy transfer between the reaction coordinate s and the transverse vibrational modes $\{Q_k\}$. The coupling functions $B_{k,3N}(s)$ of Eq. (2.4a) are determined by the geometry of the reaction path; i.e., they characterize how the <u>total</u> curvature $\kappa(s)$ of the reaction path,

$$\kappa(s) = \left[\sum_{k=1}^{F-1} B_{k,3N}(s)^2 \right]^{1/2} , \qquad (2.5)$$

projects onto the different vibrational modes k.

For applications it is often useful to use an adiabatic representation of the Hamiltonian. A canonical transformation is thus carried out to replace the cartesian coordinates and momenta (P_k, Q_k) by the adiabatic action-angle variables (n_k, q_k). It is also useful to introduce a zeroth order rotation Hamiltonian which is that of a symmetric top; i.e., from Eq. (2.2b) one has

$$\varepsilon^J_{rot} = \frac{1}{2} [A(s)+B(s)](J^2-K^2) + C(s)K^2 + \Delta\varepsilon^J_{rot}\,(K,q_K) , \qquad (2.6a)$$

where

$$\Delta\varepsilon^J_{rot}(K,q_K) \doteq \frac{1}{2} [A(s)-B(s)](J^2-K^2) \cos(2q_K) . \qquad (2.6b)$$

The asymmetric rotor coupling term $\Delta\varepsilon^J_{rot}$ is included in the perturbation term H_2, so that in the adiabatic representation the zeroth order Hamiltonian is that of a one-dimensional system (because J, K, and $\underset{\sim}{n}$ are conserved)

$$H^{(0)}_{JK\underset{\sim}{n}}(P_s, s) = \frac{1}{2} P_s^2 + V^{eff}_{JK\underset{\sim}{n}}(s) \qquad (2.7a)$$

where

$$V^{eff}_{JK\underset{\sim}{n}}(s) = V_0(s) + \sum_{k=1}^{F=1} \hbar\omega_k(s)(n_k+\tfrac{1}{2})$$

$$+ \frac{1}{2} [A(s)+B(s)](J^2-K^2) + C(s)K^2 . \qquad (2.7b)$$

Other aspects of the reaction path Hamiltonian and its application have been discussed in earlier papers to which the reader is referred.[5-12]

3. Dynamical Models.

With a Hamiltonian one can begin to describe dynamics, and this section considers some of the dynamical models that have been based on the reaction-path Hamiltonian, beginning with the simplest approaches and proceeding to more rigorous ones.

3.1 Transition-State Theory and Related Models.

The simplest dynamical models are statistical ones. The microcanonical flux, i.e., average flux for a given total energy, through a "dividing surface" that is perpendicular to the reaction path at distance s_0 along it, is given by

$$N(E,s_0) = \frac{2\pi\hbar}{(2\pi\hbar)^F} \int ds \int dP_s \int d\underset{\sim}{P} \int d\underset{\sim}{Q}\ \delta(E-H)\ \delta(s-s_0)\dot{s}h(\dot{s}) \tag{3.1}$$

where $\dot{s}$, the velocity along the reaction path, is given by Hamilton's equations

$$\dot{s} = \partial H/\partial P_s \quad ; \tag{3.2}$$

the step function $h(\dot{s})$

$$h(\dot{s}) = \begin{cases} 1 & \dot{s} > 0 \\ 0 & \dot{s} < 0 \end{cases}$$

ensures that Eq. (3.1) is the "one-way flux" through the dividing surface. With the reaction path Hamiltonian it is a straightforward calculation to show that Eq. (3.1) and (3.2) give

$$N(E,s) = [E-V_0(s)]^{F-1}/\{(F-1)!\prod_{k=1}^{F-1} \hbar\omega_k(s)\} \tag{3.3}$$

where s_0 has now been replaced simply by s. Remarkably, therefore, none of the coupling functions $B_{k,k'}(s)$ appear in the microcanonical flux; they have not been neglected in the calculation, they simply do not appear in the final result.

Transition-state theory corresponds to looking for the minimum (with respect to s) flux,[13] the main bottleneck to the reactions; i.e.,

$$N_{TST}(E) = \min_s N(E,s) \tag{3.4}$$

The thermal rate constant is given in terms of N(E), the cumulative reaction probability--which is approximated in transition-state theory by the minimum of the microcanonical flux--by

$$k(T) = (2\pi\hbar Q_0)^{-1} \int^{\infty} dE\ N(E)\ \exp(-E/kT)$$

$$= \frac{kT}{h}\frac{1}{Q_0}\int^{\infty} dE\ N'(E)\ \exp(-E/kT) \tag{3.5}$$

where Q_0 is the partition function (per unit volume) of reactants.

If there are several local minima of N(E,s) as a function of s, then this corresponds to several "bottlenecks" of the reactive flux. If one assumes that microcanonical equilibrium is established locally in the regions between these bottlenecks--e.g., by existence of long-lived intermediates--then one can derive a "unified" statistical model.[14,5] This model approximates the cumulative reaction probability as

$$N(E) = \left[\sum_{k=1,3...}^{2M+1} \frac{1}{N_k(E)} - \sum_{k=2,4...}^{2M} \frac{1}{N_k(E)}\right]^{-1} \tag{3.6}$$

where for k = 1,3,5,..., 2M+1, $\{N_k(E)\}$ are the local minima of N(E,s), and for k = 2,4,..., 2M they are the local maxima separating the local minima.

To apply these statistical theories one thus only needs the potential energy along the reaction path $V_0(s)$ and the frequencies $\{\omega_k(s)\}$ of the transverse modes. To locate the extrema of the flux N(E,s) as a function of s one can show[5] that the equation

$$\partial N(E,s)/\partial s = 0 \tag{3.7}$$

is equivalent to the following one

$$E = \mathcal{V}(s) \tag{3.8a}$$

which involves the energy-independent function

$$\mathcal{V}(s) = V_0(s) - V_0'(s)(F-1) \Big/ \sum_{k=1}^{F-1} \frac{\omega_k'(s)}{\omega_k(s)} \quad . \qquad (3.8b)$$

To find the various extrema of the microcanonical flux one thus needs only to plot the function $\mathcal{V}(s)$ and look to see where it is equal to the energy E. This is a simple way to see how the reaction "mechanism" changes with energy. Typically, for example, at low energy E there is only one bottleneck, i.e., one minimum in the flux, so that ordinary transition-state theory is a good approximation, while at higher energies there may be several minima. This latter situation is a herald, even within this statistical description, of more complex dynamics, i.e., "recrossing trajectories", which cause the breakdown of simple transition-state theory.[13]

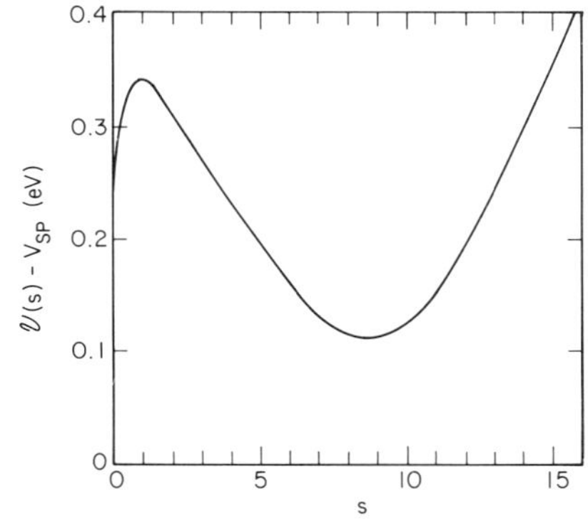

FIGURE 1. The quantity $\mathcal{V}(s) - V_{sp}$, where the former is defined by Eq. (3.8b) and the latter is the saddle point height, as a function of the (mass-weighted) reaction coordinate, for the collinear $H+H_2 \rightarrow H_2+H$ reaction.

To illustrate the above ideas, Fig. 1 shows the quantity $\mathcal{V}(s)$ of Eq. (3.8b) for the collinear $H+H_2$ reaction on the Porter-Karplus[15] H_3 potential energy surface. In the plot the saddle point height V_{sp} is subtracted from $\mathcal{V}(s)$. There are several different energy regions to distinguish.

(a) $0 \leq E - V_{sp} < 0.11$ eV. For energies less than 0.11 eV above the barrier the only extremum in the flux occurs at s = 0, and it is a minimum. Thus Eq. (3.6) with M = 0 applies, and this is the case of simple transition state theory. It is known that transition state theory gives essentially exact agreement with classical trajectory calculations for this energy region.

(b) 0.11 eV < $E - V_{sp}$ < 0.25 eV. Fig. 1 shows that for this energy region there will be two roots of Eq. (3.8a) for $s > 0$, and there will be two symmetrically related roots for $s < 0$, as well as the one at $s = 0$, for a total of five extrema in $N(s,E)$, two maxima and three minima, so that Eq. (3.6) applies with $M = 2$.

(c) 0.25 eV < $E - V_{sp}$ < 0.34 eV. Fig. 1 shows that here there are three roots to Eq. (3.8a) for $s > 0$, and with the three symmetrically related ones for $s < 0$ and the one at $s = 0$, there are seven flux extrema altogether so that Eq. (3.6) applies with $M = 3$.

(d) 0.34 eV < $E - V_{sp}$. For energies more than 0.34 eV above the saddle point Fig. 1 indicates that there is just one root to Eq. (3.8a) for $s > 0$, and with the symmetrically related one for $s > 0$ and the one at $s = 0$, there are three flux extrema, so that Eq. (3.6) applies with $M = 1$.

The appearance and disappearance of various flux extrema as a function of energy is a manifestation of how the reaction mechanism changes as a function of energy; e.g., at low energies other bottlenecks appear. These same phenomena have been seen by Pechukas and Pollak[16] in a more dynamically based theory and by Garrett and Truhlar[17] in a calculation similar in some respects to the one presented here. In the more dynamical theory of Pechukas and Pollak the dividing surfaces at the flux extrema of the present statistical model become periodic classical trajectories that oscillate across the potential valley, i.e., "trapped trajectories".[16] It is interesting to see that many aspects of this more detailed dynamical treatment appear, albeit approximately, in a purely statistical model.

3.2 Semiclassical Perturbation--Infinite Order Sudden Approximation.

Going beyond statistical approximations to more dynamically based treatments opens the door to a wide variety of possibilities, from simple approximate models to more accurate treatments that are capable (with sufficient effort) of arbitrary accuracy. Here I note one particularly simple approximate model that has been developed and applied to a variety of different dynamical phenomena, namely the semiclassical perturbation-infinite order sudden (SCP-IOS) model.[18]

The SCP-IOS model is the semiclassical approximation of Miller and Smith[19] applied to the reaction-path Hamiltonian. It has the appealing feature that it behaves qualitatively correctly both in the adiabatic limit, which is the situation if the transverse vibrational motion is much faster than motion along the reaction coordinate, and also in the sudden limit, which is the case if reaction-coordinate motion is much faster than transverse vibrational motion. For the case of a collinear atom-diatom reaction it becomes the Hofacker-Levine model.[20]

To illustrate how simple it is to apply, e.g., the probability of the vibrational transition $(n_1,\ldots,n_{F-1}) \rightarrow (n'_1,\ldots,n'_{F-1})$ in the transverse vibrational modes during motion from s_1 to s_2 along the reaction path is given by

$$P_{\underset{\sim}{n}_2 \leftarrow \underset{\sim}{n}_1}(E) = \prod_{k=1}^{F-1} |J_{\Delta n_k}(\gamma_k)|^2 \tag{3.9a}$$

where $J_{\Delta n_k}$ is the regular Bessel function of order Δn_k, and the "collision integrals" γ_k are given by

$$\gamma_k = \left| \int_{s_1}^{s_2} ds\ B_{k,3N}(s)\ \{2[E-V_{\underset{\sim}{n}}(s)]\}^{\frac{1}{2}}\ [(2n_k+1)/\omega_k(s)]^{\frac{1}{2}}\ e^{i\delta_k(s)} \right|$$

with

$$V_{\underset{\sim}{n}}(s) = V_0(s) + \sum_{k=1}^{F-1} (n_k + \tfrac{1}{2})\ \omega_k(s)$$

$$\delta_k(s) = \int_{s_1}^{s} ds'\ \omega_k(s')/\{2[E-V_{\underset{\sim}{n}}(s')]\}^{\frac{1}{2}}$$

$$\underset{\sim}{n} = 1/2(\underset{\sim}{n}_1 + \underset{\sim}{n}_2) \quad . \tag{3.9b}$$

The collision integral γ_k is a measure of how much vibrational excitation is induced in mode k during motion from s_1 to s_2: the Bessel function $J_{\Delta n_k}(\gamma_k)^2$ has its maximum at $\Delta n_k \approx \gamma_k$, so that γ_k is the most probable vibrational quantum number change. A typical application of this expression would be to predict the product state distribution of an exothermic chemical reaction: with $\underset{\sim}{n}_1 = 0$, $s_1 = 0$, and $s_2 = \infty$, Eq. (3.9) gives the distribution of product internal degrees of freedom. Clearly the modes with the larger coupling element $B_{k,3N}$ will be the ones excited most during motion from the transition state ($s_1 = 0$) to products ($s_2 = \infty$). Conversely, $s_1 = -\infty$ and $s_2 = 0$ and $\underset{\sim}{n}_2 = 0$ corresponds to the time-reversed situation. In this case the modes k for which γ_k is large are the most effective promoting modes for the reaction; i.e., vibrational energy initially in such a mode will be converted with high probability into energy along the reaction coordinate at the transition state.

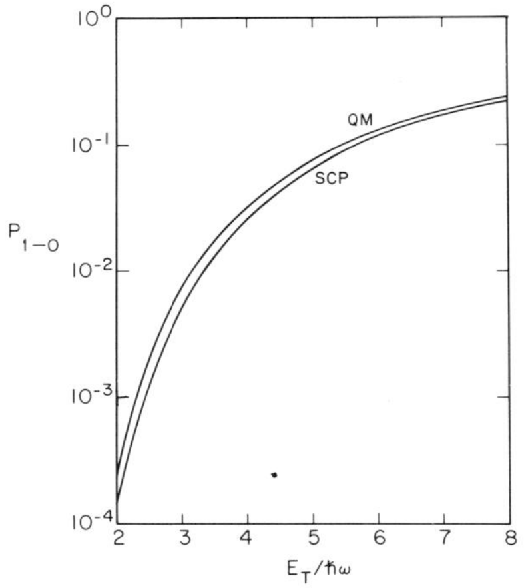

FIGURE 2. Transition probability for the 0 → 1 vibrational excitation of H_2 by collision with He, as a function of total energy. H_2 is modelled as a Morse oscillator. QM and SCP denote the essentially exact quantum mechanical results computed for this collinear system, and the present results of the SCP-IOS reaction path model.

Fig. 2 shows an example of the SCP-IOS model, i.e., Eq. (3.9), applied to vibrational excitation of H_2 by collision with He atoms.[18] One sees that this simple dynamical model based on the reaction path Hamiltonian does an extremely good job of describing vibrational inelasticity.

The SCP-IOS approximation has also been used to describe the effects of the curvature coupling elements on tunneling probabilities in chemical reactions. For example, the probability of tunneling through a simple barrier is given within the SCP-IOS model by

$$P = P_0 \prod_{k=1}^{F-1} I_0(\theta_k)^2 \qquad (3.10a)$$

where

$$P_0 = e^{-2\theta_0} \qquad (3.10b)$$

$$\theta_0 = \int ds \sqrt{2[V_{eff}(s)-E]} \qquad (3.10c)$$

$$\theta_k = \int ds \sqrt{2[V_{eff}(s)-E]} \; \frac{B_{k,3N}(s)}{\sqrt{\omega_k(s)}} \; \cosh \delta_k(s) \quad . \qquad (3.10d)$$

P_0 is the usual one-dimensional WKB tunneling probability, and the effect of curvature coupling is contained in the multiplicative factors $I_0(\theta_k)^2$, one for each mode k. I_0 is the Bessel function of imaginary argument which is an exponentially increasing function; i.e.,

$$I_0(0)^2 = 1$$

$$I_0(\theta_k)^2 \sim e^{2\theta_k}/(2\pi\theta_k)$$

for $\theta_k >> 0$. Curvature coupling thus <u>increases</u> the tunneling probability.

For the well-studied test case, the collinear H + H_2 reaction, for example, Fig. 3 shows the reaction probability as a function of initial translational energy.[6] One sees that P_0 (i.e., VAZC) is a factor of ~50-100 too small, but the SCP-IOS model, i.e., Eq. (3.10) brings it to within a factor of 2 of the correct value.

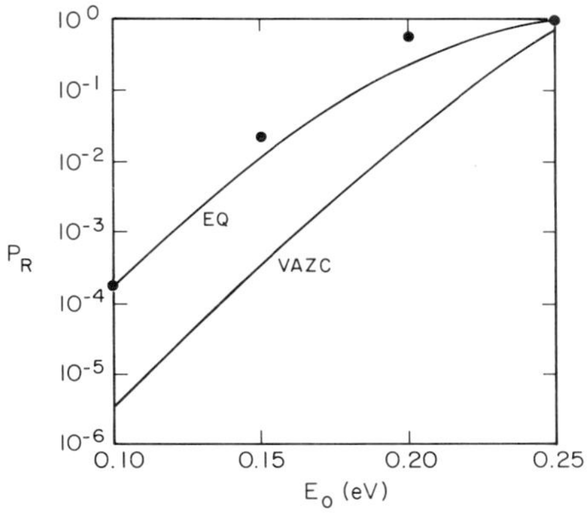

FIGURE 3. Reaction probability for collinear $H+H_2 \rightarrow H_2+H$ on the Porter-Karplus potential energy surface. EQ denotes the exact quantum mechanical values, VAZC the results of the vibrationally adiabatic zero curvature approximation, and the points the results of the present SCP-IOS reaction path model.

The SCP-IOS model has also been used to determine the degree of mode specificity in state-selected unimolecular decomposition.[21] If there were no coupling between the various modes of the polyatomic system, then the unimolecular decomposition would clearly be mode specific: i.e., different initial states with essentially the same total energy would decay at different rates because they would have various amount of energy in the reaction coordinate and there would be no energy transfer among the various degrees of freedom. Conversely, to the extent that coupling between the modes causes efficient energy transfer among them, one expects more statistical behavior, i.e., the rate of decomposition depending essentially only on the total energy of the initial state and not on the particular initial state that is prepared. The degree of mode specificity in the state-specific unimolecular decay rates is thus a sensitive measure of the intermode coupling and thus a direct test of the way that the SCP-IOS, or any other model, is able to describe this.

3.3 Rigorous Rate Constants via Path Integrals.

The reaction path Hamiltonian also provides a very useful framework for the rigorous calculation of the Boltzmann (i.e., thermally averaged) rate constant for a chemical reaction using the path integral methods described by Miller, Schwartz, and Tromp.[22] In that paper it is shown that the rate constant can be expressed as the time integral of a flux-flux autocorrelation function

$$k = Z_R^{-1} \int_0^\infty dt\ C_f(t) \quad , \tag{3.11}$$

where Z_R is the partition function (per unit volume) for reactants, and the correlation function is defined by

$$C_f(t) = \mathrm{tr}[F\ e^{-(\beta/2-it/\hbar)H}\ F\ e^{-(\beta/2+it/\hbar)H}] \quad ; \tag{3.12}$$

"tr" denotes a quantum mechanical trace, $\beta = (k_BT)^{-1}$, H is the full Hamiltonian of the system, and F is the symmetrized flux operator

$$F = \frac{1}{2}\,[\delta(s)(p_s/m) + (p_s/m)\delta(s)] \quad , \tag{3.13}$$

where s is the coordinate normal to the "dividing surface" (s=0 is the equation of the surface) through which the flux is computed. Dynamics enters in this expression for the rate via the time evolution operators, and the anticipated efficiency of this "direct" approach comes about because it is necessary to determine the quantum dynamics only for relatively short times.

To make it possible to deal with systems with many degrees of freedom, the Boltzmann operator/time evolution operators $e^{-(\beta/2 \pm it/\hbar)H}$ are represented by a Feynman path integral[23] and the path integral evaluated by a Monte Carlo random walk method. It is in general not feasible to do this for real values of the time t, however, because the integrand of the path integral would be oscillatory. We thus first calculate C_f for real values of $\tau \equiv it$, i.e., pure imaginary time,

$$C_f(t) \equiv tr[F\, e^{-(\beta/2-\tau/\hbar)H}\, F\, e^{-(\beta/2+\tau/\hbar)H}] \quad , \tag{3.14}$$

and then use these calculated values to generate a numerical analytic continuation to obtain C_f for real t. Since C_f is an even function of t, and thus of τ, one uses the values computed for real τ to construct a Padé approximant in the variable τ^2; τ^2 is then replaced by $-t^2$ to obtain $C_f(t)$. (This approach is similar in spirit, though not detail, to methods used by Berne et al.[24] to calculate quantum mechanical dipole-dipole autocorrelation functions.)

The reaction path Hamiltonian is particularly useful for evaluating the path integral representation of the trace, Eq. (3.14), because the flux operator F does not involve the "bath" degrees of freedom (i.e., the transverse vibrational modes $\{Q_k\}$), and since they are harmonic oscillators the path integrals over them can be carried out analytically.[23] All that remains to be done numerically is the path integral over only the reaction coordinate degrees of freedom itself.

The specific form one obtains for the flux correlation function at complex times $\tau_n = \hbar\beta(n/N - 1/2)$, n=1,..., N-1, is[25]

$$C_f(i\tau_n) = \left(\frac{\hbar}{2m}\right)^2 \left(\frac{\partial^2}{\partial s_N \partial s_n} + \frac{\partial^2}{\partial s_0 \partial s_n} - 4\frac{\partial^2}{\partial s_N \partial s_0}\right) \left(\frac{mN}{2\pi\hbar^2\beta}\right)^{N/2}$$

$$\times \int ds_1 \ldots \int ds_{n-1} \int ds_{n+1} \ldots \int ds_{N-1} \exp\{-\frac{1}{\hbar} \int_{-\frac{\hbar\beta}{2}}^{\frac{\hbar\beta}{2}} d\tau\, [\tfrac{1}{2}\dot{s}(\tau)^2 + V_0(s(\tau))]\}$$

$$\times Z[s(\tau)] \tag{3.15}$$

where $Z[s(\tau)]$ is the partition functional of all the degrees of freedom other than the reaction coordinate. If rotation-vibration coupling is ignored, one has

$$Z[s(\tau)] = Z_{vib}[s(\tau)]\; Z_{rot}[s(\tau)] \quad , \qquad (3.16a)$$

$$Z_{rot}[s(\tau)] = \left(\frac{\pi(kT)^2}{ABC}\right)^{1/2} \qquad (3.16b)$$

$$Z_{vib}[s(\tau)] = e^{\Delta} \prod_{k=1}^{F-1} [2\sinh(u_k/2)]^{-1} \qquad (3.16c)$$

with

$$u_k \equiv u_k[s(\tau)] = \int_{-\frac{\hbar\beta}{2}}^{\frac{\hbar\beta}{2}} d\tau\; \omega_k(s(\tau)) \qquad (3.17a)$$

$$A \equiv A[s(\tau)] = \frac{1}{\hbar\beta} \int_{-\frac{\hbar\beta}{2}}^{\frac{\hbar\beta}{2}} d\tau\; A(s(\tau)) \quad , \qquad (3.17b)$$

with B and C defined similarly. The exponent Δ in Eq. (3.16c),

$$\Delta = \sum_{k=1}^{F-1} \frac{1}{4\hbar(1-e^{-u_k})} \int_{-\frac{\hbar\beta}{2}}^{\frac{\hbar\beta}{2}} d\tau \int_{-\frac{\hbar\beta}{2}}^{\frac{\hbar\beta}{2}} d\tau' \frac{\dot{s}(\tau)^2 B_{k,3N}(s(\tau))\dot{s}(\tau')^2 B_{k,3N}(s(\tau'))}{[\omega_k(s(\tau))\;\omega_k(s(\tau'))]^{1/2}}$$

$$\times \exp\left[-\int_{\tau_<}^{\tau_>} d\tau''\; \omega_k(s(\tau''))\right] \quad , \qquad (3.18)$$

is the quantum analog of a friction that the reaction coordinate motion experiences due to coupling to the bath modes. The path integral over the reaction coordinate degree of freedom [cf. Eq. (3.15)] is carried out by a Monte Carlo random walk algorithm.[22]

Eqs. (3.15)-(3.18) are a very powerful result and should be very useful. They show that, based on the reaction path Hamiltonian, one can determine the thermal rate constant for a general polyatomic reaction by carrying out an essentially one-dimensional quantum mechanical calculation (i.e., the path integral over the s-degree of freedom). True, this one-dimensional quantum mechanical problem is somewhat more complicated than a one-dimensional Schrödinger equation--because of the partition functional in the integrand--but when evaluating the path integral by Monte Carlo it is only marginally more difficult.

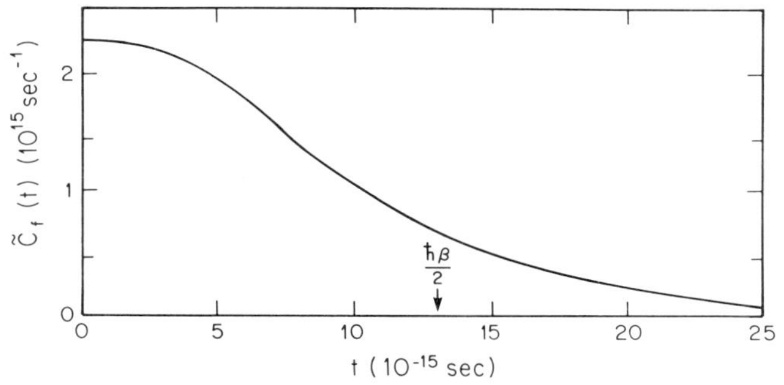

FIGURE 4. The flux-flux autocorrelation function for the 3-d $H+H_2$ reaction at T = 300°K.

Figure 4 shows the quantity $\tilde{C}_f(t)$,

$$\tilde{C}_f(t) \equiv C_f(t)/k_{TST} \quad ,$$

i.e., the ratio of the real time flux correlation function (obtained by analytic continuation of the values computed at imaginary time) to the (time-independent) transition state theory rate constant for the three-dimensional version of the $H+H_2$ reaction. The integral of $\tilde{C}_f(t)$ is thus the correction factor κ which multiplies the transition state theory rate constant to give the exact result,

$$k = \kappa \, k_{TST} \quad .$$

κ includes the effects of tunneling through the barrier, non-separability of the reaction coordinate from the other degrees of freedom, and recrossing of the transition state dividing surface.

One sees from Fig. 4 that it is necessary to determine the

quantum dynamics of this system for a time of ~ 20 fsec in order to obtain the rate constant. The integral of $C_f(t)$ gives a value of $\kappa \simeq 22$, which is the same value obtained from the quantum scattering calculations of Schatz and Kuppermann.[26]

4. Concluding Remarks.

I have attempted in this paper to illustrate the wide variety of dynamical treatments that can be usefully based on the reaction path Hamiltonian model, from simple "back of the envelope" statistical approximations (TST, RRKM, etc.) all the way to rigorous computational methods that can be practically applied to polyatomic systems. Given the necessary "input" which characterizes the model -- i.e., the quantum chemistry calculations of the reaction path, and the energy and force constant matrix along it -- the example applications that have been discussed show that it provides a quantitative *ab initio* approach to reaction dynamics in polyatomic molecular systems.

In this paper I have not reviewed the various chemical applications that have utilized the reaction path model. These include work in my research group on the unimolecular isomerization of hydrogen isocyanide (HNC → HCN), the unimolecular decomposition of formaldehyde ($H_2CO \rightarrow H_2 + CO$), the vinylidene-acetylene rearrangement ($H_2C{=}C: \rightarrow HC{\equiv}CH$), and hydrogen atom transfer in malonaldehyde,

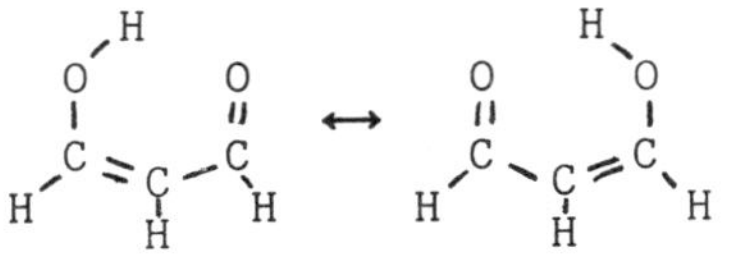

A number of interesting applications have also been carried out by Fukui, *et al.*, including hydrogen abstraction from methane,[2d] by Morokuma *et al.*, including HF elmination from CHFO[27] and C_2H_5F,[28] and by Yamabe *et al.*[29] including unimolecular decomposition of thioformaldehyde,[30] and double proton exchange in the formamidine-water system.[31] In addition, a number of atom-diatom reaction systems have been treated using reaction path descriptions by Truhlar *et al.*[12]

Since the *ab initio* determination of a minimum energy reaction path (and its necessary properties) is becoming feasible for increasingly complex polyatomic systems, one expects to see applications of reaction path methods to increasingly interesting chemical processes. This further motivates one to develop dynamical methods based on these models.

Acknowledgments

This work has been supported by the Director, Office of Energy Research, Office of Basic Energy Sciences, Chemical Sciences Division of the U.S. Department of Energy under Contract Number DE-AC03-76SF00098.

References

1. For early work on reaction paths and reaction coordinates, see (a) S. Glasstone, K. J. Laidler, and H. Eyring, The Theory of Rate Processes (McGraw-Hill, New York, 1941); (b) R. A. Marcus, J. Chem. Phys. **45**, 4493, 4500 (1966); **49**, (1968); (c) G. L. Hofacker, Z. Naturforsch. Teil A **18**, 607 (1963); J. Chem. Phys. **43**, 208 (1965); (d) S. F. Fischer, G. L. Hofacker, and R. Seiler, J. Chem. Phys. **51**, 3941 (1969).
2. Some of the recent papers by other workers on reaction path models are (a) S. F. Fischer and M. A. Ratner, J. Chem. Phys **57**, 2769 (1972); (b) P. Russegger and J. Brickman, ibid. **62**, 1086 (1975); **60**, 1 (1977); (c) M. V. Basilevsky, Chem. Phys. **24**, 81 (1977); **67**, 337 (1982); M. V. Bailevsky and A. G. Shamov, ibid. **60**, 349 (1981); (d) K. Fukui, S. Kato, and H. Fujimoto, J. Am. Chem. Soc. **97**, 1 (1975); K. Yamashita, T. Yamabe, and K. Fukui, Chem. Phys. Lett. **84**, 123 (1981); A. K. Fukui, Acc. Chem. Res. **14**, 363 (1981); (e) K. Ishida, K. Morokuma, and A. Komornicki, J. Chem. Phys. **66**, 2153 (1977); (f) A. Nauts and X. Chapuisat, Chem. Phys. Lett. **85**, 212 (1982); X. Chapuisat, A. Nauts, and G. Durrand, Chem. Phys. **56**, 91 (1981); (g) J. Pancir, Collect. Czech. Commun. **40**, 1112 (1975); **42**, 16 (1977); (h) G. A. Natanson, Mol. Phys. **46**, 481 (1982).
3. (a) P. Pulay, in Applications of Electronic Structure, edited by H. F. Schaefer (Plenum, New York, 1977), p. 153; (b) J. W. McIver, Jr. and A. Komornicki, J. Am. Chem. Soc. **94**, 2625 (1972); (c) B. R. Brooks, W. E. Laidig, P. Saxe, J. D. Goddard, Y. Yamaguchi, and H. F. Schaefer, J. Chem. Phys. **72**, 4652 (1980); (d) J. A. Pople, R. Krishnan, H. B. Schlegel, J. S. Binkley, Int. J. Quant. Chem. Symp. **13**, 225 (1970); (e) Y. Osamura, Y. Yamaguchi, P. Saxe, M. A. Vincent, J. F. Gaw, and H. F. Schaefer, Chem. Phys. **72**, 131, (1982); Y. Yamaguchi, Y. Osamura, G. Fitzgerald, H. F. Schaefer, J. Chem. Phys. **78**, 1607 (1983); (f) P. Pulay, ibid. **78**, 5043 (1983).
4. W. H. Miller, N. C. Handy, and J. E. Adams, J. Chem. Phys. **72**, 99 (1980).
5. W. H. Miller, in Potential Energy Surfaces and Dynamics Calculations, edited by D. G. Truhlar (Plenum, New York, 1981), p. 265.
6. C. J. Cerjan, S.-h. Shi, and W. H. Miller, J. Phys. Chem. **86**, 2244 (1982).
7. S. K. Gray, W. H. Miller, Y. Yamaguchi, and H. F. Schaefer, J. Chem. Phys. **73**, 2733 (1980).

8. S. K. Gray, W. H. Miller, Y. Yamaguchi, and H. F. Schaefer, J. Am. Chem. Soc. **103**, 1900 (1981).
9. Y. Osamura, H. F. Schaefer, S. K. Gray, and W. H. Miller, J. Am. Chem. Soc. **103**, 1904 (1981).
10. B. A. Waite, S. K. Gray, and W. H. Miller, J. Chem. Phys. **78**, 259 (1983).
11. W. H. Miller, J. Phys. Chem. **87**, 3811 (1983).
12. See also, (a) R. T. Skodje, D. G. Truhlar, and B. C. Garrett, J. Phys. Chem. **85**, 3019 (1981); (b) J. Chem. Phys. **77**, 5955 (1982); (c) A. D. Isaacson and D. G. Truhlar, ibid. **76**, 1380 (1982); (d) D. G. Truhlar, N. J. Kilpatrick, and B. C. Garrett, ibid. **78**, 2438 (1983); (e) R. T. Skodje and D. G. Truhlar, ibid. **79**, 4882 (1983); (f) R. T. Skodje, D. W. Schwenke, D. G. Truhlar, and B. C. Garrett, J. Phys. Chem. **88**, 628; (g) B. C. Garrett and D. G. Truhlar, J. Chem. Phys. **81**, 309 (1984).
13. For an excellent review and recent developments see (a) P. Pechukas, in 'Dynamics of Molecular Collisions' Part B, Vol. 2 of Modern Theoretical Chemistry, edited by W. H. Miller (Plenum, New York, 1970), p. 285 et seq.; Annu. Rev. Phys. Chem. **32**, 159 (1981); (b) D. G. Truhlar, W. L. Hase, and J. T. Hynes, J. Phys. Chem. **87**, 2664 (1983).
14. W. H. Miller, J. Chem. Phys. **65**, 2216 (1976).
15. R. N. Porter and M. Karplus, J. Chem. Phys. **40**, 1105 (1964).
16. E. Pollak and P. Pechukas, J. Chem. Phys. **70**, 325 (1979).
17. B. C. Garrett and D. G. Truhlar, J. Phys. Chem. **83**, 1052 (1979); **83**, 3058(E) (1979).
18. W. H. Miller and S.-h. Shi, J. Chem. Phys. **75**, 2258 (1981).
19. (a) W. H. Miller and F. T. Smith, Phys. Rev. A **17** 939 (1978); see also related works by: (b) R. J. Cross, Jr., J. Chem. Phys. **49**, 1753 (1968); **58**, 5178 (1973); (c) I. L. Beigman, L. A. Vainshtein, I. I. Sobel'man, Zh. Eksp. Teor. Fiz. **57**, 1465 (1969) (Sov. Phys.--JETP (Engl. Transl.) **30**, 920 (1973); (d) R. D. Levine and B. R. Johnson, Chem. Phys. Lett. **7**, 404 (1970); (e) I. C. Percival and D. Richards, J. Phys. B **3**, 1035 (1970); (f) A. P. Clark, A. S. Dickinson, and D. Richards, Adv. Chem. Phys. **36**, 63 (1977).
20. G. L. Hofacker and R. D. Levine, Chem. Phys. Lett. **9**, 617 (1971).
21. B. A. Waite and W. H. Miller, J. Chem. Phys. **76**, 2412 (1982).
22. W. H. Miller, S. D. Schwartz, and J. W. Tromp, J. Chem. Phys. **79**, 4889 (1983).
23. R. P. Feynman and A. R. Hibbs, Quantum Mechanics and Path Integrals, (McGraw-Hill, New York, 1965) p. 273 et seq.
24. (a) D. Thirumalai and B. J. Berne, J. Chem. Phys. **79**, 5029 (1983); (b) D. Thirumalai, E. J. Bruskin, and B. J. Berne, J. Chem. Phys. **79**, 5063 (1983); (c) D. Thirumalai and B. J. Berne, J. Chem. Phys. **81**, 2512 (1984).
25. K. Yamashita and W. H. Miller, J. Chem. Phys. **82**, 0000 (1985).
26. G. C. Schatz and A. Kuppermann, J. Chem. Phys. **65**, 4668 (1976).
27. K. Morokuma, S. Kato, and K. Hirao, J. Chem. Phys. **72**, 6800 (1980).
28. S. Kato and T. Yamabe, Int. J. Quant. Chem. **17**, 177 (1983).

30. A. Tachibana, I. Okazaki, M. Koizumi, K. Hori, and T. Yamabe, J. Am. Chem. Soc. **107**, 1190 (1985).
31. T. Yamabe, K. Yamashita, M. Kaminiyama, M. Koizumi, A. Tachibana, and K. Fukui, J. Phys. Chem. **88**, 1459 (1984).

REDUCED DIMENSIONALITY THEORIES OF QUANTUM REACTIVE SCATTERING: APPLICATIONS TO MU+H_2, H+H_2, O(^{3}P)+H_2, D_2 AND HD

Joel M. Bowman
Department of Chemistry
Illinois Institute of Technology
Chicago, IL 60616

Albert F. Wagner
Chemistry Division
Argonne National Laboratory
Argonne, IL 60439

A hierarchy of reduced dimensionality exact quantum theories of reactive scattering is presented for the vibrational state-to-state cumulative reaction probability and vibrational state-to-state thermal rate constant. The central approximation in these theories is the adiabatic treatment of the bending motion of the reactive species in the strong interaction region of configuration space. Applications of the theories are made to the reactions Mu+H_2, H+H_2, O(^{3}P)+H_2, D_2 and HD

1. Introduction

Quantum reactive scattering calculations are hampered by several difficulties which have severely impeded progress. One major difficulty is in choosing a system of coordinates which can describe all the possible reaction channels. Progress in this area has been achieved, at least conceptually, by using hyperspherical coordinates. These are described in the chapters by Kuppermann, Romelt and others. Other techniques use different coordinates for each reaction channel. This results in the problem of matching the wavefunctions in each coordinate system, as described in the chapter by Schatz, or in dealing with nonlocal potentials, as described in the chapters by Kouri and by Miller. Progress in overcoming this difficulty will certainly be made by these people and others in the near future.

The other major difficulty which cannot be overcome by a judicious choice of coordinates or coupling scheme is the very large number of coupled differential or integro-differential equations to be solved by any basis set approach. This number is given (at least) by the number of energetically accessible vibrational-rotational states. This number can vary from roughly one hundred to thousands for many reac-

D. C. Clary (ed.), The Theory of Chemical Reaction Dynamics, 47–76.

tions of interest and the vast majority of these states are the rotational states within the sparse manifold of vibrational states. The major component of the rotational states is the degenerate rotational projection states. From a computational point of view a very useful approximation is to decouple these projection states. Such an approximation exists and is called the centrifugal sudden (CS) approximation. It was first formulated for non-reactive scattering[1,2] and it has been used by several groups for reactive scattering.[3-6] In this book, Schatz applies it to studies of the $H+H_2$ reaction and it s isotopic variations. Although this method extends the range of systems which can be studied, it cannot be routinely applied to many atom-diatom systems of interest where the number of coupled rotational states may still be prohibitively large.

We have developed an approximation which goes one step further than the CS approximation. In addition to that approximation we assume (for atom plus diatom systems) that the three-atom bending motion in the strong interaction region can be treated by an uncoupled adiabatic representation.[7-12] As a result of this additional approximation, only two degrees of freedom are explicitly coupled and the number of coupled equations to be integrated is determined by the number of vibrational states open in the reaction, a substantial reduction over the number of coupled equations in the CS and the full coupled channel approaches. This approximation was motivated by results of accurate quantum studies of the collinear, coplanar and three-dimensional $H+H_2$ reaction. In going from the collinear, coplanar and three-dimensional cases the reaction threshold energies shift by the zero-point bend energy at the transition state.[13-16] This observation clearly demonstrated the existence of zero-point bending energy of the H_3 transition state. However, it was not clear how to build an approximate quantitative theory based on this observation. We were able to do this and as we show in detail in Section 2 this approximation allows us to obtain the so-called cumulative reaction probability which is summed over all initial and final rotational states.

We term this adiabatic bending approach reduced dimensionality exact quantum (RDEQ) theory because a reduced number of degrees of freedom are explicitly coupled. Independently, and somewhat later Walker and Hayes[16,17] introduced a related approach which they termed the bending corrected rotating linear model (BCRLM). This theory is an extension of the earlier rotating linear model of Wyatt[18] and Connor and Child.[19] In the BCRLM method only the ground bending state has been considered and the identification of the resulting probability as the cumulative probability was not made. Walker and Pollak have recently revised the BCRLM to make this identification.[20] (Also, see the chapter by Walker and Hayes in this book.) With that revision of the original BCRLM, that theory and the RDEQ theory are now quite similar. Recently, Carrington and Miller have generalized the reduced dimensionality approach to polyatomic systems, where two degrees of freedom are explicitly coupled.[21]

We have given a detailed review of the RDEQ theory and applications recently[12] and so here we present a brief review of the theory and suggest some new extensions of it in Section 2. Some new tests of the theory are presented for the $Mu+H_2$ reaction as well as preliminary results for excited bending state resonances in $H+H_2$ in Section 3. A review of calculations on the $O(^3P)+H_2$, D_2 and HD reactions is also given there.

2. Theory

The reduced dimensionality quantum theory of reactive scattering we have developed focuses on the so-called cumulative reaction probability. In order to motivate and review the importance of this quantity we begin this section with the rigorous collisional expressions for the thermal rate constants for an A+BC -> AB+C reaction.

Specifically, consider first the thermal rate constant which is vibrationally state-to-state but rotationally summed and averaged. This rate constant is given by

$$k_{v->v'}(T) = \frac{\exp(-E_v/kT)}{Q_{vib-rot}(T)} \sum_{j=0} (2j+1)\exp(-E_j/kT)\left(\frac{2}{kT}\right)^{3/2} \frac{1}{(\pi\mu)^{1/2}}$$

$$\times \int_0^\infty dE_t \; E_t \sigma_{vj->v'}(E_t)\exp(-E_t/kT), \tag{1}$$

where E_j is the rotational energy of BC, E_v is the BC vibrational energy, μ is the A,BC reduced mass, E_t is the initial relative kinetic energy, $\sigma_{vj->v'}(E_t)$ is the degeneracy averaged reaction cross section, summed over final AB rotational states and $Q_{vib-rot}(T)$ is the BC vibrational-rotational partition function. The cross section $\sigma_{vj->v'}(E_t)$ is given by

$$\sigma_{vj->v'}(E_t) = \frac{\pi}{k_{vj}^2(2j+1)} \sum_{\Omega j'}\sum_{\Omega'}\sum_{J=0} (2J+1)P^J_{vj\Omega->v'j'\Omega'}, \tag{2}$$

where $P^J_{vj\Omega->v'j'\Omega'}$ is the partial wave probability for the detailed vibrational-rotational state-to-state transition. The quantum numbers Ω and Ω' are the projection quantum numbers of j and j' on the body-fixed z-axes, J is the total angular momentum quantum number, and $|\Omega|$ and $|\Omega'|$ are less than or equal to J, and

$$k_{vj}^2 = \frac{2\mu}{\hbar^2} E_t = \frac{2\mu}{\hbar^2} (E-E_v-E_j) \tag{3}$$

where E is the total energy. Note, we have adopted the common approximation of expressing the vibrational-rotational energy of BC as E_v + E_j. Inserting eqs. (2) and (3) into (1), we have

$$k_{v->v'}(T) = \frac{1}{Q_{vib-rot}(T)} \frac{h^2}{(2\pi kT\mu)^{3/2}} \sum_{j=0} \exp[-(E_j+E_v)/kT]$$

$$x \int_o^\infty dE_t \sum\sum_{\Omega j'} \sum_{\Omega'} \sum_{J} (2J+1)P^J_{vj\Omega->v'j'\Omega'}(E_t)\exp(-E_t/kT). \quad (4)$$

Noting that the total energy E is just the sum $E_t+E_v+E_j$ and that

$$\frac{h^2}{(2\pi kT\mu)^{3/2}} = \frac{1}{hQ_{trans}(T)}, \quad (5)$$

where $Q_{trans}(T)$ is the A+BC translational partition function, (4) can be rewritten as

$$k_{v->v'}(T) = \frac{1}{hQ_{vib-rot}(T)Q_{trans}(T)} \sum_{j} \int_o^\infty d(E-E_v-E_j)$$

$$x \sum\sum_{\Omega j'} \sum_{\Omega'} \sum_{J}(2J+1)P^J_{vj\Omega->v'j'\Omega'}(E-E_v-E_j) \exp[-(E_t+E_j+E_v)/kT]. \quad (6)$$

Because $P^J_{vj\Omega->v'j'\Omega'}$ has the property that it vanishes for negative values of $E-E_v-E_j$, (6) can be rewritten as

$$k_{v->v'}(T) = \frac{1}{hQ_{vib-rot}(T)_{trans}(T)} \int_o^\infty dE \sum\sum\sum_{j\Omega j'} \sum_{\Omega'} \sum_{J}(2J+1)$$

$$x \quad P^J_{vj\Omega->v'j'\Omega'}(E-E_v-E_j)\exp(-E/kT) \quad (7)$$

We now define the partial wave vibrational state-to-state cumulative probability $P^J_{v->v'}(E)$ as follows

$$P^J_{v->v'}(E-E_v) = \sum\sum\sum_{j\Omega j'} \sum_{\Omega'} P^J_{vj\Omega->v'j'\Omega'}(E-E_v-E_j). \quad (8)$$

That is, $P^J_{v->v'}(E-E_v)$ is the sum of the partial wave rotational state-to-state probabilities evaluated at the appropriate translational energy for a given fixed total energy E. Because the quantum number indices are sufficient to determine this translational energy the notation $E-E_v-E_j$ and $E-E_v$ is redundant and so we can simply express (8) as

$$P^J_{v->v'}(E) = \sum\sum\sum_{j\Omega j'} \sum_{\Omega'} P^J_{vj\Omega->v'j'\Omega'}(E) \quad (9)$$

Finally, $k_{v\to v'}(T)$ can be written compactly as

$$k_{v\to v'}(T) = \frac{1}{hQ_{vib-rot}(T)Q_{trans}(T)} \int_0^\infty dE \sum_{J=0} (2J+1)P^J_{v\to v'}(E)e^{-E/kT}. \quad (10)$$

Also, it should be obvious that usual thermal rate constant k(T),

$$k(T) = \sum_{vv'}\sum k_{v\to v'}(T), \quad (11)$$

is given by

$$k(T) = \frac{1}{hQ_{vib-rot}(T)Q_{trans}(T)} \int_0^\infty dE \sum_{J=0} (2J+1)P^J(E)e^{-E/kT}, \quad (12)$$

where

$$P^J(E) = \sum_{vv'}\sum P^J_{v\to v'}(E). \quad (13)$$

Some additional quantities are useful to define. Let the cumulative vibrational state-to-state and total reaction probabilities, $P^J_{v\to v'}(E)$ and P(E), be

$$P^J_{v\to v'} = \sum_{J=0} (2J+1)P^J_{v\to v'}(E) \quad (14a)$$

and

$$P(E) = \sum_{vv'}\sum P_{v\to v'}(E) \quad (14b)$$

Then, the most compact expressions for $k_{v\to v'}(T)$ and k(T) are

$$k_{v\to v'}(T) = \frac{1}{hQ_{vib-rot}(T)Q_{trans}(T)} \int_0^\infty dE\ P_{v\to v'}(E)\exp(-E/kT) \quad (15)$$

$$k(T) = \frac{1}{hQ_{vib-rot}(T)Q_{trans}(T)} \int_0^\infty dE\ P(E)\exp(-E/kT). \quad (16)$$

At this point it is important to recall that $k_{v\to v'}(T)$ is the vibrational state-to-state rate constant from a thermal distribution of initial vibrational (and rotational) states of the reactant diatom. In some experiments the vibrational state v may not be from a thermal distribution and in this case the rate constant assuming all initial vibrational states are v can be obtained trivially from eq. (15) by multiplying that equation by $Q_{vib}(T)/e^{-E_v/kT}$.

With these rigorous expressions for $k_{v\to v'}(T)$ and k(T) in terms of cumulative reaction probabilities, a framework for approximations has been established. (We should also note that these rate constants can also be expressed in terms of rotationally averaged integral cross sections.[7,8]) We now wish to review approximations we have introduced for $P^J_{v\to v'}(E)$. The basis for these is the exact body-frame formulation of reactive scattering.[22] Within this formulation the CS approximation is made first. As noted earlier, in this approximation Ω-coupling is ignored, i.e., Ω is assumed to be a good quantum number. However, j-coupling is retained (with the attendant large number of coupled equations). We make a further approximation in which the rotational motion of the diatom is described in the strong interaction region by an uncoupled adiabatic description. The details of this formulation have been given elsewhere[12] and so we proceed with only a brief outline of the approximation. The reduced dimensionality hierarchy of approximations is then presented.

2.1 Adiabatic bend theory

We start with the full three-dimensional quantum formulation for the A+BC reaction in body-fixed coordinates[23] as given in detail by Schatz and Kuppermann.[22] The Schroedinger equation in terms of the mass-scaled Delves vectors $\mathbf{r}$ (the BC relative position vector) and $\mathbf{R}$ (the position vector of A to the center-of-mass of BC) is

$$\left[-\frac{\hbar^2}{2\mu}(\nabla_R^2 + \nabla_r^2) + V(\mathbf{R},\mathbf{r}) - E\right]\Psi(\mathbf{R},\mathbf{r}) = 0, \qquad (17)$$

where

$$\mu = [m_A m_B m_C/(m_A+m_B+m_C)]^{\frac{1}{2}} \qquad (18)$$

A standard partial wave decomposition of $\Psi(\mathbf{R},\mathbf{r})$ is done in terms of the good quantum numbers J and M, the total angular momentum and its projection on a space-fixed axis respectively, and then each component is expressed as

$$\Psi^{JM}(\mathbf{R},\mathbf{r}) = \sum_{\Omega=-J}^{J} D^J_{M\Omega}(\phi,\Theta,0)\Psi^{J\Omega}(r,R,\gamma,\psi), \qquad (19)$$

in the body-fixed coordinate system where the z-axis is along $\mathbf{R}$. The angles γ and ψ are the polar and azimuthal angles of $\mathbf{r}$ in the body-fixed system and Ω is the projection quantum number of $\mathbf{J}$ on the body-fixed z-axis. The resulting partial differential equation for $\Psi^{J\Omega}$ has been given previously.[22] In the CS approximation it is,

$$\{\frac{-\hbar^2}{2\mu}\left(\frac{\partial^2}{R\partial R^2}R + \frac{\partial^2}{r\partial r^2}r\right) + \frac{j_{op}^2}{2\mu r^2} + \frac{[J(J+1)\hbar^2 + j_{op}^2 - 2\Omega j_z\hbar]}{2\mu R^2}$$

$$+ V(r,R,\gamma) - E\}\Psi^{J\Omega}(r,R,\gamma,\psi) = 0. \qquad (20)$$

where j_{op}^2 is the square of the angular momentum operator associated with $\mathbf{r}$, expressed in terms of γ and ψ.

Instead of proceeding with the coupled channel expansion of $\Psi^{J\Omega}$ in terms of eigenfunctions of j_{op}^2 we consider an adiabatic approximation to the γ-motion. Thus, we shall express $\Psi^{J\Omega}$ as

$$\Psi^{J\Omega}(r,R,\gamma,\psi) = \exp(i\Omega\psi)\xi_{n\Omega}(\gamma;r,R)U_{n\Omega}^J(r,R)/(rR\sqrt{2\pi}). \qquad (21)$$

Inserting this into (20) we have

$$\{\frac{-\hbar^2}{2\mu}\left(\frac{\partial^2}{\partial R^2} + \frac{\partial^2}{\partial R^2}\right) - \frac{\hbar^2}{2I(r,R)}\left(\frac{\partial^2}{\partial\gamma^2} + \cot\gamma\frac{\partial}{\partial\gamma} - \frac{\Omega^2}{\sin^2\gamma} + \frac{[J(J+1)-2\Omega^2]\hbar^2}{2\mu R^2}\right.$$

$$+ V(r,R,\gamma) - E\}\xi_{n\Omega}(\gamma;r,R)U_{n\Omega}^J(r,R) = 0, \qquad (22)$$

where $I(r,R)$ equals $\mu(r^{-2} + R^{-2})^{-1}$. As it stands this equation is no less exact than eq. (20). The adiabatic approximation is now made. Let $\xi_{n\Omega}(\gamma;r,R)$ be an eigenfunction of the r and R-fixed Schroedinger equation

$$[\frac{-\hbar^2}{2I(r,R)}\left(\frac{\partial^2}{\partial\gamma^2} + \cot\gamma\frac{\partial}{\partial\gamma} - \frac{\Omega^2}{\sin^2\gamma}\right) + V_b(\gamma,r,R) - \varepsilon_{n\Omega}(r,R)]\xi_{n\Omega}(\gamma;r,R) = 0, \qquad (23)$$

where $V_b(\gamma,r,R)$ is

$$V_b(\gamma,r,R) = V(r,R,\gamma) - V(r,R,\gamma=0), \qquad (24)$$

and $V(r,R,\gamma=0)$ is the full potential for the collinear reaction A+BC. (Note, $V(r,R,\gamma=\pi)$ would be the full potential for the collinear A+CB reaction.) $\varepsilon_{n\Omega}(r,R)$ is the adiabatic bending eigenvalue. Making the adiabatic approximation now results in the following Schroedinger equation for the unknown function $U_{n\Omega}^J(r,R)$:

$$\{\frac{-\hbar^2}{2\mu}\left(\frac{\partial^2}{\partial R^2} + \frac{\partial^2}{\partial r^2}\right) + \frac{[J(J+1)-2\Omega^2]\hbar^2}{2\mu R^2} + \varepsilon_{n\Omega}(r,R) + V(r,R,\gamma=0) - E\}U_{n\Omega}^J(r,R) = 0 \qquad (25)$$

Before proceeding, we note that for J=0 (Ω=0) and ignoring the bending energy the above equation looks like the standard Schroedinger equation for a two-mathematical dimensional collinear reaction.

As discussed previously the quantity to be calculated in this reduced dimensionality theory is the partial wave cumulative reaction probability,

$$P^J_{v\to v'} = \sum_{jj'}\sum_{\Omega\Omega'} P^J_{vj\Omega\to v'j'\Omega'} \qquad (26)$$

Note that $P^J_{vj\Omega\to v'j'\Omega'}$ equals $|S^J_{vj\Omega\to v'j'\Omega'}|^2$, where $S^J_{vj\Omega\to v'j'\Omega'}$ is the full scattering matrix element. Recall also that the full scattering matrix encompasses both reactions A+BC and A+CB, whereas the above approximate approach treats these separately. This implies that the bending eigenfunctions are localized for either the ABC or ACB configurations. Referring now to eq. (25) let us introduce the associated scattering matrix, $S^{Jn\Omega}_{v\to v'}$. In this adiabatic treatment n is a good quantum number in addition to J (which is rigorously a good quantum number) and Ω (which is a good quantum number within the CS approximation). If it were our objective to obtain an approximation to the full scattering matrix $S^J_{vj\Omega\to v'j'\Omega'}$ it would be necessary to determine a correlation between the bending state $\xi_{n\Omega}(\gamma,r,R\to\infty)$ and the free-rotor states $|j\Omega\rangle$. We shall return to this point later. However, because our objective is an approximation to the cumulative reaction probability $P^J_{v\to v'}$, we can write

$$P^J_{v\to v'} = \sum_{n\Omega}\sum P^{Jn\Omega}_{v\to v'}, \qquad (27)$$

where

$$P^{Jn\Omega}_{v\to v'} = |S^{Jn\Omega}_{v\to v'}|^2. \qquad (28)$$

That is, the summation over n and Ω spans the same space as the summation over Ω,Ω',j and j' in eqs. (26) and (27). Equations (23) through (28) constitute the reduced dimensionality adiabatic bend theory. They also form the basis for a hierarchy of reduced dimensionality theories based on the centrifugal sudden and adiabatic bend approximation.

The solutions to eq. (25) with no further approximations are formally straightforward. For each n and Ω the two-mathematical dimensional equations are solved for each partial wave J. Quite often in applications it is sufficient to solve that equation for the ground bending state only and then to obtain the cumulative probability from the ground bend probability. This is because the bending eigenvalues $\varepsilon_{n\Omega}(r,R)$ can be fairly large in the neighborhood of the transition state and for total energies up to several kcal/mole above the transition state energy, excited bending states make a negligible contribution to the cumulative probability.

2.2 Reduced dimensionality hierarchy

We have introduced by a different set of arguments an approximate solution to eq. (25) which reduces the computational effort relative to the full adiabatic solution considerably.[11] In making these approximations we have been guided by transition state theory.

We now review the hierarchy of approximations to $P^{Jn\Omega}_{v->v'}$, which ultimately lead to transition state theory itself.

2.2.1 CEQB theory

In the first approximation we replace the centrifugal potential $[J(J+1)-2\Omega^2]\hbar^2/2\mu R^2$ by a constant $E^{\ddagger}_J$, which for collinear transition states is the usual linear-molecule rotational energy at the transition state configuration. With this replacement eq. (25) becomes

$$\left\{\frac{-\hbar^2}{2\mu}\left(\frac{\partial^2}{\partial R^2}+\frac{\partial^2}{\partial r^2}\right)+\varepsilon_{n\Omega}(r,R)+V(r,R,\gamma=0)-(E-E^{\ddagger}_J)\right\}U^J_{n\Omega}(r,R)=0. \quad (29)$$

This equation is formally like the two-mathematical dimensional Schroedinger equation for a collinear reaction with the addition of the constant rotational energy $E^{\ddagger}_J$ and the adiabatic bending energy. The solutions without the constant $E^{\ddagger}_J$ are denoted CEQB (collinear exact quantum with adiabatic bending energy). Thus, the solutions $U^J_{n\Omega}(r,R)$ can be related to the CEQB ones as follows,

$$U^J_{n\Omega}(E) = U^{CEQB}(E-E^{\ddagger}_J|n\Omega), \quad (30)$$

In practice, we also replace the exact bending energy eigenvalue $\varepsilon_{n\Omega}(r,R)$ by the harmonic/quartic-corrected valence bending energy $\varepsilon_{n\Omega}(r_{AB},r_{BC})$,[24] where r_{AB} and r_{BC} are the AB and BC internuclear distances. This replacement is quite accurate at least in the vicinity of tight transition states for the ground bending state.[25] This valence bending energy vanishes asymptotically for all bending states. As a result of eq. (30)

$$S^{Jn\Omega}_{v->v'}(E) = (-1)^J S^{CEQB}_{v->v'}(E-E^{\ddagger}_J|n\Omega) \quad (31)$$

where $S^{CEQB}_{v->v'}(E|n\Omega)$ is the scattering matrix element associated with $U^{CEQB}(E)$. The factor -1^J in eq. (31) appears because the phase factors $\exp(iJ\pi/2)$ and $\exp(-iJ\pi/2)$ which appear in the incoming and outgoing parts of the radial part of $U^J_{n\Omega}(E)$ must be accounted for in relating $S^{Jn\Omega}_{v->v'}(E)$ to $S^{CEQB}_{v->v'}(E-E^{\ddagger}_J|n\Omega)$.[12] This factor has important consequences for the CEQB differential cross section but it is of no significance for the probability.

Thus, the CEQB theory gives

$$P^{Jn\Omega}_{v->v'}(E) = P^{CEQB}_{v->v'}(E-E^{\ddagger}_J|n\Omega), \quad (32)$$

and for the cumulative vibrational state-to-state probability

$$P_{v->v'}(E) = \sum_{J=0}\sum_{n}\sum_{\Omega}(2J+1)P^{CEQB}_{v->v'}(E-E^{\ddagger}_J|n\Omega). \quad (33)$$

2.2.2 CEQB/G theory

A very useful additional approximation is to relate excited bending-state probabilities to the ground bending-state probability by an energy shift. Thus,

$$P^{CEQB}_{v->v'}(E-E^{\ddagger}_J|n\Omega) = P^{CEQB}_{v->v'}(E-E^{\ddagger}_J-\Delta\varepsilon^{\ddagger}_{n\Omega}|00), \quad (34)$$

where

$$\Delta\varepsilon^{\ddagger}_{n\Omega} = \varepsilon^{\ddagger}_{n\Omega} - \varepsilon^{\ddagger}_{00} \quad (35)$$

is the difference in energy between the excited and ground-bending energy at the transition state (appropriate to the ground bend). We term this approximation the CEQB/G theory. As noted above, for many reactions near threshold the ground bend makes the dominant contribution to the CEQB cumulative probability and for these cases the CEQB/G theory is essentially the same as the CEQB theory.

2.2.3 CEQ theory

An additional simplification to CEQB theory can be made, again within the spirit of transition state theory. It is to replace $\varepsilon_{n\Omega}(r,R)$ by $\varepsilon^{\ddagger}_{n\Omega}$, the ABC bending energy at the transition state. Thus, eq. (29) becomes

$$\{\frac{-\hbar^2}{2\mu}(\frac{\partial^2}{\partial R^2} + \frac{\partial^2}{\partial r^2}) + V(r,R,\gamma=0)-(E-E^{\ddagger}_J-\varepsilon^{\ddagger}_{n\Omega})\}U^J_{n\Omega}(r,R) = 0. \quad (36)$$

The scattering wavefunction $U^J_{n\Omega}$ and its corresponding scattering matrix element can be directly related to those for the collinear exact quantum problem as follows:

$$U^J_{n\Omega}(E) = U^{CEQ}(E-E^{\ddagger}_J-\varepsilon^{\ddagger}_{n\Omega}) \text{ and } S^{Jn\Omega}_{v->v'}(E) = -1^J S^{CEQ}_{v->v'}(E-E^{\ddagger}_J-\varepsilon^{\ddagger}_{n\Omega}). \quad (37)$$

Thus, in the CEQ theory

$$P^{Jn\Omega}_{v->v'}(E) = P^{CEQ}_{v->v'}(E-E^{\ddagger}_J-\varepsilon^{\ddagger}_{n\Omega}), \quad (38)$$

and for the cumulative vibrational state-to-state probability

$$P_{v->v'}(E) = \sum_{J=0} \sum_{n\Omega}\sum (2J+1)P^{CEQ}_{v->v'}(E-E^{\ddagger}_J-\varepsilon^{\ddagger}_{n\Omega}) \quad (39)$$

2.2.4 One-dimensional reaction path theories

All of the above theories require solving a two-mathematical dimensional partial differential equation which requires a fairly extensive, though routine, computational effort. A major reduction in effort results if the variables r and R are replaced by reaction coordinates s and x,[26-28] the reaction path and the displacement orthogonal to it, respectively. If the vibrational x-motion is treated adiabatically the following equation is obtained from eq. (36).

$$[T_s + V(s) + \varepsilon_v(s) - (E-E_J^{\ddagger}-\varepsilon_{n\Omega}^{\ddagger})]U_{n\Omega v}^{J}(s) = 0, \tag{40}$$

where T_s is the kinetic energy operator for the s-motion (including curvature terms), V(s) is the potential along s and $\varepsilon_v(s)$ is the adiabatic vibrational energy which correlates with the BC vth vibrational state as s->-∞. In this theory the scattering matrix $S_{v->v'}^{Jn\Omega}(E)$ is given by

$$S_{v->v'}^{Jn\Omega}(E) = \delta_{v'v}S_v^{1D}(E-E_J^{\ddagger}-\varepsilon_{n\Omega}^{\ddagger}-\varepsilon_v^{\ddagger}), \tag{41}$$

where $S_v^{1D}(E)$ is the scattering matrix corresponding to the one-dimensional Schroedinger equation

$$(T_s + V(s) + \varepsilon_v(s) - E)\, U_v^{1D}(s) = 0. \tag{42}$$

A more sophisticated reaction path approach is to replace $\varepsilon_{n\Omega}^{\ddagger}$ in eq. (40) by $\varepsilon_{n\Omega}(s)$. This is the essence of the approach taken by Garrett and Truhlar[29] and generalized by Miller et al.[28] and Skodje and Truhlar[30] for polyatomic reactions. Truhlar and coworkers have proposed one-dimensional paths which deviate from the reaction path in order to compute accurate tunneling probabilities from which transmission coefficients (see below) are then used to correct their version of variational transition state theory, the so-called improved canonical variational [transition state] theory (ICVT)[31,32] (also see below).

2.2.4 Transition state theory

Transition state theory is obtained by replacing the scattering matrix $S_v^{1D}(E)$ by a unit step function, $\Theta(E-V_0-\varepsilon_v^{\ddagger})$, where V_0 and $\varepsilon_v^{\ddagger}$ are the values of V(s) and $\varepsilon_v(s)$ at the transition state. Thus,

$$P_{v->v'}^{Jn\Omega}(E) = \delta_{v'v'}\Theta(E-V_0-\varepsilon_v^{\ddagger}-E_J^{\ddagger}-\varepsilon_{n\Omega}^{\ddagger}) \tag{43}$$

and for the cumulative vibrational state-to-state probability

$$P_{v->v'}(E) = \delta_{v'v}\sum_{J=0}\sum_{n}\sum_{\Omega}(2J+1)\Theta(E-V_0-\varepsilon_v^{\ddagger}-E_J^{\ddagger}-\varepsilon_{n\Omega}^{\ddagger}) \tag{44}$$

We have deferred up to this point giving an explicit definition of the transition state. Several definitions are of course possible. For example, the conventional transition state is the saddle point configuration of the potential surface V(r,R,γ=0). Another and better choice is obtained from a classical variational criterion in which the transition state is determined by maximizing V(s) + $\varepsilon_v(s) + E_J(s) + \varepsilon_{n\Omega}(s)$ with respect to s. This would be a bit cumbersome because of the need to reoptimize (in the general case) for each v, J, Ω, and n independently for each value of the total energy. This approach would comprise a complete microcanonical variational optimization. Another, simpler approach is to optimize in a canonical sense rather than in the microcanonical way just outlined. This canonical approach is well known[32] and is equivalent to maximizing the free energy of the transition state with respect to s for each temperature. Improvements to this method which account approximately for microcanonical threshold energies have been made by Truhlar and co-workers and the improved theory is termed ICVT.[31,32]

Pechukas and Pollak[33,34] made a very important conceptual breakthrough by defining the classically optimum transition state dividing surface without reference to any particular set of reaction coordinates. That surface is called the PODS for periodic dividing surface. This approach can also be viewed as the best method to determine a coordinate system in which to apply adiabatic theory.[35]

2.3 Transmission coefficients

It is instructive to express thermal rate constants as transmission coefficients times the transition state theory rate constant. First, recall that any rate constant can be written in this way. To see this, consider the exact rate constant $k_{v->v'}(T)$, (see eq. (15)) which we rewrite as

$$k_{v->v'}(T) = \frac{1}{hQ_{vib-rot}Q_{trans}}\langle N_v^{\ddagger}(E)\rangle\frac{\langle P_{v->v'}(E)\rangle}{\langle N_v^{\ddagger}(E)\rangle} \tag{45}$$

where $N_v^{\ddagger}(E)$ is the total number of states open at the transition state dividing surface, excluding the state which correlates with the vth reactant (see below) and where we have used the short-hand notation $\langle\ \rangle$ to indicate the thermal trace, e.g., $\langle N_v^{\ddagger}(E)\rangle$ equals $\int_0^\infty dEN_v^{\ddagger}(E)\exp(-E/kT)$. Equation (45) can be rewritten as

$$k_{v->v'}(T) = \frac{kTQ_v^{\ddagger}(T)}{hQ_{vib-rot}Q_{trans}}\Gamma_{v->v'}(T), \tag{46}$$

where

$$kTQ_v^{\ddagger}(T) = \langle N_v^{\ddagger}(E)\rangle \tag{47}$$

and

$$\Gamma_{v->v'}(T) = \frac{\langle P_{v->v'}(E)\rangle}{\langle N_v^{\ddagger}(E)\rangle} \tag{48}$$

is the transmission coefficient. In the usual separable approximation

$$N_v^{\ddagger}(E) = \sum_{J=0}\sum_{n}\sum_{\Omega}(2J+1)\Theta(E-V_0-E_J^{\ddagger}-\varepsilon_v^{\ddagger}-\varepsilon_{n\Omega}^{\ddagger}), \tag{49}$$

where $\varepsilon^{\ddagger}_{n\Omega}$ is the transition state bending energy and $\varepsilon^{\ddagger}_{v}$ is the energy of the transition state vibration which correlates adiabatically with the vth vibrational state of the reactant. It follows from this separable form for $N^{\ddagger}_{v}(E)$ and the simple (but unnecessary) assumption that location of the transition state is independent of E that

$$Q^{\ddagger}_{v}(T) = Q^{\ddagger}_{rot}Q^{\ddagger}_{bnd}\exp[-(V_0+\varepsilon^{\ddagger}_{v})/kT], \tag{50}$$

where $Q^{\ddagger}_{rot}$ and $Q^{\ddagger}_{bnd}$ are the rotational and bending partition functions and V_0 is the potential energy of the transition state. (In canonical variational transition state theory the partition function $Q^{\ddagger}_{v}(T)$ is computed along the reaction path s and minimized with respect to s.)

The transmission coefficients, $\Gamma_{v\to v'}(T)$, corresponding to the reduced dimensionality hierarchy are straightforward to obtain; they are given by the various approximations to the cumulative reaction probability $P_{v\to v'}(E)$. Rather than derive the expression for each approximation we shall give a detailed derivation for the CEQB theory and merely quote the results for the other approximations in Table 1. Before deriving the CEQB transmission coefficient we note that the adiabatic bend transmission coefficient is simply obtained by replacing the exact cumulative probability $P_{v\to v'}$ by eqs. (27) and (28).

The CEQB approximation to $\Gamma_{v\to v'}(T)$ is obtained as follows. First, using eqs. (27) and (33), the CEQB approximation to $\langle P_{v\to v'}(E)\rangle$ can be expressed as

$$\langle P_{v\to v'}(E)\rangle = \int_0^\infty dE \sum_{J=0}\sum_{n\Omega}\sum P^{CEQB}_{v\to v'}(E-E^{\ddagger}_J|n\Omega)\exp(-E/kT) \tag{51}$$

$$= \sum_{J=0}(2J+1)\exp(-E^{\ddagger}_J/kT)\int_0^\infty dE\sum_{n\Omega}\sum P^{CEQB}_{v\to v'}(E|n\Omega)\exp(-E/kT) \tag{52}$$

$$= Q^{\ddagger}_{rot}(T)\int_0^\infty dE\sum_{n\Omega}\sum P^{CEQB}_{v\to v'}(E|n\Omega)\exp(-E/kT). \tag{53}$$

Equation (52) follows from a change of integration variable from E to $E-E^{\ddagger}_J$ and the fact that $P^{CEQB}_{v\to v'}(E|n\Omega)$ vanishes for negative values of E. Also, it is easy to prove that

$$\langle N^{\ddagger}_{v}(E)\rangle = Q^{\ddagger}_{rot}(T)\int_0^\infty dE\sum_{n\Omega}\sum\Theta(E-V_0-\varepsilon^{\ddagger}_{v}-\varepsilon^{\ddagger}_{n\Omega}) \tag{54}$$

Finally from eqs. (53) and (54) the CEQB transmission coefficient is given by

$$\Gamma_{v\to v'}(T) = \frac{\langle \sum_{n\Omega}\sum P^{CEQB}_{v\to v'}(E|n\Omega)\rangle}{\langle \sum_{n\Omega}\sum\Theta(E-V_0-\varepsilon^{\ddagger}_{v}-\varepsilon^{\ddagger}_{n\Omega})\rangle} \tag{55}$$

This result is quite transparent. The consequence of the energy shift $E_J^{\ddagger}$ in the thermal trace of the CEQB expression for the cumulative probability is to factor out the corresponding transition state partition function, $Q_{rot}^{\ddagger}(T)$. In the expression for $\Gamma_{v->v'}(T)$ that partition function exactly cancels with the same transition state partition function which appears in the thermal trace of $N_v^{\ddagger}$. Continued energy shifting as in the CEQB/G and CEQ theories results in additional cancellation of partition function factors in the numerator and denominator. Finally, in the TST approximation to $P_{v->v'}(E)$ the numerator cancels exactly with the denominator and the TST transmission coefficient is unity, as of course it must be. The transmission coefficients from the hierarchy of reduced dimensionality quantum theories are summarized in Table 1. The adiabatic bend theory approximation with no energy shifting is abbreviated ADB in that table. Also, as noted just after eq. (16), if the vth BC vibrational state is assumed to be populated with unit probability then the expression for $k_{v->v'}(T)$ is modified by a factor given earlier. For convenience we incorporate that factor into the transmission coefficient in Table 1 for "non-Boltzmann v-states".

Table 1. Expressions for the partial wave vibrational state-to-state cumulative reaction probabilty $P_{v->v'}^J$, and the corresponding transmission coefficients $\Gamma_{v->v'}(T)$.

THEORY	$P_{v\to v'}^J(E)$	$\Gamma_{v\to v'}(T)$
EXACT	$\sum_{j\Omega}\sum_{j'\Omega'} P_{vj\Omega\to v'j'\Omega'}^J(E)$	$\langle P_{v\to v'}^{EX}(E)\rangle / \langle N_v^{\ddagger}(E)\rangle$
ADB	$\sum_{n\Omega} P_{v\to v'}^J(E\mid n\Omega)$	$\langle P_{v\to v'}^{ADB}(E)\rangle / \langle N_v^{\ddagger}(E)\rangle$
CEQB	$\sum_{n\Omega} P_{v\to v'}^{CEQB}(E-E_J^{\ddagger}\mid n\Omega)$	$\langle \sum_{n\Omega} P_{v\to v'}^{CEQB}(E\mid n\Omega)\rangle / \langle \sum_{n\Omega} \theta(E-V_0-E_n^{\ddagger}-E_v^{\ddagger})\rangle$
CEQB/G	$\sum_{n\Omega} P_{v\to v'}^{CEQB}(E-E_J^{\ddagger}-\Delta E_n^{\ddagger}\mid 00)$	$\langle P_{v\to v'}^{CEQB}(E\mid 00)\rangle / \langle \theta(E-V_0-E_{n=0}^{\ddagger}-E_v^{\ddagger})\rangle$
CEQ	$\sum_{n\Omega} P_{v\to v'}^{CEQ}(E-E_J^{\ddagger}-E_n^{\ddagger})$	$\langle P_{v\to v'}^{CEQ}(E)\rangle / \langle \theta(E-V_0-E_v^{\ddagger})\rangle$
TST	$\sum_{n\Omega} \theta(E-V_0-E_J^{\ddagger}-E_n^{\ddagger}-E_v^{\ddagger})\,\delta_{v',v}$	UNITY

$$N_v^{\ddagger}(E) = \sum_{J=0}\sum_{n\Omega}(2J+1)\theta(E-V_0-E_J^{\ddagger}-E_n^{\ddagger}-E_v^{\ddagger})$$

Note for non-Boltzmann v-states:

$$\Gamma_{v\to v'}(T) \longrightarrow \Gamma_{v\to v'}(T)Q_v(T)\exp(E_v/kT)$$

2.4 Correlation with asymptotic rotational states

The centrifugal sudden-adiabatic bend theory we just reviewed represents approximations to the partial wave vibrational state-to-state cumulative reaction probability, $P^J_{v->v'}$. Because that quantity is summed over initial and final rotational states it was not necessary to specify a correlation between the adiabatic bending states in the strong interaction region and the asymptotic reactant or product rotational states. However, to extend the theory to describe rotational state-to-state processes, some correlation procedure must be adopted. We discuss several possible procedures now.

In fact a partial correlation does exist between the Ω-quantum number of the bending states $|n\Omega\rangle$ and free rotor states $|j\Omega\rangle$. As a consequence of the centrifugal sudden approximation Ω is assumed to be a good quantum number. This partial correlation itself leads to some interesting (and qualitatively accurate) predictions. It implies that in the energy range where $P^J_{v->v'}$ is given predominantly by ground state bending state probability, for which Ω equals zero (see eq. (33)) that only initial and final rotational states with Ω equal to zero have significant reaction probabilities. That result is indeed found in the coupled channel calculations of Schatz and Kuppermann for $H+H_2(v=0)$ for the energies they considered, i.e., E less than or equal to 0.7 eV.[14] For these energies the ground bend state does dominate the cumulative reaction probability.[7]

A completely adiabatic correlation between $|n\Omega\rangle$ and $|j\Omega\rangle$ states would assume that n equals j, a result, which unlike the partial Ω-correlation, is not borne out by coupled channel calculations.[14] Physically, this adiabatic correlation is untenable because the spatial character of free-rotor and bending wavefunctions is quite different and so non-adiabatic coupling is certain to be large. Thus, the change between the free-rotor and bending wavefunctions is better described by a sudden correlation. It is possible to incorporate a sudden correlation within the present adiabatic bend theory and to achieve the objective of extending that theory to describe rotational state-to-state processes. A requirement of any such extension is that the cumulative probability $P^J_{v->v'}$ obtained from it agree with one from the adiabatic bend theory.

One obvious way to extend the adiabtic bend theory which preserves $P^J_{v->v'}$ is to relate $S^{Jn\Omega}_{v->v'}$ to some approximate $S^J_{vj\Omega->v'j'\Omega}$ by an orthogonal transformation. Because Ω is assumed to be a good quantum number, the transformation is between n-space and j-space. Thus, in matrix notation we have

$$\mathbf{S}^{J\Omega} = \mathbf{C}^t \mathcal{S}^{J\Omega}\mathbf{C}, \tag{56}$$

where $\mathbf{S}^{J\Omega}$ is a diagonal matrix in the n-representation, i.e., $S^{J\Omega}_{nm} = S^{Jn\Omega}_{v->v'}\delta_{m,n}$ and in general $\mathcal{S}^{J\Omega}$ is non-diagonal in j-space, i.e., $\mathcal{S}^{J\Omega}_{jj'} = S^J_{vj\Omega->v'j'\Omega}$. It is straightforward to show that such a transformation preserves the cumulative probability $P^J_{v->v'}$. The proof of this is as follows. Note that

$$\sum_n |S^{Jn\Omega}_{v\to v'}|^2 = \mathrm{Tr}[\mathbf{S}^{J\Omega}(\mathbf{S}^{J\Omega})^t], \qquad (57)$$

From eq. (54), eq. (57) can be rewritten as

$$\mathrm{Tr}[\mathbf{C}^t \mathcal{A}^{J\Omega}\mathbf{C}(\mathbf{C}^t\mathcal{A}^{J\Omega}\mathbf{C})^t] = \mathrm{Tr}[\mathbf{C}^t \mathcal{A}^{J\Omega}(\mathcal{A}^{J\Omega})^t\mathbf{C}], \qquad (58)$$

which equals $\mathrm{Tr}[\mathcal{A}^{J\Omega}(\mathcal{A}^{J\Omega})^t]$ from the cyclic property of the trace operation. Thus, $\mathrm{Tr}[\mathbf{S}^{J\Omega}(\mathbf{S}^{J\Omega})^t]$ equals $\mathrm{Tr}[\mathcal{A}^{J\Omega}(\mathcal{A}^{J\Omega})^t]$, i.e.,

$$\sum_n |S^{Jn\Omega}_{v\to v'}|^2 = \sum_{jj'}\sum |S^J_{vj\Omega\to v'j'\Omega'}|^2. \qquad (59)$$

Both sides of eq. (59) can then be summed over Ω completing the proof that $P^J_{v\to v'}$ is preserved under a unitary transfomation of $\mathbf{S}^{J\Omega}$.

If a strict adiabatic correlation between the n and j-spaces is made then $C_{nj} = \delta_{nj}$ in which case

$$S^J_{vj\Omega\to v'j'\Omega} = \delta_{j'j}\delta_{jn}S^{Jn\Omega}_{v\to v'}. \qquad (60)$$

As discussed above this correlation is not reasonable and so we do not pursue it further.

A physically reasonable correlation is a sudden one. This implies that $C^{\Omega}_{jn} = \langle j\Omega|n\Omega\rangle$ and so

$$S^J_{vj\Omega\to v'j'\Omega'}(E) = \langle j'\Omega|n\Omega\rangle S^{Jn\Omega}_{v\to v'}(E)\langle n\Omega|j\Omega\rangle. \qquad (61)$$

(Note that **C** depends on Ω.) This result is similar in spirit to the Franck-Condon theory of reactions,[36-38] and eq. (61) is quite similar to one given by Schatz and Ross for three-dimensional reactions.[38] An important difference is that eq. (61) is meant to be a quantitative approximation for the scattering matrix. Previous Franck-Condon theories contained an unknown electronic coupling interaction and so were capable of giving relative final state distributions. The present expression, though intended to be quantitative, does share a problem with those from Franck-Condon theories. Namely, it violates conservation of energy and so transitions to energetically closed rotational states can occur. A simple **ad hoc** procedure to eliminate energetically closed transitions is to restrict the matrix equation, eq. (56), to only energetically open values of j and j' and reorthogonalize the resulting non-orthogonal C-matrix.

An important issue in obtaining $S^J_{vj\Omega\to v'j'\Omega}$ by this projection method is where to take the projections $\langle j'\Omega|n\Omega\rangle$ and $\langle n|j\Omega\rangle$. Clearly, the final results will depend on this. Physically, the projection should be made in the region where the free-rotor basis undergoes the most rapid change in character. (That is the prescription in the Franck-Condon theory of reactions.) It may be possible to eliminate this dependence to a large extent by a procedure which we now outline. This extension will also eliminate transitions to energetically closed channels.

The basic idea is quite simple. It is to use the uncoupled adiabatic $|n\Omega\rangle$-basis in the strong interaction region and to do the arrangement channel matching in that basis. However, instead of continuing the integration into the asymptotic regions of space with that basis, a transformation to the asymptotic $|j\Omega\rangle$-basis would be made in each arrangement channel. The wavefunction in the new free rotor basis would then be propagated in each arrangement channel in a standard non-reactive scattering mode into the asymptotic regions of space. Scattering boundary conditions will be imposed as usual and the full rotational state-to-state S-matrix will be obtained.

These ideas and many obvious modifications of them for extending the adiabatic bend theory to obtain rotational state-to-state reaction probabilites are, we believe, a promising new area of research in the quantum theory of reactive scattering.

In the next section, we present several applications of the reduced dimensionality theory just reviewed. The applications are to the reaction of muonium+H_2, resonances in H+H_2 and the reactions $O(^3P)+H_2$, D_2 and HD where comparisons with experiment are given.

3. Applications

3.1 Muonium plus H_2

Schatz has recently performed three-dimensional centrifugal sudden (CS) calculations of the reaction of muonium with H_2[39] (also see his chapter) using the **ab initio** LSTH potential energy surface.[40,41] Muonium (Mu) is an unstable, very light isotope of H and as a result very substantial tunneling is expected in this reaction. Schatz's calculations offer another opportunity to test the reduced dimensionality theory. Previous tests for the H+H_2(v=0,1) reactions demonstrated the high accuracy of the reduced dimensionality theory.[7,10] We performed CEQB/G calculations for this reaction using the LSTH surface. (For the total energies considered the ground bending state makes the dominant contribution to the cumulative reaction probability and so the CEQB/G results would be essentially identical to the full CEQB ones.) The J=0 cumulative reaction probabilities are compared with the CS ones of Schatz in Figs. 1 and 2. The semi-log plot in Fig. 1 shows the accuracy of the present calculations over a wide range of values of the cumulative probability. In Fig. 2 the comparison is shown for both the reaction with H_2(v=0) and H_2(v=1). In both cases only MuH(v=0) is energetically open for the energy range shown.

In Figure 3 we compare the partial wave opacities for H_2(v=0) versus J for a total energy of 0.75 eV. As discussed in the previous section the CEQB/G approximation to $P^J_{v\to\Sigma'}$ is obtained from the J=0 probability by a simple energy shift $E^\ddagger_J$, where $E^\ddagger_J$ is the rotational energy of the transition state. For this reaction the variational transition state is shifted considerably from the saddle

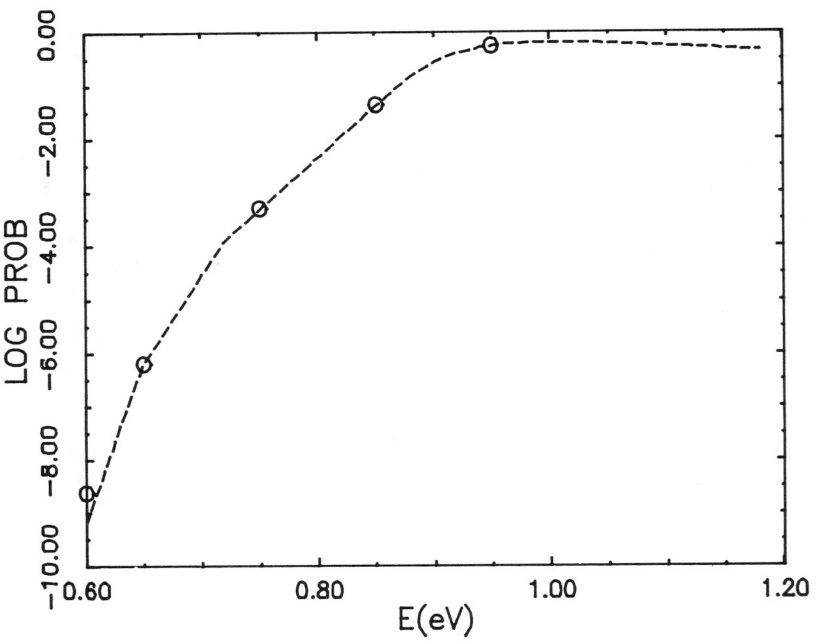

Figure 1. Semi-log plot of the J=0 cumulative reaction probability for Mu + H_2(v=0) on the LSTH surface versus the total energy E. The circles are the centrifugal sudden (CS) results of Schatz and the dashed line is the present reduced dimensionality (CEQB/G) result.

point.[42] For the effective potential including the ground state bend the transition state is located at R_{MuH} = 1.52 bohrs and R_{HH} = 2.10 bohrs. The saddle point distances are R_{MuH} = R_{HH} = 1.757 bohrs. The agreement between the simple energy-shifted CEQB/G opacities and the CS ones of Schatz is quite good. The sum of the CEQB/G opacities equals 0.86 the sum of the CS opacities. If the rotational energy of MuH_2 at the saddle point is used to calculate the opacities the agreement is considerably worse with the sum of the opacities equal to 0.69 the CS summed opacities. Thus, it is very important to use the variational transition state $E_J^{\ddagger}$ for this reaction.

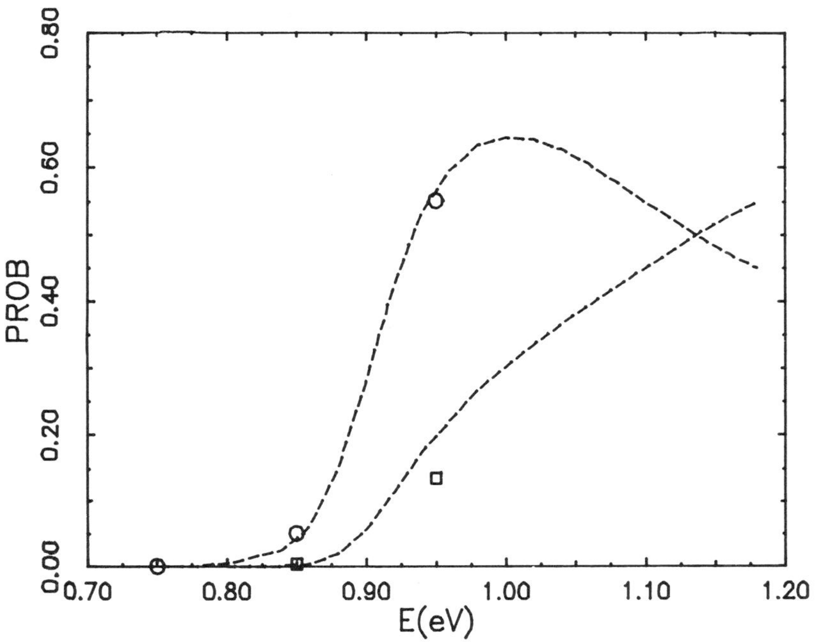

Figure 2. J=0 cumulative reaction probability for Mu + H_2(v=0) and H_2(v=1) on the LSTH surface versus the total energy E. The circles and squares are the centrifugal sudden (CS) results of Schatz for H_2(v=0) and H_2(v=1), respectively and the dashed lines are the present reduced dimensionality (CEQB/G) results.

3.2 Resonances in $H+H_2$

The $H+H_2$ reaction has played a central role in the theory of gas phase reactions. Very recently, Schatz reported CS calculations on $H+H_2$ on the LSTH surface for total energies up to 1.2 eV[43] (also see his chapter in this book). A number of features, identified as resonances were observed. We decided to perform reduced dimensionality CEQB calculations on this system with special attention to these resonant features.

Reduced dimensionality CEQB reaction probabilities for the ground, first, and second excited bending states were calculated. The calculations for the ground bend state were done with the quartic correction to the harmonic bending energy[24] and those for the excited state bends were done with the harmonic approximation to the bend energy. We found that for excited bending states of H_3 the harmonic approximation for the adiabatic bending energy at the transition state is roughly as accurate as the quartic-corrected harmonic energy but for the ground bending state the quartic-corrected harmonic energy is considerably more accurate than the

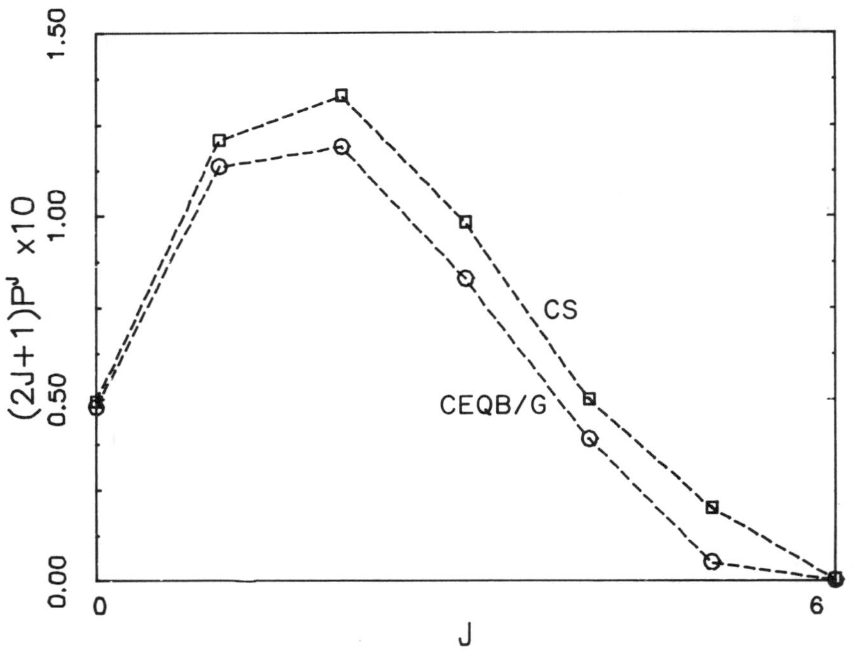

Figure 3. Comparison of energy shifted CEQB/G and centrifugal sudden (CS) partial wave cumulative opacities for $Mu+H_2(v=0)$ for a total energy of 0.75 eV.

harmonic one.[25] Thus, the extra effort in doing the quartic correction was made only for the ground-bend state. The results for $H+H_2(v=0) \rightarrow H_2(v'=0)+H$ are shown in Fig. 4 as a function of the total energy E. As discussed in detail below, the ground and second excited bend probability is for J = 0 whereas the one for the first excited bend is for J = 1. Resonance features are seen in all the probability curves. As expected, the probabilities for the excited bending states look qualitatively like the ground-bend probability shifted roughly by the bend energy at the transition state. However, excited bending state resonance features, although also roughly energy-shifted from the ground bend ones, are broader than the corresponding ground-bend ones. The resonance energies for the various bending state probabilities are given in Table 2. These energies were simply determined by the energy where the probability has a local minimum. This is a rather crude method to determine resonance energies, especially for such broad resonances. As discussed in the previous section, the bending states are labeled by the quantum numbers n and Ω, the latter being the projection of the total rotor momentum on the body-fixed z-axis, with the restriction that |Ω| must be less than or equal to the

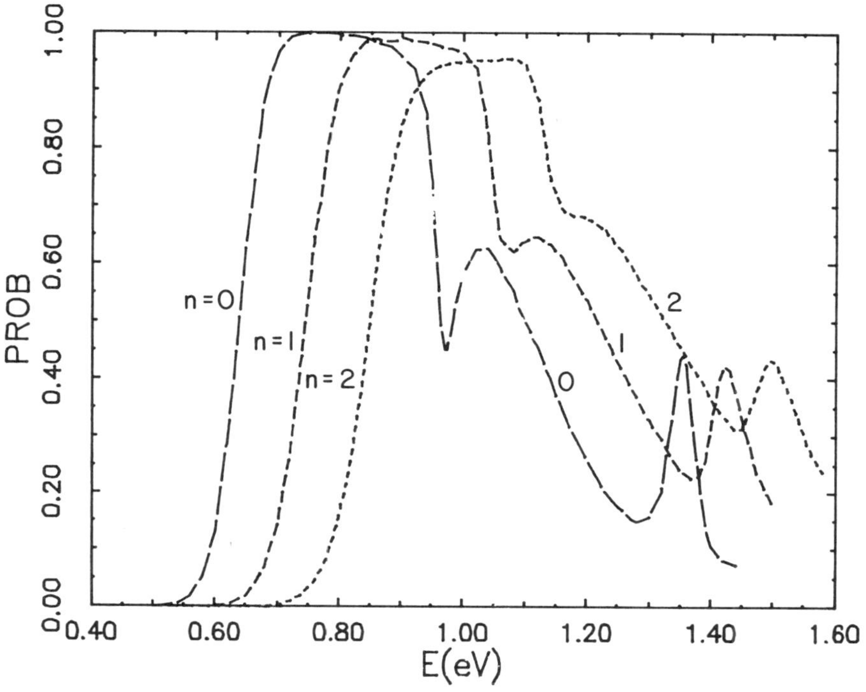

Figure 4. Reduced dimensionality CEQB reaction probabilities for $H+H_2(v=0) \rightarrow H_2(v'=0)+H$ for the ground (n=0), first (n=1) and second (n=2) excited bending states on the LSTH surface.

total angular momentum quantum number J. In the harmonic approximation, which, as noted, was used for the excited state bends, the bend energy for a given n is the same for all allowable values of Ω and Ω is restricted to the range −n to n in steps of two.[44] In addition, the value of J must be greater than or equal to $|\Omega|$. Thus, for n equal to 0 and 1 $|\Omega|$ equals 0 and 1 respectively and so the minimum values of J are also 0 and 1, respectively. For the second exicited bending state n equals 2 and so $|\Omega|$ equals 0 or 2. The cumulative probabilities shown in Fig. 4 are for J = 0 for n equal to 0 and 2 and for J = 1 for n equal to 1. In Table 2, the resonance energies for n = 2 are given for the components $|\Omega|$ equal to 0 and 2. They differ by the energy shift $E_J^{\ddagger}$ that is used in the CEQB theory to obtain the partial wave cumulative probabilities. We should note that for the bending states considered here a greater splitting is due to anharmonic correction to the harmonic energy. These are discussed in detail elsewhere.[45]

Table 2. Reduced dimensionality (CEQB) resonance energies for the ground and two excited bending states of $H+H_2(v=0)\rightarrow H_2(v'=0)+H$ on the LSTH potential surface.

n	$\lvert\Omega\rvert$	E_{res} (eV)
0	0	0.973
1	1	1.09
2	0	1.20
2	2	1.21
0	0	1.35
1	1	1.42
2	0	1.50
2	2	1.51

Several points should be noted. First, the first four resonance energies listed in Table 2 are all in good agreement with the CS results of Schatz[43] (also, his chapter in this book). Second, not all possible resonances in the energy range shown are given. A resonance for the third excited bending state could occur at a total energy of roughly 1.34 eV.

The agreement noted between the ground and excited bend-state reduced dimensionality resonances and the CS ones of Schatz is quite significant. It indicates that the adiabatic bend approximation for this reaction continues to be a realistic description of the dynamics even for total energies up to 1.20 eV. This should stimulate further inquiry about the realism of the adiabatic bend approximation in reactive scattering.

A more complete discussion of these results along with comparisons with approximate stabilization predictions of the resonances in this system is given elsewhere.[45]

3.3 $O(^3P)+H_2$, D_2 and HD

The reactions of $O(^3P)$ with H_2, D_2, and HD have become one of the most intensively studied set of reactions, both experimentally and theoretically. Theoretically, there have been several three-dimensional quasiclassical trajectory studies,[46-49], collinear exact quantum studies,[50-52] several reduced dimensionality quantum[50-52] and quasiclassical studies,[53-55] variational transition state theory studies with tunneling corrections,[56,57] and two distorted wave Born studies.[58,59] We recently reviewed our previous reduced dimen-

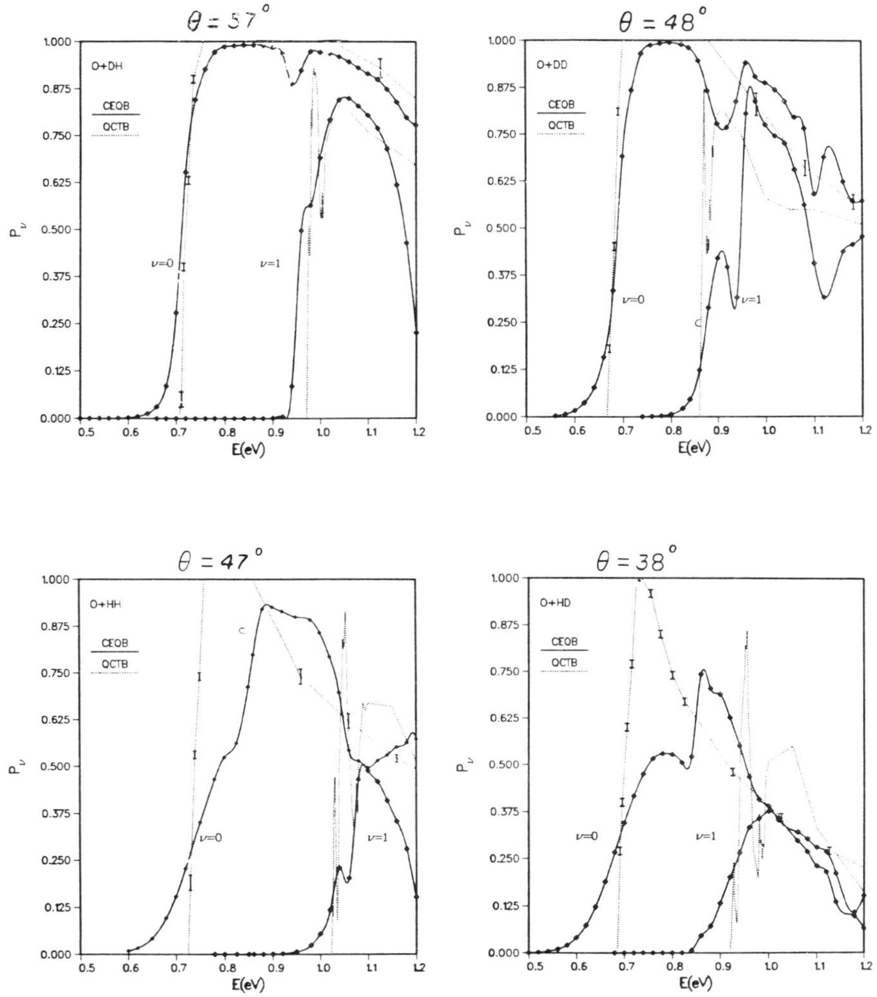

Figure 5. **Reduced dimensionality exact quantum and quasiclassical total reaction probabilities for the ground bending state (CEQB/G and QCTB/G, respectively) for the O(^{3}P)+DH, D_2, H_2 and HD reactions versus the total energy E. The potential surface used is the *ab initio* MODPOLCI surface (refs. 52 and 54).**

sionality calculations of the $O+H_2$ and D_2 reactions.[11] Our initial study used the CEQ reduced dimensionality theory to obtain rate constants for the thermal and vibrationally excited H_2 reaction using five potential energy surfaces.[52,53] Based on that study, we decided to continue with a CEQB/G study of the H_2, D_2 and HD reactions[54] using one of the **ab initio** potential surfaces, the MODPOLCI surface.[52,54] Here we wish to review and unify these studies and to present the latest comparisons with experiment.

In Fig. 5 we present the CEQB/G and corresponding quasiclassical trajectory total reaction probabilities for the reactant molecule initially in the ground and first excited vibrational states. The angle given at the top of each frame is the skew angle for the reaction in mass-weighted coordinates. Several obvious trends are seen with respect to the skew angle. First, both the quantum and quasiclassical results show decreasing reactivity with decreasing skew angle. This is due presumably to increased re-crossing of the transition state with decreasing skew angle. Essentially, the incoming reactant flux has greater difficulty in turning the corner to form products as the skew angle decreases. Second, the difference between the quantum and quasiclassical thresholds for reaction increases with decreasing skew angle. This is due to increasing "corner-cutting" tunneling in the quantum calculation as the skew angle decreases. That is, the quantum mechanical tunneling can occur over extended regions of space especially for smaller skew angles.

These threshold energies are very important quantitities, as they determine to a large degree the behavior of the corresponding thermal rate constants. We decided to apply vibrational adiabatic (VA) theory in an attempt to analyze both the quantum and quasiclassical threshold energies. For the four reactions the steepest descent path from the saddle point was calculated in mass-weighted coordinates and the vibrational eigenvalues were obtained numerically (by a finite difference algorithm) for the motion transverse to this path. For the purpose of determining adiabatic threshold energies all that is required is the maximum in the corresponding VA potential, which is simply the sum of the bare potential (which in this case includes the adiabatic bending energy) and the local VA energy. The approach to calculate VA potentials we have outlined has well-known difficulties in the vicinity of the saddle point, where transverse cuts can intersect both the reactant and product channels. This cutting across channels turns out to be a difficulty mainly for vibrationally exicted states. Fortunately, for these states, the maximum in the VA potentials occurs away from this region. For the ground vibrational state, we found that the maximum in the VA potential occurs essentially at the saddle point for each reaction. The VA barrier heights are given in Table 3 for both the forward and reverse directions of each reaction. Consider first the predicted threshold energies for the vibrational ground state reactions. In general these are in good agreement with the observed quasiclassical threshold energies in Fig. 5. As expected, the quantum reaction probabilities exhibit thresholds considerably below the VA barrier heights, due to tunneling. For the vibrationally

Table 3. Vibrational adiabatic barrier heights (in eV) for $O(^3P)+H_2$, D_2, HD and DH forward and reverse reactions.

Reaction	vibrational state 0	1	2
$O+H_2$	0.738	1.02	1.39
OH+H	0.738	0.911	1.21
$O+D_2$	0.685	0.855	1.10
OD+D	0.685	0.818	0.945
O+HD	0.706	0.921	1.23
OH+D	0.706	0.841	1.21
O+DH	0.720	0.968	1.28
OD+H	0.729	0.884	1.03

excited state reactions the quasiclassical threshold energies are in excellent agreement with the predictions based on the VA barrier heights. As already noted, for these reactions the VA barriers occur away from the saddle point where the reaction path curvature is small. Thus, the VA assumption is expected to be quite realistic and so the accuracy of the predicted VA threshold energies is not surprising. The quantum threshold energies for the vibrational ground and excited reactions are below the VA (and quasiclassical) threshold energies due to tunneling, of course.

Thermal rate constants have been measured for the $O(^3P)+H_2$ reaction by numerous groups over a wide temperature range.[61-66] The rate constant for vibrational excited H_2 has also been measured at 298 K.[67] We calculated reduced dimensionality CEQB/G and CEQ rate constants for these reactions, using the MODPOLCI surface,[52,54] and the comparison with available expermients is given in Fig. 6. For the thermal rate, which is completely dominated by the ground vibrational state of H_2, the CEQB/G and the simpler CEQ results are quite similar and in good agreement with experiment over the entire temperature range. The inset, showing the highest temperatures, indicates better agreement between experiment and the CEQB/G result than with the CEQ one. For the reaction with $H_2(v=1)$ there is a substantial difference between the CEQ and CEQB/G rate constants,

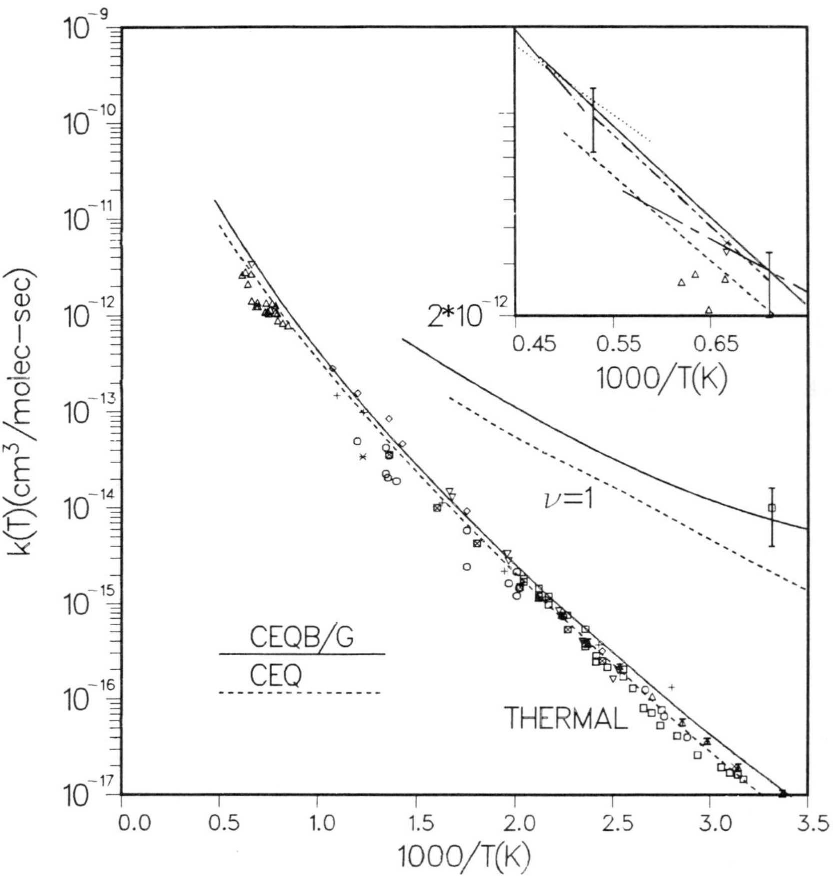

Figure 6. Reduced dimensionality CEQB/G and CEQ rate constants versus 1/T for the thermal and vibrationally excited H_2. The inset is an expanded scale of the high temperature end of the thermal rate constant. The experimental thermal results are taken from refs. 61-66 and for the vibrationally excited measurment from ref. 67. The recent high high temperature measurments (— – —) are from ref. 65.

with the latter one in very good agreement with the one experimental measurment. The main reason for the difference between the CEQ and CEQB/G rate constants is in the way that the adiabatic bending energy is treated. Recall that in the CEQ theory the adiabatic bending eigenvalue is replaced by a constant value equal to its energy at the transition state. In these previous CEQ calculations the transition state was taken at the saddle point for <u>both</u> the ground and vibrationally excited reaction. While this is correct for the ground state reaction it is not correct for the vibrationally excited one. In fact, the bending energy at the transition state for that reaction is considerably less than at the saddle point and so using the higher (incorrect) saddle point bending

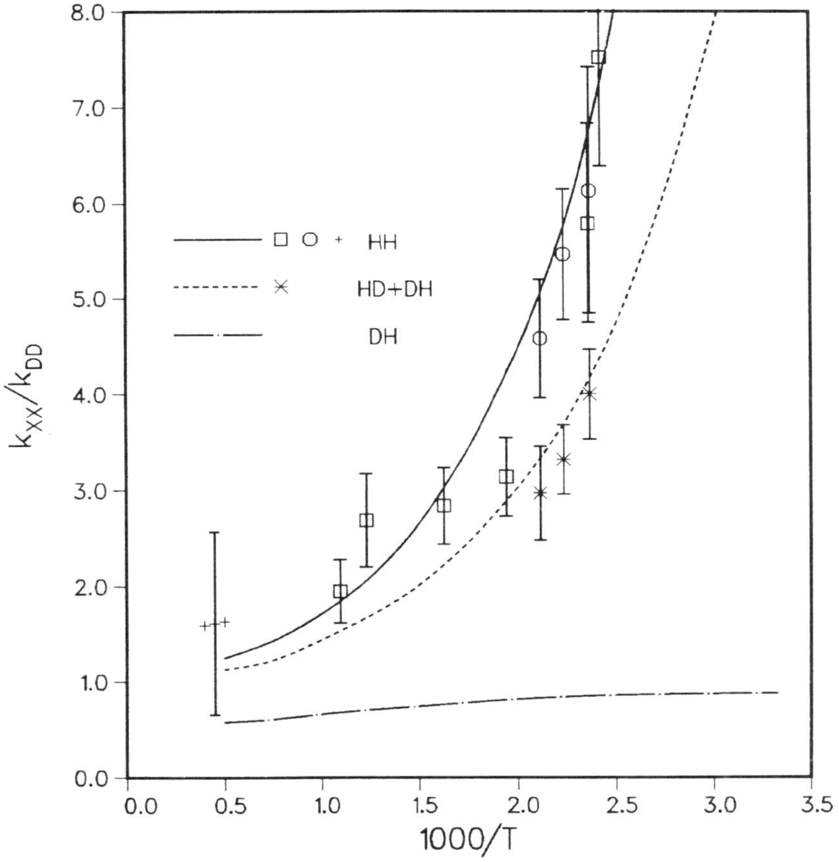

Figure 7. The ratio of CEQB/G rate constants for the reactions of $O(^3P)$ with H_2 (solid curve) and HD, summed over the OH and OD reaction channels, (dotted curve) and to form OH (dashed-dotted curve) to the rate constant for the reaction with D_2 versus 1/T. The corresponding experimental results are from refs. 64 and 65 (+), 62 and 63 (□), and ref. 66 (0,*)

energy excessively reduces the CEQ approximation to the cumulative reaction probablity which in turns results in a lower rate constant. The more accurate CEQB/G theory does not make this error as the ground state adiabatic bending energy is added everywhere to the potential energy surface. (For the temperature range shown for the H_2(v=1) reaction the ground bending state dominates the summation over bending states for the cumulative reaction probability. Thus, there is in practice no difference between the CEQB/G and the full CEQB rate constant.) Indeed, a simple estimate[53] of the correct bending shift to apply in the CEQ theory would bring the CEQ rate constant into close agreement with the CEQB/G one, as expected.

Recently measurments of isotope effects in this reaction have been reported by several groups.[64-66,68] These effects, expressed as ratios of thermal rate constants, are compared to the calculated CEQB/G ones on Fig. 7. The agreement between theory and experiment is excellent. The branching ratio in the O+HD reaction to form either OH or OD has not yet been measured, although experiments to do so are in progress.[68]

4. Acknowledgments

This work was supported by the U.S. Department of Energy, Division of Chemical Sciences under grant DOEDE-AC0291ER10900 (JMB) and contract W-31-109-Eng-38 (AFW). We are grateful to Professor George Schatz for supplying unpublished resuts and helpful discussions. We also thank Professor Donald G. Truhlar for helpful comments on notation.

5. References

1. For a review, see D.J. Kouri in Atom-Molecule Collision Theory: A Guide for the Experimentalist, R.B. Bernstein, Editor, (Plenum, N.Y., 1979), Chap. 9.
2. D. Secrest, in Molecular Collision Dynamics, J.M. Bowman, Editor, (Springer-Verlag, Berlin, 1983), Chap. 2.
3. A.B. Elkowitz and R.E. Wyatt, Mol. Phys. 31, 189 (1976).
4. A. Kuppermann, G.C. Schatz, and J.P. Dwyer, Chem. Phys. Lett. 45, 71 (1977).
5. G.C. Schatz, Chem. Phys. Lett. 94, 183 (1983).
6. G.C. Schatz, Chem. Phys. Lett. 108, 532 (1984).
7. J.M. Bowman, G-Z. Ju, and K-T. Lee, J. Chem. Phys. 75, 5199 (1981).
8. J.M. Bowman, G-Z. Ju, and K-T. Lee, J. Phys. Chem. 86, 2238 (1982).
9. J.M. Bowman, K.-T. Lee, and G-Z. Ju, Chem. Phys. Lett. 86, 5199 (1981).
10. J.M. Bowman and K-T. Lee, Chem. Phys. Lett. 74, 363 (1983)

11. J.M. Bowman, K-T. Lee, H. Romanowski and L.B. Harding, in Resonances in Electron-Molecule Scattering, van der Waal's Complexes, and Reactive Chemical Dynamics, D.G. Truhlar, Editor, (ACS, Washington D.C., 1984), Chap. 4.
12. J.M. Bowman, Adv. Chem. Phys. 61, xxx (1985).
13. A.B. Elkowitz and R.E. Wyatt, J. Chem. Phys. 62, 2504 (1975).
14. G.C. Schatz and A Kuppermann, J. Chem. Phys. 65, 4668 (1976). (1983).
15. R.B. Walker, E.B. Stechel, and J.C. Light, J. Chem. Phys. 69, 2922 (1978).
16. R.B. Walker and E.F. Hayes, J. Phys. Chem. 87, 1255 (1983).
17. E.F. Hayes and R.B. Walker, in Resonances in Electron-Molecule Scattering, van der Waal's Complexes, and Reactive Chemical Dynamics, D.G. Truhlar, Editor, (ACS, Washington D.C., 1984), Chap. 26.
18. R.E. Wyatt, J. Chem. Phys. 51, 3489 (1969).
19. J.N.L. Connor and M.S. Child, Mol. Phys. 18, 653 (1970).
20. R.B. Walker and E. Pollak, J. Chem. Phys. 83, xxxx (1985).
21. T. Carrington and W.H. Miller, J. Chem. Phys. 81, 3942 (1984).
22. G.C. Schatz and A. Kuppermann, J. Chem. Phys. 65, 4642 (1976).
23. R.T Pack, J. Chem. Phys. 60, 633 (1974).
24. B.C. Garrett and D.G. Truhlar, J. Phys. Chem. 83, 1915 (1979).
25. J.M. Bowman, unpublished.
26. R.A. Marcus, J. Chem. Phys. 45, 4493, 4500 (1966).
27. G.L. Hofacker, Z. Naturforsch. A18, 607 (1963).
28. W.H. Miller, N.C. Handy, and J.E. Adams, J. Chem. Phys. 72, 99 (1980).
29. B.C. Garrett and D.G. Truhlar, Proc. Natl. Acad. Sci. U.S.A. 76, 4755 (1979).
30. R.T. Skodje and D.G. Truhlar, J. Chem. Phys. 79, 4882 (1983).
31. B.C. Garrett, D.G. Truhlar, R.S. Grev and A.W. Magnuson, J. Phys. Chem. 84, 1730 (1980).
32. For a forthcoming review, see D.G. Truhlar, A.D. Isaacson and B.C. Garrett, in The Theory of Chemical Reaction Dynamics, Ed. M. Baer (CRC Press, Boca Raton, FL), in press.
33. P. Pechukas and E. Pollak, J. Chem. Phys. 67, 5976 (1977).
34. For a review, see P. Pechukas, Ann. Rev. Phys. Chem. 32, 159 (1981).
35. E. Pollak, Chem. Phys. 61, 305 (1981).
36. M. Berry, Chem. Phys. Lett. 27, 73 (1974).
37. U. Halavee and M. Shapiro, J. Chem. Phys. 64, 2826 (1976).
38. G.C. Schatz and J. Ross, J. Chem. Phys. 66, 1037 (1977).
39. G.C. Schatz, J. Chem. Phys., to be published.
40. B. Liu and P. Siegbahn, J. Chem. Phys. 68, 2457 (1978).
41. D.G. Truhlar and C.J. Horowitz, J. Chem. Phys. 68, 2466 (1978); ibid., 71, 1514E (1979).
42. D.K. Bondi, D.C. Clary, J.N.L. Connor, B.C. Garrett and D.G. Truhlar, J. Chem. Phys. 76, 4986 (1982).
43. G.C. Schatz, Chem. Phys. Lett., to be published.
44. See for example, C.H. Townes and A.L. Schalow, Microwave Spectroscopy, (McGraw-Hill, New York, 1955), Chap. 3.
45. J.M. Bowman, Chem. Phys. Lett., to be published.

46. B.R. Johnson and N.W. Winter, J. Chem. Phys. 66, 4116 (1977).
47. P.A. Whitlock, J.T. Muckerman and E.P. Fisher, 'Theoretical Investigations of the Energetics and Dynamics of the Reaction of $O(^3P), ^1D)+H_2$ and $C(^1D)+H_2$', (Research Institute for Engineering Sciences, Wayne State Univ., Detroit, 1976).
48. R. Schinke and W.A. Lester, J. Chem. Phys. 70, 4893 (1979).
49. M. Broida and A. Persky, J. Chem. Phys. 80, 3687 (1984).
50. D.C. Clary, J.N.L. Connor and C.J. Edge, Chem. Phys. Lett. 68, 154 (1979).
51. N.W. Winter, unpublished.
52. K-T. Lee, J.M. Bowman, A.F. Wagner and G.C. Schatz, J. Chem. Phys. 76, 3563 (1982).
53. K-T. Lee, J.M. Bowman, A.F. Wagner and G.C. Schatz, J. Chem. Phys. 76, 3583 (1982).
54. J.M. Bowman, G-Z. Ju, K-T. Lee, A.F. Wagner and G.C. Schatz, J. Chem. Phys. 75, 141 (1981).
55. J.M. Bowman, A.F. Wagner, S.P. Walch, and T.H. Dunning, Jr., J. Chem. Phys. 81, 1739 (1984).
56. B.C. Garrett, D.G. Truhlar, R.S. Grev and A.W. Magnuson, J. Phys. Chem. 84, 1730 (1980).
57. B.C. Garrett and D.G. Truhlar, Proc. 20th Intl. Symp. of Comb. (Univ. of Mich., Ann Arbor, 1984), to be published.
58. D.C. Clary and J.N.L. Connor, Mol. Phys. 41, 689 (1980).
59. G.C. Schatz, J. Chem. Phys., to be published.
60. See, for example, J.M. Bowman, A. Kuppermann, J.T. Adams and D.G. Truhlar, Chem. Phys. Lett. 20, 229 (1973).
61. K. Westberg and N. Cohen, J. Phys. Chem. Ref. Data 12, 531 (1983).
62. A.A. Westenberg and N. de Haas, J. Chem. Phys. 47, 4241 (1967).
63. A.A. Westenberg and N. de Haas, J. Chem. Phys. 50, 2512 (1969).
64. K.M. Pamidimukkala and G.B. Skinner, J. Chem. Phys. 76, 311 (1982).
65. K.M. Pamidimukkala and G.B. Skinner, 13th Intl. Symp. on Shock Tubes and Waves, (Niagara Falls, N.Y., 1981), p. 585
66. N. Presser and R.J. Gordon, J. Chem. Phys. 82, 1291 (1985).
67. G.C. Light, J. Chem. Phys. 68, 2831 (1978).
68. N. Presser and R.J. Gordon, to be published.

CALCULATIONS ON COLLINEAR REACTIONS USING HYPERSPHERICAL COORDINATES

J. Römelt
Lehrstuhl für Theoretische Chemie
Universität Bonn
Wegelerstraße 12
D-5300 Bonn 1, FRG

ABSTRACT. A number of new effects in reaction dynamics discovered by theoretical investigations using hyperspherical coordinates, e.g.

- vibrational adiabaticity
- oscillating reactivity
- resonances and their mode specific decay
- vibrational bonding

are presented and discussed with respect to their experimental consequences. Finally it is shown that the important class of organic S_N2-reactions may be investigated in an approximate collinear treatment using a combination of molecular structure and reaction dynamics theory.

1. INTRODUCTION

Considering the literature it has to be realized that the hyperspherical coordinates (sometimes called Delves coordinates or mass weighted polar coordinates, too) have quite a long standing tradition in describing three body problems in a variety of physical fields. Originally they seem to have appeared in studies of the helium atom (1932) [1] and since then a continuous stream of publications indicate their application to the treatment of two electron atoms [2] and the H_2^+ molecule [3]. Clapp and Delves rediscovered them for the field of nuclear physics (1949) [4] and within the last two decades the set of hyperspherical coordinates was introduced to the theory of molecular dynamics (i.e. molecular collisions (1960) [5], collision induced dissociation (1971) [6], photolysis of molecules (1974) [7], chemical reactions (1975) [8], electron molecule scattering (1980) [9], molecular vibrations (1981) [10]). In 1980 the first numerical examples for a quantum mechanically exact treatment of a collinear reaction were published [11,12] and since then the hyperspherical coordinates have provided a very useful tool to handle the dynamics of collinear three-body reaction processes

$$A + BC(v) \rightarrow A + BC(v) \quad \text{(elastic)} \tag{1}$$

$$\rightarrow A + BC(v') \quad \text{(inelastic)} \tag{2}$$

D. C. Clary (ed.), The Theory of Chemical Reaction Dynamics, 77–104.

$\rightarrow AB(v'') + C$ (reactive) (3)
$\rightarrow A + B + C$ (dissociative) (4)

Traditionally, chemical reactions have been treated using the natural reaction coordinates [13] following the chemist's view leading from the reactants to the products. Unfortunately it is an inherent problem of this picture in accurate quantum calculations that this scheme excludes the description of branching ratios (cf. [14]), dissociative processes (cf. [15]) or reactions associated with small skewing angles in mass weighted coordinates (i.e. heavy-light-heavy reactions, cf. [16]). Compared to these restrictions the hyperspherical coordinates allow a simultaneous treatment of all processes (1) - (4) for all mass combinations of the atoms A, B and C. Up to now a considerable amount of work has been published applying hyperspherical coordinates to various reactions and revealing some new and interesting phenomena (cf. Ref. [17], too). This contribution is meant to review and to summarize the essential results and conclusions of this work (including some very recent and still unpublished parts of it). Finally it is tried to outline some future aspects emerging from all the effects which - at least from the author's point of view - might be fruitful for the field of reaction dynamics.

2. COLLINEAR REACTIONS DESCRIBED IN HYPERSPHERICAL COORDINATES AND THE CORRESPONDING ADIABATIC APPROXIMATION

The purpose of the following resume is to provide the essential equations of the method and to introduce the adiabatic approximation in terms of hyperspherical coordinates, which turns out to be a helpful and illustrative picture for an understanding of the numerical results. Considering the collinear reaction

$$A + BC \rightarrow AB + C$$

the Hamilton operator in mass weighted coordinates x,y [18] (with the center of mass already factored out) is given by (in atomic units):

$$X = (m_{A,BC}/m_{BC})^{1/2}\ r_{A,BC} \tag{5}$$

$$Y = r_{BC} \tag{6}$$

$$H = -\frac{1}{2m_{BC}}\left\{\frac{\partial^2}{\partial x^2} + \frac{\partial^2}{\partial y^2}\right\} + V(x,y) \tag{7}$$

($m_{A,BC}$ and m_{BC} represent the reduced masses for the configuration A+BC and the BC molecule, respectively, M denotes the total mass. Note that the mass scaling is slightly different compared to Kuppermann's notation). The transformation to the hyperspherical coordinates r,φ (cf. Fig. 1)

$$r = (x^2+y^2)^{1/2} \tag{8}$$

$$\varphi = \arctan\,(x/y) \tag{9}$$

(with a skewing angle $\varphi_{max} = \arctan\left(\left(\frac{m_B M}{m_A m_C}\right)^{1/2}\right)$

yields the following Schrödinger equation [11,12]:

$$\left\{-\frac{1}{2m_{BC}}\left(\frac{1}{r}\frac{\partial}{\partial r} r \frac{\partial}{\partial r} + \frac{1}{r^2}\frac{\partial^2}{\partial\varphi^2}\right) + V_{(r,\varphi)} - E\right\} \Psi_{(r,\varphi)} = 0 \tag{10}$$

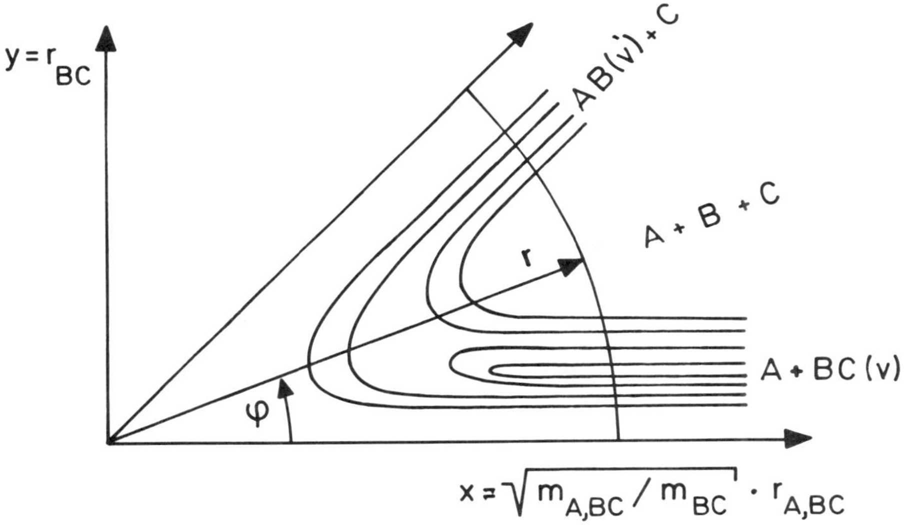

Fig. 1 Schematic contour diagram of the potential energy surface for a collinear reaction A+BC → AB+C in mass weighted coordinates x,y. The hyperspherical coordinates are shown as the radius r and the angular variable φ .

Assuming the ansatz

$$\Psi_{(r,\varphi)} = r^{-1/2} \sum_i \chi_i(r) \; \Phi(r,\varphi), \tag{11}$$

substitution of this expansion (11) into the Schrödinger equation (10) and integration over the angular variable φ produces two separated parts [19]: the angular equation along φ at a constant $r=\bar{r}$

$$\left\{-\frac{1}{2m_{BC}}\frac{\partial^2}{\partial\varphi^2} + V_{(\bar{r},\varphi)}\right\} \Phi_i(\bar{r},\varphi) = \varepsilon_i(\bar{r}) \; \Phi_i(\bar{r},\varphi) \tag{12}$$

and a set of coupled radial equations

$$\left\{-\frac{1}{2m_{BC}}\frac{\partial^2}{\partial r^2} - \left(E - \varepsilon_i(r) + \frac{1}{8m_{BC}r^2}\right)\right\} \chi_i(r) = \frac{1}{2m_{BC}} \sum_j \left\{2 P_{ij}(r)\frac{\partial}{\partial r} + Q_{ij}(r)\right\} \chi_j(r) \tag{13}$$

with the non-adiabatic coupling elements

$$P_{ij}(r) = \langle \Phi_i(r,\varphi) \frac{\partial}{\partial r} \Phi_j(r,\varphi) \rangle \qquad (14)$$

$$Q_{ij}(r) = \langle \Phi_i(r,\varphi) \frac{\partial^2}{\partial r^2} \Phi_j(r,\varphi) \rangle \qquad (15)$$

While the angular equation (12) can be solved numerically (cf. [12]) or by expansion techniques (cf. [11]) the set of coupled radial equations (13) has been treated by propagating quantum wavefunctions (cf. [11,20,21]) or, equivalently, S-matrices (cf. [12]) or R-matrices (cf. [22,23]) along the hyperspherical radius r. Due to the potential boundaries at $\varphi = 0$ and $\varphi = \varphi_{max}$ the angular wavefunctions $\Phi_i(\bar{r},\varphi)$ form a discrete and complete set for the entire range of collision energies E, even in the domain above the dissociation limit

$$A + BC \rightarrow A + B + C.$$

This fact provides the enormous advantage that reactions probabilities may be calculated for high energy collisions (cf. [20,22,23]) and even collision induced dissociation can be included into the description of reaction processes (cf. [24]). These results are qualitatively beyond the capabilities of the traditional techniques. Furthermore it has been shown that using hyperspherical coordinates the convergence of reaction probabilities with respect to the number of channels included in the numerical treatment is much better compared to the conventional methods of propagation in natural reaction coordinates [12,25]. Parallel to the development of exact quantum techniques semiclassical approaches have been elaborated and applied quite successfully [19,26-28] providing illuminating information for the understanding of reaction mechanisms. In the scheme of hyperspherical coordinates the angular eq. (12) describes the internal (vibrational) motion at a fixed value of $r=\bar{r}$ while the set of radial equations (13) represents the translational motion coupled by the vibrational mode. Here the hyperspherical radius provides a generalized translational coordinate including in its asymptotic region ($r\rightarrow\infty$) all possible reactant and product states of the processes (1) - (4) (cf. Fig. 1). The angular (vibrational) energies $\varepsilon_i(r)$ determined as angular eigenenergies in subsequent potential profiles along φ at a constant radius $\bar{r}$ (cf. eq. (12)) provide potential energy curves very similar to the familiar Born-Oppenheimer potentials of a diatomic molecule. This formal analogy in separating the angular and radial motion for a collinear reaction to the Born-Oppenheimer separation of electronic and nuclear motion [29] allows to introduce corresponding experience of molecular structure theory into this type of approximation. In fact the Born-Oppenheimer type energy curves $\varepsilon_i(r)$ give a very reasonable zero-order picture for the interpretation of energy transfer processes and resonance phenomena. For those systems with non-neglegible coupling elements P_{ij} and Q_{ij} it turned out useful to go beyond the pure Born-Oppenheimer type potential $\varepsilon(r)$ to the adiabatic potential $U_i(r)$ [30] which includes the diagonal coupling term $Q_{ii}(r)$ [20,31,32]

$$U_i(r) = \varepsilon_i(r) - \frac{1}{8m_{BC}r^2} + Q_{ii}(r) \qquad (16)$$

To illustrate this type of adiabatic potential $U_i(r)$ in hyperspherical coordinates let us consider the reactions

$$F + HBr \rightarrow FH + Br \qquad (17)$$

$$I + HI \rightarrow IH + I \qquad (18)$$

which will serve as typical representatives for the classes of asymmetric and symmetric reactions, respectively. Here they will be used as examples to demonstrate the essential findings emerging from a rather extended series of theoretical investigations in the field. Although both have to be regarded as reactions with the mass combination heavy-light-heavy, the author would like to emphasize that processes with different mass combinations have been treated equally well in the frame of hyperspherical coordinates (cf. [11,12,20]). In Fig. 2 and 3 the adiabatic angular (vibrational) potentials $U_i(r)$ are displayed for the reactions (17) and (18), respectively. In fact it has to be realized that they provide rather similar energy curves as the traditional electronic potential energy curves of diatomic molecules. In the asymptotic limit they represent the different reactant and product states and at smaller values of r their shape is closely related to the variation of the angular potential profile along the hyperspherical radius r. For symmetric type of reactions the potential energy surface is symmetric with respect to the angle $\varphi_{max}/2$. Hence the angular wavefunction can be classified as g and u according to their parity against reflection at the symmetry axis. In the asymptotic limit reactant and product states are degenerate and they split in the interaction region (cf. Fig. 3).

3. VIBRATIONAL ADIABATICITY AND OSCILLATING REACTIVITY IN COLLINEAR HYDROGEN TRANSFER REACTIONS

Hydrogen transfer reactions usually provide a system with a mass combination heavy-light-heavy associated with a small mass angle φ_{max} in mass weighted coordinates. For these reactions the hyperspherical coordinates are extremely powerful as they provide almost the normal coordinates of the system. Here the angular variable φ represents the exchange vibration of the light atom (asymmetric stretch vibration of the AHB complex), which is fast compared to the motion of the heavy nuclei and nearly uncoupled to their translational motion (i.e. a scan of φ from zero to φ_{max} produces only little change in the distance of the two heavy atoms). Consequently there is only a very weak coupling in the radial equation (13) and the adiabatic approximation describing the reaction as an uncoupled one-dimensional process on the potential $U_i(r)$ is justified. Furthermore the diagonal adiabatic correction is small and therefore $U_i(r)$ is mainly determined by the shape of the Born-Oppenheimer type angular energy curve $\varepsilon_i(r)$. For symmetric type reactions (i.e. I+HI → IH+I, cf. eq. (18)) an oscillative structure of the reaction probability with energy is found (cf. Fig. 4) [23],

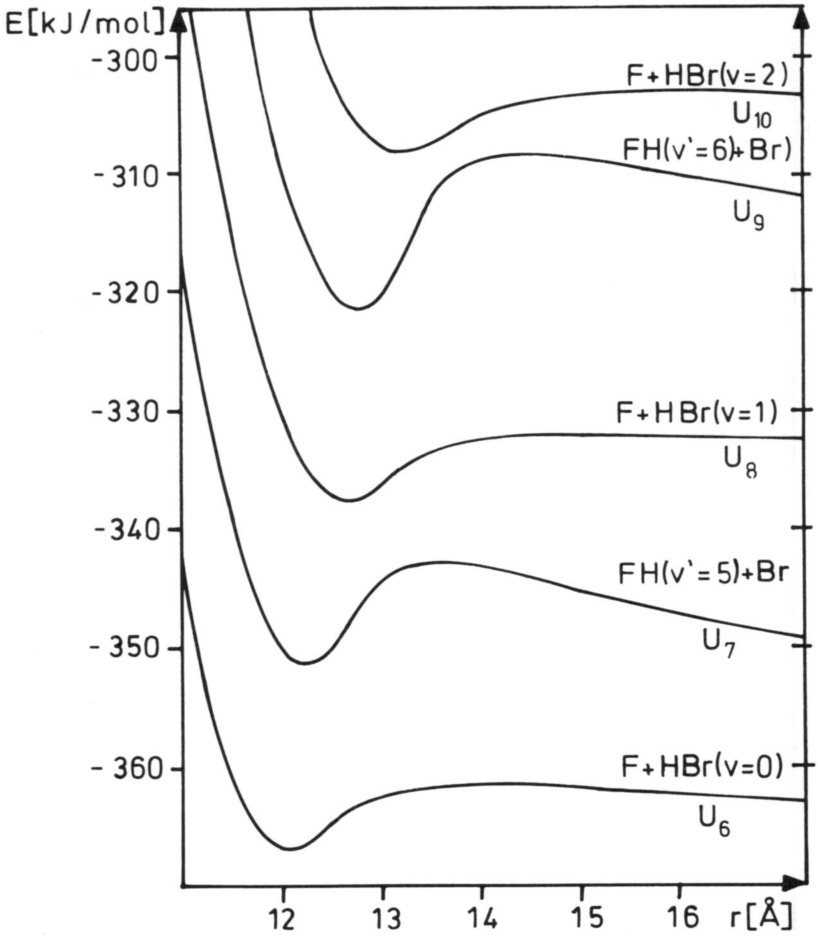

Fig. 2 Adiabatic angular (vibrational) potentials $U_6(r)$ - $U_{10}(r)$ with respect to the hyperspherical radius r for the collinear reaction $F+HBr(v) \rightarrow FH(v')+Br$. The potential energy surface used is a LEPS potential with its parameters adjusted to experimental kinetic data by Jonathan et al. [33].

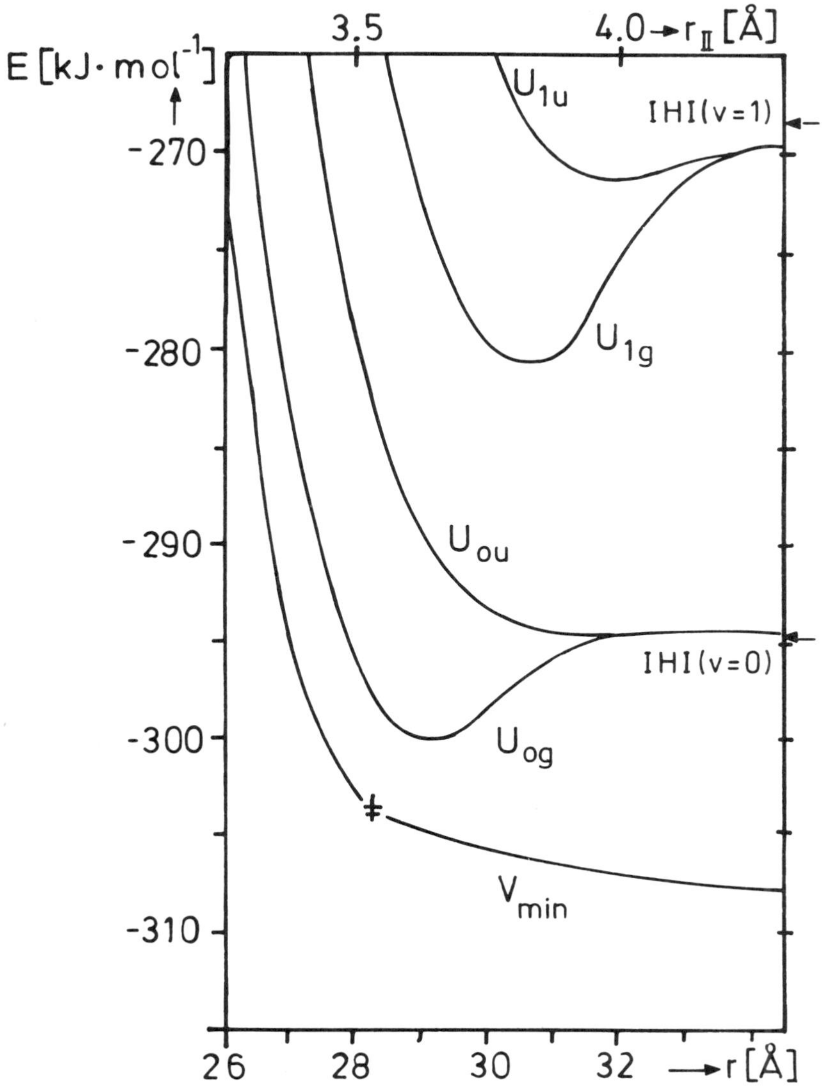

Fig. 3 Adiabatic angular (vibrational) potentials U_{og}, U_{ou}, U_{1g} and U_{1u} with respect to the hyperspherical radius r for the reaction I+HI → IH+I. The top abscissa displays the corresponding r_{I-I} distance. The V_{min} curve provides the potential energy minimum curve with the point of transition state indicated by ‡. The arrows at the right hand side denote the asymptotic energies of the corresponding reactant/product state. The LEPS-A potential used to describe the reaction is given in Ref. [34].

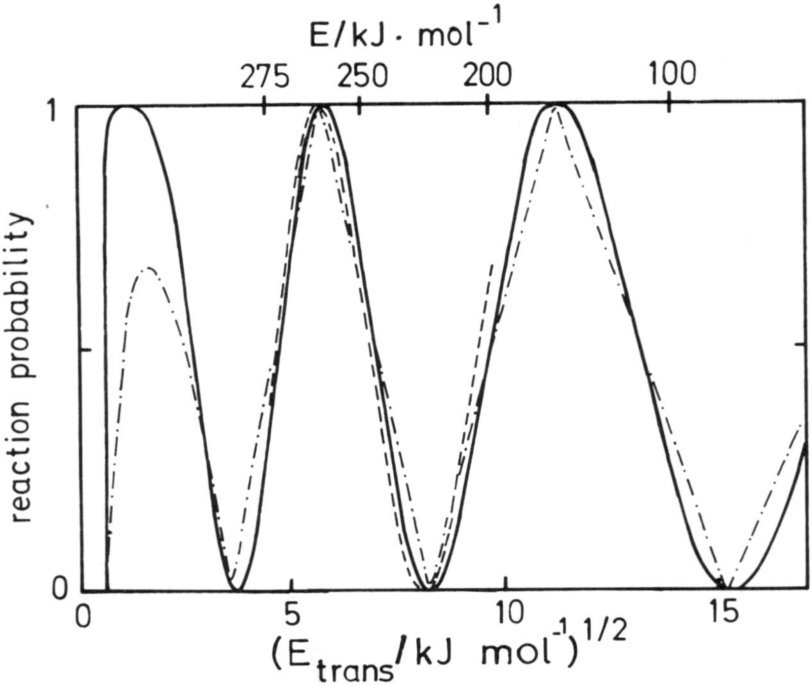

Fig. 4 Oscillating reaction probability for the process I+HI(v=0) → IH(v'=0)+I versus $(E_{trans.})^{1/2}$ (--- quantum, —— semiclassical, .-.-. classical treatment). At $(E_{trans.})^{1/2}$ < 5 kJ/mol the quantum and semiclassical calculations provide identical results.

[26-28,34] which may be explained as quantum mechanical interference pattern of processes occurring on the independent and asymptotic degenerate potentials U_{og} and U_{ou} (cf. Fig. 3). Semiclassically this structure corresponds to the phenomenon of classical multiple encounters increasing in numbers with energy [19,28,35].

In recent model calculations these findings have been confirmed and verified for a full three-dimensional treatment [36]. Due to the weak couplings in these systems almost no V-T energy transfer occurs during the reaction process; i.e. the translational energy (mainly stored in the heavy atoms) is conserved [37]. Hence these type of reactions provide classical examples of the traditional "spectator stripping model" [38].

A similar situation is found for the asymmetric type of hydrogen transfer processes (i.e. F+HBr → FH+Br; cf. eq. (17)). Again there exists only very weak coupling of the heavy atom translational motion through the light atom exchange vibration (i.e. translational energy is conserved). Since for the asymmetric case the degeneracies of asymptotic reactant and product states are removed according to the adiabatic approximation no reaction should happen at all. But a closer inspection of the adiabatic potentials $U_i(r)$ (cf. Fig. 2) reveals that avoided crossings occur between curves $U_i(r)$ and $U_{i+1}(r)$ providing regions with increased non-adiabatic interactions which allow probability flow in different reaction channels (cf. Fig. 5).

Hence for asymmetric reactions the probability pattern is mainly governed by the structure of the avoided crossings between the different adiabatic channel potentials [27,39,40]. Dominant elastic scattering probabilities are found for F+HBr(v=0) → F+HBr(v=0) while the reaction probability for F+HBr(v=0) → FH(v'=4)+Br is quite small in the region about 150 KJ/mol above the reaction threshold (cf. Fig. 5) [41]. For its isotope reactions F+DBr(v=0) → FD(v'=6)+Br and F+MuBr(v=0) → FMu(v'=3)+Br the corresponding probabilities are even smaller because of smaller non-adiabatic couplings and larger energy splittings in the avoided crossing region. Again the probabilities exhibit the oscillative pattern found for other asymmetric hydrogen transfer reactions [39,42] and hence the picture of the classical multiple encounters also can be applied. Stressing the analogy to the diatomic molecules, again, these oscillations can be regarded as formal counterparts to the so- called Landau-Zener oscillations in the theory of electronic non-adiabatic transitions [43]. In total the following hydrogen transfer reactions have been investigated by various groups obtaining similar results:

I + HI(v)	→IH(v') + I	[23,26-28,34]
Cl + HCl(v)	→ClH(v') + Cl	[32, 23]
F + HF(v)	→FH(v')+F	[26,27]
H_3C + HCH_3(v)	→H_3CH(v')+CH_3	[26]
Br + HCl(v)	→BrH(v')+CH	[27,36,39,42]
F + HBr(v)	→FH(v') + Br	[41]

On the experimental side a considerable amount of kinetic data for hydrogen transfer reactions is available. The conclusions emerging from these data are in a remarkable agreement with the theoretical findings just presented. Considering the F+HBr reaction explicitly the experiments exhibit an inverted FH vibrational population produced in the course of the reaction [33,44]. This result is in qualitative agreement (since the theoretical calculation dealt with a collinear system a quantitative comparison is not appropriate) with the theoretical finding of translational energy conservation. In this exoergic reaction nearly all the potential energy is converted into vibrational motion thus producing a highly inverted product vibrational distribution. Furthermore the experiments reveal a surprising non-Arrhenius behaviour of k(T) and σ(E) for the F+HBr [45] and similar hydrogen transfer reactions [46] that tentatively has been explained in terms of competing direct and migratory reaction paths (microscopic branching). However, this interpretation is in conflict with the analysis of rotational and vibrational product populations [44]. The theoretical investigation of collinear reactions suggests that the interference effect presented above may contribute to the experimental oscillating reactivity of the hydrogen transfer reactions corresponding to multiple encounters in classical trajectory simulations [33].

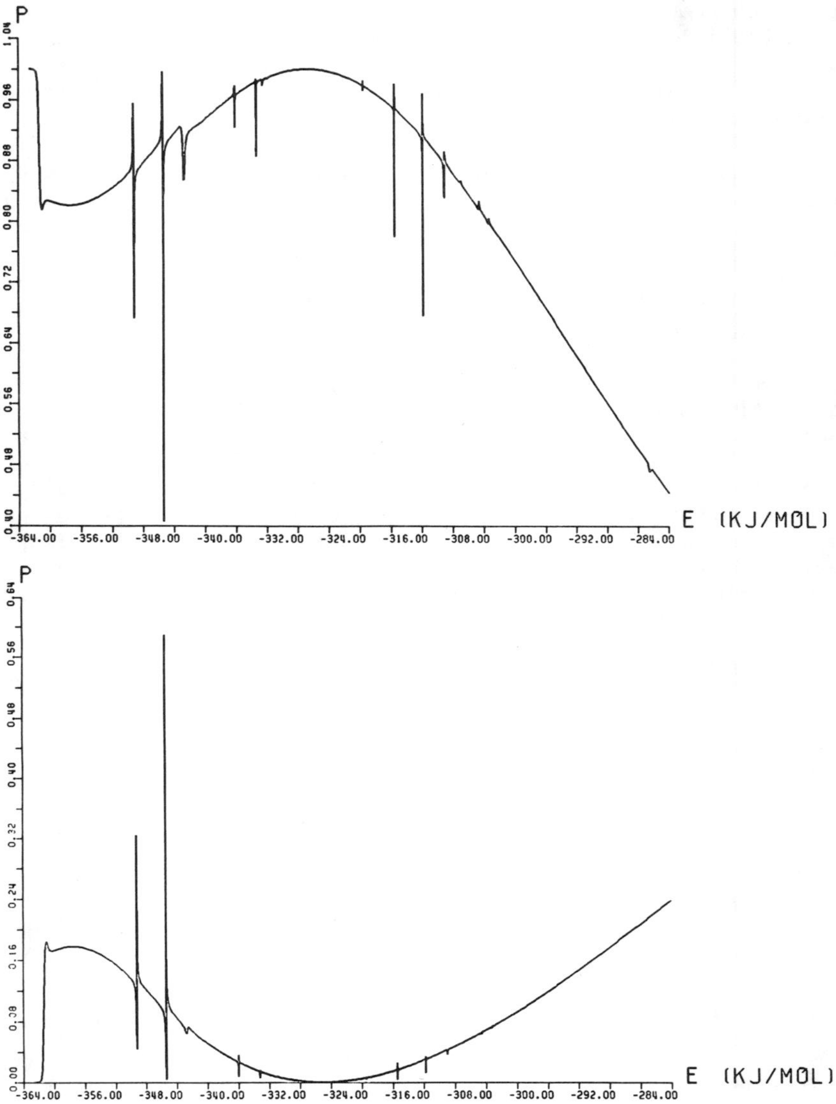

Fig. 5 Reaction probabilities P for the processes
a) F+HBr(v=0) → F+HBr(v=0)
b) F+HBr(v=0) → FH(v'=4)+Br
with respect to the total energy of the system. The LEPS surface used for the calculations is described in Ref. [33]. Note, that the scale of the probabilities is different in panel a) and b).

4. REACTIVE SCATTERING RESONANCES AND VIBRATIONAL BONDING

Considering the set of coupled radial equations (13) with the assumption

$$P_{ij}(r) = 0 \quad ; \quad Q_{ij}(r) = 0$$

we are left with the Diagonal Corrected Vibrational Adiabatic Hyperspherical (DIVAH) model [20,31,32] and equation (13) reduces to

$$\{- \frac{1}{2m_{BC}} \cdot \frac{\partial^2}{\partial r^2} + U_i(r) - E\} \chi_i(r) = 0 \qquad (17)$$

(with $U_i(r)$ defined in eq. (16)). Within this model the problem of reactive scattering is reduced to an elastic scattering treatment in an effective potential $U_i(r)$ (cf. eq. (16)) along the radial coordinate r. An inspection of these effective potentials $U_i(r)$ (cf. Figs. 2,3) reveals the existence of quasi-bound and bound states (analogous to the vibrational states of diatomic molecules). However, the existence of these states on a potential $U_i(r)$ implies distortions of the continua belonging to potentials $U_j(r)$ $(j<i)$, depending mainly on the magnitude of the corresponding non-adiabatic coupling elements $P_{ij}(r)$, $Q_{ij}(r)$ and leading to resonance features of Feshbach type which should appear in $P^R_{ij}(E)$ and $P^R_{ii}(E)$ reaction probabilities. Furthermore, quasi-bound states (behind a potential barrier) are causing shape type resonances in $P^R_{ii}(E)$ probabilities. As a matter of fact this simple model turned out to be a very effective and accurate tool for the prediction and interpretation of resonance features in reactive scattering probabilities [31,47]. In Fig. 6 it is shown for the F+HBr case that this type of analysis accounts for the entire resonance spectrum found in a quantum exact two-dimensional treatment. From the shape of the resonances at -349 kJ/mol, -345 kJ/mol and -342 kJ/mol (being induced by quasi-bound states in the FH(v'=5)+Br channel potential altogether) can be seen, that their decay mechanism is changing with increasing energy from a predominantly Feshbach type (at -349 kJ/mol, -345 kJ/mol) to a predominantly shape type decay (at -342 kJ/mol). The lifetime of these three resonances (τ_1=0.8 ps, τ_2=0.5 ps, τ_3 = 5.7 fs) exceed thermal collision times as well as reactant or product hydride vibrational periods (τ_{HBr}= 2.0 fs, τ_{FH}=1.3 fs). Although it is expected that in a full three-dimensional (experimental) treatment their lifetime will be reduced due to additional bending mode decay channels still they should exist longer than a vibrational period; hence a new type of experimental transition state spectroscopy should be possible (cf. [48]).

Another interesting aspect connected with resonances and their decay is the effect of mode selectivity of bi- and unimolecular reactions [49] in light-heavy-light atom processes. Here the central heavy atom may block the energetic flux and exchange of light atoms; i.e. for the Rosen-Thiele-Wilson model of two coupled Morse-oscillators

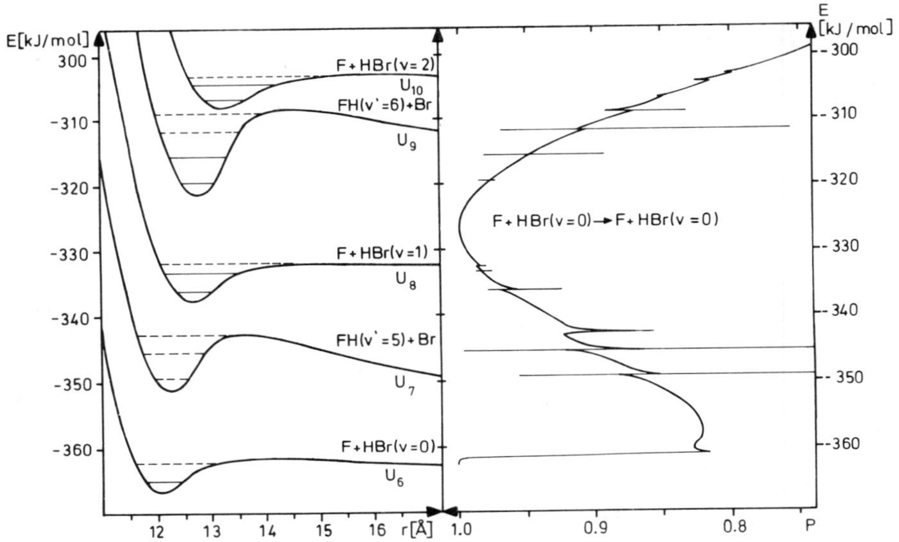

Fig. 6 a) Diagonal Corrected Vibrational Adiabatic Hyperspherical (DIVAH) potentials $U_i(r)$ for the FHBr system including the bound (——) and quasi-bound (---) states compared to

b) the elastic scattering probability P(E) for the process $F+HBr(v=0) \rightarrow F+HBr(v=0)$.

[50,51] the process

$$A + BA(v) \rightarrow A + BA(v') \qquad m_B >> m_A$$

is almost completely non-reactive in off-resonance energy regions (cf. Fig. 7). Only the resonance states provide a kind of doorway for reactive processes [49,52,53]

$$A + BA(v) \rightarrow ABA^{\#} \rightarrow AB(v') + A. \tag{18}$$

Interestingly their widths and amplitudes vary non-monotonously with energy implying non-RRKM variations of the corresponding lifetimes [49] (cf. the resonances at E=0.123 eV and E=0.138 eV in Fig. 7). Using a time dependent fast-Fourier-Transform propagation the decay of a variety of resonance states has been investigated [49] and it was shown that the corresponding lifetime depends critically on the mode structure associated with the resonance state: local mode resonances [51] decay is fast, hyperspherical modes [53] decay much slower (cf. Fig. 7). This type of mode selectivity may be rationalized by the structure of resonance wavefunctions in a rather similar way as pointed out for the Hénon-Heiles model [54]. On the experimental side it is a challange to discover this type of hyperspherical versus local mode specifity via spectroscopic investigation of line widths in highly excited $ABA^{\#}$ molecules (possible candidates are dihydrides like H_2O [10,50,51,55] or other systems like SO_2 [56]).

Finally, a most interesting and surprising phenomenon was discovered in the I+HI system: vibrational bonded molecules. As shown in Fig. 8, for this system even the lowest DIVAH potential curve U_{og} supports four bound states, which, due to the absence of a lower lying continua, are not destroyed by non-adiabatic interactions and hence they have to be regarded as stable, dynamically bonded molecular complexes IHI [57]. Stimulated by a classical analysis of oscillating I+HI reactivity [35,58] (cf. Section 3) the possibility of infinite encounters was found corresponding to a stable IHI molecule. Quantummechanically their existence is based on a decrease of the angular, exchange vibrational zero point energy outweighting the increase in potential energy along the hyperspherical radius (cf. Fig. 8) [57,59]. Their bonding mechanism is dynamical in contrast to all other bonding types known in chemistry. This mechanism has been generalized to a full three-dimensional treatment and although it is found that the bonding effect is considerably decreased by the additional bending mode a single bound state has been discovered [59]. The bond strength is comparable to van der Waals type interactions and an extensive analysis of its spectroscopic properties reveal an abnormal isotope effect; i.e. the bond is weakened replacing the hydrogen by a deuterium atom [60]. A complete review of the literature on vibrational bonding is given in Ref. [61]. Experimentally this new type of chemical bonding provides a real challenge, although a verification is not easy, since it will be mixed up with the always present van der Waals interactions. Among several candidates the $CS^+ \cdot BrHBr^-$ complex, being observed just recently, seem to have a small contribution ($\sim$ 2 kJ/mol) of vibrational bonding [62].

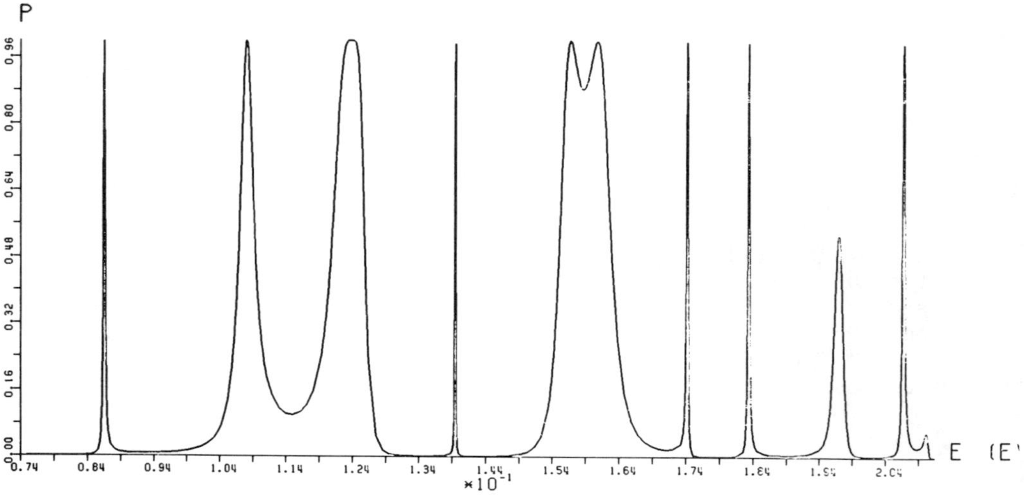

Fig. 7 Reaction probabilities for the process A+BA(v=0) → AB(v'=0)+A for the Rosen-Thiele-Wilson model of two coupled Morse-oscillators ($m_A << m_B$). The resonance at E=0.123 eV is a local mode one (τ =0.17 ps), the resonance at E=0.139 eV is of hyperspherical character (τ =4.7 ps).

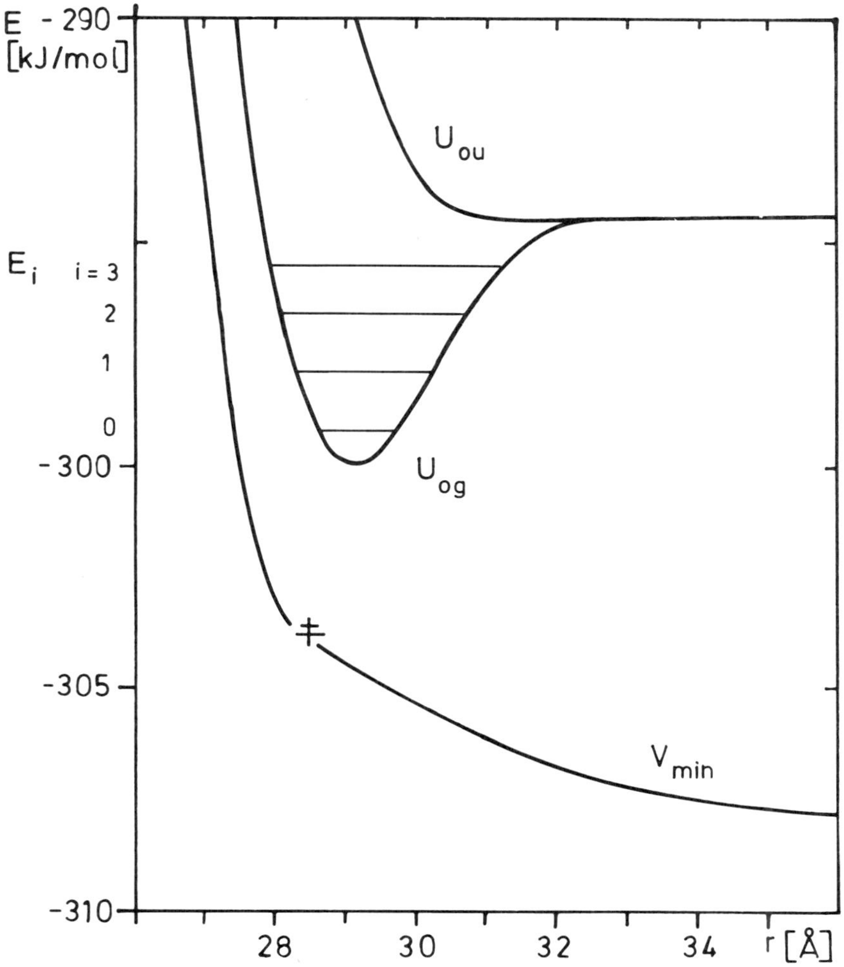

Fig. 8 Minimal potential energy $V_{min}(r)$ and the DIVAH potentials $U_{og}(r)$ and $U_{ou}(r)$ for the collinear I+HI(v=0)→ IH(v'=0)+I reaction. The LEPS potential used is described in Ref. [34]. The bound states in the $U_{og}(r)$ potential are indicated by horizontal lines.

5. FUTURE ASPECTS

As shown in section 3 and 4, the theory of molecular dynamics in hyperspherical coordinates revealed a number of new and exciting effects (e.g. vibrational adiabaticity, oscillating reactivity, resonance features and their mode specificity vibrational bonding). They provide real challenges for experimentalists as well as theoreticians to verify and to investigate these phenomena in more detail. But, as emphasized before, it is necessary to bear in mind that their theoretical prediction and interpretation in many instances still suffer from the restriction to collinearity (physical 1D) in the theoretical treatments. There have been great efforts to extent the basic ideas to the full three-dimensional (experimental) situation in terms of classical, semiclassical or even approximate quantum methods. Nevertheless it is highly desirable to adopt the dynamical treatment in hyperspherical coordinates to the full three-dimensional situation. As far as theory and computational methodology is concerned, several strategies have been developed [63-66], but, due to the enormous amount of computational power involved, only preliminary results are available [66,67]. But the author is hopeful that in the near future a physical 3D extended computer code will be generated, giving the opportunity for more realistic studies, which will provide even more insight into the intimate mechanisms of elementary reactions. Despite this natural tendency to a 3D extension, the collinear methods remain important for a rather large class of reactions: in organic chemistry there is an enormous field which due to electronic structure or steric forces are kept in a perfect collinear arrangement. An example for this type of reaction is represented by the class of bimolecular nucleophilic substitutions [68] (S_N2-reaction; cf. Fig. 9):

$$Y + H_3CX \rightarrow [Y \ldots CH_3 \ldots X]^{\#} \rightarrow YCH_3 + X \qquad (19)$$

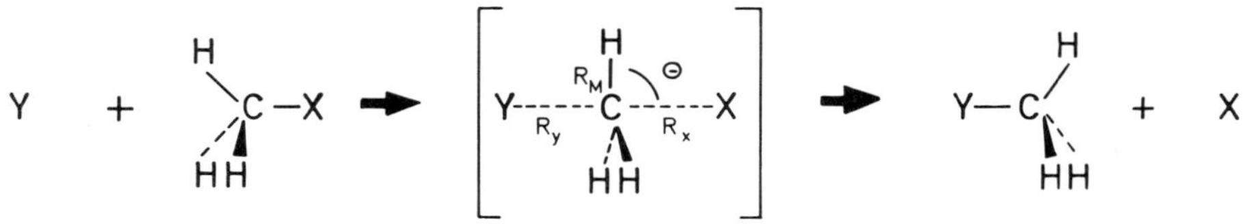

Fig. 9 Schematic mechanism of a bimolecular nucleophilic substitution (S_N2-reaction).

The attacking nucleophile approaches the H_3C-X molecule perpendicular to the H_3-plane at the central carbon atom and under inversion of the CH_3-umbrella (Walden inversion [69]) the system reacts via a necessarily collinear arranged Y ... C ... X transition complex (due to the electronic sp^2-hybridization at the central C atom) to the product configuration Y-CH_3+X. Hence in this reaction the bond-breaking and the bond-forming distance are fixed to a collinear configuration. In a zero-order approximation this bimolecular reaction can be treated as

$$Y + R\text{-}X \rightarrow [Y \ldots R \ldots X]^{\#} \rightarrow Y\text{-}R + X \qquad (20)$$

(the central CH_3 group is considered as structureless particle R) in complete analogy to the conventional collinear atom plus diatom processes [70].

Identifying Y and X with a hydrogen negative ion and a hydrogen atom, respectively

$$H^- + HCH \rightarrow [H \ldots CH_3 \ldots H]^- \rightarrow HCH_3 + H^- \qquad (21)$$

a complete two-dimensional potential energy surface has been calculated with a very good accuracy (cf. Table I) using ab initio electronic structure methods [70]. The CH_3 umbrella inversion has been taken into account in a minimal energy path approximation; e.g. the optimal inversion angle Θ (cf. Fig. 9) was calculated at each point of the two-dimensional surface. In Fig. 10 the potential energy together with the inversion angle is plotted along the minimal energy reaction path. At large distances there is a shallow well accompanied by a slight decrease of the angle Θ, due to the static multipole interactions, which is followed by a steep barrier (~ 56 kcal/mol) and a sudden inversion of the central CH_3 group. This mechanism deduced from the calculations is in contrast to the experimental findings [71], assuming a much smoother reaction process. Obviously this discrepancy is due to solvent effects [72]. Preliminary dynamical calculations on the ab initio surface indicate the tendency that increasd translational energy enhance vibrational non-adiabaticity; e.g. the product vibrational distribution approach more and more a statistical distribution with increasing translational energy.

Table I: Comparison of experimental and ab initio calculated molecular energies and structure parameters for the S_N2-reaction $H^- + H_3CH \rightarrow [H-CH_3-H]^- \rightarrow H-CH_3 + H^-$.

$H^- + H_3CH$		R_y/a_o	R_H/a_o	Θ/degree	D_e(C-H)/kcal·mol^{-1}	E/h
experimental	a)	∞	2.06	109.3	107	-
ab initio	b)	20	2.09	109.3	106	-40.8926
	c)	∞	2.06[d)]	109.5	-	-40.8963

$[H - CH_3 - H]^-$		$R_x=R_y/a_o$	R_H/a_o	Θ/degree	$E^{\#}$/kcal·mol^{-1}	E/h
experimental	a)	-	-	90	> 6	-
ab initio	b)	3.22	2.04	90	56.1	-40.8032
	c)	3.28	2.01	90	56.4	-40.8064

a) The experimental values are taken from:
D.L. Gray and A.G. Robiette, Mol. Phys. **37** (1979) 1901
K.E. McCulloh and V.H. Dibeler, J. Chem. Phys. **64** (1976) 445
J.D. Paysant, K. Tanaka, L.D. Betowski and D.K. Bohme, J. Am. Chem. Soc. **98** (1976) 894

b) H. Dohmann and J. Römelt, to be published

c) F. Keil and R. Ahlrichs, J. Am. Chem. Soc. **98** (1976) 4787

d) The experimental R_H distance has been used in the calculations.

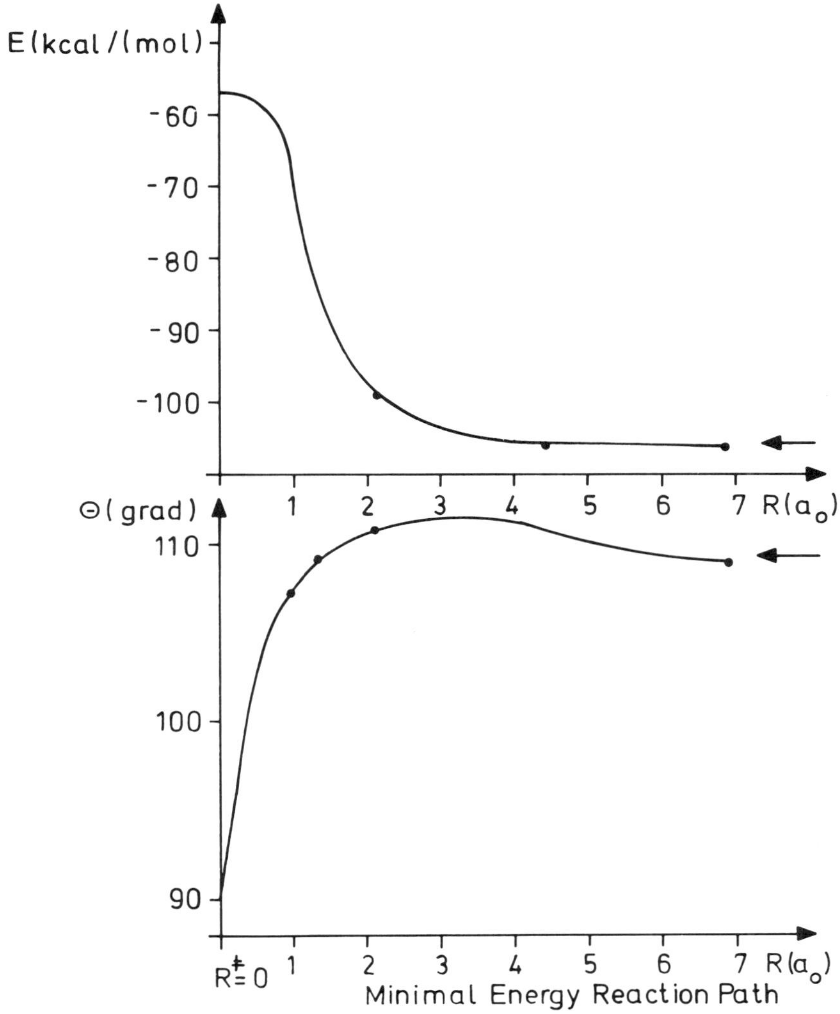

Fig. 10 Ab initio calculated electronic potential energy E and inversion angle Θ (angle H-C-X, cf. Fig. 9) along the minimal energy reaction path for the S_N2-reaction $H^- + H_3C-H \rightarrow H-CH_3 + H^-$. The arrows at the right hand side denote the corresponding asymptotic values ($R \rightarrow \infty$); $R^{\#} = R = 0$ corresponds to the location of the transition state complex.

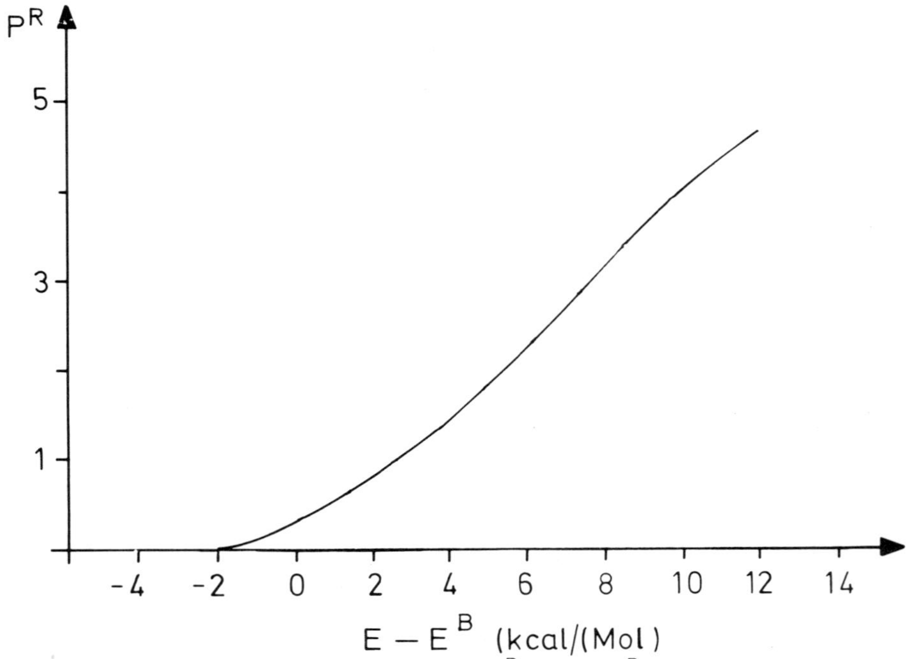

Fig. 11 Total reaction probability $P^R = \sum_{v'} \sum_{v} P^R_{v'v}$ for the model S_N2 reaction $H^- + R\text{-}H(v) \rightarrow H\text{-}R(v') + H^-$ as a function of the energy difference $E\text{-}E^B$. E and E^B denote the total energy and the value of the electronic energy barrier $E^B = E^{\#}(CH_5^-)$, respectively.

Fig. 11 displays the total reaction probability (from all reactant to all product states energetically accessible) for the S_N2 model reaction (cf. eqs. (20), (21))

$$H^- + R{+}H(v) \rightarrow H\text{-}R(v') + H^- \qquad (22)$$

as a function of energy. The onset of the probability is found to be lower by about 2 kcal/mol compared to the electronic potential energy barrier due to vibrational bonding and tunnelling effects. This result demonstrates, that it is not sufficient just to know the potential energy surface but dynamical effects have to be incorporated in the theoretical treatment in order to calculate such an important parameter like the reaction barrier of a chemical process. As in the present model only the bond-breaking and bond-forming C-H distances are taken into account explicitly the vibrational bonding effect will be even greater considering all degrees of freedom in the system.
More studies are in progress considering different attacking and leaving groups Y and X, respectively, in order to investigate their influence

on the mechanism and dynamics of the system. It is absolutely clear that this kind of treatment is only a very crude approximation, but nevertheless it should provide a useful and illuminating look into the details of these important organic chemistry reactions. In the light of classical trajectory calculations [73] as well as gradient analysis [74] the approximation to fix the C-H distances seems to be justified, whereas the angular degree of freedom is considerably involved in the reaction process, which contradicts our assumption of a structureless central CH_3 group. Therefore a generalization of the hyperspherical treatment to polyatomic reactions [75] or a combination of two hyperspherical (representing the bond-breaking and bond-forming degrees of freedom) and 3N-8 ordinary normal coordinates (serving as a kind of energy storage medium) [76] would improve the quality of these investigations considerably. Furthermore this project represents another example [77], that ab initio potential energy surfaces can be used in dynamical studies. Since the quality of electronic structure calculations is improving more and more, it is the author's opinion, that both fields, electronic structure theory and molecular dynamics theory, should come into much closer contact to design efficient and, for the purpose of dynamical treatments, satisfying strategies to improve molecular potential energy surfaces. Similar collaboration would be useful on other problems, too; e.g. ab initio calculations are able to provide electronically non-adiabatic coupling elements [78] and, just recently, pioneering work has been presented including more than one electronic potential energy surface in collinear reaction processes [63,79].

6. CONCLUSION

A variety of new and exciting phenomena have been presented resulting from molecular reaction dynamics studies in the frame of hyperspherical coordinates. The rule of translational energy conservation in heavy-light-heavy atom reactions based on the effect of vibrational adiabaticity could be related to highly inverted product vibrational distributions in corresponding exogetic processes. Oscillating reactivity, interpreted as classical multiple encounters, may contribute to the remarkable non-Arrhenius behavior of reaction rates for some hydrogen transfer reactions in a close analogy to oscillatory electron transfer reactions. Resonance features have been discovered and consequences of their mode specific decay have been discussed providing a challenge for experimentalists to develop methods and experiments for the fascinating field of "spectroscopy of the transition state", which to the authors opinion will be one of the most exciting and fruitful problems in future reaction dynamics. A new dynamical kind of chemical bonding, vibrational bonding, was discovered which still needs to be verified and investigated in more detail by experiments as well as theory. Finally a litle step has been undertaken in the direction of investigating the chemically very important class of bimolecular nucleophilic substitutions (S_N2-reactions) by combining molecular structure and molecular dynamics theory.
The author would be glad, if this little review will be able to transfer some of his own fascination of molecular reaction dynamics theory onto the reader and it is his greatest hope, that it may stimulate experimentalists to new and interesting experiments.

7. ACKNOWLEDGEMENT

I would like to express my thanks to all my coauthors for their fruitful and illuminating cooperation, especially to Dr. J. Manz, with whom I am collaborating since many years. I am indebt to Prof. S.D. Peyerimhoff for her constant and encouraging support, to the Heisenberg Foundation for a fellowship and to many colleagues providing their results often prior to publication. Financial support of the Deutsche Forschungsgemeinschaft and the Fonds der Chemischen Industrie is also gratefully acknowledged.

REFERENCES

1. T.H. Gronwall, Annals. Math. **33** (1932) 279
   T.H. Gronwall, Phys. Rev. **51** (1937) 655
   J.H. Bartlett, Phys. Rev. **51** (1937) 661
   P.M. Morse and H. Feshbach, "Methods of Theoretical Physics", (McGraw-Hill, 1953), p. 1729

2. U. Fano, Rep. Prog. Phys. **46** (1983) 97 and Ref. therein
   C.D. Liu and J.H. Macek, Phys. Rev. **A30** (1984) 2161 and Ref. therein
   C.H. Greene and C.W. Clark, Phys. Rev. **A30** (1984) 2161 and Ref. therein
   H. Klar, in "Electronic and Atomic Collisions", eds. J. Eichler, I.V. Hertel and N. Stolterfoht (Elsevier, 1984), p. 767 and Ref. therein

3. R.C. Whitten and J.S. Sims, Phys. Rev. **A9** (1974) 1586
   C.H. Greene, Phys. Rev. **A26** (1982) 2974

4. R.E. Clapp, Phys. Rev. **76** (1949) 873
   L.H. Delves, Nucl. Phys. **9** (1959) 391; **20** (1960) 275

5. F.T. Smith, Phys. Rev. **120** (1960) 1058
   F.T. Smith, in "Kinetic Processes in Gases and Plasmas", ed. A.R. Hochstim (Academic Press, 1969), p. 321

6. R.D. Levine and R.B. Bernstein, Chem. Phys. Letters **11** (1971) 552
   C. Rebick and R.D. Levine, J. Chem. Phys. **58** (1973) 3942

7. M. Tamir, U. Halavee and R.D. Levine, Chem. Phys. Letters **25** (1974) 38

8. A. Kuppermann, Chem. Phys. Letters **32** (1975) 374
   R.A. Marcus, Faraday Symp. Chem. Soc. **10** (1975) 60

9. C.W. Clark and J. Siegel, J. Phys. B **13** (1980) L 31
   Vo Ky Lan, M. Le Dourneuf and J.M. Launay, in "Electron-Atom and Electron-Molecule Collisions", ed. J. Hinze (Plenum Press, 1983), p. 161

10. C.H. Greene and Ch. Jungen, Abstract of Papers, 11th Int. Conf. Physics of Electronic and Atomic Collision, ed. S. Datz (Gatting-burg, Tennesse, USA, 1981), p. 1019
    J.G. Frey, Chem. Phys. Letters **102** (1983) 421

11. A. Kuppermann, J.A. Kaye and J.P. Dwyer, Chem. Phys. Letters **74** (1980) 257

12. G. Hauke, J. Manz and J. Römelt, J. Chem. Phys. **73** (1980) 5040
    J. Römelt, Chem. Phys. Letters **74** (1980) 263

13. G.L. Hofacker, Z. Naturforsch. A **18** (1963) 607
R.A. Marcus, J. Chem. Phys. **45** (1966) 4432, 4600

14. J. Manz, Chem. Phys. Letters **15** (1972) 136

15. D.J. Diestler, in "Atom-Molecule Collision Theory", ed. R.B. Bernstein (Plenum Press, 1979), p. 655

16. M. Baer, J. Chem. Phys. **62** (1975) 305

17. J. Manz, Comment on At. Mol. Phys. (1985), in press

18. F.T. Smith, J. Chem. Phys. **31** (1959) 1352

19. V.K. Babamov and R.A. Marcus, J. Chem. Phys. **74** (1981) 1790

20. J.M. Launay and M. Le Dourneuf, J. Phys. B **15** (1982) L 455

21. F. Mrugala and J. Römelt, to be published

22. D.K. Bondi and J.N.L. Connor, Chem. Phys. Letters **92** (1982) 570
D.K. Bondi and J.N.L. Connor, J. Chem. Phys., in press

23. C.L. Shoemaker, N. AbuSalbi and D.J. Kouri, J. Chem. Phys. **87** (1983) 5389

24. J. Manz and J. Römelt, Chem. Phys. Letters **77** (1981) 172
J.A. Kaye and A. Kuppermann, Chem. Phys. Letters **78** (1981) 546; **115** (1985) 158

25. J. Römelt, Chem. Phys. Letters **87** (1982) 259

26. V. Aquilanti, G. Grossi and A. Laganà, Chem. Phys. Letters **93** (1982) 174, 179
V. Aquilanti, S. Cavalli, G. Grossi and A. Laganà, J. Mol. Struc. **93** (1983) 319; **107** (1984) 95

27. V.K. Babamov, V. Lopez and R.A. Marcus, J. Chem. Phys. **78** (1983) 5621; **80** (1984) 1812; Chem. Phys. Letters **101** (1983) 507

28. C. Hiller, J. Manz, W.H. Miller and J. Römelt, J. Chem. Phys. **78** (1983) 3850

29. M. Born and J.R. Oppenheimer, Ann. Physik **84** (1927) 457

30. J.H. van Vleck, J. Chem. Phys. **4** (1936) 327

31. J. Römelt, Chem. Phys. **79** (1983) 197

32. D.K. Bondi, J.N.L. Connor, J. Manz and J. Römelt, Mol. Phys. **50** (1983) 467

33. N.B.H. Jonathan, P.V. Sellers and A.J. Stace, Mol. Phys. **43** (1981) 215

34. J.A. Kaye and A. Kuppermann, Chem. Phys. Letters **77** (1981) 573
J. Manz and J. Römelt, Chem. Phys. Letters **81** (1981) 179

35. E. Pollak, J. Chem. Phys. **78** (1983) 1228

36. N. AbuSalbi, S.-H. Kim, D.J. Kouri and M. Baer, Chem. Phys. Letters **112** (1984) 502
D.C. Clary, Mol. Phys. **53** (1984) 3
E. Pollak, M. Baer, N. AbuSalbi and D.J. Kouri, Chem. Phys., in press

37. M. Baer, J. Chem. Phys. **62** (1975) 305

38. R.D. Levine and R.B. Bernstein, in "Molecular Reaction Dynamics" (Clarendon Press, 1974)

39. N. AbuSalbi, D.J. Kouri, V. Lopez, V.K. Babamov and R.A. Marcus, Chem. Phys. Letters **103** (1984) 458

40. P.V. Coveney, M.S. Child and J. Römelt, to be published

41. J. Manz and H.H.R. Schor, Chem. Phys. Letters **107** (1984) 549
P.L. Gertitschke, J. Manz, J. Römelt and H.H.R. Schor, J. Chem. Phys. **83** (1985) 208
J. Manz, J. Römelt and H.H.R. Schor, to be published

42. J.A. Kaye and A. Kuppermann, Chem. Phys. Letters **92** (1982) 579

43. E.E. Nikitin, in "Theory of Elementary Atomic and Molecular Processes" (Oxford Univ. Press, 1974)

44. J.P. Sung and D.W. Setser, Chem. Phys. Letters **48** (1977) 413
P. Beadle, M.R. Dunn, N.B.H. Jonathan, J.P. Liddy and J.C. Naylor, J. Chem. Soc. Faraday Trans. 2, **74** (1978) 2170
K. Tamagake, D.W. Setser and J.P. Sung, J. Chem. Phys. **73** (1980) 2203
L.S. Dzelzkalns and F. Kaufmann, J. Chem. Phys. **79** (1983) 3836

45. E. Würzberg and P.L. Houston, J. Chem. Phys. **72** (1980) 5915
J.W. Hepburn, K. Liu, R.G. Macdonald, F.J. Northrup and J.C. Polanyi, J. Chem. Phys. **75** (1981) 3353

46. C.C. Mei and C.B. Moore, J. Chem. Phys. **67** (1977) 3936, **70** (1979) 1759

47. B.C. Barrett, D.W. Schwenke, R.T. Skodje, D. Thirumalai, T.C. Thompson and D.G. Truhlar, in "Resonances in Electron-Molecule Scattering, van der Waals Complexes and Reactive Chemical Dynamics", ed. D.G. Truhlar (Plenum Press, 1984), p. 421

48. T. Carrington Jr., J.C. Polanyi and J.R. Wolf, in "Physics of Electronic and Atomic Collisions", ed. S. Datz (North Holland, 1982), p. 393 and Ref. therein
H.R. Mayne, R.A. Poinier and J.C. Polanyi, J. Chem. Phys. **80** (1984) 4025
W. Kamke, B. Kamke, I. Hertel and A. Gallagher, J. Chem. Phys. **80** (1984) 4879

49. R.H. Bisseling, R. Kosloff, J. Manz and H.H.R. Schor, Ber. Bunsenges. Phys. Chem. **89** (1985) 270
R.H. Bisseling, R. Kosloff and J. Manz, J. Chem. Phys., in press
R.H. Bisseling, R. Kosloff, J. Manz and J. Römelt, to be published

50. N. Rosen, J. Chem. Phys. **1** (1933) 319
E. Thiele and D.J. Wilson, J. Chem. Phys. **35** (1961) 1256
R.M. Hedges Jr. and W.P. Reinhardt, Chem. Phys. Letters **91** (1982) 241; J. Chem. Phys. **78** (1983) 3964

51. M.S. Child and L. Halonen, Adv. Chem. Phys. **57** (1984) 1 and Refs. therein

52. K.C. Kulander, J. Chem. Phys. **79** (1983) 1279

53. K.C. Kulander, J. Manz and H.H.R. Schor, J. Chem.Phys., in press
J. Manz and H.H.R. Schor, Chem. Phys. Letters **107** (1984) 542

54. G. Hose and H.S. Taylor, J. Chem. Phys. **76** (1982) 5356
Phys. Rev. Letters **51** (1983) 947
G. Hose, H.S. Taylor and Y.Y. Bai, J. Chem. Phys. **80** (1984) 4363

55. T.C. Thompson and D.G. Truhlar, Chem. Phys. Letters **101** (1983) 235
V. Lopez and R.A. Marcus, Chem. Phys. Letters **93** (1982) 232
K.N. Swamy and W.L. Hase, J. Chem. Phys. **82** (1985) 123
H. Shyldkrot and M. Shapiro, J. Chem. Phys. **79** (1983) 5927

56. M.D. Feit and J.A. Fleck Jr., J. Chem. Phys. **78** (1983) 301

57. J. Manz, R. Meyer, E. Pollak and J. Römelt, Chem. Phys. Letters **93** (1982) 184

58. E. Pollak, in "Intramolecular Dynamics", eds. J. Jortner and B. Pullmann (D. Reidel, 1982), p. 1

59. D.C. Clary and J.N.L. Connor, Chem. Phys. Letters **94** (1983) 81
E. Pollak, Chem. Phys. Letters **94** (1983) 85
J. Manz, R. Meyer and J. Römelt, Chem. Phys. Letters **96** (1983) 607

60. J. Manz, R. Meyer, E. Pollak, J. Römelt and H.H.R. Schor, Chem. Phys. **83** (1984) 333
J. Manz, R. Meyer and H.H.R. Schor, J. Chem. Phys. **80** (1984) 1562
D.C. Clary and J.N.L. Connor, J. Phys. Chem. **88** (1984) 2758

61. E. Pollak, Comment on At. Mol. Phys. **15** (1984) 73
J. Römelt and E. Pollak, in "Resonances in Electron-Molecule Scattering, van der Waals Complexes and Reactive Chemical Dynamics", ed. D.G. Truhlar (Plenum Press, 1984), p. 353
J. Manz and J. Römelt, Nachr. Chem. Techn. Lab. **33** (1985) 210

62. B.S. Ault and J. Manz, Chem. Phys. Letters, in press

63. A. Kupperman, in "Potential Energy Surfaces and Dynamics Calculations", ed. D.G. Truhlar (Plenum Press, 1981), p. 375

64. B.R. Johnson, J. Chem. Phys. **73** (1983) 1906, 1916

65. R.T. Pack, Chem. Phys. Letters **108** (1984) 333

66. H. Mishra and J. Linderberg, Mol. Phys. **50** (1983) 91
M. Mishra, J. Linderberg and Y. Öhrn, Chem. Phys. Letters **111** (1984) 439

67. A. Kuppermann, private communication

68. C.K. Ingold, "Structure and Mechanism in Organic Chemistry" (Cornell Univ. Press, 1953)

69. P. Walden, Chem. Ber. **29** (1896) 133; **30** (1897) 3146

70. S.D. Peyerimhoff and J. Römelt, Phys. Bl. **40** (1984) 300
H. Dohmann and J. Römelt, to be published

71. e.g. K. Müller and L.D. Brown, Theor. Chim. Acta (Berl.) **53** (1979) 75 and Refs. therein

72. K. Morokuma, S. Kato, K. Kitaura, S. Obara, K. Ohta and M. Hanamura, in "Proceedings of the 4th Int. Congress of Quantum Chemistry", eds. P.-O. Löwdin and B. Pullmann (D. Reidel Publ. Comp., 1983), p. 221

73. C. Leforestier, J. Chem. Phys. **68** (1978) 4406

74. K. Ishida, K. Morokuma and A. Kormonicki, J. Chem. Phys. **66** (1977) 2153

75. J. Linderberg, in "Intramolecular Dynamics", eds. J. Jortner and B. Pullmann (D. Reidel, 1982), p. 325
J. Linderberg, in "Proceedings of the 4th Int. Congress on Quantum Chemistry", eds. P.-O. Löwdin and B. Pullmann (D. Reidel, 1983) p. 7
J. Avery, B. Christensen-Dalsgaard, P.S. Larsen and Shen Hengyi, Intern. J. Quant. Chem. S **18** (1984) 321

76. T. Carrington Jr. and W.H. Miller, J. Chem. Phys. **81** (1984) 3942

77. "Potential Energy Surfaces and Dynamics Calculations", ed. D.G. Truhlar (Plenum Press, 1981) and Ref. therein

78. R.J. Buenker, G. Hirsch, S.D. Peyerimhoff, P.J. Bruna, J. Römelt, M. Bettendorff and C. Petrongolo, in "Studies in Physical and Theoretical Chemistry", Vol. **21** (Elsevier Publ. Comp., 1982), p. 81 and Refs. therein

79. B.C. Garrett and D.G. Truhlar, Theor. Chem.: Adv. Persp. **6A** (1981) 215
J.M. Launay and M.L. Dourneuf, poster presented at Molec V, Jerusalem 03.-07.09.1984

REACTIVE SCATTERING IN THE BENDING-CORRECTED ROTATING LINEAR MODEL

Robert B. Walker
Group T-12, MS J569
Theoretical Division
Los Alamos National Laboratory
Los Alamos, NM 87545

Edward F. Hayes
Controllers Office
National Science Foundation
Washington, DC 20550

ABSTRACT. We review the theory and applications of the Bending-Corrected Rotating Linear Model (BCRLM) to problems in the quantum description of reactions between atoms and diatomic molecules.

1. INTRODUCTION TO BCRLM

The Bending-Corrected Rotating Linear Model (BCRLM) is a straightforward extension of the Rotating Linear Model (RLM) proposed in the late 1960's by Child,[1] Wyatt,[2] and Connor and Child.[3] The RLM constrains the dynamics of three dimensional (3D) collisions by requiring the molecular species to maintain an orientation collinear with the atomic species during the course of collision. The classical dynamics of three particles on a line was considered prior to this by Jepsen and Hirschfelder,[4] and more recently by Agmon,[5] but the BCRLM is an outgrowth of the model presented by the authors of Refs. 1-3.

By neglecting the two internal rotational (or bending) degrees of freedom, the mathematical description of the rearrangement collision event is simplified so extensively that the computational treatment of reaction dynamics within this model is routinely possible. This computational simplication arises because the rotational motion of the line of collision is treated analytically by a partial wave expansion of the scattering wavefunction. Consequently, the computational effort reduces to that of a family of collinear reactive scattering calculations, one for each partial wave term in the wavefunction expansion.

The obvious shortcoming of the RLM is its neglect of the internal rotational degrees of freedom. In comparison to the asymptotic vibrational degrees of freedom, the asymptotic rotational degrees of freedom impose a relatively modest constraint on the energetics of collision,

D. C. Clary (ed.), The Theory of Chemical Reaction Dynamics, 105–134.

but they correlate adiabatically to higher energy bending states when the collision partners are close together. The results of the earliest accurate 2D and 3D coupled-channel calculations[6-14] for the $H+H_2$ reaction showed that these bending degrees of freedom are important in determining the energetic position of the reaction threshold. Consequently, Walker and Hayes[15] implemented the suggestion made in Wyatt's[2] paper, and supplemented the RLM with an ad hoc correction to include the adiabatic effects of the lowest energy bending degrees of freedom, producing the bending-corrected RLM, or BCRLM. Including the bending degrees of freedom as an effective potential within a collinear reactive framework was first described by Mortensen and Pitzer[16,17] and is now widely used by Bowman and coworkers[18-24] in reduced dimensionality theories of reaction, and by Truhlar and coworkers[25-32] in variational transition state theories of reactions.

In practice, all BCRLM calculations to date have been done so that only the lowest energy (i.e., zero point) bending state has been explicitly treated. At this level, the additional computational effort for a BCRLM calculation instead of an RLM calculation is minimal --- it is necessary only to compute an effective collinear potential energy surface which is the sum of the usual collinear potential and the bending zero point energy determined at each collinear geometry. In principle, however, a full treatment of the bending degrees of freedom within the adiabatic approximation would require a family of RLM calculations, one for each bending state.

Another obvious defect of both the RLM and BCRLM models is that they assume a collinearly dominated reaction intermediate. While the potential energy surfaces for many collision systems do favor collinear geometries, there are of course many reactions which do not. Extensions of the BCRLM model are therefore needed to treat noncollinear systems, perhaps along the lines defined by the Carrington and Miller[33] reaction surface Hamiltonian theory.

In the next section (Sec. 2), we will develop the theory of the BCRLM. We discuss the solution of the coupled-channel equations in both natural collision coordinates[34-38] and hyperspherical coordinates.[39-47] Both coordinate systems are widely used to treat collinear reactive scattering processes. We will discuss the projection[45,48] of the hyperspherical equations on coordinate surfaces appropriate for applying scattering boundary conditions and review the definition of integral and differential scattering cross sections in this model.

In Sec. 3, we will briefly review applications of BCRLM calculations to reactive systems and discuss in Sec. 4 some possible future developments which may be made through extensions of the method. Sec. 5 then concludes with a summary.

2. THEORY

In this section, we will present a mathematical description of the BCRLM. We will define the classical and quantum mechanical Hamiltonian for the translational, vibrational, tumbling, and bending degrees of freedom for the system. After expanding the scattering wavefunction in a total angular momentum representation, we obtain coupled-channel equations which may be solved numerically subject to reactive scattering boundary conditions. The solution of these coupled-channel equations at a fixed total scattering energy E and angular momentum J determines the scattering matrix, $\underline{\underline{S}}^J(E)$. From the scattering matrix, we can then compute reaction probabilities, integral and differential cross sections, and reaction rate constants.

2.1. Internal Coordinate Systems

We restrict ourselves here to the atom-diatom reactive collision process defined chemically by the equation

$$A + BC(m) \rightarrow AB(n) + C, \tag{1}$$

in which A and BC are the reactant atom and molecule respectively, and AB and C are the product molecule and atom. The vibrational quantum numbers of the reactant and product molecules are m and n respectively. We further assume that the collision dynamics is represented by the motion of the A, B, and C nuclei on a single Born-Oppenheimer electronic potential energy surface, at energies below the threshold for collision induced dissociation. The atomic masses are defined as m_A, m_B, and m_C.

Coupled-channel equations arise in scattering dynamics when all but one of the degrees of freedom of the system are expanded in a square integral basis (of "channels"). The coupled-channel equations are then solved numerically and describe motion in the unbound, or scattering coordinate. The principal difficulty of any reactive scattering calculation is that the coordinate system which best describes the asymptotic motions of reactants differs from the coordinate system best suited for products. Consequently, computational methods commonly use different coordinate systems in different parts of configuration space. Boundary conditions are expressed in terms of Jacobi coordinates (sometimes referred to as "cartesian coordinates"), where in the A+BC arrangement r_{BC} is the internuclear separation of the BC molecule,

$$r_{BC} = |\vec{r}_{BC}| = |\vec{r}_B - \vec{r}_C|, \tag{2}$$

and $R_{A,BC}$ is the distance between the atom and the center of mass of the BC molecule,

$$R_{A,BC} = |\vec{R}_{A,BC}| = |\vec{r}_C - \frac{m_A\vec{r}_A + m_B\vec{r}_B}{m_A + m_B}|. \tag{3}$$

In Eqs. (2) and (3), the vectors $\vec{r}_A$, $\vec{r}_B$, and $\vec{r}_C$ locate the atoms A, B, and C, respectively, relative to an origin of a space-fixed Cartesian reference frame, and $|\vec{r}|$ denotes the length of the vector $\vec{r}$. Equations analogous to (2) and (3) are obtained for the B + AC and C + AB arrangements by cyclically permuting the A, B, and C labels, and define the appropriate Jacobi coordinates for other asymptotic configurations. Because the RLM and BCRLM consider only collinear or near-collinear reaction intermediates, only a single arrangement of product species is possible (as in Eq. (1)), and so we need to consider only the Jacobi coordinates for A+BC geometries (the α arrangement) and AB+C geometries (the γ arrangement). We then define mass-scaled Jacobi coordinates so that motion in both r and R occurs with the same effective reduced mass. These coordinates are

$$
\begin{aligned}
R_\alpha &= C_\alpha^{-1} R_{A,BC}, \qquad & R_\gamma &= C_\gamma^{-1} R_{C,AB}, \\
r_\alpha &= C_\alpha\, r_{BC}, & r_\gamma &= C_\gamma\, r_{AB}, \\
C_\alpha^{4} &= \left[\frac{(m_A + m_B + m_C) m_B m_C}{m_A (m_B + m_C)^2}\right] .
\end{aligned}
\tag{4}
$$

Early treatments of collinear reaction dynamics addressed the coordinate problems associated with different asymptotic arrangement channels by using natural collision coordinates.[34-38] The generic

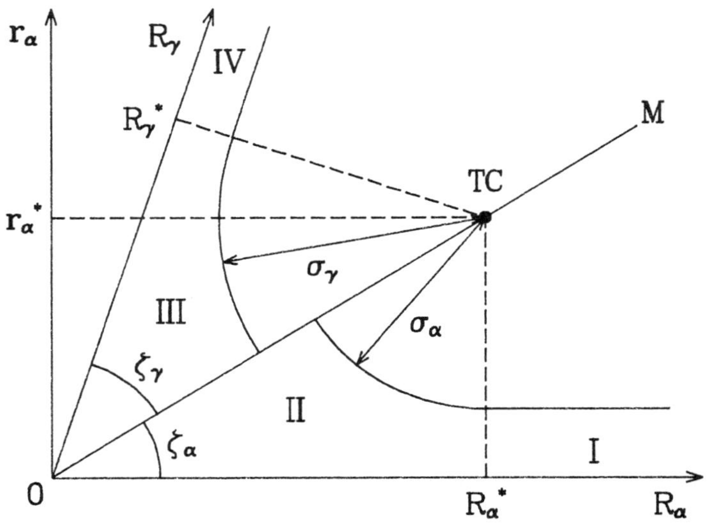

Figure 1. Collinear configuration space, subdivided into regions (I-IV) in which different coordinate systems are used. Regions I and II are for reactants, and III and IV are for products. M is a matching line between reactants and products, and TC is the origin of the polar natural collision coordinates used in Regions II and III.

feature these curvilinear coordinate systems share is that they deform smoothly from the Jacobi coordinates of reactants to the Jacobi coordinates of products. In practice, BCRLM calculations[15] have used natural collision coordinates[49] which can be visualized with the aid of Fig. 1. The NCC are actually plane polar coordinates with an origin located at a turning center labelled TC in Figure 1. The TC has projections R^*_α and r^*_α on the α axes and R^*_γ and r^*_γ on the γ axes. For computational purposes, the collinear configuration space (between the R_α and R_γ axes in Fig. 1) is divided into four regions. Regions I and II (reactants) are separated by a matching surface M from regions III and IV (products). In region I, containing geometries in which $R_\alpha > R^*_\alpha$, and $\tan^{-1}(r_\alpha/R_\alpha) < \varphi_\alpha$, Jacobi coordinates R_α and r_α are used. Similarly, in Region IV, Jacobi coordinates R_γ and r_γ are used. Natural collision coordinates u_α and v_α are used for configurations in Region II, within the triangle defined by (O, TC, R^*_α); coordinates u_γ and v_γ are used in Region III, within the triangle defined by (O, TC, R^*_γ). In terms of R_α and r_α, u_α and v_α are

$$
\begin{aligned}
R_\alpha &= R^*_\alpha - \eta_\alpha \sigma_\alpha \sin\tau_\alpha, \\
r_\alpha &= r^*_\alpha - \eta_\alpha \sigma_\alpha \cos\tau_\alpha, \\
\eta_\alpha &= 1 + v_\alpha/\sigma_\alpha, \\
\tau_\alpha &= \pi/2 - \varsigma_\alpha - u_\alpha/\sigma_\alpha.
\end{aligned}
\tag{5}
$$

Equations analogous to Eq. (5) define R_γ and r_γ in terms of u_γ and v_γ.

Natural collision coordinates defined in this way are convenient for many reactive systems but have the drawback that one must decide where to locate the turning center TC. Physical considerations require that it be placed far away from the origin, in a region of sufficiently high potential energy that the scattering wavefunction, once determined, will be negligibly small there. This requirement immediately implies that these coordinates are unsuitable at scattering energies above the threshold for collision induced dissociation.

A second problem is encountered for "heavy-light-heavy" (HLH) systems in which the mass of the transferred atom B is small in comparison to the masses of A and C. In such cases, the skew angle ($\varsigma_{\alpha\gamma}=\varsigma_\alpha+\varsigma_\gamma$, see Fig. 1) becomes very small, and tunneling between the reactant and product valleys may occur at large distances, requiring that TC be located far from the origin. When this is done, the vibrational motion of the system is poorly represented by the v coordinate. Consequently, slices of the potential at fixed values of u generate broad effective vibrational wells whose shape changes rapidly with u. As a result, a large basis of target functions in the v coordinate is required in the coupled-channel equations at each of a large number of integration steps in the coordinate u.

It is more economical to use hyperspherical coordinate systems[39-47] for HLH systems. For collinear configurations, these coordinates are also plane polar coordinates, but the turning center is located at the origin. These coordinates have had a wide application to collinear reactions,[50-61] especially those of the HLH variety. The hyperspherical radius ρ is independent of the arrangement channel index

$$\rho^2 = R_\alpha^{\ 2} + r_\alpha^{\ 2} = R_\gamma^{\ 2} + r_\gamma^{\ 2}, \qquad 0 \le \rho < \infty, \tag{6}$$

and the hyperspherical angle depends in a simple way on α or γ

$$\begin{aligned} \tan\varphi &= r_\alpha/R_\alpha, \\ \tan(\varsigma_{\alpha\gamma}-\varphi) &= r_\gamma/R_\gamma, \qquad 0 \le \varphi \le \varsigma_{\alpha\gamma}. \end{aligned} \tag{7}$$

Whether the numerical problem is solved in natural collision coordinates or in hyperspherical coordinates, we still must express boundary conditions in the appropriate asymptotic Jacobi coordinates. In the natural collision coordinates of Fig. 1, there is a common boundary between Regions I and II in the α coordinates, between Regions III and IV in γ coordinates, and between Regions II and III separating arrangement channels. In a hyperspherical approach, however, the boundaries between regions which employ different coordinates do not match, as in Fig. 2. Consequently, we must numerically project the solutions of Schrodinger's equation inside the hyperspherical region onto constant R_α and R_γ surfaces. This projection is more complicated in comparison to the analagous but analytic projection procedures[49] required in the NCC approach. This asymptotic matching requirement may be regarded as a minor disadvantage of hyperspherical coordinates.

2.2. The Classical Kinetic Energy

The classical kinetic energy of an A+BC system in a 3D center-of-mass frame may be written in mass-scaled α Jacobi coordinates as

$$T_\alpha = \mu/2\left[\dot{R}_\alpha^2 + \dot{r}_\alpha^2 + R_\alpha^2(\dot{\theta}_\alpha^{\ 2} + \sin^2\theta_\alpha\dot{\phi}_\alpha^2) + r_\alpha^2(\dot{\xi}_\alpha^2 + \sin^2\xi_\alpha\dot{\nu}_\alpha^2)\right], \tag{8}$$

where μ is a reduced mass common to all arrangements (because of the mass scaling of Eq. (4)),

$$\mu = \left[\frac{m_A m_B m_C}{(m_A+m_B+m_C)^2}\right]^{1/2} . \tag{9}$$

In Eq. (8), Θ_α and Φ_α are the spherical polar angles of the $\vec{R}_\alpha$ vector in a space-fixed coordinate frame (see Eqs. (3) and (4)), and ξ_α and ν_α are the corresponding spherical polar angles of the $\vec{r}_\alpha$ vector.

The essence of the approximation in the RLM is to require that both $\vec{r}_\alpha$ and $\vec{R}_\alpha$ be parallel, and hence their spherical polar angles are equal. The atoms A, B, and C now lie on a line in 3D, whose spherical polar angles are defined as Θ and Φ, so that

$$\begin{aligned} \Theta &= \Theta_\alpha = \Theta_\gamma = \xi_\alpha = \xi_\gamma, \\ \Phi &= \Phi_\alpha = \Phi_\gamma = \nu_\alpha = \nu_\gamma. \end{aligned} \tag{10}$$

Consequently, the RLM kinetic energy is simpler than Eq. (8), namely,

$$\begin{aligned} T &= \mu/2\left[\dot{R}_\alpha^2 + \dot{r}_\alpha^2 + (R_\alpha^2 + r_\alpha^2)(\dot{\Theta}^2 + \sin^2\Theta\,\dot{\Phi}^2)\right], \\ &= \mu/2\left[\dot{R}_\gamma^2 + \dot{r}_\gamma^2 + (R_\gamma^2 + r_\gamma^2)(\dot{\Theta}^2 + \sin^2\Theta\,\dot{\Phi}^2)\right]. \end{aligned} \tag{11}$$

In natural collision coordinates, the RLM kinetic energy becomes

$$T = \mu/2\left[\eta^2\dot{u}^2 + \dot{v}^2 + \rho^2(\dot{\Theta}^2 + \sin^2\Theta\,\dot{\Phi}^2)\right], \tag{12}$$

and in hyperspherical coordinates we obtain

$$T = \mu/2\left[\dot{\rho}^2 + \rho^2\dot{\varphi}^2 + \rho^2(\dot{\Theta}^2 + \sin^2\Theta\,\dot{\Phi}^2)\right]. \tag{13}$$

2.3. The Quantum Mechanical Kinetic Energy Operator

To obtain the quantum kinetic energy operator, we first rewrite the classical expression in terms of momenta conjugate to the coordinates, and then follow the prescription described by Podolsky[62] or Margenau and Murphy.[63] In α-channel Jacobi coordinates, we obtain

$$\hat{T}_{RLM} = -\frac{\hbar^2}{2\mu}\left[\frac{1}{\rho^2}\frac{\partial}{\partial R_\alpha}\rho^2\frac{\partial}{\partial R_\alpha} + \frac{1}{\rho^2}\frac{\partial}{\partial r_\alpha}\rho^2\frac{\partial}{\partial r_\alpha}\right] + \frac{\hat{J}^2}{2\mu\rho^2}, \tag{14}$$

where $\hat{J}$ is the total angular momentum operator for the system

$$\hat{J}^2 = -\hbar^2\left[\frac{1}{\sin\Theta}\frac{\partial}{\partial\Theta}\sin\Theta\frac{\partial}{\partial\Theta} + \frac{1}{\sin^2\Theta}\frac{\partial^2}{\partial\Phi^2}\right]. \tag{15}$$

In natural collision coordinates, $\hat{T}$ becomes

$$\hat{T}_{RLM} = -\frac{\hbar^2}{2\mu}\left[\frac{1}{\eta^2\rho^2}\frac{\partial}{\partial u}\rho^2\frac{\partial}{\partial u} + \frac{1}{\eta\rho^2}\frac{\partial}{\partial v}\eta\rho^2\frac{\partial}{\partial v}\right] + \frac{\hat{J}^2}{2\mu\rho^2}, \tag{16}$$

and in hyperspherical coordinates we obtain

$$\hat{T}_{RLM} = -\frac{\hbar^2}{2\mu}\left[\frac{1}{\rho^3}\frac{\partial}{\partial\rho}\rho^3\frac{\partial}{\partial\rho} + \frac{1}{\rho^2}\frac{\partial^2}{\partial\varphi^2})\right] + \frac{\hat{J}^2}{2\mu\rho^2}. \tag{17}$$

In the RLM, the Hamiltonian operator is simply

$$\hat{H}_{RLM} = \hat{T}_{RLM} + V_{1D}(R,r), \tag{18}$$

where $V_{1D}(R,r)$ is the electronic potential energy hypersurface, for collinear geometries. Of course, we assume V_{1D} may be evaluated as needed in any of the required coordinate systems.

2.4. The Bending Hamiltonian

We next elaborate upon the RLM to account approximately for the neglected bending degrees of freedom. Bending is treated as if it is adiabatically separable from motion in the R and r coordinates, as if bending time scales were faster than time scales for translational and vibrational motion. The true time scales associated with these motions almost never satisfy these conditions (especially asymptotically), except for some reactions with highly constrained linear intermediates and at collision energies near the reaction threshold. Nevertheless, we include the bending approximation to improve the threshold behavior of reactions, hoping that in some average sense, it may recover some of the features expected from the internal rotational degrees of freedom in a more accurate 3D theory. However, the two degenerate bending modes correlate to zero-frequency modes asymptotically, and not to the proper diatomic rotational levels. Consequently, we cannot identify the results of a BCRLM calculation for a specific set of bending states with those of a 3D theory for specific rotational transitions. However, as we describe later, we may identify bending averaged[64] BCRLM results with rotationally averaged 3D results.

Following Garrett and Truhlar,[25] we define the angle γ (not to be confused with the arrangement channel index) as the bond angle between the $-\vec{r}_{BC}$ and $\vec{r}_{AB}$ vectors defined by Eq. (2). For small displacements in the γ angle, we may define a bending Hamiltonian for each (R,r) or (ρ,φ)

$$\hat{H}_{bend} = \frac{-\hbar^2}{2I_b}\frac{\partial^2}{\partial\gamma^2} + V_{bend}(\gamma;R,r), \tag{19a}$$

where I_b is a moment of inertia,

$$I_b^{-1} = [m_A R_{AB}^2]^{-1} + [m_C R_{BC}^2]^{-1} + m_B^{-1}[R_{AB}^{-1} + R_{BC}^{-1}]^2. \quad (19b)$$

The eigenvalues of this Hamiltonian are $\epsilon_\lambda(R,r)$,

$$\hat{H}_{bend}\Psi_\lambda^{bend} = \epsilon_\lambda(R,r)\Psi_\lambda^{bend} \quad (20)$$

and form an effective potential which, when added to the collinear potential surface, forms the BCRLM potential. We have therefore

$$V_{\lambda_1\lambda_2}(R,r) = V_{1D}(R,r) + \epsilon_{\lambda_1}(R,r) + \epsilon_{\lambda_2}(R,r), \quad (21)$$

$$\hat{H}_{BCRLM}^{\lambda_1\lambda_2} = \hat{T}_{RLM} + V_{\lambda_1\lambda_2}(R,r). \quad (22)$$

The bending eigenvalue functions appear twice in Eq. (21) because of the degeneracy of the two bending modes of a linear triatomic molecule. In practice, BCRLM calculations have been reported[15,64-71] only for an approximate form of the bending eigenvalue function, and for $\lambda_1=\lambda_2=0$. The approximation used[15] is expressed in natural collision coodinates,

$$\tilde{\epsilon}_\lambda(u,v) = \epsilon_\lambda(u,v_0), \quad (23)$$

where v_0 is the value of v where the potential $V_{1D}(u,v)$ has a minimum at fixed u. This approximation has been used for computational convenience but may have several disadvantages. The first problem[15] arises because the approximation is tied to the definition of the natural collision coordinates. This dependency arises because the position of the vibrational minimum v_0 depends slightly on the location of TC, and lines of constant u are not perpendicular to the minimum energy path from the saddle point toward reactants (or products). A second problem arises in hyperspherical coordinates, because Eq. (23) becomes quite cumbersome to implement, and indeed, the effective potential becomes multivalued at TC. A third problem arises at subthreshold collision energies, where collinear calculations show that significant corner-cutting of reactive flux occurs to the concave side of the minimum energy path. It has been pointed out[72] that in this region, the approximate potential is likely to be larger than ϵ_λ; consequently, the barrier to tunnelling may be overestimated. The simple solution to each of these problems is to avoid the approximation Eq. (23) altogether.

2.5. The Coupled-Channel Equations

The angular momentum operator in Eq. (15) suggests that the overall rotational degrees of freedom can be expanded in partial waves using spherical harmonics $Y_J^0(\Theta,\Phi)$, so that

$$\Psi_m^{\lambda_1\lambda_2}(R,r,\theta,\Phi) = \sum_{J=0}^{\infty} A_m^J \Psi_m^{J\lambda_1\lambda_2}(R,r)\ Y_J^0(\theta,\Phi), \tag{24}$$

where λ_1 and λ_2 label adiabatic bending states, J is the total angular momentum quantum number, and m labels the initial vibrational state. The coefficient A_m^J is chosen to satisfy asymptotic boundary conditions in Sec. 2.8. We next expand the coefficient functions $\Psi_m^{J\lambda_1\lambda_2}(R,r)$ as appropriate for each coordinate system. In Jacobi coordinates, we have

$$\Psi_m^{J\lambda_1\lambda_2}(R,r) = \rho^{-1} \sum_{i=1}^{M} \sum_{n=1}^{N} f_{nm}(R;iJ\lambda_1\lambda_2)\ F_n(r;iJ\lambda_1\lambda_2), \tag{25}$$

and in natural collision coordinates, we have

$$\Psi_m^{J\lambda_1\lambda_2}(R,r) = \rho^{-1}\eta^{1/2} \sum_{i=1}^{M} \sum_{n=1}^{N} g_{nm}(u;iJ\lambda_1\lambda_2)\ G_n(v;iJ\lambda_1\lambda_2), \tag{26}$$

and in hyperspherical coordinates, we have

$$\Psi_m^{J\lambda_1\lambda_2}(R,r) = \rho^{-3/2} \sum_{i=1}^{M} \sum_{n=1}^{N} h_{nm}(\rho;iJ\lambda_1\lambda_2)\ H_n(\varphi;iJ\lambda_1\lambda_2). \tag{27}$$

In Eqs. (25)-(27), we subdivided configuration space into sectors, each labeled by the index i; the boundary between sectors in each coordinate system is formed by curves on which the propagation variables (R, u, and ρ, respectively) are constant. Since the wavefunction expansion may change from sector to sector, the functions f, F, g, G, h, and H depend parametrically on the i index, as well as the total angular momentum index J and the adiabatic bend quantum numbers λ_1 and λ_2.

The functions F, G, and H are determined by solving a reference vibrational Hamiltonian defined at the center of each sector,

$$\left[-\frac{\hbar^2}{2\mu}\frac{d^2}{dr^2} + \tilde{V}_F(r;iJ\lambda_1\lambda_2) - \epsilon_n^F(iJ\lambda_1\lambda_2)\right]F(r;iJ\lambda_1\lambda_2) = 0, \tag{28}$$

$$\left[-\frac{\hbar^2}{2\mu}\frac{d^2}{dv^2} + \tilde{V}_G(v;iJ\lambda_1\lambda_2) - \epsilon_n^G(iJ\lambda_1\lambda_2)\right]G(v;iJ\lambda_1\lambda_2) = 0, \tag{29}$$

$$\left[-\frac{\hbar^2}{2\mu\rho^2}\frac{d^2}{d\varphi^2} + \tilde{V}_H(\varphi;iJ\lambda_1\lambda_2) - \epsilon_n^H(iJ\lambda_1\lambda_2)\right]H(\varphi;iJ\lambda_1\lambda_2) = 0. \tag{30}$$

The actual choice of the reference vibrational potential depends on the particular application. In the RXN1D program,[73] a quadratic reference potential is chosen[49] in the NCC and Jacobi coordinate systems, and the functions F and G form a harmonic oscillator basis. In hyperspherical coordinates, we use the entire potential and determine the basis H by a finite difference approach.

Combining Eqs. (24)-(30) with Eqs. (14)-(18), we obtain the coupled-channel equations for the propagation functions $f_{nm}(R)$, $g_{nm}(u)$, and $h_{nm}(\rho)$, which after suppressing the parametric labels $(iJ\lambda_1\lambda_2)$ are,

$$\frac{d^2}{dR^2} f_{nm}(R) = \sum_{n'=0}^{N} (\underline{\underline{D}}_F)_{nn'} f_{n'm}(R), \tag{31}$$

$$\frac{d^2}{du^2} g_{nm}(u) = \sum_{n'=0}^{N} (\underline{\underline{D}}_G)_{nn'} g_{n'm}(u), \tag{32}$$

$$\frac{d^2}{d\rho^2} h_{nm}(\rho) = \sum_{n'=0}^{N} (\underline{\underline{D}}_H)_{nn'} h_{n'm}(\rho), \tag{33}$$

where in Eqs. (31)-(33) the coupling matrices (we denote matrices by a double underline) are

$$\frac{\hbar^2}{2\mu}(\underline{\underline{D}}_F)_{nn'} = (\epsilon_n - E)\delta_{nn'} + + \langle F_n|V_{\lambda_1\lambda_2} - \tilde{V}_F + \frac{\hbar^2}{2\mu\rho^2}[J(J+1)+1]|F_{n'}\rangle, \tag{34}$$

$$\frac{\hbar^2}{2\mu}(\underline{\underline{D}}_G)_{nn'} = \frac{3}{4\sigma^2}\delta_{nn'} + + \langle G_n|\eta^2\left\{V_{\lambda_1\lambda_2} - \tilde{V}_G + \frac{(\epsilon_n+\epsilon_{n'})}{2} - E + \frac{\hbar^2}{2\mu\rho^2}[J(J+1)+1]\right\}|G_{n'}\rangle, \tag{35}$$

$$\frac{\hbar^2}{2\mu}(\underline{\underline{D}}_H)_{nn'} = \left[\epsilon_n - E + \frac{\hbar^2}{2\mu\rho^2}\left[J(J+1) + \frac{3}{4}\right]\right]\delta_{nn'} + + \langle H_n|V - \tilde{V}_H|H_{n'}\rangle. \tag{36}$$

When we change the target basis (Eqs. (28)-(30)) between two adjacent sectors, we must ensure that the wavefunction and its derivative are continuous across the sector boundary. Enforcing this requirement defines overlap matrices $\underline{\underline{\sigma}}$ in each coordinate system, and for Jacobi coordinates we obtain

$$f_{nm}(R_i^+;i) = \sum_{n'} [\underline{\underline{\sigma}}_F(i,i+1)]_{nn'} \, f_{n'm}(R_{i+1}^-;i+1), \tag{37}$$

$$[\underline{\underline{\sigma}}_F(i,i+1)]_{nn'} = \langle F_n(r;i)|F_{n'}(r;i+1)\rangle, \tag{38}$$

where we have suppressed the labels $(J\lambda_1\lambda_2)$ on the f's, F's, and T's. In Eq. (37), R_i^- and R_i^+ are the values of the propagation coordinate at the inner and outer boundaries of sector i. Equations analagous to Eqs. (37)-(38) also hold in the NCC and hyperspherical coordinate systems.

2.6. Solving the Coupled-Channel Equations

The coupled-channel equations (Eq. (31), (32), or (33)), may be solved in a variety of ways, but we use the R-matrix propagation method of Light and Walker.[49,74-76] We will review this method briefly in this section, as applied to the coupled-channel equations in Jacobi coordinates. The approach is essentially the same in other coordinates. The coupling matrices $\underline{\underline{D}}$ (Eqs. (34)-(36)) are evaluated at the center of each sector, and are assumed to be constant across the sector. The real symmetric $\underline{\underline{D}}$ matrices are diagonalized by a real orthogonal matrix $\underline{\underline{U}}$,

$$\underline{\underline{U}}^T(i)\cdot\underline{\underline{D}}(i)\cdot\underline{\underline{U}}(i) = \underline{\underline{\lambda}}^2(i), \tag{39}$$

where $\underline{\underline{U}}^T$ is the transpose of $\underline{\underline{U}}$. The matrix $\underline{\underline{U}}$ transforms to a locally uncoupled representation, and defines new propagation functions $\tilde{f}_{nm}(R;i)$ in each sector,

$$\underline{\underline{\tilde{f}}}(R;i) = \underline{\underline{U}}(i)\cdot\underline{\underline{f}}(R;i). \tag{40}$$

The global R matrix, between the initial sector and sector i, is

$$\begin{bmatrix} \underline{\underline{\tilde{f}}}(R_0^-;0) \\ \underline{\underline{\tilde{f}}}(R_i^+;i) \end{bmatrix} = \begin{bmatrix} \underline{\underline{R}}_1(i) & \underline{\underline{R}}_2(i) \\ \underline{\underline{R}}_3(i) & \underline{\underline{R}}_4(i) \end{bmatrix} \begin{bmatrix} -\underline{\underline{\tilde{f}}}'(R_0^-;0) \\ \underline{\underline{\tilde{f}}}'(R_i^+;i) \end{bmatrix}. \tag{41}$$

The sector R matrix relating the values of the locally uncoupled functions to derivatives within sector (i+1) is

$$\begin{bmatrix} \underline{\underline{\tilde{f}}}(R_{i+1}^-;i+1) \\ \underline{\underline{\tilde{f}}}(R_{i+1}^+;i+1) \end{bmatrix} = \begin{bmatrix} \underline{\underline{r}}_1(i+1) & \underline{\underline{r}}_2(i+1) \\ \underline{\underline{r}}_3(i+1) & \underline{\underline{r}}_4(i+1) \end{bmatrix} \begin{bmatrix} -\underline{\underline{\tilde{f}}}'(R_{i+1}^-;i+1) \\ \underline{\underline{\tilde{f}}}'(R_{i+1}^+;i+1) \end{bmatrix}, \tag{42}$$

where for open channels ($\lambda^2 \leq 0$) we have

$$[\underline{\underline{r}}_1(i)]_{nn'}=[\underline{\underline{r}}_4(i)]_{nn'} = \delta_{nn'}\left[-|\lambda_n(i)|^{-1}\cot\{\Delta R_i|\lambda_n(i)|\}\right],$$
$$[\underline{\underline{r}}_2(i)]_{nn'}=[\underline{\underline{r}}_3(i)]_{nn'} = \delta_{nn'}\left[-|\lambda_n(i)|^{-1}\csc\{\Delta R_i|\lambda_n(i)|\}\right], \quad (43)$$

and for closed channels ($\lambda^2 \geq 0$) the sector R matrix is

$$[\underline{\underline{r}}_1(i)]_{nn'}=[\underline{\underline{r}}_4(i)]_{nn'} = \delta_{nn'}\left[|\lambda_n(i)|^{-1}\coth\{\Delta R_i|\lambda_n(i)|\}\right],$$
$$[\underline{\underline{r}}_2(i)]_{nn'}=[\underline{\underline{r}}_3(i)]_{nn'} = \delta_{nn'}\left[|\lambda_n(i)|^{-1}\mathrm{csch}\{\Delta R_i|\lambda_n(i)|\}\right]. \quad (44)$$

In Eqs. (43)-(44), ΔR_i is the width of sector i. The transformation matrix from the locally uncoupled representation of sector i to the locally uncoupled representation of sector i+1 is

$$\underline{\underline{T}}(i,i+1) = \underline{\underline{U}}^T(i)\cdot\underline{\underline{\sigma}}(i,i+1)\cdot\underline{\underline{U}}(i+1). \quad (45)$$

Assuming we know the global R matrix of Eq. (41), we can now compute the global R matrix for sector i+1 using the sector R matrix of Eq. (42) and the overlap matrix of Eq. (45). The R-matrix recursion relations are[77]

$$\underline{\underline{R}}_1(i+1) = \underline{\underline{R}}_1(i) - \underline{\underline{R}}_2(i)\cdot\underline{\underline{T}}(i,i+1)\cdot\underline{\underline{Z}}(i+1)\cdot\underline{\underline{T}}^T(i,i+1)\cdot\underline{\underline{R}}_3(i), \quad (46)$$

$$\underline{\underline{R}}_2(i+1) = \underline{\underline{R}}_3^T(i+1) = \underline{\underline{R}}_2(i)\cdot\underline{\underline{T}}(i,i+1)\cdot\underline{\underline{Z}}(i+1)\cdot\underline{\underline{r}}_2(i+1), \quad (47)$$

$$\underline{\underline{R}}_4(i+1) = \underline{\underline{r}}_4(i+1) - \underline{\underline{r}}_3(i+1)\cdot\underline{\underline{Z}}(i+1)\cdot\underline{\underline{r}}_2(i+1), \quad (48)$$

$$\underline{\underline{Z}}(i+1) = [\underline{\underline{r}}_1(i+1) - \underline{\underline{T}}^T(i,i+1)\cdot\underline{\underline{R}}_4(i)\cdot\underline{\underline{T}}(i,i+1)]^{-1}. \quad (49)$$

By repeatedly applying Eqs. (46)-(49), the coupled-channel equations are solved by propagating towards asymptotic regions of configuration space. We also note[76] that, if desired, we may propagate the R-matrix inverse (the log-derivative or L matrix) with equations essentially the same as Eqs. (46)-(49), where only the definition of the sector L matrix is changed. At the conclusion of the propagation, we compute the scattering matrix $\underline{\underline{S}}$ by enforcing boundary conditions.

In the RXN1D program,[73] both the NCC and Jacobi coordinate systems are used. We begin at the collinear matching surface (M in Fig. 1) with a sector R matrix as the first "global" R matrix, and propagate all four blocks of the R matrix outwards toward the α-channel asymptotic region, and then switch to α-Jacobi coordinates when $R_\alpha = R^*_\alpha$. For asymmetric systems ($m_A \neq m_C$), propagation resumes at the matching surface, and proceeds toward the γ-channel asymptotic region, switching to γ-Jacobi coordinates when $R_\gamma = R^*_\gamma$. The α- and γ-channel R matrices are then combined and boundary conditions enforced.

In hyperspherical coordinates, propagation begins at a small hyperspherical radius, and continues to larger hyperspherical radii. Because the potential is repulsive at small radii, only regular functions at the origin are physically allowed, and in this case it is necessary to propagate only the $\underline{R}_4$ block of the R matrix. As ρ increases, the angular potential evolves from a single well to a double well, one each for the reactant and product molecules. At energies below dissociation, the barrier between the two wells becomes large and broad enough that the eigenstates of the angular potential are completely localized within each well. For symmetric systems ($m_A = m_C$), we may obtain degenerate pairs of delocalized functions, but these are easily localized (i.e., $\Psi(\text{local}) = 2^{-1/2}[\Psi_1 \pm \Psi_2]$). Once the angular eigenstates are localized, we may continue propagating in Jacobi coordinates, or if appropriate, we may enforce boundary conditions. However, in either case we first project the hyperspherical solutions onto constant R_α (and R_γ) surfaces.

2.7. Hyperspherical Projection

Asymptotic boundary conditions are most conveniently expressed in Jacobi coordinates, and so if we solve the coupled-channel equations in hyperspherical coordinates, we first express our solutions, defined on a hyperspherical radius, on the appropriate Jacobi surfaces (see Fig. 2).

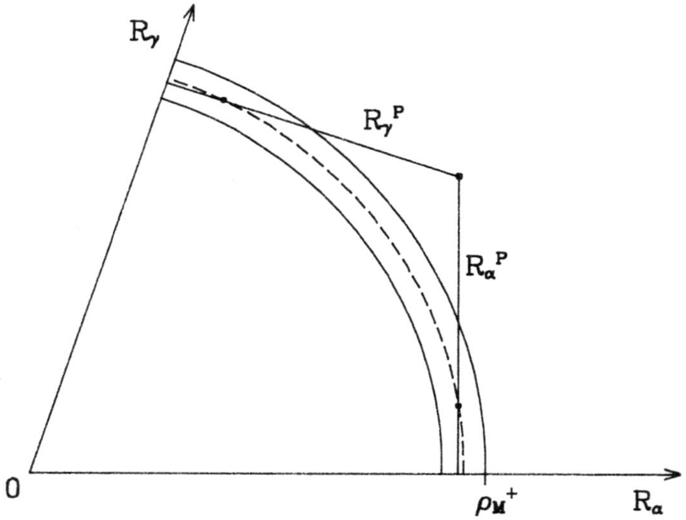

Figure 2. Collinear configuration space, showing the projection of hyperspherical solutions onto Jacobi surfaces. The solid arcs are the inner and outer boundaries of the last hyperspherical sector, and the dashed arc is the center of the sector. The Jacobi surfaces R_α^P and R_γ^P intersect the dashed arc at the vibrational minima.

We describe in this section a procedure which determines a two-surface R matrix (four blocks) from the single surface hyperspherical R matrix. The procedure we describe is essentially that of Bondi and Connor,[45,48] except for minor differences in strategy (they evaluate asymptotic boundary conditions directly on the final hyperspherical radius). We begin by recalling the definition of the final hyperspherical R matrix,

$$\underline{\underline{h}}(\rho_M^+;M) = \underline{\underline{R}}^H \cdot \underline{\underline{h}}'(\rho_M^+;M), \tag{50}$$

where here M labels the final hyperspherical sector (see Eq. (27)), and ρ_M^+ is the value of ρ at the outer boundary of this sector. Within the last sector, the propagation functions (and their derivatives) may be expanded in sine- and cosine-like solutions, so that

$$\underline{\underline{h}}(\rho;M) = \underline{\underline{s}}(\rho;M) \cdot \underline{\underline{A}}_M + \underline{\underline{c}}(\rho;M) \cdot \underline{\underline{B}}_M, \qquad \rho_M^- \le \rho \le \rho_M^+, \tag{51}$$

$$\underline{\underline{h}}'(\rho;M) = \underline{\underline{s}}'(\rho;M) \cdot \underline{\underline{A}}_M + \underline{\underline{c}}'(\rho;M) \cdot \underline{\underline{B}}_M, \qquad \rho_M^- \le \rho \le \rho_M^+, \tag{52}$$

where $\underline{\underline{A}}_M$ and $\underline{\underline{B}}_M$ are undetermined coefficient matrices, constant within the final sector, which depend on asymptotic boundary conditions. The diagonal matrices $\underline{\underline{s}}(\rho;M)$ and $\underline{\underline{c}}(\rho;M)$ are

$$\begin{aligned} s_n(\rho) &= \sin[k_n^M(\rho-\rho_M)], && \text{channel n open,} \\ &= \sinh[k_n^M(\rho-\rho_M)], && \text{channel n closed,} \end{aligned} \tag{53}$$

$$\begin{aligned} c_n(\rho) &= \cos[k_n^M(\rho-\rho_M)], && \text{channel n open,} \\ &= \cosh[k_n^M(\rho-\rho_M)], && \text{channel n closed,} \end{aligned} \tag{54}$$

where ρ_M is the value of ρ at the center of the final sector, and

$$\left[k_n^M\right]^2 = |(\underline{\underline{D}}_H)_{nn}|. \tag{55}$$

By substituting Eqs. (51) and (52) into Eq. (50), we can relate the coefficient matrices $\underline{\underline{A}}$ and $\underline{\underline{B}}$,

$$\begin{aligned} \underline{\underline{A}}_M &= \underline{\underline{X}}_M \cdot \underline{\underline{B}}_M, \\ \underline{\underline{X}}_M &= [\underline{\underline{s}}(\rho_M^+) - \underline{\underline{R}}^H \cdot \underline{\underline{s}}'(\rho_M^+)]^{-1} \cdot [\underline{\underline{c}}(\rho_M^+) - \underline{\underline{R}}^H \cdot \underline{\underline{c}}'(\rho_M^+)]. \end{aligned} \tag{56}$$

We now require the right hand sides of Eqs. (25) and (27) to agree on the projection surface $R_\alpha = R_\alpha^P$ (see Fig. 2),

$$\rho^{-1/2} \sum_{n'} h_{n'm}(\rho;M)\, H_{n'}(\varphi;M) = \sum_n f_{nm}(R_\alpha^P)\, F_n(r_\alpha), \tag{57}$$

and a similar equation must also hold for the derivatives

$$\frac{\partial}{\partial R_\alpha}\left[\rho^{-1/2} \sum_{n'} h_{n'm}(\rho;M)\, H_{n'}(\varphi;M)\right] = \sum_n f'_{nm}(R_\alpha^P)\, F_n(r_\alpha). \tag{58}$$

Two additional equations, similar to Eqs. (57) and (58), also hold on the product surface R_γ^P. We next multiply by $F_n^*(r_\alpha)$ and integrate over r_α (and over r_γ on the product surface), and use Eqs. (51) and (52) to obtain equations for the propagation functions in Jacobi coordinates,

$$\begin{bmatrix} \underline{\underline{f}}(R_\alpha^P) \\ \underline{\underline{f}}(R_\gamma^P) \end{bmatrix} = \begin{bmatrix} \underline{\underline{I}}_\alpha^{(1)} & \underline{\underline{0}} \\ \underline{\underline{0}} & \underline{\underline{I}}_\gamma^{(1)} \end{bmatrix} \cdot \underline{\underline{A}}_M + \begin{bmatrix} \underline{\underline{I}}_\alpha^{(2)} & \underline{\underline{0}} \\ \underline{\underline{0}} & \underline{\underline{I}}_\gamma^{(2)} \end{bmatrix} \cdot \underline{\underline{B}}_M, \tag{59}$$

$$\begin{bmatrix} \underline{\underline{f}}'(R_\alpha^P) \\ \underline{\underline{f}}'(R_\gamma^P) \end{bmatrix} = \begin{bmatrix} \underline{\underline{I}}_\alpha^{(3)} & \underline{\underline{0}} \\ \underline{\underline{0}} & \underline{\underline{I}}_\gamma^{(3)} \end{bmatrix} \cdot \underline{\underline{A}}_M + \begin{bmatrix} \underline{\underline{I}}_\alpha^{(4)} & \underline{\underline{0}} \\ \underline{\underline{0}} & \underline{\underline{I}}_\gamma^{(4)} \end{bmatrix} \cdot \underline{\underline{B}}_M, \tag{60}$$

where the α-channel matching matrices are defined (suppressing the M label on the s, c, and H functions)

$$\left[\underline{\underline{I}}_\alpha^{(1)}\right]_{nn'} = \int_0^\infty F_n(r_\alpha)\, \rho^{-1/2}\, s_{n'}(\rho)\, H_{n'}(\varphi)\, dr_\alpha, \tag{61}$$

$$\left[\underline{\underline{I}}_\alpha^{(2)}\right]_{nn'} = \int_0^\infty F_n(r_\alpha)\, \rho^{-1/2}\, c_{n'}(\rho)\, H_{n'}(\varphi)\, dr_\alpha, \tag{62}$$

$$\begin{aligned}\left[\underline{\underline{I}}_\alpha^{(3)}\right]_{nn'} = \int_0^\infty F_n(r_\alpha)\, \rho^{-1/2} \{ & s'_{n'}(\rho)\, H_{n'}(\varphi) \cos\varphi - \\ & - (2\rho)^{-1}\, s_{n'}(\rho)\, H_{n'}(\varphi) \cos\varphi \\ & - \rho^{-1}\, s_{n'}(\rho)\, H'_{n'}(\varphi) \sin\varphi \}\, dr_\alpha,\end{aligned} \tag{63}$$

$$\begin{aligned}\left[\underline{\underline{I}}_\alpha^{(4)}\right]_{nn'} = \int_0^\infty F_n(r_\alpha)\, \rho^{-1/2} \{ & c'_{n'}(\rho)\, H_{n'}(\varphi) \cos\varphi - \\ & - (2\rho)^{-1}\, c_{n'}(\rho)\, H_{n'}(\varphi) \cos\varphi \\ & - \rho^{-1}\, c_{n'}(\rho)\, H'_{n'}(\varphi) \sin\varphi \}\, dr_\alpha.\end{aligned} \tag{64}$$

The off-diagonal blocks of the $\underline{\underline{I}}^{(1)}$, $\underline{\underline{I}}^{(2)}$, $\underline{\underline{I}}^{(3)}$ and $\underline{\underline{I}}^{(4)}$ matrices in Eqs. (59)-(60) are zero because we have assumed that the hyperspherical angular eigenfunctions have been localized in the reactant and product potential wells. Referring back to Eq. (41), we can define an R matrix for the Jacobi coordinates

$$\begin{bmatrix} \underline{\underline{f}}(R_\alpha^P) \\ \underline{\underline{f}}(R_\gamma^P) \end{bmatrix} = \begin{bmatrix} \underline{\underline{R}}_1^F & \underline{\underline{R}}_2^F \\ \underline{\underline{R}}_3^F & \underline{\underline{R}}_4^F \end{bmatrix} \begin{bmatrix} \underline{\underline{f}}'(R_\alpha^P) \\ \underline{\underline{f}}'(R_\gamma^P) \end{bmatrix}. \tag{65}$$

Combining Eqs. (59) and (60) with Eqs. (56) and (65), the Jacobi R matrix is determined in terms of the matching matrices as

$$\underline{\underline{R}}^F = \left[\underline{\underline{I}}^{(1)} \cdot \underline{\underline{X}}_M + \underline{\underline{I}}^{(2)} \right] \cdot \left[\underline{\underline{I}}^{(3)} \cdot \underline{\underline{X}}_M + \underline{\underline{I}}^{(4)} \right]^{-1}, \tag{66}$$

where in Eq. (66) we have implicitly arranged the rows and columns of $\underline{\underline{X}}$ to agree with the labeling implied by Eqs. (59)-(60). Having determined the R matrix in Jacobi coordinates, we can now either continue the propagation or apply asymptotic boundary conditions.

2.8. Boundary Conditions

The coupled-channel equations (Eqs. (31) and (34)) decouple at large values of R_α (or R_γ), because in the limit that $\rho \to R$, we obtain

$$\frac{\hbar^2}{2\mu} (\underline{\underline{D}}_F)_{nn'} = \left[\epsilon_n - E + \frac{\hbar^2}{2\mu R^2}[J(J+1)+1]\right]\delta_{nn'} \ . \tag{67}$$

The form of Eq. (67) implies that the functions $f_{nm}(R;J)$ will approach a linear combination of Bessel functions of unusual order, because of the $J(J+1)+1$ term. In our calculations, we have ignored the additional $1/R^2$ centrifugal potential in applying boundary conditions, in order to use the more familiar spherical Bessel functions. Our experience has been, and others have shown,[70] that this approximation has a small effect on the magnitudes and phases of the final S-matrix elements. We therefore require the functions $f_{nm}(R;J)$ to go asymptotically[3] as

$$f_{nm}(R;J) \sim -ik_n R\left[h_J^{(1)}(k_n R)\delta_{nm} + h_J^{(1)}(k_n R)\ (k_m/k_n)^{-1/2} S_{nm}^J\right], \tag{68}$$

where k_n is the channel wavenumber $\hbar^2 k_n^2 = 2\mu(E-\epsilon_n)$, S_{nm}^J is an element of the S matrix, and the functions h_J are spherical Hankel functions of the first and second kind, which themselves have the asymptotic behavior

$$\pm i h_J^{(1),(2)}(z) = z^{-1}\exp[\pm i(z - J\pi/2)]. \tag{69}$$

The wavefunction of Eq. (24) must satisfy the boundary condition[3]

$$\Psi_m^{\lambda_1\lambda_2}(R,r,\theta,\Phi) \sim F_m^\infty(r_\alpha)\exp[ik_m R_\alpha \cos\theta_\alpha] + $$

$$+ R_\alpha^{-1} \sum_n \exp[ik_n R_\alpha]\ F_n^\infty(r_\alpha)\ \mathcal{A}_{nm}(\theta,\Phi), \tag{70}$$

when R_α is large,

$$\sim \quad R_\gamma^{-1} \sum_{n'} \exp[ik_{n'} R_\gamma]\ F_{n'}^\infty(r_\gamma)\ \mathcal{A}_{n'm}(\theta,\Phi), \tag{71}$$

when R_γ is large.

In Eqs. (70)-(71), the functions $F^\infty(r)$ are the eigenfunctions of the asymptotic vibrational Hamiltonian, and $\mathcal{A}(\theta,\Phi)$ is the scattering amplitude. Using the asymptotic form of f_{nm} defined in Eqs. (68) and (69) in the right-hand-side of Eq. (24), we determine the expansion coefficients A_m^J in Eq. (24) and the scattering amplitude $\mathcal{A}_{nm}$ by equating with Eqs. (70) and (71). The coefficients A_m^J are determined by equating the coefficients of the incoming spherical waves, obtaining

$$A_m^J = k_m^{-1}\ i^{J+1}\ [\pi(2J+1)]^{1/2}. \tag{72}$$

The scattering amplitude is similarly determined by equating the coefficients of the outgoing spherical waves, after first expanding $\mathcal{A}_{nm}$ in Legendre polynomials. We obtain[3]

$$\mathcal{A}_{nm}(\theta,\Phi) = i(4k_n k_m)^{-1/2} \sum_{J=0}^{\infty} (2J+1)\ (\delta_{nm} - S_{nm}^J)\ P_J(\cos\theta). \tag{73}$$

The calculation of the S matrix from the final R matrix is accomplished by rewriting Eq. (68) and its derivative in matrix form,

$$\underline{\underline{f}}(R;J) = \underline{\underline{\mathcal{I}}}(J) - \underline{\underline{\mathcal{O}}}(J)\cdot\underline{\underline{k}}^{-1/2}\cdot\underline{\underline{S}}^J\cdot\underline{\underline{k}}^{1/2}, \tag{74}$$

$$\underline{\underline{f}}'(R;J) = \underline{\underline{\mathcal{I}}}'(J) - \underline{\underline{\mathcal{O}}}'(J)\cdot\underline{\underline{k}}^{-1/2}\cdot\underline{\underline{S}}^J\cdot\underline{\underline{k}}^{1/2}, \tag{75}$$

where $\underline{\underline{f}}(R)$ and $\underline{\underline{f}}'(R)$ are matrices of the values of the propagation functions and their derivatives on the final R-matrix boundaries in both the α and γ arrangement channels. Here the diagonal $\underline{\underline{\mathcal{I}}}$ and $\underline{\underline{\mathcal{O}}}$ matrices (and their derivatives) are the spherical Hankel functions of Eq. (68),

$$[\underline{\underline{\mathcal{O}}}(J)]_{nn} = ik_n R\, \mathcal{h}_J^{(1)}(k_n R), \tag{76}$$

$$[\underline{\underline{\mathcal{I}}}(J)]_{nn} = [\underline{\underline{\mathcal{O}}}(J)]^*_{nn}, \tag{77}$$

where * denotes the complex conjugate. The $\underline{\underline{\mathcal{O}}}$ matrix defined here should not be confused with the overlap matrix used in Eqs. (37), (38), and (45). Defining the final R matrix as $\underline{\underline{R}}^{\infty}$, and combining the definition of the R matrix (see Eq. (65)) with Eqs. (74)-(75), the S matrix is

$$\underline{\underline{S}}^J = \underline{\underline{k}}^{1/2}\cdot\left[\underline{\underline{\mathcal{O}}}(J) - \underline{\underline{R}}^{\infty}\cdot\underline{\underline{\mathcal{O}}}'(J)\right]^{-1}\cdot\left[\underline{\underline{\mathcal{O}}}(J) - \underline{\underline{R}}^{\infty}\cdot\underline{\underline{\mathcal{O}}}'(J)\right]^*\cdot\underline{\underline{k}}^{-1/2}. \tag{78}$$

2.9. Differential and Integral Cross Sections, Thermal Rate Constants

The differential scattering cross section in the RLM is defined as usual, the ratio of the spherically scattered flux into final state n originating from an incident plane wave in molecular state m,

$$\frac{d}{d\Omega}\sigma_{nm}(\theta,\Phi;E) = (k_n/k_m)\,|\mathcal{A}_{nm}(\theta,\Phi)|^2, \tag{79}$$

where $\mathcal{A}_{nm}(\theta,\Phi)$ is defined in Eq. (73). The integral cross section is obtained by integrating over the polar angles, giving the familiar form

$$\sigma_{nm}(E) = \pi k_m^{-2}\sum_{J=0}^{\infty}(2J+1)\,|\delta_{nm} - S^J_{nm}|^2. \tag{80}$$

From the integral cross section we can compute a state-to-state thermal rate constant in the standard way,[78]

$$K_{nm}(T) = N\,(2/k_B T)^{3/2}\,(\pi\mu_{A,BC})^{-1/2}\times$$
$$\times\int_0^{\infty} E_t\,\sigma_{nm}(E_t+\epsilon_m)\,\exp[-E_t/k_B T]\,dE_t, \tag{81}$$

where N is Avogadro's number, k_B is Boltzmann's constant, E_t is the initial translational energy of reactants in vibrational state m, and $\mu_{A,BC}$ is the reduced mass of the initial collision partners.

$$\mu_{A,BC} = m_A(m_B + m_C)/(m_A + m_B + m_C). \tag{82}$$

Definitions similar to Eqs. (79)-(81) also hold for differential cross sections, integral cross sections, and rate constants in the BCRLM, except that each is obtained for every choice of bending states λ_1 and λ_2.

2.10. The Relationship between RLM/BCRLM and 3D -- Rotational Averaging

Although cross sections and rate constants in the RLM are well defined quantities, it is nevertheless difficult to compare directly to full three-dimensional calculations because the RLM neglects internal bending and rotational degrees of freedom. Philosophically, it is better to ask how one should sum or average the results of full 3D calculations in order to obtain quantities which best compare with the models. Since in the RLM or BCRLM, the diatomics do not rotate before or after the collision, we may be tempted to compare 3D $(n,j=0) \rightarrow (n',j'=0)$ processes with RLM $n \rightarrow n'$ processes. In cases where such comparisons can be made, the RLM probabilities, cross sections, and rate constants are larger than the corresponding 3D quantities at all energies and temperatures. The comparison is improved for 3D processes from $(n,j=0)$ to $(n',\text{all } j')$; but even here, RLM results are too large, especially at reaction thresholds, where at least for the cases where detailed 3D results are available,[6-8,13] threshold behavior is strongly influenced by the bending zero point energy of the collision complex in the strong interaction region (i.e., the transition state). It is this latter effect which we address to some extent by augmenting the RLM with a bending Hamiltonian.

The inclusion of effective potentials into the BCRLM in order to account approximately for the neglected bending degrees of freedom in the RLM should make it possible to compare more directly with 3D calculations. Such comparisons are difficult because bending motion is relevant only when the collision partners are close together, and not asymptotically, where boundary conditions are imposed and where the angular motion becomes that of a free rotor. Although the lowest bending states $\lambda_1=\lambda_2=0$ do correllate with the lowest free rotor states $j=j'=0$, we cannot generally define a mapping between higher bending states and higher free rotor states. Consequently, comparisons between BCRLM and full 3D calculations require that we average both sets of results.[64,79] In the BCRLM, we average over the bending degrees of freedom labelled by λ_1 and λ_2, and compare to 3D calculations averaged over the analogous rotational degrees of freedom j, j', l, and l', where l is a label for orbital angular momentum.

The appropriate kind of rotational averaging has been discussed for several years by Bowman and coworkers,[18-24] in connection with a hierarchy of dimensionality reducing theories of reactions. Although the BCRLM differs in origin from these dimensionality reducing theories, it resembles them in spirit, and in detail at some levels. Specifically, the application of microcanonical rotational averaging to BCRLM has been presented by Walker and Pollak,[64] and we will review only the final results here.

In this section, we use square brackets ([]) to indicate quantities which have been microcanonically summed, braces ({}) to indicate quantities which have been microcanonically averaged, and angle brackets (<>) to indicate quantities which have been thermally averaged. The appropriate 3D microcanonically averaged rotational cross section with which we wish to compare is[64,80]

$$\{\sigma_{nm}(E)\} = \pi\ [\mathcal{P}_{nm}(E)]\ /\ [k_m^2], \tag{83}$$

where $[k_m^2]$ is a cumulative translational wavenumber for reactants,

$$[k_m^2] = 2\mu_{A,BC}\hbar^{-2} \sum_{j=0}^{\infty} (2j+1)\ (E-\epsilon_{mj})\ \Theta(E-\epsilon_{mj}), \tag{84}$$

and ϵ_{mj} is the internal energy of the initial molecule in vibrational state m and rotational state j. The Heaviside function $\Theta(x)$ in Eq. (84) indicates that the summation runs over only open channels at total energy E. In Eq. (83), $[\mathcal{P}_{nm}(E)]$ is a cumulative rotational probability, which for a full 3D calculation is defined as

$$[\mathcal{P}_{nm}(E)] = \sum_{J=0}^{\infty} (2J+1) \sum_{j,j'=0}^{\infty} \sum_{l,l'} P^{J}_{nj'l'mjl}(E-\epsilon_{mj}), \tag{85}$$

where the sums over l and l' run over the triangle inequality with J and j (or j'), and $P^{J}_{nj'l'mjl}$ is the reaction probability, the absolute square of an element of the 3D S matrix.

In the BCRLM, the cumulative reaction probability is one in which we sum over the bending degrees of freedom, approximating Eq. (85) as

$$[\mathcal{P}_{nm}(E)] = \mathcal{M} \sum_{\lambda=0}^{\infty} \sum_{\Lambda=-\lambda}^{\lambda}{}_{2} \sum_{J=|\Lambda|}^{\infty} (2J+1)\ P^{J\lambda_1\lambda_2}_{nm}(E), \tag{86}$$

where λ is a principal bending quantum number ($\lambda = \lambda_1+\lambda_2$), Λ is an internal bending angular momentum ($\Lambda = \lambda_1-\lambda_2$), and the notation Σ_2 indicates that the summation over Λ goes in steps of two. The reaction path multiplicity factor $\mathcal{M}$ in Eq. (86) assumes values of one or two, the latter for the case of an initial homonuclear diatomic. In practice,[64] we have further approximated Eq. (86) by writing reaction probabilities for higher bending states in terms of those for the lowest bending state, using transition state theory arguments.[19,24]

Given the cumulative reaction probabilities, we can compute thermally averaged rate constants,

$$K_{nm}(T) = \hbar^2 N(2\pi)^{1/2}(\mu_{A,BC}k_B T)^{-3/2}\ \langle\mathcal{P}_{nm}(E)\rangle\ /\ Q_m(T), \tag{87}$$

where Q_m is the rotational partition function

$$Q_m(T) = \sum_{j=0}^{\infty} (2j+1)\ \exp(-\epsilon_{mj}/k_B T), \tag{88}$$

and the thermal average of the cumulative reaction probability is

$$\langle \mathcal{P}_{nm}(E)\rangle = \int_0^\infty \exp(-E/k_BT)\ [\mathcal{P}_{nm}(E)]\ dE. \quad (89)$$

3. REVIEW OF APPLICATIONS OF THE BCRLM

To date, the BCRLM has been applied to a handful of chemically reactive systems, namely the hydrogen exchange reaction $H + H_2$ and its isotopic counterparts,[15,64,65] and to the $F + H_2$ reaction and its isotopic counterparts.[66-69] Some preliminary results have also been presented[67] for the $He + H_2^+$ reaction as well. Lagana[70] has extended the calculations for the $H+H_2$ reaction to higher collision energies, and de Haar, Balint-Kurti, and Wyatt[71] have considered the $H + Cl_2$ reaction. We have already discussed in the previous section an extension[64] of the BCRLM in which we define averaged cross sections and rate constants; when applied to the $D+H_2(m=0,1)$ reaction, we obtained an excellent comparison with shifted sudden calculations of Abu-Salbi, Kouri, Shima, and Baer.[81,82]

The primary concern of the first BCRLM paper[15] was to investigate the extent to which the 300K rate constants (RLM and BCRLM) for the reactions $H + H_2(m=1)$ and $D + H_2(m=1)$ are determined by collisions at energies below the height of the adiabatic reaction barrier. The rate constants determined were compared to experiment[83,84] and to a classical trajectory calculation,[85] but since they are not rotationally averaged, these rates are certainly an upper limit to the true rates on the potential surface[86,87] we used.

Our interest turned then to the relationship between scattering resonances and the angular distribution (differential cross section) predicted by the BCRLM. We showed[65] that for both reactions mentioned above, the angular distribution moves from backwards peaked at low collision energies to more sideways peaked at higher energies, even though the reaction dynamics at threshold is dominated in the $H+H_2(m=1)$ case by a resonance and no resonance appears in the $D+H_2(m=1)$ case. Our interest in this relationship was sparked by the $F+H_2$ reaction, which also shows[88-95] a shift in the angular distribution, and has a definite threshold resonance contribution. We therefore analyzed[66] the BCRLM angular distribution for the $F+H_2$ reaction, and concluded that while the presence of a threshold resonance does contribute to the sideways shift in the angular distribution, it is probably not the only source of the feature. Pursuing this idea further, we attempted to separate the resonant and background contributions to the angular distribution in a following paper,[67] using isolated narrow resonance approximations.

We have also used the BCRLM as a tool to investigate the relationship between parameters which define potential energy surfaces and dynamic features such as resonances, angular distributions,[68] and the position of reaction thresholds.[69] This work has concentrated on the $F+D_2$ reaction, and has aided the development of improved potential energy surfaces.[69,96]

4. EXTENSIONS AND FURTHER APPLICATIONS OF THE METHOD

There is currently work in progress which will extend some of the ideas of the BCRLM either to improve the quantitative reliability of the method, or to enlarge the range of problems to which it is applicable.

4.1. Hybrid Sudden and Adiabatic Methods

As is evidenced by recent literature[97-100] and in other contributions in this volume, there is considerable interest in understanding the nature of the rotation-bending dynamics of reactions in the energy regime near the reaction threshold. For many reactions, and perhaps for most reactions, the dynamics of bending motion at threshold is adiabatic, but above threshold energies, the motion seems to switch over to a sudden type of behavior.[99] Consequently, work in progress[100-1] would define a hybrid sudden-adiabatic theory which would produce reaction cross sections in agreement with adiabatic thresholds (e.g., BCRLM) and in agreement with reactive sudden cross sections[81-2,102-3] at higher collision energies. If this work proves fruitful, and we learn how to model the crossover between adiabatic and sudden reaction dynamics, then we may hope to considerably improve the predictive nature of approximate theories of reactions.

4.2. Non-collinearly Dominated Reactions

The BCRLM is by its very nature constrained to treating collinearly dominated reaction processes. One could extend the method to non-collinear systems by including effective potential terms and more complicated kinetic energy operators to represent the motion of the reacting system along its (bent) minimum energy path from reactants to products. This is indeed an example of the Carrington and Miller[33] reaction surface Hamiltonian theory, which at present is probably the most fruitful approach for noncollinear systems.

4.3. Coupling the Bending Degrees of Freedom

A fairly straightforward extension of the method would be to include the coupling between the lowest and higher bending degrees of freedom while solving the coupled-channel equations. Such an approach may have the beneficial effect of lowering the overall reactivity characteristic of the BCRLM at post-threshold energies, since inelastic bending transitions may reflect otherwise reactive flux prior to reaching the reaction barrier. Unfortunately, there is no significance to individual bend-state to bend-state cross sections, because bend states have no well defined asymptotic meaning; we would therefore still need to rotationally average (i.e., bending average) our results. Furthermore, any improvements would come at the cost of substantially increasing the computer requirements of the method, to a level comparable to 3D centrifugal sudden (CS) reactive calculations. If such a level of computer effort is available, it would therefore seem appropriate to do the 3D CS calculation instead.

4.4. Photodissociation of Linear Triatomics

A promising extension of the BCRLM to new problems lies in the photodissociation of some triatomics. The application of quantum half-scattering methods to problems of photodissociation is well known,[104-8] and for molecules whose ground and excited electronic surfaces are linearly dominated, the approximations inherent in BCRLM are quite appropriate. To treat photodissociation, we must compute the overlap of the scattering wavefunction on the excited electronic surface with the initial bound state wavefunction on the ground surface. Methods for computing the required overlaps as the R-matrix solution of the coupled-channel equations progresses have been described by Kulander and Light[106] and by Schneider and Taylor.[108] In addition to its relevance to photodissociation, these techniques provide a way to recover the scattering wavefunction from an R-matrix calculation, since the "bound state" wavefunction can be a delta function or a narrow gaussian. The technique for accumulating these overlaps resembles the hyperspherical-to-Jacobi projection described earlier, and so we will review it here.

Using the Jacobi coordinate system as an example, we seek the overlap $\mathcal{X}_{nm}$ of the scattering wavefunction Ψ_m in Eq. (24) with a bounded function $\mathcal{F}_n(R,r)$. We begin by expanding the propagation functions in each sector with sine- and cosine-like functions,

$$\begin{aligned}
s_n(R;i) &= \sin[k_n^i(R-R_i^+)], && \text{channel n open,}\\
s_n(R;i) &= \sinh[k_n^i(R-R_i^+)], && \text{channel n closed,}\\
c_n(R;i) &= \cos[k_n^i(R-R_i^+)], && \text{channel n open,}\\
c_n(R;i) &= \cosh[k_n^i(R-R_i^+)], && \text{channel n closed,}
\end{aligned} \tag{90}$$

where k_n^i is the local channel wavenumber in sector i. Note specifically that we are expanding about the right-hand-side of each sector, because it simplifies the propagation of the overlaps. Next we define primitive overlap integrals of the bound function(s) in sector i,

$$[\underline{\underline{N}}_s^i]_{nm} = \int_{R_i^-}^{R_i^+} \mathcal{F}_n(R,r)\ s_m(R;i)\ F_m(r;i)\ dR, \tag{91}$$

$$[\underline{\underline{N}}_c^i]_{nm} = \int_{R_i^-}^{R_i^+} \mathcal{F}_n(R,r)\ c_m(R;i)\ F_m(r;i)\ dR. \tag{92}$$

Note that if $\mathcal{F}_n(R,r) = \delta(R-R_o^n)\ \delta(r-r_o^n)$, or is a narrow gaussian[109,110] centered at (R_o^n,r_o^n), then the integrals above are easily evaluated.

If $\underline{\underline{R}}(i)$ is the global R matrix accumulated after propagating through sector i, we then compute from it a local overlap in sector i,

$$\underline{\underline{d}}(i) = \underline{\underline{N}}_s^i\cdot[\underline{\underline{R}}(i)\cdot\underline{\underline{k}}^i]^{-1} + \underline{\underline{N}}_c^i, \tag{93}$$

and a matrix relating overlaps in sector i to those in sector i+1,

$$\underline{\underline{Q}}(i,i+1) = \underline{\underline{T}}(i,i+1)\cdot\left[\underline{\underline{s}}(R_-^{i+1})\cdot[\underline{\underline{R}}(i+1)\cdot\underline{\underline{k}}^{i+1}]^{-1} + \underline{\underline{c}}(R_-^{i+1})\right], \tag{94}$$

where $\underline{\underline{T}}$ is defined in Eq. (45), and the diagonal $\underline{\underline{s}}$ and $\underline{\underline{c}}$ matrices are defined in Eq. (90). The accumulated overlap through sector i+1 is now given by

$$\underline{\underline{D}}(i+1) = \underline{\underline{d}}(i+1) + \underline{\underline{D}}(i)\cdot\underline{\underline{Q}}(i,i+1). \tag{95}$$

The overlap propagation begins with $\underline{\underline{D}}(1) = \underline{\underline{d}}(1)$ and continues through the sector where boundary conditions are imposed. At this point we compute the desired overlap matrix by taking the same linear combination of the propagated overlaps as for asymptotic boundary conditions (see Eq. (74)),

$$\underline{\underline{\mathfrak{X}}} = \underline{\underline{D}}(\text{final})\cdot\left[\underline{\underline{\mathcal{I}}} - \underline{\underline{\sigma}}\cdot\underline{\underline{k}}^{-1/2}\cdot\underline{\underline{S}}^J\cdot\underline{\underline{k}}^{1/2}\right]. \tag{96}$$

It should also be clear that a similar approach could be employed, if desired, to propagate overlaps of bound state wavefunctions with the gradient of the scattering wavefunction. Note from Eqs. (93)-(94) that the propagation of overlaps requires at each step an inversion of the R matrix; if one propagates the log-derivative matrix[76] instead of the R matrix, then the propagation of overlaps by this scheme requires only a few additional matrix multiplications at each step.

5. SUMMARY

The Bending Corrected Rotating Linear Model should be useful as a tool for insight into the importance of some of the three-dimensional features expected of collinearly dominated atom-diatom reactions. The Rotating Linear Model, defined by Child,[1] Connor and Child,[3] and by Wyatt,[2] augments the collinear world naturally with an impact parameter, making it possible to compute integral and differential cross sections. Adding a bending correction then improves the quantitative predictive ability of the method, and permits a more direct comparison with 3D rotationally averaged integral cross sections and rate constants.

Additional theoretical refinement is needed before we can quantitatively compare BCRLM differential cross sections with 3D. The shape of the BCRLM differential cross section contains information about the impact parameter dependence of the reaction probability, and whenever the 3D angular distribution retains only this level of dynamical detail, we would expect the BCRLM differential cross section to compare nicely. Consequently, it is likely that BCRLM will fare best when compared to 3D differential cross sections from the ground state of reactants to all product rotational states, since this type of cross section retains the least amount of rotational information.

REFERENCES

1. M. S. Child, Mol. Phys. **12**, 401 (1967).
2. R. E. Wyatt, J. Chem. Phys. **51**, 3489 (1969).
3. J. N. L. Connor and M. S. Child, Mol. Phys. **18**, 653 (1970)
4. D. W. Jepsen and J. O. Hirschfelder, J. Chem. Phys. **30**, 1032 (1959)
5. N. Agmon, Chem. Phys. **61**, 189 (1981)
6. G. C. Schatz and A. Kuppermann, J. Chem. Phys. **62**, 2502 (1975)
7. A. B. Elkowitz and R. E. Wyatt, J. Chem. Phys. **62**, 2504 (1972)
8. A. B. Elkowitz and R. E. Wyatt, J. Chem. Phys. **63**, 702 (1975)
9. R. E. Wyatt, ACS Symp. Ser. No. 56, edited by P. R. Brooks and E. F. Hayes (American Chemical Society, Washington, D. C., 1977), p. 185.
10. G. C. Schatz and A. Kuppermann, J. Chem. Phys. **65**, 4642 (1976)
11. G. C. Schatz and A. Kuppermann, J. Chem. Phys. **65**, 4668 (1976)
12. R. E. Wyatt, in *Atom-Molecule Collision Theory*, edited by R. B. Bernstein, (Plenum, New York, 1979), p. 567.
13. R. B. Walker, E. B. Stechel, and J. C. Light, J. Chem. Phys. **69**, 2922 (1978)
14. R. B. Walker, J. C. Light, and A. Altenberger-Siczek, J. Chem. Phys. **64**, 1166 (1976)
15. R. B. Walker and E. F. Hayes, J. Phys. Chem. **87**, 1255 (1983)
16. E. M. Mortensen and K. S. Pitzer, Chem. Soc. (London), Spec. Publ. **16**, 57 (1962)
17. E. M. Mortensen, J. Chem. Phys. **48**, 4029 (1968)
18. J. M. Bowman, G. Z. Ju, and K. T. Lee, J. Chem. Phys. **75**, 5199 (1981)

19. J. M. Bowman, G. Z. Ju, and K. T. Lee, J. Phys. Chem. **86**, 2232 (1982)
20. J. M. Bowman, K. T. Lee, and G. Z. Ju, Chem. Phys. Lett. **86**, 384 (1982)
21. K. T. Lee and J. M. Bowman, J. Phys. Chem. **86**, 2289 (1982)
22. J. M. Bowman and K. T. Lee, Chem. Phys. Lett. **94**, 363 (1983)
23. J. M. Bowman, K. T. Lee, and R. B. Walker, J. Chem. Phys. **79**, 3742 (1983)
24. J. M. Bowman, Adv. Chem. Phys. (to be published)
25. B. C. Garrett and D. G. Truhlar, J. Am. Chem. Soc. **101**, 4534 (1979)
26. B. C. Garrett and D. G. Truhlar, J. Phys. Chem. **83**, 1915 (1979)
27. B. C. Garrett and D. G. Truhlar, Proc. Natl. Acad. Sci. USA **76**, 4755 (1979)
28. B. C. Garrett and D. G. Truhlar, J. Chem. Phys. **72**, 3460 (1980)
29. B. C. Garrett, D. G. Truhlar, R. S. Grev, and A. W. Magnuson, J. Phys. Chem. **84**, 1730 (1980)
30. B. C. Garrett, D. G. Truhlar, and R. S. Grev, J. Phys. Chem. **84**, 1749 (1980)
31. D. G. Truhlar, A. D. Isaacson, R. T. Skodje, and B. C. Garrett, J. Phys. Chem. **86**, 2252 (1982)
32. B. C. Garrett and D. G. Truhlar, "Generalized Transition State Theory and Least-Action Tunneling Calculations for the Reaction Rates of H(D) + H_2(n=1) → H_2(HD) + H," to be published, J. Phys. Chem.
33. T. Carrington and W. H. Miller, J. Chem. Phys. **81**, 3942 (1984)
34. R. A. Marcus, J. Chem. Phys. **45**, 4493 (1966)
35. R. A. Marcus, J. Chem. Phys. **49**, 2610 (1968)
36. R. E. Wyatt, J. Chem. Phys. **56**, 390 (1972)
37. C. C. Rankin and J. C. Light, J. Chem. Phys. **51**, 1701 (1969)
38. G. Miller and J. C. Light, J. Chem. Phys. **54**, 1635 (1971)
39. L. M. Delves, Nucl. Phys. **9**, 391 (1959); **20**, 275 (1960)
40. A. Kuppermann, Chem. Phys. Lett. **32**, 374 (1975)
41. G. Hauke, J. Manz, and J. Romelt, J. Chem. Phys. **73**, 5040 (1980)
42. A. Kuppermann, J. A. Kaye, and J. P. Dwyer, Chem. Phys. Lett. **74**, 257 (1980)
43. J. Romelt, Chem. Phys. Lett. **74**, 263 (1980)
44. B. R. Johnson, J. Chem. Phys. **73**, 5051 (1980)
45. D. K. Bondi and J. N. L. Connor, Chem. Phys. Lett. **92**, 570 (1982)
46. V. Aquilanti, G. Grossi, and A. Lagana, Chem. Phys. Lett. **93**, 174 (1982)
47. J. M. Launay and M. Le Dourneuf, J. Phys. B**15**, L455 (1982)
48. D. K. Bondi, Ph.D. Thesis, University of Manchester, 1985, preprint kindly provided by J. N. L. Connor.
49. J. C. Light and R. B. Walker, J. Chem. Phys. **65**, 4272 (1976)
50. J. Manz and J. Romelt, Chem. Phys. Lett. **76**, 337 (1980)
51. J. Manz and J. Romelt, Chem. Phys. Lett. **77**, 172 (1981)
52. J. Manz and J. Romelt, Chem. Phys. Lett. **81**, 179 (1981)
53. J. Manz, E. Pollak, and J. Romelt, Chem. Phys. Lett. **86**, 26 (1982)
54. J. Romelt, Chem. Phys. Lett. **86**, 26 (1982)
55. J. Romelt, Chem. Phys. Lett. **87**, 259 (1982)

56. J. A. Kaye and A. Kuppermann, Chem. Phys. Lett. **77**, 573 (1981)
57. J. A. Kaye and A. Kuppermann, Chem. Phys. Lett. **78**, 546 (1981)
58. V. Aquilanti, S. Cavalli, and A. Lagana, Chem. Phys. Lett. **93**, 179 (1982)
59. V. Aquilanti, Hyperfine Interactions **17-19**, 739 (1984)
60. V. Aquilanti, S. Cavalli, G. Grossi, and A. Lagana, J. Molec. Structure **93**, 319 (1983)
61. V. Aquilanti, S. Cavalli, G. Grossi, and A. Lagana, J. Molec. Structure **107**, 95 (1984)
62. B. Podolsky, Phys. Rev. **32**, 812 (1928)
63. H. Margenau and G. M. Murphy, *The Mathematics of Physics and Chemistry*, 2nd Edition (D. Van Nostrand, Inc., Princeton, 1956), pp. 192-7.
64. R. B. Walker and E. Pollak, J. Chem. Phys. **83**, 0000 (1985)
65. R. B. Walker and E. F. Hayes, J. Phys. Chem. **88**, 1194 (1984)
66. E. F. Hayes and R. B. Walker, J. Phys. Chem. **88**, 3318 (1984)
67. E. F. Hayes and R. B. Walker, ACS Symp. Ser. No. 263, edited by D. G. Truhlar (American Chemical Society, Washington, D. C., 1984), p. 493
68. R. B. Walker, N. C. Blais, and D. G. Truhlar, J. Chem. Phys. **80**, 246 (1984)
69. R. Steckler, D. G. Truhlar, B. C. Garrett, N. C. Blais, and R. B. Walker, J. Chem. Phys. **81**, 5700 (1984)
70. A. Lagana, private communication.
71. B. M. D. D. Jansen op de Haar, G. G. Balint-Kurti, and R. E. Wyatt, "An Approximate Three-Dimensional Quantum Mechanical Calculation of Reactive Scattering Cross Sections for the H + $Cl_2(v) \rightarrow HCl(v')$ + Cl Reaction," preprint.
72. W. H. Miller, private communication.
73. R. B. Walker, QCPE Program No. 352, Department of Chemistry, Indiana University, Bloomington, IN
74. E. B. Stechel, R. B. Walker, and J. C. Light, J. Chem. Phys. **69**, 3518 (1978)
75. J. C. Light, R. B. Walker, E. B. Stechel, and T. G. Schmalz, Computer Phys. Comm. **17**, 89 (1979)
76. J. C. Light, contribution in this volume.
77. D. J. Zvijac and J. C. Light, Chem. Phys. **12**, 237 (1976)
78. R. D. Levine and R. B. Bernstein, *Molecular Reaction Dynamics*, (Oxford University Press, New York, 1974), p. 108.
79. E. Pollak and R. E. Wyatt J. Chem. Phys. **78**, 4464 (1983)
80. R. A. Marcus, J. Chem. Phys. **45**, 2138 (1966); K. Morokuma, B. C. Eu, and M. Karplus, J. Chem. Phys. **51**, 5193 (1969)
81. N. Abu-Salbi, D. J. Kouri, Y. Shima, and M. Baer, J. Chem. Phys. **82**, 2650 (1985)
82. N. Abu-Salbi, D. J. Kouri, Y. Shima, and M. Baer, Chem. Phys. Lett. **105**, 472 (1984)
83. E. B. Gordon, B. I. Ivanov, A. P. Perminov, V. E. Balalaev, A. N. Ponomarev, and V. V. Filatov, Chem. Phys. Lett. **58**, 425 (1978)
84. G. P. Glass and B. K. Chaturbedi, J. Chem. Phys. **79**, 3478 (1982)

85. H. R. Mayne and J. P. Toennies, J. Chem. Phys. **75**, 1794 (1981)
86. P. Siegbahn and B. Liu, J. Chem. Phys. **68**, 2457 (1978); B. Liu, *ibid*. **58**, 1925 (1973)
87. D. G. Truhlar and C. J. Horowitz, J. Chem. Phys. **68**, 2466 (1978); **71**, 1514 (E) (1979)
88. M. J. Redmon and R. E. Wyatt, Chem. Phys. Lett. **63**, 209 (1979)
89. R. E. Wyatt, in *Horizons in Quantum Chemistry*, edited by K. Fukui and B. Pullman, (Reidel, Dordrecht, 1980), p. 63.
90. R. E. Wyatt, J. F. McNutt, and M. J. Redmon, Ber. Bunsenges. Phys. Chem. **86**, 437 (1982)
91. R. E. Wyatt and M. J. Redmon, Chem. Phys. Lett. **96**, 284 (1983)
92. M. Baer, J. Jellinek, and D. J. Kouri, J. Chem. Phys. **78**, 2962 (1983)
93. S. H. Suck, Chem. Phys. Lett. **77**, 390 (1981)
94. S. H. Suck and R. W. Emmons, Phys. Rev. A **24**, 129 (1981)
95. R. W. Emmons and S. H. Suck, Phys. Rev. A **25**, 178 (1982)
96. D. G. Truhlar, B. C. Garrett, and N. C. Blais, J. Chem. Phys. **80**, 232 (1984)
97. N. Abusalbi, D. J. Kouri, Y. Shima, and M. Baer, Chem. Phys. Lett. **105**, 472 (1984)
98. G. C. Schatz, J. Chem. Phys. **79**, 5386 (1983)
99. E. Pollak and R. E. Wyatt, Chem. Phys. Lett. **110**, 340 (1984)
100. E. Pollak, J. Chem. Phys. **83**, 1111 (1985)
101. G. G. Balint-Kurti, private communication.
102. E. Pollak, N. Abusalbi, and D. J. Kouri, Chem. Phys. Lett. **113**, 585 (1985)
103. N. Abu-Salbi, D. J. Kouri, M. Baer, and E. Pollak, J. Chem. Phys. **82**, 4501 (1985)
104. Y. B. Band, K. F. Freed, and D. J. Kouri, J. Chem. Phys. **74**, 4380 (1980)
105. G. G. Balint-Kurti and M. Shapiro, Chem. Phys. **61**, 137 (1981)
106. K. C. Kulander and J. C. Light, J. Chem. Phys. **73**, 4337 (1980)
107. M. Shapiro and R. Bersohn, Ann. Rev. Phys. Chem. **33**, 409 (1982)
108. B. I. Schneider and H. S. Taylor, J. Chem. Phys. **77**, 379 (1982)
109. J. C. Light, I. P. Hamilton, and J. V. Lill, J. Chem. Phys. **82**, 1400 (1985)
110. I. P. Hamilton and J. C. Light, "On Distributed Gaussian Bases for Multi-dimensional Vibrational Problems," preprint.

PERIODIC ORBITS AND REACTIVE SCATTERING: PAST, PRESENT AND FUTURE

Eli Pollak
Chemical Physics Department
Weizmann Institute of Science
Rehovot 76100
Israel

ABSTRACT. The study of periodic orbits embedded in the continuum has provided a new tool for understanding the dynamics of molecular collisions. The application of periodic orbit theory to classical variational transition state theory, quantal threshold and resonance effects is presented. Special emphasis is given to the stability analysis of periodic orbits in collinear and three dimensional systems. Future applications of periodic orbit theory are outlined.

1. INTRODUCTION

The basic question posed in reactive scattering theory today is simple. Given a potential energy surface for the interacting molecules, what will be the outcome of a collision between them. The outcome is described in various terms such as a reaction probability, distribution of final states, differential cross section, rate constant etc. All of these quantities depend on the initial state of the colliders.

The laws of physics needed to answer the question are well known. Since the potential energy surface is given, one knows the masses of the colliders and so one only needs to solve the Schrödinger equation. The problem of course is that the number of coupled equations that need to be solved is enormous and not yet within reach of present day computers. Necessarily then the theorist is restricted to studying model systems and construction of approximations. One type of approximation is to solve the exact classical mechanical equations of motion.[1] One selects initial conditions which correspond to the experimental initial state, integrates the equations of motion forward in time till the process is 'over' and then obtains cross sections, product distributions etc. In essence, Hamilton's equations of motion serve as a 'black box', whose structure is determined by the masses and the potential energy surface. This black box provides the necessary transformation from initial conditions to final conditions.

The study of periodic orbits originated from an attempt to uncover the 'black box' and understand how the potential energy surface and masses affect the outcome of a collision. Even before the advent of

D. C. Clary (ed.), The Theory of Chemical Reaction Dynamics, 135–165.

computers it was realised that the saddle point region of the surface was critical in determining thermal rates of reaction. Transition state theory formulated by Eyring[2] and Wigner,[3] is based on this observation. Wigner showed that classical transition state theory provides a variational upper bound to the classical rate. The variational theory was then developed by Horiuti[4] Keck[5] and Koeppl.[6] In all cases though, the theory was formulated with the aid of expansions around the saddle point region. Pechukas,[7] was the first to give an exact solution to the variational problem. He proved that the variational transition state must be itself a classical trajectory. If in an A+BC collision the atoms are forced to be collinear then the variational transition state is a periodic orbit. Periodic orbits may be found also far away from the saddle point. Therefore a study of their properties reveals many aspects of the global dynamics. In section two we review briefly the classical variational transition state theory. Special attention is given to the stability[8] (in the Lyapunov sense) of the variational transition states. We show that stability analysis is useful for predicting the energetic threshold at which the variational theory is no longer exact.

Classical mechanically, we know today, that periodic orbits govern the flow of trajectories in a collinear collision.[9] In a sense, one may say that classical collinear collisions are well understood. However the world is quantal and it is of greater interest to also try and understand the quantum mechanics of collinear collisions. In the past decade various numerical techniques have been devised which enable a relatively fast and cheap evaluation of exact quantal collinear reaction probabilities.[10,11] Here too, a study of classical periodic orbits has provided insight into the quantum mechanics. In section III we show how periodic orbits may be used as an analytic tool for understanding quantal phenomena such as Feshbach resonances[12] and tunneling.[13]

The main objective of any theory is to be able to understand and predict the results of experiments. Since the world is three dimensional one cannot limit oneself to the study of collinear systems. In section IV we show how the collinear analysis based on periodic orbits may be generalised to three dimensional systems.[14] We provide a 3D adiabatic transition state theory which is used to analyse numerical computations as well as experimental results. A 3D analysis of quantal resonances predicts that one should hope that quantal resonances[15] will provide a new spectroscopy of transition states. A discussion of the future role of periodic orbits and reactive scattering is given in section V.

II. PERIODIC ORBIT DIVIDINGS SURFACES - PODS

a. Classical Variational Transition State Theory

In this section we consider the reaction A+BC→AB+C where A, B and C are atoms with masses m_A, m_B, m_C respectively. The classical Hamiltonian H of the system given in terms of the reactant diatom internuclear separation vector $\underline{r}_{BC}$, the atom to center of mass of diatom separation vector $\underline{R}_{A-BC}$ and the conjugate momenta $\underline{p}_{BC}$, $\underline{P}_{A-BC}$ is

$$H = \frac{p_{BC}^2}{2\mu_{BC}} + \frac{p_{A-BC}^2}{2\mu_i} + V(\underline{r}_{BC}, \underline{R}_{A-BC}) \quad . \tag{1}$$

Here μ_{BC} and μ_i are reduced masses

$$\mu_{BC} = \frac{m_B m_C}{m_B + m_C} \qquad \mu_i = \frac{m_A(m_B+m_C)}{m_A+m_B+m_C} \tag{2}$$

and V is the potential energy surface.

The structure of the potential energy surface is assumed to be such that in the reactant region, that is for large enough $\underline{R}_{A,BC}$ and small $\underline{r}_{BC}$, $V \to V^o(\underline{r}_{BC})$ where $V^o(\underline{r}_{BC})$ is the BC diatomic potential. Similarly in the product region, that is for small $\underline{r}_{AB}$ and large enough $\underline{R}_{C-AB}$, $V \to V^o(\underline{r}_{AB})$. Correspondingly, the reactants (products) Hamiltonian is denoted as $H^o_{BC}(H^o_{AB})$. Furthermore, we assume that at energy E there are two equipotential surfaces, defined by

$$V(\underline{r}_{BC}, \underline{R}_{A,BC}) = E \tag{3}$$

running from reactants to products and enclosing between them a band of configuration space in which V<E. Outside these surfaces V>E. (A typical potential energy surface is plotted in Fig. 1, cf. p.140). A trajectory is a line in phase space defined by the variation in time (t) of a set of coordinates and momenta $[\underline{p}(t), \underline{q}(t)]$ under the influence of the Hamiltonian H. A reactive trajectory is one on which as $t \to -\infty$ $\underline{q}(t)$ is in the reactants region, while as $t \to +\infty$ $\underline{q}(t)$ is in the products region. With each point in phase space one can associate a characteristic function of reactive phase points $\chi_r(\underline{p},\underline{q})$ such that

$$\chi_r(\underline{p},\underline{q}) = \begin{cases} 1 \text{ if } (\underline{p},\underline{q}) \text{ lies on a reactive trajectory} \\ 0 \text{ otherwise} \end{cases} \tag{4}$$

A dividing surface $S(\underline{q})=0$ is defined as a (5-dimensional) surface in configuration space, connecting the two V=E equipotentials, without loops, so that the surface divides the band V<E into two pieces, one containing the reactants region and the other the products region. Let $\underline{n}_S(\underline{q})$ denote a unit vector (at $\underline{q}$) perpendicular to the surface and pointing in the direction of products. For a microcanonical distribution in phase space $[\delta(E-H)]$, the number of reactive phase points crossing this surface per unit time per unit energy E in the direction of products is[5] (remember Liouville's equation)

$$F_R(E) = \frac{1}{h^6} \int_{-\infty}^{\infty} d\underline{p}\, d\underline{q}\, \delta[S(\underline{q})](\dot{\underline{q}} \cdot \underline{n}_s) \cdot \delta(E-H) \chi_r(\underline{p},\underline{q}) \tag{5}$$

Here $\delta(x)$ is the Dirac delta function and $\dot{\underline{q}}$ is the velocity at $(\underline{p},\underline{q})$ $(\dot{\underline{q}} = \frac{\partial H}{\partial \underline{p}})$. Note that by virtue of Liouville's theorem, $F_R(E)$ is independent of the choice of dividing surface. As is evident from Eq. (5) all the dynamical information lies in the characteristic function $\chi_r(\underline{p},\underline{q})$.

The transition state approximation to the reactive flux $F_{TS}(E)$ is obtained by replacing $\chi_r(\underline{p},\underline{q})$ in Eq. (5) by the function

$$\chi_+(\underline{p},\underline{q}_s) = \begin{cases} 1 \text{ if } (\dot{\underline{q}}\cdot\underline{n}_s) > 0 \\ \\ 0 \text{ otherwise} \end{cases} \qquad (6)$$

χ_+ 'counts' all trajectories passing through the dividing surface S in the products direction. Since any reactive trajectory must ultimately cross the dividing surface it is clear that

$$(\dot{\underline{q}}\cdot\underline{n}_s)\chi_+(\underline{p},\underline{q}_s) > (\dot{\underline{q}}\cdot\underline{n}_s)\chi_r(\underline{p},\underline{q}_s) \qquad (7)$$

for any dividing surface. The transition state flux $F_{TS}(E)$ is <u>for any dividing surface</u> an upper bound to the reactive flux. Variational transition state theory is the problem of finding that dividing surface for which $F_{TS}(E)$ is minimal. Note, that by substituting χ_r by χ_+ we have done away with the dynamics. As shall be shown shortly, given a dividing surface, $F_{TS}(E)$ is just the number of internal states at the dividing surface and so is easy to evaluate. The more difficult problem to solve is the variational one.

Under what conditions does $F_{TS}(E)=F_R(E)$? All that is needed is an equality in Eq. (9) for almost any point $(\underline{p},\underline{q}_s)$ at the surface S at energy E. If <u>any</u> reactive trajectory at energy E crosses S only once then $\chi_+=\chi_r$ and transition state theory (TST) is exact. This necessary and sufficient condition was formulated by Wigner almost fifty years ago.[3] This condition can be phrased in a different form. TST is exact at energy E, if no trajectory of energy E that leaves the dividing surface, in either products or reactants direction, ever returns.

Thus far, the discussion of transition state theory has been very general. In the following the variational problem will be discussed for the specific case of a collinear collision. The configuration space of a collinear collision consists of two independent coordinates q_1, q_2 defined such that the Hamiltonian of the system can be written as

$$H = \frac{1}{2m}(p_1^2 + p_2^2) + V(q_1,q_2) \qquad (8)$$

where (p_1,p_2) are the momenta conjugate to (q_1,q_2) and m is a mass. For example one might choose

$$q_1 = \alpha R_{A-BC}, \quad q_2 = \frac{1}{\alpha} r_{BC}, \quad \alpha^4 = \frac{\mu_i}{\mu_{BC}}, \quad m^2 = \mu_i\mu_{BC} \qquad (9)$$

Any dividing surface or equipotential is now a line in configuration space. The dividing surface $S(q_1,q_2)=0$ is parametrized by arc length s, so that $[q_1(s), q_2(s)]$ is the point on the line a distance s from the start. We now define a new set of canonical orthogonal (curvilinear) coordinates $u(q_1,q_2)$, $v(q_1,q_2)$ such that on the dividing surface

$$u[q_1(s), q_2(s)] = u_o (=\text{constant}) \qquad (10)$$

i.e., the v coordinate at $u=u_o$ coincides with the dividing line. By definition then, the normal to the dividing surface $\underline{n}_s$ is just the unit vector $\hat{u}_o$. The volume element in phase space is invariant under canonical transformation. Also, one can show that the transformed Hamiltonian is[16]

$$H = \frac{1}{2\mu(u,v)} [\lambda(u,v)p_u^2 + p_v^2] + V(u,v) \; . \tag{11}$$

Here

$$\mu(u,v) = m[(\partial q_1/\partial v)_u^2 + (\partial q_2/\partial v)_u^2] \tag{12}$$

$$\lambda(u,v) = (\partial q_2/\partial v)_u^2/(\partial q_1/\partial u)_v^2 \quad , \tag{13}$$

and (p_u,p_v) are the momenta conjugate to (u,v).
With these preliminaries, one can show that the transition state theory estimate (cf. Eq. 6) for the reactive flux is given in the new coordinate system as

$$F_{TS}(E,u_o) = \frac{1}{h^2}\int_{-\infty}^{\infty} dp_v dv \; \theta[E - \frac{p_v^2}{2\mu(u_o,v)} - V(u_o,v)] \tag{14}$$

where $\theta(x)$ is the unit step function. The form, given in Eq. (14) is familiar, it expresses the well known fact that the transition state estimate for the reactive flux crossing an arbitrary dividing surface at energy E is just the number of states of the internal Hamiltonian h at that energy.

The integration over p_v in Eq. (14) is straightforward and one finds that collinearly the transition state flux is just proportional to an action integral. The variational problem is to vary the dividing surface u_o so as to minimise the flux. Maupertuis' principle of least action implies that u_o must be the configuration space path of a classical trajectory of energy E.[17] However, this trajectory starts at the turning point $v_<$ with zero momentum [remember that $V(u_o,v_<)=E$], heads out perpendicular to the equipotential reaching after time T/2 the other equipotential at the turning point $v_>$ where it stops and heads back for $v_<$ along a trajectory which is just the time reversal of the motion from $v_<$ to $v_>$. Therefore, the trajectory must be periodic with period T. We have therefore shown that the variational solution to TST is a <u>periodic orbit dividing surface</u> or in short a pods.

The first system on which variational transition state theory has been tested is the (symmetric) hydrogen exchange reaction on the Porter-Karplus (II) potential energy surface.[18] This surface has a saddle point at $E_{SP}=0.396$eV relative to the bottom of the asymptotic reactants and products well. Because of symmetry, the symmetric stretch is a pods at all energies greater than E_{SP} and lower than the three atom dissociation limit.

The pods of the system are shown in Fig. 1. One finds that for $E_{SP}<E<0.603$eV there is at each energy only a single pods - the symmetric stretch. For $0.603<E<0.722$eV one finds five pods, two on each side of the symmetric stretch. The pair on each side (not shown in the Figure) appears at E=0.603eV, as energy is increased one pods migrates outwards,

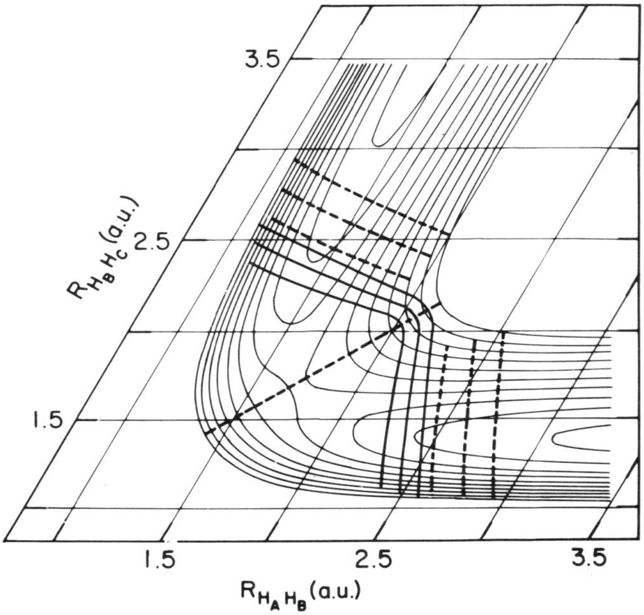

Figure 1. Potential energy surface, (PKII)[18] pods and RPO's of the H_3 system. The dashed lines are pods, heavy solid lines RPO's. Equipotential contours are at E=0.1-1.0eV with 0.1eV intervals. Adapted from Ref. 19.

towards reactants or products, the other pods migrates towards the symmetric stretch pods and coalesces with it at 0.723eV. Above 0.723eV one finds three pods, the symmetric stretch and one on each side. For almost all energies the pods with minimal flux, that is, the variational solution to TST is the outermost pods. In the bottom panel of Fig. 2 we plot the exact reaction probability (defined as the ratio of reactive flux to incident flux) and the variational estimate. Also shown in the top half of the figure is the temperature dependence of the ratio of the variational to the thermal rate constant as a function of temperature. Here, the collinear rate is defined as

$$k(T) = (2\pi\mu_i\hat{k}T)^{-1/2}\int_0^\infty dE\ P(E)e^{-E/\hat{k}T} \qquad (15)$$

where $\hat{k}$ is Boltzmann's constant.

There are a few important characteristic features associated with the results presented in Fig. 2, which are actually not specific to the H_3 exchange reaction. Foremost, we find, that in the threshold region transition state theory seems to be exact. Actually, it has been proven[20] that a sufficient condition for TST to be exact is that only one pods exists at a given energy. Secondly, although the variational result fails qualitatively at high energy this has little effect on the thermal rate constant. Even at 2500°K the variational rate overestimates

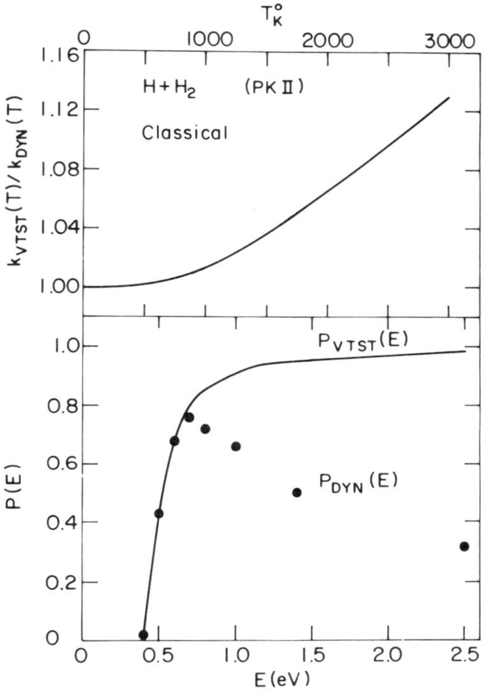

Figure 2. Classical variational TST for the collinear $H+H_2$ reaction. For details, see text.

the exact rate by only fifteen percent. Since much of chemistry is thermal this is a very gratifying result. Variational TST is at room temperature highly accurate.

b. Linear Stability Analysis of Pods

As noted earlier, a necessary and sufficient condition for TST to be exact at a dividing surface is that any classical trajectory crossing the surface will never recross it. It is of practical interest to determine when TST must fail.[21] Usually, the complete potential energy surface is not available so that one would want to develop local criteria for the failure of TST. Here we will show that the stability properties of periodic orbits can be used to answer this question. Loosely speaking, a periodic orbit is defined as stable in the sense of Lyapunov[8] if every trajectory originating at t=0 close enough to the periodic orbit remains close to the orbit for all time t. Obviously, TST cannot be exact if the pods is stable, since there are, by definition, trajectories crossing the pods that stay in its vicinity forever. For these trajectories $\chi_r \neq \chi_+$ and so TST is not exact.

One determines the stability of periodic orbits using linear stability analysis. We denote the phase space of the periodic orbit by (p_u, u, p_v, v). A trajectory starting close to the orbit and at the same total energy shall be denoted as $(p_u+\delta p_u, u+\delta u, p_v+\delta p_v, v+\delta v)$. Using the form of the Hamiltonian given in Eqs. (11-13) and keeping only terms linear in $(\delta p_u,\delta u,\delta p_v,\delta v)$ we note that the linearised (Hamilton) equations of motion for $(\delta p_u,\delta u)$ are

$$(\delta\dot{u}) = [\lambda(u_o,v)/\mu(u_o,v)](\delta p_u) \tag{16}$$

$$(\delta\dot{p}_u) = -(\partial^2 H/\partial u_o^2)(\delta u) \tag{17}$$

Here, the (u,v) coordinate system is defined such that the periodic orbit lies on a line of constant u, that is $u=u_o$, $p_u=0$. Thus the coefficients is Eqs. (16,17) are time dependent but periodic with the same period T as that of the periodic orbit.

Equations (16,17) can be used to define the matrix equation

$$\delta\dot{\underset{\sim}{q}} = \underset{\sim}{A}\cdot\delta\underset{\sim}{q} \quad , \quad \underset{\sim}{A} \equiv \begin{pmatrix} 0 & \lambda/\mu \\ -H_{uu} & 0 \end{pmatrix} \quad , \quad \delta\underset{\sim}{q}(0) = \underset{\sim}{I} \tag{18}$$

where $\underset{\sim}{I}$ is the (2x2) identity matrix and as noted $\underset{\sim}{A}(t+T)=\underset{\sim}{A}(t)$. One may solve for $\delta\underset{\sim}{q}$ from time t=0 to t=T and define a constant matrix $\underset{\sim}{B}$ as $\delta\underset{\sim}{q}(T)=\underset{\sim}{B}$. The eigenvalues of $\underset{\sim}{B}$; b_1,b_2 can be shown[8] to obey the relation $b_1b_2=1$. We may therefore use the notation

$$b_1 = e^{-i\beta T} \quad , \quad b_2 = e^{i\beta T} \tag{19}$$

where β is defined as the stability frequency of the periodic orbit. It is a well known theorem in stability theory[8] that if β is purely real then the orbit is stable, if it has an imaginary component then the orbit is unstable.

This result can be understood intuitively. If β is real, then $\delta\underset{\sim}{q}$ oscillates indefinitely but will stay small. Thus a small perturbation in the vicinity of the orbit will leave us in this vicinity. However, if β has an imaginary component then the exponent will always cause $\delta\underset{\sim}{q}$ to diverge so that one cannot stay indefinitely in the vicinity of the periodic orbit. Hence the orbit is unstable.

In principle, the numerical solution of Eq. (19) is not more difficult than the solution of Hamilton's equations of motion. All that one needs are second derivatives of the Hamiltonian evaluated at the pods. Moreover, it is often possible to evaluate the stability frequency of the pods analytically. For example in a symmetric exchange reaction the potential energy may be expanded in the vicinity of the saddle point, to third order such that

$$H = \frac{1}{2m_A}(p_u^2 + p_v^2) + \frac{1}{2}[\lambda\cdot v^2 - ku^2 - bu^2v + \ldots] \tag{20}$$

where λ,k and b are parameters, independent of u and v.
Here we anticipate two properties in the saddle point region. u - the anti-symmetric stretch, is the translational coordinate so that the potential along u is basically that of an inverted harmonic oscillator and k>0. More importantly though, inspection of Fig. 1 shows (as can also be seen explicitly from Fig. 4 of ref. 22) that on the symmetric

stretch $\partial^2V/\partial u^2$ is a monotonically decreasing function of the vibrational coordinate v so that also b>0.

Disregarding all fourth order and higher contributions to the expansion of the potential we find that the vibrational motion on the symmetric stretch (u=0) is purely harmonic. If the energy relative to V(0,0) is E then Eqs. (16,17) may be rewritten as

$$(\ddot{\delta u}) = \frac{1}{m_A}\,[k+b(2E/\lambda)^{1/2}\cos\omega t]\delta u \tag{21}$$

where ω is the harmonic frequency of the symmetric stretch. Eq.(21) is precisely the Mathieu equation whose stability properties are well known. In the limit E→0, Eq. (21) is just that of an inverted harmonic oscillator with imaginary frequency $(k/m_a)^{1/2}$. This of course makes sense, at low energies, the harmonic expansion about the saddle point suffices and motion about an inverted harmonic oscillator is of course unstable. As energy increases the oscillatory term in Eq. (21) becomes more important and one can show[23] that there exists a lowest critical energy at which δu(t) will become a purely periodic function. At this point TST will of course break down. It is not too difficult to obtain an analytical estimate for the critical energy. In Fig. 3 we compare

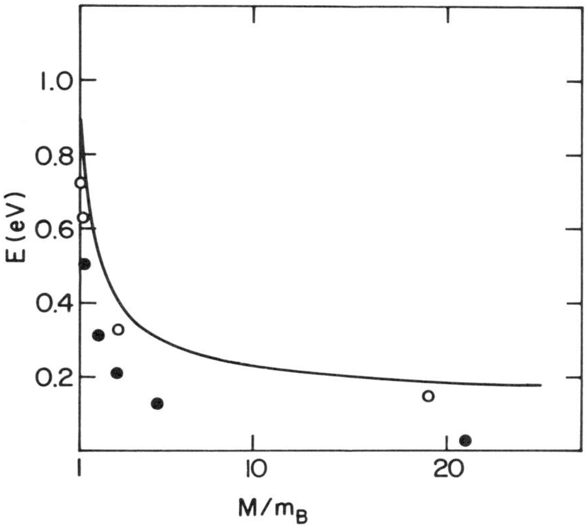

Figure 3. The onset of stability and breakdown of TST in the collinear $H+H_2$ reaction. Solid line - estimate by analysis of Mathieu Equation. Open circles - the exact onset of stability; full circles - the breakdown of TST. Here $M=2m_A+m_B$ for the symmetric A+BA→AB+A reaction.

the critical energy obtained from such an estimate with exact results for the PK(II) potential energy surface and a variety of symmetric mass combinations. Note the reasonable qualitative agreement. The onset of stability is a sufficient condition for breakdown of TST. Interestingly, it is, as seen from the figure, not a necessary condition. As the mass of the middle atom gets higher, stability sets in earlier and TST breaks down at a lower energy. Thus we expect even variational TST to be quite poor for a light atom transfer reaction and very good for heavy atom transfer.

Lack of space prevents us from describing in any detail other properties of pods. It should be mentioned though that pods may be characterized as repulsive or attractive according to the local behaviour of trajectories in their vicinity.[9] Loosely speaking, repulsive pods repel trajectories in their vicinity, attractive pods attract them. These properties have enabled the development of a lower bound to the classical reaction probability,[9] the construction of a new theory of direct reactions[24] as well as new numerical methods for computation of classical product state distributions.[25,26] The repulsive attractive properties provide a global picture of the classical flow of trajectories at a given energy.

III. PERIODIC ORBITS AND COLLINEAR QUANTUM SCATTERING

a. An Optimal Coordinate System Defined by Pods

One of the nice aspects of classical variational TST is that the theory provides upper and lower bounds to the exact classical rate of reaction. To date, there is no such quantal theory. Quantally, transition state theory is an approximation[27] - one of the objectives of the theorist is to optimize the approximation. The concept of a transition state remains though as an important guide towards understanding the quantal structure of the reaction probability. In this section we shall see that quantal barriers and wells and their associated properties determine both threshold behaviour and resonance phenomena in quantal reactive scattering.

To do this one must introduce the concept of vibrational adiabaticity during a collision process. This notion is not new. The foundation was laid by Marcus[28] from both a classical and a quantal point of view. The basic idea is that one picks out a reaction coordinate parametrized by u and assumes that motion along this coordinate is much slower than motion along the other degrees of freedom denoted by v. Thus the translational u motion is governed by the vibrational motion, just as in the Born-Oppenheimer approximation the motion of the nuclei is governed by the electronic motion. The analogy is complete, the nuclear motion does not change the electronic state - the translational motion cannot change the vibrational state. The energy of the vibrational state E_n depends parametrically on the translational coordinate u and so it effectively acts as a potential energy $E_n(u)$ for the translational motion. Classically, one assumes that the translational motion cannot change the classical action, n, of the vibrational motion. $E_n(u)$ is called the vibrationally adiabatic potential energy surface.

These surfaces are used extensively in collision theory. Kuppermann[27] and Truhlar[29,30] have demonstrated their importance in quantum mechanical transition state theory. Wyatt and coworkers[31] have used them as analytic tools, to show that the existence of wells in these surfaces is a necessary condition for the appearance of quantal resonances in the reaction probability.

All of these applications suffer from one major defect. In the Born-Oppenheimer approximation, the difference in mass of the nuclei and electrons is large enough to admit a natural set of adiabatic coordinates. This is not the case in the collision of an atom with a diatom. For any curvilinear coordinate system (u,v) one obtains different adiabatic surfaces. To do away with these ambiguities we will use the pods to define the (u,v) coordinate system. We will then find that a pods is necessarily a vibrationally adiabatic barrier or well and that at the pods the adiabatic approximation is exact.

Suppose that we find a periodic orbit, it defines a line in configuration space. We will use this line to define the coordinate system: a periodic orbit defines a line of constant u. We further suppose that the location of the periodic orbit is a continuous function of an external parameter - the total energy of the system. Then we find a continuous set of lines of constant u and in this manner define an orthogonal curvilinear coordinate system.[16] For each value of u, there exists a value of the energy g(u) for which we find at u a periodic orbit with energy E=g(u).

Consider now a line of constant $u=u_o$ at the energy $E=g(u_o)$. At this energy we have a periodic orbit at u_o so that $p_u=\dot{p}_u=0$. Via Hamilton's equations of motion, this implies (provided that $(\partial\mu/\partial u)\neq 0$) that

$$g(u) = V(u,v) + \mu(u,v)\,\frac{\partial V/\partial u}{\partial \mu/\partial u} \tag{22}$$

where we have dropped the superscript since Eq. (22) holds for all u in the vicinity of the pods. Note that for Eq. (22) to hold we do not have to assume that periodic orbits cover all of configuration space. If the periodic orbit is a continuous function of the energy for a finite interval it will cover a finite connected band of configuration space. Accordingly, the special curvilinear coordinate system we constructed will also cover this band. Within this band, which may be small or large, Eq. (22) holds exactly.

Quantally, the vibrationally adiabatic potential energy surface, $E_n(u)$ is defined by the energy eigenvalue of the n'th vibrational state at u. Semiclassically, the n'th eigenvalue is determined by the condition that the action in the v direction, at fixed u be equal to (n+1/2)h. This determines an energy $E_{n+1/2}(u)$. Thus the action m(u) is defined as

$$m(u) \equiv 2\int_{v_<}^{v_>} \{2\mu[E_m(u) - V]\}^{1/2} dv\ . \tag{23}$$

$v_<$ and $v_>$ are the turning points $V(u,{}^{v_>}_{v_<})=E_m(u)$ and they are functions of the action m and the 'translational' coordinate u. The vibrationally adiabatic potential energy surface $E_m(u)$ is defined by the demand that m(u) be independent of u. By definition therefore

$$\frac{dm}{du} = \frac{d^2m}{du^2} = \dots = 0 \tag{24}$$

Equation (24) enables us to study the properties of the function $E_m(u)$. If at u_o we find a pods whose energy $g(u_o)$ is equal to $E_m(u_o)$ then it is easy to see (by differentiation of Eq. (23) and insertion of Eq. (22)) that

$$\left.\frac{dE_m}{du}\right|_{u=0} = 0 \; ,$$

in other words a periodic orbit of action m is an extremum of a vibrationally adiabatic potential energy surface. Using similar reasoning, it is possible to obtain an explicit expression for d^2E_m/du^2 at u_o. One finds after some manipulation that[32]

$$\left.\frac{d^2E_m}{du^2}\right|_{u=u_o} = \frac{dg}{du}\,\frac{1}{T}\oint \frac{1}{\mu}\frac{\partial\mu}{\partial u}\,dt\,\bigg|_{u=u_o} \tag{25}$$

where T is the period of the periodic orbit at u_o.

Thus far we have dealt purely with definitions. We have shown that a periodic orbit is an extremum of a classical adiabatic potential energy surface and that given the periodic orbit and its immediate vicinity of energy dependent periodic orbits one can compute the coefficients of a Taylor expansion of the surface around the extremum. One must now turn to dynamics. In fact, it is not too difficult to show that if translational motion is much slower than vibrational then the Hamiltonian governing the translational motion may be written as

$$H = p_u^2/(2M(u)) + E_m(u) \tag{26}$$

where M(u) is an effective mass defined as

$$M^{-1}(u) = \frac{1}{T}\int_0^T dt\,\frac{\lambda}{\mu} \tag{27}$$

Consider now a periodic orbit at u_o at energy $E_m(u_o)$. By definition, $p_u=0$ so that at u_o Eq. (26) holds exactly. We have thus proved that classically, not only does a periodic orbit correspond to an adiabatic barrier or well but also that at the barrier or well the adiabatic approximation is exact. It is also easy to see that if the adiabatic approximation is exact at the extremum of $E_m(u)$ then at the extremum u_o one must find a periodic orbit of energy $E_m(u_o)$. From Eq. (26) we find that $p_u=0$, and since $dE_m/du=0$ at u_o necessarily $\dot{p}_u=0$. We have thus proved that a necessary and sufficient condition for the classical adiabatic approximation to be exact at an adiabatic barrier or well is the existence of a periodic orbit at the barrier or the well. Periodic orbits though are defined uniquely by the potential energy surface and the masses. Classically, we have removed the ambiguity in the definition of the extrema of $E_m(u)$.

Given the adiabatic Hamiltonian (Eq. (26)) in the vicinity of the pods, one may define an adiabatic stability frequency ω_m for motion perpendicular to the orbit as

$$\omega_m^2 = \frac{d^2E_m/du_o^2}{M(u_o)} = \frac{1}{T}\oint \frac{\lambda}{\mu}\,dt \cdot \frac{1}{T}\frac{dg}{du_o} \oint \frac{1}{\mu}\frac{\partial\mu}{\partial u_o}\,dt \quad . \tag{28}$$

Within the adiabatic approximation if $\omega_m^2>0$ then the periodic orbit is an adiabatic well, if $\omega_m^2<0$ then the orbit is an adiabatic barrier.

There is a close connection between the adiabatic frequency and the stability frequency (cf. Eq. 19) of a periodic orbit. Since at the pods the time average of H_{uu} is exactly d^2E_m/du^2 one finds that the adiabatic stability frequency is just the first order Magnus approximation to the exact stability frequency.[32] From a practical point of view it is actually easier to compute the stability frequency. Finding the adiabatic frequency implies actual construction of the (u,v) coordinate system. The stability frequency may be computed by integrating along the pods for one period, but using the Cartesian set of coordinates. Note that the stability frequency is of course independent of the coordinate system used.

From the properties of the Magnus approximation, one can show[32] that if $\beta T<<1$ than the adiabatic approximation is excellent. Thus, given a pods, one may easily compute its period and stability frequency and so assess the validity of the adiabatic approximation in the vicinity of the pods.

b. Semiclassical Adiabatic TST for Collinear Systems

The vibrationally adiabatic approximation is coordinate dependent. However one may formulate the quantal adiabatic approximation using the coordinate system defined by pods. One can then show that up to terms of order $\hbar^2$, that if at u_o there exists a pods with action $(n+1/2)h$, and energy $E_n(u_o)$ then also quantally u_o is an adiabatic barrier or well of the n-th vibrationally adiabatic potential energy surface.[33] Furthermore, the quantal adiabatic frequency for motion perpendicular to the pods is excellently approximated by the adiabatic frequency of the pods. Finally, one can show, that to order $\hbar^2$, all quantal nonadiabatic coupling elements are identically zero at u_o.[33] In other words, one should expect that just as in the classical case, the coordinate system defined by the pods is also quantally, the optimal coordinate system for the vibrationally adiabatic approximation. One should also expect that the semiclassical barrier and well energies and frequencies computed via the pods are an excellent approximation to the quantal energies and frequencies.

Given the n-th adiabatic barrier height $E_n^{\neq}$ one can easily formulate an adiabatic transition state theory for the reaction probability, from the n-th reagents vibrational state $P_n^{1D}(E_T)$. The simplest estimate is unit transmission probability for translational energies (E_T) greater than the barrier height and zero otherwise:

$$P_n^{1D}(E_T) = \theta(E_T - E_n^{\neq}) \tag{29}$$

Here $\theta(x)$ is the unit step function, ΔE_n the n-th adiabatic barrier height relative to the n-th asymptotic state of reactants. When the adiabatic barrier is evaluated via semiclassical quantization of periodic orbits we denote the result as SemiClassical ADiabatic (SCAD) theory. The collinear rate constant $k_n^{1D}(T)$ at temperature T is given in terms of the reaction probability as

$$k_n^{1D}(T) = (2\pi\mu_i kT)^{-1/2} \int_0^{\infty} dE_T P_n^{1D}(E_T) \exp[-E_T/kT] \tag{30}$$

where k is Boltzmann's constant and μ_i the reduced mass of reactants. Insertion of Eq. (29) into Eq. (30) gives

$$k_n^{1D}(T) = (kT/2\pi\mu)^{1/2} \exp(-\Delta E_n/kT) \tag{31}$$

The next simple estimate is to make use of the imaginary adiabatic or stability frequency $\tilde{\omega}_n$ of the n-th adiabatic barrier to estimate the tunneling correction. Thus one replaces Eq. (29) with the well known expression:

$$P_n^{1D}(E_T) = \{1+\exp[2\pi(\Delta E_n - E_T)/|\hbar\tilde{\omega}_n|]\}^{-1} \tag{32}$$

To obtain the thermal rate one inserts the tunneling probability into Eq. (31).

Remarkably, the SCAD estimate is very good. In Fig. 4 we analyse the exact quantal reaction probabilities and rate constants of the collinear $D+H_2 \rightarrow DH+H$ reaction on the LSTH[34] potential energy surface. The dots on the right hand panel are the exact quantal reaction probabilities. The solid line is the SCAD estimate of the reaction probability based on Eq. (32). The dashed line is based on the SCAD approximation without tunneling (Eq. 29). For the ground state reaction we find that the SCAD estimate with tunneling gives a reasonably accurate description of the exact quantal reaction probability in the threshold region. In fact the SCAD rate is in almost perfect agreement with the quantal rate (cf Table I). However, upon close scrutiny one notes that the SCAD

Table I. Rate Constants for the Collinear $D+H_2$ Reaction[a]

	$D+H_2$ (v=0)		$D+H_2$ (v=1)		
T	$k_{ex}(T)$	$k_{SCAD}(T)$[b]	$k_{ex}(T)$	$k_{SCAD}(T)$	$k_{ex}(T)+k_{ex}^{NR}(T)$[c]
200	9.18(-2)	2.46(-1)	1.64(2)	2.52(2)	1.91(2)
250	1.07	1.32	4.02(2)	5.96(2)	4.69(2)
300	6.19	6.17	8.02(2)	1.19(3)	9.38(2)
400	6.32(1)	5.87(1)	2.15(3)	3.24(3)	2.53(3)
600	7.46(2)	7.25(2)	6.71(3)	1.03(4)	8.05(3)
1000	6.17(3)	6.57(3)	1.89(4)	2.92(4)	2.33(4)
1500	1.90(4)	2.19(4)	3.24(4)	5.07(4)	4.10(4)
2400	4.59(4)	5.95(4)	4.72(4)	7.43(4)	6.09(4)

a. Exact quantal reaction rates are based on collinear quantal computations of Ref. 35. Rate constants are in units of cm/(sec molec.).
b. Rate constants based on Eqs. (31,32) with parameters provided in Table 2.
c. k_{ex}^{NR} is the rate for the inelastic 1→0 transition.

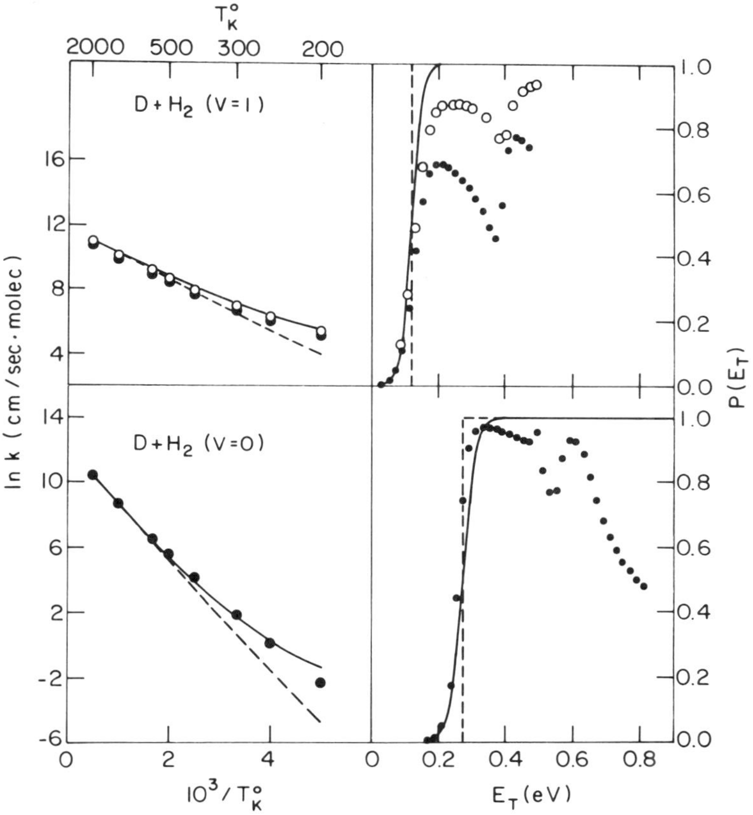

Figure 4. Probabilities and rate constants for the collinear D+H_2 reaction on the LSTH potential energy surface.

estimate is for some temperatures, somewhat lower than the exact rate. This result is worrisome. TST should always overestimate the exact rate! Although the adiabatic TST isn't a rigorous upper bound, one might expect, that barring quantal resonances - which don't exist in the $D+H_2$ system - the SCAD theory would give a practical upper bound. After all, it is well known that the harmonic tunneling correction will, if anything overestimate the exact tunneling probability.[26] In fact at 200K, SCAD does overestimate the rate. Note though, that actually, the SCAD threshold is ∿0.01eV greater than the exact quantal threshold. This is probably a result of our using the WKB approximation in estimating the barrier height. As shown in Ref. 33 the exact quantal barrier energy will be slightly different from the WKB estimate as a result of curvature corrections. If in fact one lowers the SCAD estimate of the barrier height by 0.01eV one obtains perfect agreement with the exact quantal result.

The situation is somewhat different for $D+H_2(v=1)$. Here, SCAD with tunneling overestimates the exact quantal rates by 50% over the whole temperature range shown in the figure (cf. Table I). Here the adiabatic barrier is far out in the reactants region of the potential energy surface so that it is almost a straight line. However, the transmission probability through the barrier does not account for the possibility of being back reflected from the interaction region. The exact quantal reaction probability reaches only a maximum of 0.7 while for v=0 the maximal probability is 0.97. Note that the SCAD theory overestimates the rate by ∿1/0.7. This point is easily checked out by adding to the exact reaction probability the $v=1 \rightarrow v=0$ inelastic transition probability. This gives the open circles shown in the figure. The SCAD rates overestimate the quantal total inelastic rates by ∿15%.

To show that the good agreement found for $D+H_2$ is not a fluke we show in Fig.5 a comparison of the SCAD estimate with exact quantal results[37] for the $Mu+D_2$ and $Mu+H_2$ thermal rate constants. The SCAD parameters are given in Table II. For $Mu+D_2$ we find excellent agreement over a large temperature range. The overestimate at very low temperatures may be attributed to the harmonic tunneling correction. Garrett

Table II. Parameters for Collinear TST Computations on the LSTH Potential Energy Surface

system	n	E_n^a	$E_n^{\neq}$	$\tilde{\omega}_n^b$
$D+H_2$	0	.2703	.5403	927
$D+H_2$	1	.7864	.9055	783
$Mu+H_2$[c]	0	.2703	.7853	1137
$Mu+D_2$[d]	0	.1946	.7783	801

a. All energies in eV relative to the asymptotic diatom well.
b. Stability frequencies in cm^{-1}.
c. The threshold of this reaction is at E=.5945eV.
d. The threshold of this reaction is at E=.5796eV.

and Truhlar[38] using a more sophisticated tunneling approximation get slightly better results. The overestimate at 2400°K reflects the fall-off of the quantal reaction probability at high energies. For $Mu+H_2$

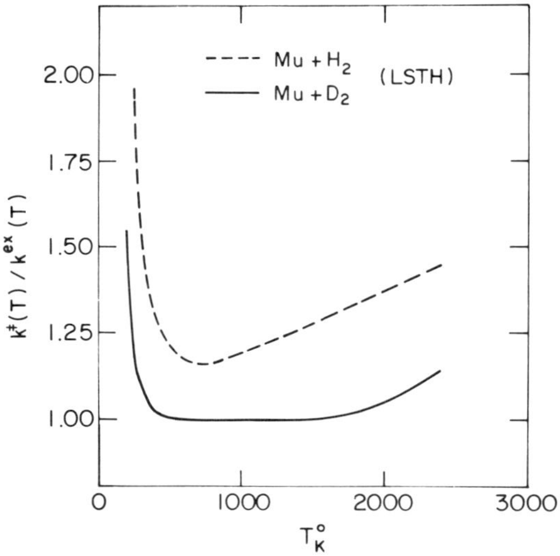

Figure 5. Ratio of SCAD rate constants to exact quantal rates for the collinear Mu+H_2 and Mu+D_2 reactions. The exact quantal rates are given in Table II of Ref. 37.

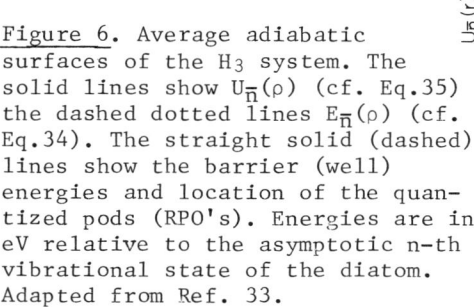

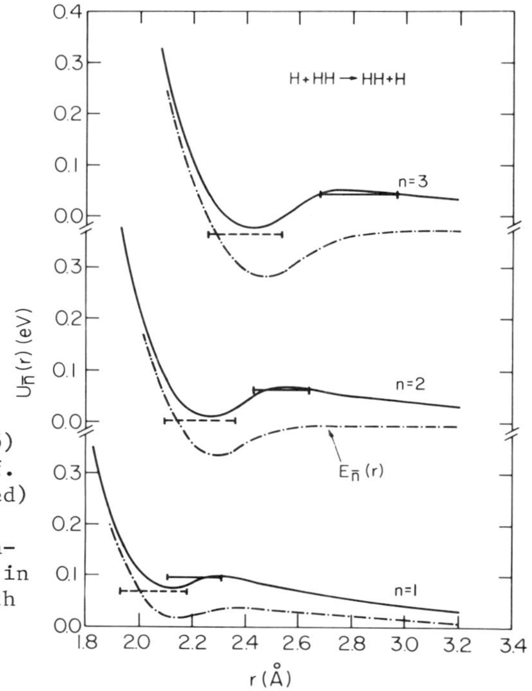

Figure 6. Average adiabatic surfaces of the H_3 system. The solid lines show $U_{\bar{n}}(\rho)$ (cf. Eq.35) the dashed dotted lines $E_{\bar{n}}(\rho)$ (cf. Eq.34). The straight solid (dashed) lines show the barrier (well) energies and location of the quantized pods (RPO's). Energies are in eV relative to the asymptotic n-th vibrational state of the diatom. Adapted from Ref. 33.

the situation is similar but with one difference. For Mu+D_2 the cumulative reaction probability rises to 1.0 in the threshold region. For Mu+H_2 it only reaches 0.85 and even then falls off rapidly. Thus for Mu+H_2 the SCAD theory is at best 15% greater than the exact quantal results.

c. Periodic Orbits and Collinear Quantal Resonances

One of the intriguing facts uncovered by numerical computations for collinear reactive scattering of an atom on a diatom was the existence of sharp spikes in the reaction probability as a function of energy.[39] It was immediately realised that these spikes may be called resonances since they are a result of a long lived intermediate. Phase shift analysis has confirmed this conclusion.

A number of empirical facts have been established about the properties of the resonances. They usually show up at the threshold of vibrational channels. Their width can vary by orders of magnitude. Their existence, location and width can be a sensitive function of the potential energy surface.

Although the resonances show up on potential energy surfaces exhibiting a single saddle point separating reactants and products, the mechanism for the formation of a long lived intermediate is today quite well understood. Usually vibrationally adiabatic potential energy surfaces of excited vibrational states have wells in the interaction region. These wells may support quasibound states. Of course, the quantitative determination of the energy and width of a resonance using such an adiabatic model is in principle a tricky problem. The adiabatic surface is coordinate dependent so that different choices of the vibrational and translational coordinate will give different quantitative results. Recently though, Launay and LeDorneuf[41] and Römelt[42] have shown that if one uses the radial Delves coordinate system and solves the diagonal part of the Schrödinger equation including all 'diagonal non-adiabatic' terms, then one finds quantitative agreement with exact quantal computations.

The radial coordinate system is defined by

$$q_1 = \rho\cos\theta \quad q_2 = \rho\sin\theta \tag{33}$$

where (q_1,q_2) are defined in Eq. (9). It is easy to see that $\lambda(\rho,\theta)=\mu(\rho,\theta)/m=\rho^2$. The adiabatic approximation is obtained by considering the angular θ motion as the fast motion. If $\phi_m(\theta;\rho)$ denotes the ρ dependent eigenfunction with energy $E_m(\rho)$ then one can show[42] that the effective m-th potential energy surface including the diagonal non-adiabatic correction is

$$U_m(\rho) = E_m(\rho) - \frac{\hbar^2}{4m\rho^2} + \frac{\hbar^2}{2m}\left\langle \frac{\partial\phi_m}{\partial\rho}\middle|\frac{\partial\phi_m}{\partial\rho}\right\rangle \tag{34}$$

$U_m(\rho)$ may have wells and these wells support bound or quasibound states which correspond to the resonances.

As stressed earlier the classical signature of an adiabatic barrier or well is a periodic orbit. Adiabatic wells, in the Delves coordinate system should correspond to periodic orbits that follow the angular motion at constant ρ values.[33] In fact such 'resonant' periodic orbits (RPO's)[19] exist, they are shown in Fig. 1 for the $H+H_2$ exchange reaction. To find the semiclassical well energies one should quantise the action of the orbits semiclassically. However note that the orbits in Fig. 1 move over a double well potential and their energy is quite close to the potential energy of the barrier separating the two potential wells. In principle, it is necessary to include above barrier reflection terms,[43,44] a simple quantisation of the action won't do.

Consider though a symmetric exchange reaction. Here the angular wave function must be either symmetric (g) or antisymmetric (u) with respect to reflection about the symmetric stretch. For the pair of surfaces $U_n^g(\rho)$ and $U_n^u(\rho)$ one finds that as $\rho\to\infty$ $U_n^g \simeq U_n^p$. Semiclassically, the doubly degenerate energies at large ρ are given by an $(n+1/2)h$ action condition in each well. Since one has two (symmetric) wells, this condition may be interpreted as $(2n+1)h$ action taken over the double well. If one doesn't include any tunneling or above barrier reflection contributions, classical adiabaticity would imply that the action $(2n+1)h$ stays constant for all ρ. Thus one should quantise the resonant orbit with a $(2n+1)h$ action condition. The energy of the orbit would correspond to the adiabatic well of the averaged adiabatic potential energy surface

$$U_{\bar{n}}(\rho) \equiv \frac{1}{2}\ [U_n^g(\rho) + U_n^u(\rho)] \tag{35}$$

since semiclassically, the splitting of the eigenvalues is symmetric about $U_{\bar{n}}$.

In Fig. 6 we show for the hydrogen exchange reaction a comparison between the well energies and locations as computed from resonant periodic orbits (RPO's) and the exact quantal averaged adiabatic surfaces $E_{\bar{n}}(\rho)$ and $U_{\bar{n}}(\rho)$. Note that there is quantitative agreement with the wells of $U_{\bar{n}}(\rho)$.

Because of the splitting due to the double well potential for fixed ρ, one knows that the real well energies of $U_n^g(\rho)$ will be lower than the well energy of the RPO. On the other hand, the resonance energies are given by adding the zero point energy for motion perpendicular to the well. Since both the splitting and the zero point energies are much smaller than the separation between the adiabatic well energies it is not surprising that the $(2n+1)h$ RPO's give a good zero order estimate of the resonance energies.[19] For example, for $H+H_2$, on the PKII surface, quantal resonances occur at E=0.875eV and 1.32eV, 3h and 5h RPO's are found at E=0.867eV and 1.29eV, respectively. In principle though, using semiclassical methods, it is possible to evaluate the splitting,[44] while the zero point energy could be assessed from the adiabatic frequency of the RPO, however such a computation has yet to be done.

A similar analysis is applicable to asymmetric reactions. RPO's have been computed for the $F+H_2$ reaction and again there is good correspondence between integer action quantized RPO's and exact quantal resonance energies.[19] It should be stressed though that the most important property of the RPO's is that they give a reasonably accurate description of the adiabatic well responsible for the resonance - its energy and location, at a very cheap price. As a result RPO's are very useful for dealing with resonances in three-dimensional systems.[45]

Before ending this section we mention a novel application of periodic orbits towards understanding the structure of resonances in reactive scattering. Exact quantal calculations of reaction probabilities are available for example, for the $H+H_2$ exchange reaction on two different potential energy surfaces - the PK(II) and the LSTH surfaces.[37] One finds, that qualitatively the quantal reaction probabilities are identical on the two surfaces. It came therefore as quite a surprise to find that for the $Mu+D_2$ reaction, the LSTH surface exhibits many narrow resonances which simply don't exist on the PKII surface. On the other hand, for $Mu+H_2$ the reaction probabilities on the two surfaces are again qualitatively similar - no resonances are found for either surface.

There is though, one qualitative difference between the more accurate LSTH surface and the PK(II) surface. The former includes the van der Waals wells in the far entrance and exit valleys. Using pods, we've computed the adiabatic well depths for different H_2 vibrational states. In addition we've computed the stability frequencies of the quantised pods. These frequencies are now of course real - the pods are truly stable. The results are summarised for Mu+DD in Table III. Within

Table III. Parameters[a] of van der Waals Wells in the $Mu+D_2$ Reaction on the LSTH Potential Energy Surface

n	E_n [b]	E_n^{vdW} [c]	$\hbar\omega_n$ [d]
0	.1946	.1904	.0218
1	.5717	.5627	.0228
2	.9329	.9202	.0229
3	1.2782	1.2631	.0235

a. All energies are in eV.
b. Energies relative to the bottom of the asymptotic DD well.
c. van der Waals vibrationally adiabatic well energies obtained from quantised pods in the entrance channel of collinear Mu+DD.
d. Real stability frequencies (in eV) of the quantised pods.

a harmonic approximation, the adiabatic bound state energies are just the adiabatic well energies plus $(m+1/2)\hbar\omega_{st}$ where ω_{st} is the stability frequency.

For Mu+DD we note that the n=2 and n=3 adiabatic surfaces support exactly one bound state each at E=0.9285 and 1.2705eV respectively. These energies should be compared with the exact results 0.9233 and 1.2646eV.[37,46] A similar analysis may be carried out for the $Mu+H_2$ reaction,[47] here the larger stability frequencies raise the resonance energy above the threshold energy so the resonance is, if at all, a virtual resonance, very broad and seemingly unobservable.

To summarise, we have shown that periodic orbits are an extremely useful analytic tool. Many aspects of exact quantal computations were explained or even predicted, based on periodic orbit analysis. Here we should mention that the stability frequency of RPO's has been used to analyse resonance widths.[48] Moreover, a study of stability of RPO's in light atom transfer systems has led to the discovery of a new type of chemical bond - the vibrational bonding of molecules.[49]

IV. PERIODIC ORBITS AND 3D QUANTAL SCATTERING

a. A Reduction Method

In principle, one may generate adiabatic potential energy surfaces for n degree of freedom systems. For each value of a translational coordinate u one must diagonalise (quantally or semiclassically) an n-1 degree of freedom internal Hamiltonian $h(\underline{v};u)$. However, such a prescription is quite expensive numerically. For example Wyatt and coworkers[50] do not diagonalise h exactly - they fit the potential to convenient numerical forms. Although, in a fully coupled reactive scattering computation, the fit is in principle not important, it may change reaction rates considerably in a transition state theory type computation. In 3D, apart from the question what is the unique coordinate system to be used, there is also the problem of given the coordinates how does one compute the adiabatic potential energy surfaces efficiently.

Garrett and Truhlar[51] circumvent the latter problem by making separable Taylor expansions around the bottom of the well of the internal Hamiltonian. For transition state theory, at low temperatures, this is a reasonable prescription, however the coordinate system remains arbitrary. Miller and coworkers[52] have derived exact expressions for the Hamiltonian of the system as a function of a reaction path. Their theory incorporates non-adiabatic and rotational coupling between the internal degrees of freedom, however, it too is based on harmonic expansions and an arbitrary reaction coordinate.

To circumvent these problems and still take advantage of the simplicity of the periodic orbit approach, we have taken a different approach which makes full use of the spirit of the adiabatic approximation. The most general motion of a triatomic molecule at an adiabatic barrier or well is a combination of stretching, bending and overall rotation. Invariably, overall rotation is much slower than bending and stretching motion. Thus, one may first solve for these fast motions, average the Hamiltonian over the fast motion and obtain an effective adiabatic Hamiltonian for the overall rotation. We are thus left with a two degree of freedom Hamiltonian - one stretch and one bend motion. We also know from spectroscopy that usually stretch frequencies are greater than bend frequencies. Thus we may freeze the bend, solve for the fast stretch and obtain an adiabatic Hamiltonian for the bend motion. Note though that by now we have reduced the total Hamiltonian to that of a two degrees of freedom system-translation and vibration. For a fixed value of the bend angle (and the rotational angles) we find an angle dependent periodic orbit and quantise its action. This gives us a semiclassical angle dependent adiabatic barrier height or well depth. We average the Hamiltonian over the period of the orbit and so obtain an adiabatic bend Hamiltonian. We then quantise the bend motion semiclassically and obtain an adiabatic Hamiltonian for the overall rotation.

In principle there are two arbitrary ingredients in this prescription. One must define rotational angles and a bend angle. The rotational angles imply a transformation from a space fixed coordinate system to a body fixed coordinate system. There are an infinity of such transformations and one must choose one arbitrarily. Similarly the bend angle is defined arbitrarily. Of course, if *a posteriori* we find that indeed there is a good separation of time scales then this arbitrariness is irrelevant from a practical point of view.

In Ref. 14 we have derived in detail a useful transformation from the space fixed to the body fixed coordinate system. Without going into detail, we provide here, the final result. The full classical Hamiltonian in 3D, in our body fixed coordinate system is

$$H = H_{2D} + T_{tumble} , \tag{36}$$

$$H_{2D} = \frac{1}{2m}[p_r^2+p_R^2] + \frac{1}{2m}(\frac{1}{r^2}+\frac{1}{R^2})p_\gamma^2 + \frac{1}{2m(r^2+R^2)}[1+\frac{4\gamma^2r^2R^2}{(r^2+R^2)^2}]J_x^2 + \frac{2\gamma rR}{m(r^2+R^2)^2}(Rp_r-rp_R)J_x + V(R,r,\gamma) . \tag{37}$$

$$T_{tumble} = \frac{1}{2}(\frac{I_{zz}}{I_{yy}I_{zz}-I_{yz}^2})J_y^2 + \frac{1}{2}(\frac{I_{yy}}{I_{yy}I_{zz}-I_{yz}^2})J_z^2 - (\frac{I_{yz}}{I_{yy}I_{zz}-I_{yz}^2})J_yJ_z \tag{38}$$

Here, r,R are (for example) mass scaled diatom and atom to center of mass of diatom distances (cf. Eqs. 1,8,9) and γ the (bend) angle between r and R. Actually r,R,γ may represent a much wider definition of coordinates. For example, <u>r</u>,<u>R</u> may be the heliocentric coordinates of Smith.[53] H_{2D} is the coplanar Hamiltonian and T_{tumble} represents the instantaneous tumbling of the body fixed nuclear plane. The components of the inertia tensor are given explicitly in Ref. 14. The x axis is oriented perpendicular to the nuclear plane.

The first step of the adiabatic separation is to separate the 'fast' variables from the slow variables. We assume that the fast motion is vibration, all the other variables are slow. This leads to the 'pseudo-collinear' translation-vibration Hamiltonian

$$h(\gamma) = \frac{1}{2m}(p_R^2 + p_r^2) + V(R,r;\gamma) \quad , \tag{39}$$

in which γ is taken to be a parameter. So far, this reduction is equivalent to the first step of the usual IOS scheme.[54,55]

As we have already seen in the previous section, finding adiabatic barriers and wells of the n-th quantal vibrational adiabatic potential surface for the γ dependent Hamiltonian $h(\gamma)$ is equivalent semiclassically to finding periodic orbits of $h(\gamma)$ with quantised action - $(n+1/2)h$ if the periodic orbit is over a simple well potential. The time dependent coordinates and momenta of the (γ dependent) periodic orbit are denoted $r_n(t;\gamma)$, $R_n(t;\gamma)$, $p_r(t;\gamma)$, and $p_R(t;\gamma)$, and the period of the orbit is $T_n(\gamma)$. We thus find for each value of γ a vibrationally adiabatic barrier or well at energy $E_n(\gamma)$, a stability frequency $\tilde{\omega}_n(\gamma)$ and effective mass $M_n(u_o)$ (cf. Eq. 27) for motion perpendicular to the periodic orbit. $h_n(\gamma)$ may then be represented as

$$h_n(\gamma) = \frac{p_u^2}{2M_n(u_o,\gamma)} + \frac{1}{2} M_n(u_o\gamma)\tilde{\omega}_n^2 (\gamma)(u-u_o)^2 + E_n(\gamma) \quad (40)$$

At this point one may make an adiabatic or a sudden assumption. If the translational motion is faster than the bend motion, one must first solve for the u motion in Eq. (40) and only then return to the full Hamiltonian. This sudden approximation will be derived later. If translational motion is slow relative to the bend then one inserts $h(\gamma)$ back in the full Hamiltonian and averages the full Hamiltonian over the period of the angle dependent periodic orbit. The adiabatic Hamiltonian that emerges from this process is

$$H_n = h_n(\gamma) + B_n(\gamma)p_\gamma^2 + B_x^n(\gamma)J_x^2 + T_n^{tumble}(\gamma) \quad (41)$$

Here the coefficients $B_n(\gamma)$ and $B_x^n(\gamma)$ are time averages over the angle dependent periodic orbit of the respective coefficients in Eq. (36), and likewise $T_n^{tumble}(\gamma)$ is obtained by averaging Eq. (38).

Since the translational and rotational motions are assumed slow compared to the bending motion they may now be treated perturbatively. The bending Hamiltonian for the n-th vibrational barrier or well is

$$H_B^n = E_n(\gamma) + B_n(\gamma)p_\gamma^2 \quad (42)$$

The energy levels associated with the γ motion may be approximated semi-classically. Depending upon $E_n(\gamma)$, this motion will be either librational (between γ_{min} and γ_{max}) or rotational (hindered, but "complete"). Usually though one will find only libration at the adiabatic barriers or wells, so we will assume this situation in the following formulation. Semiclassically, the k-th bending eigenvalue at an adiabatic barrier or well is found by demanding as usual that $\oint p_\gamma d\gamma=(k+1/2)h$. One now averages the full Hamiltonian over a period of the bend motion of the k-th bend state. If the potential energy surface is minimal for a collinear configuration then the bend vibration adiabatic Hamiltonian reduces to

$$H_{n,k}(u,J) = E_{nm} + \frac{p_u^2}{M_{nk}} + \frac{1}{2} M_{nk}\tilde{\omega}_{nk}^2(u-u_o^2) + B_x^{nk}J^2 \quad (43)$$

Here the frequency $\tilde{\omega}_{nk}$ is obtained by averaging $\tilde{\omega}_n(\gamma)$ over the γ motion. A similar averaging gives M_{nk} and B_x^{nk}. Since J is conserved one now solves for the translational u motion for each J.

The periodic reduction method has been used to generalise the RPO picture of resonances to 3D collisions.[45] The zero point bend energy of RPO's is in good agreement with exact (J=0) 3D quantal computations of the shift of the resonance energy as one goes from the collinear world to 3D. The total angular momentum dependence of the resonance energy is in good agreement with all available approximate 3D computations. The periodic reduction method as applied to RPO's has predicted the existence of a bend level substructure to 3D excited resonance states.[45] This prediction has yet to be confirmed. The method has also been used to predict the 3D structure of vibrationally bonded molecules.[49,56] In the next subsection we will show in detail how the method is applied to TST.

The periodic reduction method may be generalised. Instead of arbitrarily defining a bend angle one may start with the J=0 coplanar Hamiltonian which has three degrees of freedom. Suppose we denote an arbitrary orthogonal curvilinear coordinate system as (u,v,w) where u denotes a translational coordinate and v,w the 'internal' coordinates. For each value of u one can quantise the 'internal' Hamiltonian obtained by setting $p_u=0$. The internal Hamiltonian has two degrees of freedom and so its quantised states will carry two indices. Thus $E_{nm}(u)$ will be the n-th stretch, m-th bend, adiabatic potential energy surface. Semiclassically one must use standard semiclassical quantisation techniques for two degree of freedom systems. The following argument should by now be familiar. If $E_{nm}(u)$ has an extremum at u_o and the adiabatic approximation is exact, then at u_o we must find a bound orbit of energy $E_{nm}(u_o)$. This orbit will usually be quasiperiodic. Conversely, if one finds a quasiperiodic orbit, it will be an adiabatic barrier or well. The proof may be derived by complete analogy to the collinear case.

Given the quasiperiodic orbit embedded in the continuum one can proceed to identify in configuration space the coordinates (u,v,w). In principle then, using classical mechanics one is able to identify also in a coplanar reaction a 'best' set of adiabatic coordinates. It may also happen that the bend and stretch frequencies are similar in magnitude. In such a case the periodic reduction method cannot work. However usually the rotational motion will still be much slower than the stretches and bends and so could still be treated adiabatically. Such a scheme may be termed a quasiperiodic reduction method.[57] It has been applied successfully for adiabatic barriers[57] and wells.[58] However there is a price to pay, it is more tedious.

b. Adiabatic and Sudden TST

Given the bend state or angle dependent Hamiltonian, one may easily generalise the collinear TST, with harmonic tunneling correction to 3D systems. The rotationally averaged cross section from initial vibrational state n summed over all final vibrational (n') and rotational (j') states is defined as[59]

$$\bar{\sigma}_n(E) = \frac{\pi\hbar^2}{2\mu\langle E_t\rangle_n\langle 2j+1)\rangle_n}\,\bar{P}_n(E) \tag{44}$$

We let E_{nj} denote the vibrotational energy of the n,j reactant state,E the total energy so that

$$\langle E_t\rangle_n\langle 2j+1\rangle_n = \sum_{j=0}^{\infty}(2j+1)(E-E_{nj})\theta(E-E_{nj}) \tag{45}$$

$$\bar{P}_n(E) = \sum_{j=0}^{\infty}\sum_{J=0}^{\infty}(2J+1)\sum_{\ell=|J-j|}^{J+j}P^J_{nj\ell}(E-E_{nj}) \tag{46}$$

Here $P^J_{nj\ell}(E-E_{nj})$ is the reaction propability out of reagents state n,j with orbital angular momentum quantum number ℓ and total angular momentum J. $\bar{P}_n(E)$ is therefore the cumulative reaction probability out of reagents vibrational state n at energy E.

Adiabatic transition state theory is obtained by first quantising semiclassically the bend Hamiltonian Eq. (42). In the following we'll limit ourselves to a collinearly dominated potential energy surface. Generalisation to other cases is straightforward. With this condition, the bend motion is doubly degenerate. For the level with λ_1 quanta in one mode and λ_2 in the other one may construct the bending levels $E_n^{\lambda_1,\lambda_2}$ using the $(\lambda_i+\frac{1}{2})h$ quantization condition. For any level λ_1,λ_2 one defines averaged frequencies and rotational constants as

$$\tilde{\omega}_n^{\lambda_1\lambda_2} = \frac{1}{2}(\tilde{\omega}_{n\lambda_1}+\tilde{\omega}_{n\lambda_2}), \quad B_x^{n\lambda_1\lambda_2} = \frac{1}{2}(B_x^{n\lambda_1}+B_x^{n\lambda_2}) \tag{47}$$

The constraints imposed by conservation of total angular momentum make it useful though to employ the notation (λ,Λ) instead,[60] where λ is the principal bend quantum number $(\lambda=\lambda_1+\lambda_2)$ and Λ is an internal rotational angular momentum $(\Lambda=\lambda_1-\lambda_2)$ which takes on the $\lambda+1$ values $-\lambda,-\lambda+2,\ldots\lambda-2,\lambda$. Angular momentum conservation implies that $J>|\Lambda|$. For any bend state characterised now by $E_n^{\lambda,\Lambda}$, $\tilde{\omega}_n^{\lambda,\Lambda}$, $B_x^{n,\lambda,\Lambda}$ one defines a SCAD reaction probability as

$$P_n^{J\lambda\Lambda}(E) = \{1+\exp[\frac{2\pi}{\hbar\tilde{\omega}_n^{\lambda,\Lambda}}(E_n^{\lambda,\Lambda} + B_x^{n,\lambda,\Lambda}J(J+1) - E)]\}^{-1} \tag{48}$$

The SCAD cumulative reaction probability is now

$$\overline{P}_n^{SCAD}(E) = \sum_{\lambda=0}^{\infty}\sum_{\Lambda=-\lambda}^{\lambda}\sum_{J=|\Lambda|}^{\infty}(2J+1)P_n^{J\lambda\Lambda}(E) \tag{49}$$

A sudden rotationally averaged cross section may be derived analogously. In the sudden limit all reagent rotational states collapse to the ground rotational state and the summation over j states is replaced by integration over an angle. Thus

$$\overline{\sigma}_n^{ADRS}(E) = \frac{\pi\hbar^2}{2\mu E_T}\frac{1}{2}\int_0^{\pi}\overline{P}_n(E,\gamma)\sin\gamma d\gamma \tag{50}$$

where E_T is the translational energy $(E_T=E-E_{no})$, and ADRS stands for ADiabatic Reactive Sudden. The angle dependent reaction probability is further decomposed to

$$\overline{P}_n(E,\gamma) = \sum_{J=0}^{\infty}(2J+1)P_n^J(E,\gamma) \tag{51}$$

and $P_n^J(E,\gamma)$, is as usual evaluated using a harmonic tunneling correction

$$P_n^J(E,\gamma) = \{1+\exp[\frac{2\pi}{\hbar\tilde{\omega}_n(\gamma)}(E_n(\gamma) + B_x^n(\gamma)J(J+1)-E)]\}^{-1} \tag{52}$$

The thermal reaction rate constant from the n-th reagents vibrational state is easily computed from $\overline{\sigma}_n(E)$ via the well known relationship

$$\overline{k}_n(T)=N\left(\frac{2}{k_BT}\right)^{3/2}(\pi\mu)^{-1/2}\frac{\int_0^{\infty}dE\, e^{-E/k_BT}\langle E_T\rangle_n\langle 2j+1\rangle_n\overline{\sigma}_n(E)}{\sum_{j=0}^{\infty}(2j+1)\exp[-(E_{nj}-E_{no})/k_BT]} \tag{53}$$

Here N is Avogadro's number and k_B is Boltzmann's constant.

Notice that there really isn't too great a difference between the adiabatic and sudden TST. The only noticeable difference is in the reaction threshold. For SCAD the threshold is larger because of the zero point bend energy. One thus expects that in general, adiabatic and sudden rotationally averaged cross sections will be of similar magnitude, only shifted by the zero point bend energy relative to each other. In Fig. 7[61] we show a comparison between the SCAD theory, CS results of Schatz[62] and (ℓ_{av}) sudden results of Kouri et al.[63] (shifted by the bend frequency of 0.09eV) for the $H+H_2(n=1)$ reaction on the PKII potential energy surface. We have also shown[61] that the sudden results of Kouri et al.[63] are well approximated by the ADRS theory. Figure 7 thus shows that indeed, the sudden and adiabatic theories are almost identical, apart from the zero point bend motion.[64]

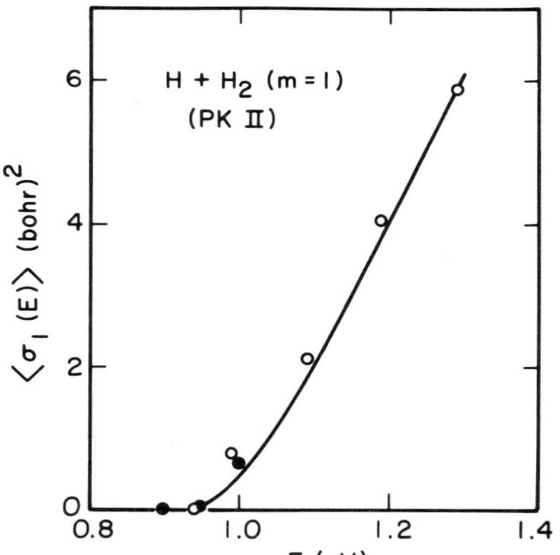

Figure 7. Rotationally averaged cross sections for the $H+H_2(n=1)$ exchange reaction. For details see text. Adapted from Ref. 61.

The figure also shows that the CS results of Schatz, which do not assume an adiabatic or a sudden approach, are compatible with the SCAD theory. In fact, the quantal cross section seems to be primarily dominated by the probability to cross the vibrational bend adiabatic barrier. This is an encouraging result since it indicates that TST, when properly applied, may be just as accurate in 3D as in 1D.

c. 3D Stability Analysis

One of the theoretical shortcomings of the periodic reduction method is the rather arbitrary definition of the bend angle. Although, as was the case to date, one may *a posteriori* verify that the adiabatic reduction is justified by comparing bend frequencies with stretch frequencies, it is still desirable to construct a method which *a priori* does not have this ambiguity. One possibility is using the quasiperiodic reduction method outlined in this section. However, as we shall point out a much simpler method may be based on stability analysis of periodic orbits in 3D.

As noted earlier, the stability frequencies of periodic orbits are invariant under canonical transformation.[65] Consider first the J=0 coplanar Hamiltonian

$$H_{2D}(J=0) = \frac{1}{2m}(p_r^2 + p_R^2) + \frac{1}{2m}\left(\frac{1}{r^2} + \frac{1}{R^2}\right)p_\gamma^2 + V(r,R,\gamma) \tag{54}$$

The generalisation of the $\underset{\sim}{A}$ matrix defined in Eq. (18) is for any number of degrees of freedom[65]

$$\underset{\sim}{A} = \begin{pmatrix} \underset{\sim}{H}_{pq} & \underset{\sim}{H}_{pp} \\ -\underset{\sim}{H}_{qq} & -\underset{\sim}{H}_{qp} \end{pmatrix} \tag{55}$$

If we deal with a 3 degree of freedom Hamiltonian than $\underset{\sim}{H}_{ij}$ are 3x3 sub-matrices whose elements are the second derivatives of the Hamiltonian with respect to the momenta and coordinates, along the periodic orbit.

If the potential energy surface is collinear in the sense that the collinear plane is everywheres a minimum then

$$\left.\frac{\partial V}{\partial \gamma}\right|_{\gamma=0} = 0 \,.$$

This means that, for a collinear periodic orbit, along the orbit, $p_\gamma = V_{r\gamma} = V_{R\gamma} = 0$. This immediately implies that the matrix stability equation (Eq. 18) decouples, into the two independent equations:

$$\delta\dot{\underset{\sim}{q}} = \begin{pmatrix} \underset{\sim}{0} & \begin{matrix} \frac{1}{m} & 0 \\ 0 & \frac{1}{m} \end{matrix} \\ \begin{matrix} -V_{rr} & -V_{rR} \\ -V_{Rr} & -V_{RR} \end{matrix} & \underset{\sim}{0} \end{pmatrix} \cdot \delta\underset{\sim}{q} \tag{56}$$

and

$$\delta\dot{\underset{\sim}{q}}' = \begin{pmatrix} 0 & \frac{1}{m}\left(\frac{1}{r^2} + \frac{1}{R^2}\right) \\ -V_{\gamma\gamma} & 0 \end{pmatrix} \cdot \delta\underset{\sim}{q}' \tag{57}$$

Equation (56) is the stability equation for the collinear Hamiltonian. The bending motion does not affect the collinear stability frequencies!

From Eq. (57) one may obtain the stability frequency for the bending degree of freedom. This frequency may then be used in a number of ways. Its value gives the harmonic bend potential and bend energy levels. Its magnitude relative to the stretch frequency sheds light on the validity of an adiabatic approximation for the bend motion. Finally, if it agrees with the bend frequency determined via periodic reduction then it validates that method. One may undertake a similar analysis using the full coplanar Hamiltonian (Eq. 37). One finds that also angular momentum won't change the collinear stability frequency nor the bend stability frequency.

V. DISCUSSION

The main purpose of this article is to show that periodic orbits are a very useful tool for investigating and predicting the dynamics of reactive scattering. In this section I would like to point out various problems for which a periodic orbit analysis will probably be quite useful. Foremost, we have only dealt with atom diatom collisions. The techniques outlined here are generalisable to atom-triatom and diatom-diatom collisions.[66,67] Since more accurate potential energy surfaces are becoming available simultaneously with more detailed and accurate experiments such a study should be worthwhile.

In practical applications periodic orbits have been used mainly for direct collisions - where the potential energy surface has a saddle point separating reactants and products. Ion molecule collisions have not been studied extensively via periodic orbits.[68] This is even somewhat surprising in view of the recent controversy over the switching from loose to tight transition states.[69-71] Loose transition states are also periodic orbits. Rynefors and Markovic[72] have found them numerically. Clary[73] has undertaken quantal studies of neutral reactions on attractive potential energy surfaces leading to rates exhibiting negative activation energies. Here too, a periodic orbit analysis could provide insight into the quantal results.

It seems to me that stability analysis of periodic orbits is a powerful tool which has yet to be fully utilised. The discussion in sec. IV.c is really more a research proposal than a review. No one has yet studied the stability of periodic orbits in 3D systems. The study presented in this paper is simplified since only collinear like potential energy surfaces were used. What happens in a system like LiFH whose minimum energy path is certainly not collinear?[74,75]

A different example is the H_3^+ system. Here the ground state of H_3^+ is an equilateral triangle. Recent experiments of Carrington and Kennedy[76] indicate that H_3^+ has a high density of quasibound states embedded in the continuum above the H_3^+ ground electronic state dissociation limit. These states or at least some of them should be representable as stable periodic and quasiperiodic orbits.

Stability analysis may be put to other uses. Any perturbation added to the Hamiltonian may change the stability parameters. For example, a radiation field may change the stability of RPO's causing changes in the resonance lifetime. An open question today is how does a bath affect the tunneling rate. Here, stability analysis in conjunction with the reaction surface Hamiltonian of Carrington and Miller[77] may prove instructive.

Finally, maybe the nicest feature of periodic orbits is the very plastic picture they provide for the reaction dynamics. Hopefully we have convinced the reader that 'one picture is worth a thousand words'.

REFERENCES

1. J.T. Muckerman in *Theoretical Chemistry: Theory of Scattering*. D. Henderson, ed., Academic Press, New York, 1981, 6A, Chap. 1.
2. H. Eyring, *J.Chem.Phys*. 3, 107 (1935).
3. E. Wigner, *Trans.Faraday Soc*. 34, 29 (1938).
4. J. Horiuti, *Bull.Chem.Soc*. Japan 13, 210 (1938).
5. J.C. Keck, *Adv.Chem.Phys*. 13, 85 (1967).
6. G.W. Koeppl, *J.Amer.Chem.Soc*. 96, 6539 (1974).
7. P. Pechukas in *Dynamics of Molecular Collisions*, Part B, W.H. Miller, ed., Plenum Press, New York, 1976, Chap. 6.
8. N. Minorsky, *Nonlinear Oscillations*, van Nostrand, Princeton, 1962, Chap. 5.
9. E. Pollak, M.S. Child and P. Pechukas, *J.Chem.Phys*. 72, 1669 (1980).
10. J.C. Light and R.B. Walker, *J.Chem.Phys*. 65, 4272 (1976).
11. G. Hauke, J. Manz and J. Römelt, *J.Chem.Phys*. 73, 5040 (1980).
12. A. Kuppermann in *Potential Energy Surfaces and Dynamics Calculations*, D.G. Truhlar, ed., Plenum Press, New York, 1981, Chap. 16.
13. E. Pollak, *J.Chem.Phys*. 75, 4435 (1981).
14. E. Pollak and R.E. Wyatt, *J.Chem.Phys*. 78, 4464 (1983).
15. E. Pollak and R.E. Wyatt, *J.Chem.Phys*. 81, 1801 (1984).
16. M.S. Child and E. Pollak, *J.Chem.Phys*. 73, 4365 (1980).
17. E. Pollak and P. Pechukas, *J.Chem.Phys*. 69, 1218 (1978).
18. R.N. Porter and M. Karplus, *J.Chem.Phys*. 40, 1105 (1964).
19. E. Pollak and M.S. Child, *Chem.Phys*. 60, 23 (1981).
20. P. Pechukas and E. Pollak, *J.Chem.Phys*. 71, 2062 (1979).
21. D.I. Sverdlik and G.W. Koeppl, *Chem.Phys.Lett*. 59, 449 (1978).
22. J. Costley and P. Pechukas, *J.Chem.Phys*. 77, 4957 (1982).
23. E. Pollak in *Theory of Chemical Reaction Dynamics*, M. Baer, ed., CRC Press, Boca Raton, FL. in press.
24. E. Pollak, *J.Chem.Phys*. 78, 1228 (1983).
25. E. Pollak and M.S. Child, *J.Chem.Phys*. 73, 4373 (1980).
26. E. Pollak and P. Pechukas, *J.Chem.Phys*. 79, 2814 (1973).
27. A. Kuppermann, *J.Phys.Chem*. 83, 171 (1979).
28. R.A. Marcus, *J.Chem.Phys*. 45, 4493, 4500 (1966).
29. D.G. Truhlar, *J.Chem.Phys*. 53, 2041 (1970).
30. B.C. Garrett and D.G. Truhlar, *J.Phys.Chem*. 83, 1079 (1979).
31. S.L. Latham, J.F. McNutt, R.E. Wyatt and M.J. Redmon, *J.Chem.Phys*. 69, 3746 (1978).
32. E. Pollak, *Chem.Phys*. 61, 305 (1981).
33. E. Pollak and J. Römelt, *J.Chem.Phys*. 80, 3613 (1984).

34. P. Siegbahn and B. Liu, *J.Chem.Phys.* 68, 2457 (1978); D.G. Truhlar and C.J. Horowitz, *J.Chem.Phys.* 68, 2466 (1978); 71, 1514(E) (1979).
35. N. Abu Salbi and D.J. Kouri, private communication.
36. H.S. Johnston and D. Rapp, *J.Amer.Chem.Soc.* 83, 1 (1961).
37. D.K. Bondi, D.C. Clary, J.N.L. Connor, B.C. Garrett and D.G. Truhlar, *J.Chem.Phys.* 76, 4986 (1982).
38. B.C. Garrett and D.C. Truhlar, *J.Chem.Phys.* 81, 309 (1984).
39. For a recent review of resonances see ACS Symposium 263 (1984), edited by D.G. Truhlar.
40. R.D. Levine and S.F. Wu, *Chem.Phys.Lett.* 11, 557 (1971).
41. J.M. Launay and M. LeDorneuf, *J.Phys.B* 15, L455 (1982).
42. J. Römelt, *Chem.Phys.* 79, 197 (1983).
43. J.N.L. Connor, *Chem.Phys.Lett.* 4, 419 (1969).
44. V. Aquilanti, S. Cavalli, G. Grossi and A. Lagana, *J.Mol.Struct.* 93, 319 (1983).
45. E. Pollak and R.E. Wyatt, *J.Chem.Phys.* 81, 1801 (1984).
46. J.N.L. Connor, private communication.
47. E. Pollak, to be published.
48. E. Pollak, *J.Chem.Phys.* 76, 5843 (1982).
49. E. Pollak, *Com.At.Mol.Phys.* 15, 73 (1984).
50. R.E. Wyatt and A.B. Elkowitz, *J.Chem.Phys.* 63, 702 (1975).
51. D.G. Truhlar, A.D. Isaacson and B.C. Garrett in *Theory of Chemical Reaction Dynamics*, M. Baer, ed., CRC Press, Boca Raton, FL. in press.
52. W.H. Miller, N.C. Handy and J.E. Adams, *J.Chem.Phys.* 72, 99 (1980).
53. F.T. Smith, *Phys.Rev.Lett.* 45, 1157 (1980).
54. M. Baer, D.J. Kouri and V. Khare, *J.Chem.Phys.* 71, 1188 (1979).
55. J.M. Bowman and K.T. Lee, *J.Chem.Phys.* 72, 5071 (1980).
56. D.C. Clary and J.N.L. Connor, *J.Phys.Chem.* 88, 2758 (1984).
57. E. Pollak, *Chem.Phys.Lett.* 91, 27 (1982).
58. C.C. Marston and R.E. Wyatt in *Resonances*, D.G. Truhlar, ed., ACS Symposium Ser. 263, 441 (1984).
59. J.M. Bowman, *Adv.Chem.Phys.*, in press.
60. L.D. Landau and E.M. Lifshitz, *Quantum Mechanics*, Pergamon Press, New York, 1977, pp. 410-420.
61. E. Pollak, *J.Chem.Phys.* 82, 106 (1985).
62. G.C. Schatz, *Chem.Phys.Lett.* 94, 183 (1983).
63. D.J. Kouri, V. Khare and M. Baer, *J.Chem.Phys.* 75, 1179 (1981).
64. R.B. Walker and E. Pollak, *J.Chem.Phys.*, in press.
65. W.H. Miller, *J.Chem.Phys.* 62, 1899 (1975).
66. G.C. Schatz, *J.Chem.Phys.* 74, 1133 (1981).
67. D.G. Truhlar and A.D. Isaacson, *J.Chem.Phys.* 77, 3516 (1982).
68. W.J. Chesnavich, T. Su and M.T. Bowers, *J.Chem.Phys.* 72, 2641(1980).
69. D.G. Truhlar, *J.Chem.Phys.* 82, 2166 (1985).
70. W.J. Chesnavich and M.T. Bowers, *J.Chem.Phys.* 82, 2168 (1985).
71. J.A. Dodd, D.M. Golden and J.I. Brauman, *J.Chem.Phys.* 82, 2169 (1985).
72. K. Rynefors and N. Markovic, *Chem.Phys.* 92, 327 (1985).
73. D.C. Clary, *Molec.Phys.* 53, 3 (1984).
74. Y. Zeiri and M. Shapiro, *Chem.Phys.* 31, 217 (1978).

75. I. Noorbatcha and N. Sathyamurthy, *Chem.Phys.* **77**, 67 (1983).
76. A. Carrington and R.A. Kennedy, *J.Chem.Phys.* **81**, 91 (1984).
77. T. Carrington, Jr. and W.H. Miller, *J.Chem.Phys.* **81**, 3942 (1984).

THE SUDDEN APPROXIMATION FOR REACTIONS

Michael Baer,
Soreq Nuclear Research Center, Yavne 70600, Israel
and
Donald, J. Kouri,
Department of Chemistry, University of Houston, Houston TX
77004, USA

ABSTRACT. In this work we have examined the relevance of the reactive infinite order sudden approximation (RIOSA). This was done by selecting from our previous published results only those that can be directly compared either with experiment or with results obtained from less approximate treatments (exact quantum mechanical, coupled states). In addition the quasiclassical trajectory (QCT) method represents the exact classical reference relation to which comparisons are to be made. The main findings are: (a) The RIOSA yields reliable total integral cross sections. (b) There is strong evidence that the RIOSA yields relevant integral vibrational state-to-state cross sections. (c) The situation is unclear with regard to differential cross section because very little is known about the angular distributions. So far there is some evidence that the RIOSA is able to yield angular distributions which are encountered in experiment and which the QCT method fails to do.

1. INTRODUCTION

The quantum mechanical treatment of a three-dimensional atom-diatom reactive system is one of the main subjects of theoretical chemistry [1]. About a decade ago when the first numerical results for the $H + H_2$ reactions appeared in print [2] it seemed that the problem was solved. However, difficulties associated with numerical instabilities and with the bifurcation into two nonsymmetric product channels slowed progress with this kind of treatment. This situation caused a change in the order of priorities; whereas previously most of the effort was directed toward developing algorithms for yielding "exact" cross sections, now it is mostly aimed at developing reliable approximations.

Such an approximation, which is a derivative of the exact formalism presented by Kuppermann Schatz and Baer (KSB) [3] is described in this paper. Use is made of the infinite order sudden approximation (IOSA) as applied in the inelastic case but contains significant changes: whereas in the inelastic case one has to treat one arrangement channel at a time, in the case of exchange at least two channels are

D. C. Clary (ed.), The Theory of Chemical Reaction Dynamics, 167–192.

to be treated simultaneously [4]. This fact imposes the treatment of the matching of the wave functions of the two arrangement channels. However, the situation is even more complicated than that. The IOSA is a fixed $\gamma(= \cos(\hat{R}\cdot\hat{r}))$ angle (see fig. 1) calculation. In the case of two arrangement channels one encounters two γ angles one in the entrance channel, i.e. γ_λ and one in the exit channel, i.e. γ_ν. The relation between the two is not obvious. The reactive IOSA (RIOSA) as presented in this work treats the two problems and delivers a reasonable solution.

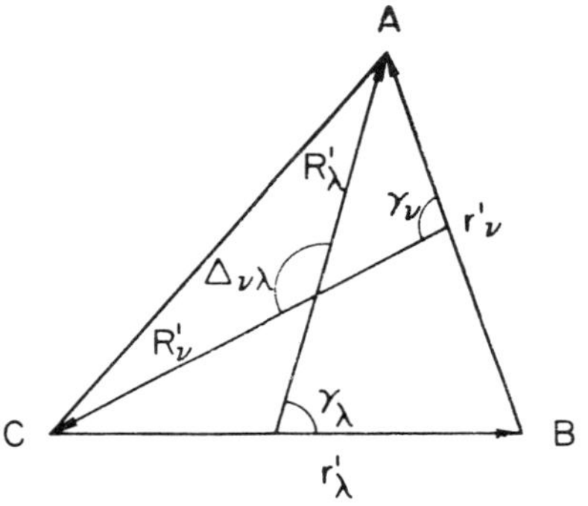

Figure 1. The coordinate system of the three-body system in three dimensions: R'_α and r'_α, $\alpha = \lambda,\nu$ are the unscaled distances. Note the angles γ_λ, γ_ν and $\Delta_{\lambda\nu}$.

The paper is arranged in the following way. In the next section a few aspects of the RIOSA which are relatively new are discussed and a detailed procedure is given for doing the actual calculations. The third section deals with numerical results as obtained by us in the last five years. However, in contrast to what has been published elsewhere, we concentrate only on those results which can be compared either with findings obtained by reliable methods, i.e. exact quantum mechanical (EQM), coupled states (CS) and quasiclassical trajectory (QCT) methods or with experiment. A discussion and a summary are given in the last section.

2. THEORY

2.1. General Remarks

The IOSA which is employed in this and in our previous publications is known as the ℓ-labeled IOSA. In general one distinguishes between two types of ℓ-labeling, namely ℓ-in in which the $\bar{\ell}$ is identified with ℓ-initial (ℓ_i) and the ℓ-av in which $\bar{\ell}$ is identified with $(\ell_i+\ell_f)/2$ (ℓ_f is the final ℓ-value). From the many studies performed for inelastic systems the ℓ-av is known to give much better results. This, as will be shown in the next section, is also the case when exchange is included.

The detailed derivation of the RIOSA was given by Khare, Kouri and Baer (KKB) [4a]. Starting with the exact formalism of KSB, they obtained the RIOSA by substituting the IOSA into the wave functions in the various arrangement channels. Then the matching and the asymptotic analysis was performed exactly as in the KSB formalism [3]. This procedure, which will not be repeated here, yields the following basic relation:

$$\sum_{j_\nu'} Y_{j_\nu'\Omega_\nu}(\gamma_\nu,0)\, S^{\lambda j_i\Omega_i}_{J_\nu j_\nu'\Omega_\nu}(B_{\lambda\nu}) = \sum_{\ell\Omega_\lambda'} \frac{2\ell+1}{2J+1} \langle \ell 0 j_i \Omega_\lambda' | J\Omega_\lambda' \rangle \langle \ell 0 j_i\Omega_i | J\Omega_i \rangle\, d^J_{\Omega_\nu\Omega_\lambda'}(\Delta_{\lambda\nu}) \quad (1)$$
$$Y_{j_i\Omega_\lambda'}(\gamma_\lambda,0)\, S^{\lambda}_{\bar{\ell}\nu}(\gamma_\nu(\gamma_\lambda,B_{\lambda\nu}))$$

This is eq. (105) in KKB [4a] except that it is written for S-matrix elements, instead of the R-matrix elements. In this expression three variables γ_λ, γ_ν and $B_{\nu\lambda}$ appear, the latter being a parameter connected with matching which be discussed later in more detail. The three parameters are interrelated in the sense that each two uniquely determine the third. Equation (1) was derived assuming $B_{\lambda\nu}$ is given and γ_ν is then determined (uniquely) for each γ_λ. We now want to change this order and make the ansatz that eq. (1) is valid for each value of γ_λ and γ_ν (each such pair determinates one value of $B_{\lambda\nu}$). Consequently, eq. (1) will be rewritten as:

$$\sum_{j_\nu'} Y_{j_\nu'\Omega_\nu}(\gamma_\nu,0)\, S^{\lambda j_i\Omega_i}_{J\nu j_\nu'\Omega_\nu}(\gamma_\lambda) = \sum_{\ell\Omega_\lambda'} \frac{2\ell+1}{2J+1} \langle \ell 0 j_i\Omega_\lambda' | J\Omega_\lambda'\rangle \quad (1')$$
$$\langle \ell 0 j_i\Omega_i | J\Omega_i\rangle\, d^J_{\Omega_\nu\Omega_\lambda}(\Delta_{\lambda\nu})\, Y_{j_i\Omega_\lambda}'(\gamma_\lambda,0)\, S^{\lambda}_{\bar{\ell}\nu}(\gamma_\nu,\gamma_\lambda)$$

In this expression $S^{\lambda j_i\Omega_i}_{J\nu j_\nu'\Omega_\nu}(\gamma_\lambda)$ for the transition $(\lambda j_i\Omega_i) \rightarrow (\nu j_\nu\Omega_\nu)$ is the physical S-matrix element as obtained for one fixed value of γ_λ but averaged over all γ_ν values, $S^{\lambda}_{\bar{\ell}\nu}(\gamma_\lambda,\gamma_\nu)$ is the fixed internal IOS S matrix element which is derived, as will be seen, for one given pair of angles from a collinear type calculation. $Y_{j\Omega}(\gamma,0)$ are the spherical harmonics, $d^J_{\Omega_\nu\Omega_\lambda}(\Delta_{\lambda\nu})$ are the Wigner rotation matrix elements and $\Delta_{\lambda\nu}$ is the angle of rotation from the $\hat{R}_\lambda$ axis to the $\hat{R}_\nu$ axis (see fig. 1).

Multiplying eq. (1') by $Y^*_{j_\nu\Omega_\nu}(\gamma_\nu,0)$ and integrating over γ_ν leads to

$$S^{\lambda j_i\Omega_i}_{J\nu j_\nu\Omega_\nu}(\gamma_\lambda) = \sum_{\ell\Omega'_\lambda}\left(\frac{2\ell+1}{2J+1}\right)\langle \ell 0 j_i\Omega'_\lambda|J\Omega'_\lambda\rangle\langle \ell 0 j_i\Omega_i|J\Omega_i\rangle Y_{j_i\Omega'_\lambda}(\gamma_\lambda,0)$$

$$2\pi\int_{-1}^{+1} d(\cos\gamma_\nu)Y^*_{j_\nu\Omega_\nu}(\gamma_\nu,0)d^J_{\Omega_\nu\Omega'_\lambda}[\Delta_{\lambda\nu}(\gamma_\lambda,\gamma_\nu)]S^\lambda_{\bar{\ell}\nu}(\gamma_\lambda,\gamma_\nu). \quad (2)$$

We now define the physical $S^{\lambda j_i\Omega_i}_{J\nu j_\nu\Omega_\nu}$ as

$$S^{\lambda j_i\Omega_i}_{J\nu j_\nu\Omega_\nu} = \frac{1}{2}\int_{-1}^{+1} d(\cos\gamma_\lambda)S^{\lambda j_i\Omega_i}_{J\nu j_\nu\Omega_\nu}(\gamma_\lambda) \quad (3)$$

and obtain

$$S^{\lambda j_i\Omega_i}_{J\nu j_\nu\Omega_\nu} = \pi\sum_{\ell\Omega'_\lambda}\left(\frac{2\ell+1}{2J+1}\right)\langle \ell 0 j_i\Omega_i|J\Omega_i\rangle\langle \ell 0 j_i\Omega'_\lambda|J\Omega'_\lambda\rangle$$

$$\int_{-1}^{1}\int_{-1}^{1} d(\cos\gamma_\lambda)d(\cos\gamma_\nu)Y^*_{j_\nu\Omega_\nu}(\gamma_\nu,0)d^J_{\Omega_\nu\Omega'_\lambda}[\Delta_{\lambda\nu}(\gamma_\lambda,\gamma_\nu)] \quad (4)$$

$$S^\lambda_{\bar{\ell}\nu}(\gamma_\lambda,\gamma_\nu)Y_{j_\lambda\Omega'_\lambda}(\gamma_\lambda,0).$$

Equation (4) is the general expression for a physical S-matrix element as obtained from the RIOSA assuming the λ channel is the initial channel. The characteristic feature of this expression is that γ_λ and γ_ν are completely independent so that fixing one does not impose any restriction on the other.

The introduction of some kind of dependence between the two angles can be done only in ad-hoc fashion by multiplying the integral by $2Q(\gamma_\lambda|\gamma_\nu)$ where $Q(\gamma_\lambda|\gamma_\nu)$ is a conditional probability function fulfilling the requirement:

$$\int_{-1}^{+1} d(\cos\gamma_\nu) Q(\gamma_\lambda|\gamma_\nu) = 1 \tag{5}$$

Consequently eq. (4) becomes:

$$S_{J\nu j_\nu \Omega_\nu}^{\lambda j_i \Omega_i} = 2\pi \sum_{\ell\Omega'_\lambda} \left(\frac{2\ell+1}{2J+1}\right) <\ell 0 j_i \Omega'_\lambda | J\Omega'_\lambda><\ell 0 j_i \Omega_i | J\Omega_i>$$

$$\int_{-1}^{1}\int_{-1}^{1} d(\cos\gamma_\nu) d(\cos\gamma_\lambda) Q(\gamma_\lambda|\gamma_\nu) Y^*_{j_\nu\Omega_\nu}(\gamma_\nu,0) d^J_{\Omega_\nu\Omega'_\lambda}[\Delta_{\lambda\nu}(\gamma_\lambda,\gamma_\nu)]$$

$$S^\lambda_{\ell\nu}(\gamma_\lambda,\gamma_\nu) Y_{j_\lambda\Omega'_\lambda}(\gamma_\lambda,0). \tag{6}$$

The reason for multiplying by $2Q(\gamma_\lambda|\gamma_\nu)$ and not by $Q(\gamma_\lambda|\gamma_\nu)$ is to avoid overnormalization. Equation (6) is in the appropriate form to obtain a single $B_{\lambda\nu}$ equation. For a given γ_λ choosing a simple fixed value for $B_{\lambda\nu}$ yields a single value γ_ν according to:

$$\gamma_\nu = \gamma_\nu(B_{\lambda\nu},\gamma_\lambda) \tag{7}$$

This can be introduced into eq. (6) by taking $Q(\gamma_\lambda|\gamma_\nu)$ to be a Dirac δ function:

$$Q(\gamma_\lambda|\gamma_\nu) = \delta(\cos\gamma_\nu - \cos\left\{\gamma_\nu(B_{\lambda\nu},\gamma_\lambda)\right\}) \tag{8}$$

Substituting eq. (8) in eq. (6) yields:

$$S_{J\nu j_\nu \Omega\nu}^{\lambda j_i \Omega_i} = 2\pi \sum_{\ell\Omega_\lambda} \left(\frac{2\ell+1}{2J+1}\right) <\ell 0 j_i \Omega'_\lambda | J\Omega'_\lambda><\ell 0 j_i \Omega_i | J\Omega_i>$$

$$\int_{-1}^{1} d(\cos\gamma_\lambda) Y^*_{j_\nu\Omega_\nu}(\gamma_\nu(B_{\lambda\nu},\gamma_\lambda),0) d^J_{\Omega_\nu\Omega'_\lambda}[\Delta_{\lambda\nu}(B_{\lambda\nu},\gamma_\lambda)] \tag{9}$$

$$S^\lambda_{\ell\nu}(B_{\lambda\nu},\gamma_\lambda) Y_{j_\lambda\Omega'_\lambda}(\gamma_\lambda,0).$$

Equation (9) is the expression for the physical S-matrix element obtained by fixing $B_{\lambda\nu}$ and choosing the λ channel to be the initial channel. It also has to be emphasized that this expression is an ℓ-in expression. The extension to the ℓ-av case is done in an ad-hoc fashion exactly like in the inelastic case.

2.2. Differential and Integral Cross Sections

The degeneracy averaged differential cross section from an initial state (v_λ, j_λ) to a final vibrational state v_ν is given in the form:

$$\frac{d\sigma(v_\nu|v_\lambda,j_\lambda|\theta,\phi)}{d\omega} = \frac{1}{[j_\lambda]} \sum_{j_\nu m_\nu m_\lambda} |f(v_\lambda,j_\lambda,m_\lambda|v_\nu,j_\nu,m_\nu|\theta,\phi)|^2 \quad (10)$$

where ω is a solid angle comprised of the scattering angle θ and the azimuthal angle ϕ, m_λ and m_ν are the p-helicity quantum numbers, $f(v_\lambda,j_\lambda,m_\lambda|v_\nu,j_\nu,m_\nu|\theta,\phi)$ is the scattering amplitude function and $[j_\lambda]$ stands for $(2j_\lambda+1)$. In what follows the v_λ and v_ν indices are dropped in order to simplify the notation.

The scattering amplitude is related to the body fixed S-matrix element according to:

$$f(j_\lambda,m_\lambda|j_\nu,m_\nu|\theta,\phi) = \frac{i^{j_\lambda-j_\nu+1}}{2k_{j_\lambda}} \sum_J [J] d^J_{m_\lambda m_\nu}(\theta) S^{j_\lambda m_\nu}_{Jj_\lambda m_\lambda} \quad (11)$$

where $d^J_{m_\lambda m_\nu}(\theta)$ is the Wigner rotation matrix (the notation of Rose [5] is employed), k_{j_λ} is the initial wave numer defined as : $k_{j_\lambda} = \left(\frac{2\mu}{\hbar^2}(E-\varepsilon_{j_\lambda})\right)^{\frac{1}{2}}$ and μ is the mass of the system. The body fixed S-matrix elements are related to the Arthurs Dalgarno (AD) [6] matrix elements according to:

$$S^{j_\nu m_\nu}_{Jj_\lambda m_\lambda} = \sum_{\ell_\lambda \ell_\nu} i^{\ell_\lambda-\ell_\nu} \sqrt{\frac{[\ell_\lambda][\ell_\nu]}{[J]}} <\ell_\lambda 0 j_\lambda m_\lambda|Jm_\lambda><\ell_\nu 0 j_\nu m_\nu|Jm_\nu> \quad (12)$$
$$S^{j_\nu \ell_\nu}_{Jj_\lambda \ell_\lambda}$$

Employing the orthogonality relations of the Clebsch-Gordan coefficients and recalling eq. (9), one can show that the AD S-matrix elements approximated within the RIOSA are:

$$S^{j_\nu \ell_\nu}_{Jj_\lambda \ell_\lambda} = i^{\ell_\lambda - \ell_\nu - 2\bar{\ell}} \sum_{\Omega_\lambda \Omega_\nu} \frac{\sqrt{[\ell_\lambda][\ell_\nu]}}{[J]} <\ell_\lambda 0 j_\lambda \Omega_\lambda | J\Omega_\lambda><\ell_\nu 0 j_\nu \Omega_\nu | J\Omega_x>$$

$$2\pi \int_{-1}^{+1} d(\cos\gamma_\lambda) Y^*_{j_\nu \Omega_\nu}(\gamma_\nu(\gamma_\lambda, B), 0) d^J_{\Omega_\lambda \Omega_\nu}(\Delta(\gamma_\lambda, B)) \tag{13}$$

$$S_{\bar{\ell}}(\gamma_\lambda, B) Y_{j_\lambda \Omega_\lambda}(\gamma_\lambda, 0) \quad .$$

For the case $4\ \ell_\lambda - (\ell_\lambda + \ell_\nu)/2$ no further simplifications are possible and thus the procedure for calculating the differential cross sections described so far should be used.

For the choice $\bar{\ell}_\lambda = \ell$-in, simplifications are possible (see Ref. 4) and the final result is (following integration over ϕ):

$$\frac{d\sigma(v_\nu | v_\lambda, j_\lambda | \theta)}{d(\cos\theta)} = \frac{\pi}{4k^2_{v_\lambda j_\lambda}} \sum_L d^L_{00}(\theta) \sum_{\ell_\lambda \ell'_\lambda} [\ell_\lambda][\ell'_\lambda] |<\ell_\lambda 0 \ell'_\lambda 0 | L0>|^2 \tag{14}$$

$$\int_{-1}^{+1} d(\cos\gamma) d^L_{00}(\Delta(\gamma_\lambda, B)) S^{v_\nu}_{\ell_\lambda v_\lambda}(\gamma_\lambda, B) S^{v_\nu}_{\ell'_\lambda v_\lambda}(\gamma_\lambda, B)$$

To obtain total cross sections we have to integrate eq. (10) over $d(\cos\theta)d\phi$, which for the general case yields the following expressions:

$$\sigma(v_\nu | v_\lambda, j_\lambda) = \frac{\pi}{2k^2_{v_\lambda j_\lambda}} \sum_{\substack{j_\nu m_\nu \\ m_\lambda J}} [J] \left| S^{v_\nu j_\nu m_\nu}_{Jv_\lambda j_\lambda m_\lambda} \right|^2 \tag{15}$$

A more explicit expression can be obtained for the case $\bar{\ell} = \ell$-in [7]. i.e.

$$\sigma(v_\nu | v_\lambda, j_\lambda) = \frac{\pi}{2k^2_{v_\lambda j_\lambda}} \sum_{\ell_\lambda} [\ell_\lambda] \int_{-1}^{+1} d(\cos\gamma_\lambda) |S^{v_\nu}_{\ell_\lambda v_\lambda}(\gamma_\lambda, B)|^2 \tag{16}$$

2.3. Determination of the $S_{\bar{\ell}}(\gamma_\lambda, B)$ Matrix Elements

The only unknown functions which appear in the above formalism are $S^{v_\nu}_{\bar{\ell} v_\lambda}(\gamma_\lambda, B)$, and these are determined by solving fixed angle (γ_λ) "collinear type" Schroedinger equations. In contrast to the inelastic case the treatment of reactive scattering takes place in two arrangement channels and therefore usually two γ angles, i.e. γ_λ and γ_ν are encountered. As mentioned earlier the two γ angles are not necessarily dependent on each other and therefore the corresponding S-matrix element $S^{v_\nu}_{\bar{\ell} v_\lambda}$ is expected to be dependent on two angles. However we showed how the general two-angle formula is reduced to a single-angle formula, as given in eq. (9). This reduction leads to the introduction of a parameter B. Consequently, each value of γ_λ uniquely determines one value of γ_ν i.e. $\gamma_\nu = \gamma_\nu(\gamma_\lambda, B)$. In what follows we describe how this is accomplished.

$(R_\lambda, r_\lambda, \gamma_\lambda)$ are coordinates of a three-dimensional space in which γ_λ is the polar coordinate. Fixing γ_λ reduces the three-dimensional space to a plane. In the same way $(R_\nu, r_\nu, \gamma_\nu)$ are coordinates of a three-dimensional space in which γ_ν is the polar coordinate. Fixing γ_ν reduces the three-dimensional ν-space to a plane (usually the γ_λ and γ_ν planes are different). The intersection of two planes is a straight line termed here the matching line. In Ref. (8) it is proved that the equation of this line is determined from the relation

$$r_\nu = B r_\lambda \tag{17}$$

where B is uniquely determined by the choice of γ_λ and γ_ν [8]:

$$B = B(\gamma_\lambda, \gamma_\nu) \tag{18}$$

In a single γ formula, as the one given in eq. (9), one fixes B and then for each given γ_λ the value of γ_ν is uniquely determined by the equation (4a):

$$\cos\gamma_\nu = -\frac{\cos\gamma_\lambda + (1-B^2)\cot\Phi_{\lambda 0}\cot\alpha}{B[1 + (1-B^2)\cot^2\Phi_{\lambda 0}]^{\frac{1}{2}}} \tag{19}$$

where

$$\cot\Phi_{\lambda 0} = \frac{\sin\alpha}{B^2 - \cos^2\alpha}\left[\cos\alpha\cos\gamma_\lambda + (B^2 - \sin^2\gamma_\lambda\cos^2\alpha)^{\frac{1}{2}}\right] \tag{20}$$

and

$$\cos\alpha = -\left\{\frac{m_A m_C}{(m_A+m_B)(m_C+m_B)}\right\}^{\frac{1}{2}}; \quad \sin\alpha = (1-\cos^2\alpha)^{\frac{1}{2}} \tag{21}$$

Here m_A, m_B and m_C are the masses of the three interacting atoms and the reaction that is studied is $A + BC \rightarrow AB + C$.

In the $H + H_2$ case [7], B was taken to be 1, a natural choice for such a highly symmetrical system. The same choice was made for $D + H_2$ [9] and Cl + HCl [10], whereas for the $F + H_2(D_2)$ system, [11,12] B was assumed to be 1.4 which yields a straight line that follows the ridge of the potential in the plateau region. For the particular choice of B = 1 eq. (19) reduces to

$$\cos\gamma_\nu = -\cos\gamma_\lambda \tag{19'}$$

or

$$\gamma_\gamma = \pi - \gamma_\lambda \tag{19''}$$

To calculate the $S^{\nu}_{\bar{\ell}v_\lambda}(\gamma_\lambda, B)$ matrix elements, one solves the collinear-type Schroedinger equations (4a):

$$\left\{-\frac{\hbar^2}{2\mu}\left[\left(\frac{\partial^2}{\partial R_\alpha^2}+\frac{\partial^2}{\partial r_\alpha^2}\right)-\frac{\ell(\ell+1)}{R_\alpha^2}\right]+V(R_\alpha,r_\alpha;\gamma_\alpha)-E\right\}\psi_\alpha(R_\alpha,r_\alpha;\gamma_\alpha)=0 \tag{22}$$

$$\alpha = \lambda, \nu$$

Here E is the total energy, $V(R_\alpha, r_\alpha; \gamma_\alpha)$ is the potential energy surface where γ_α ($\alpha = \lambda, \nu$) enters as a parameter, R_α and r_α are the scaled translational and vibrational coordinates, respectively, and μ is the reduced mass of the system;

$$\mu = \left(\frac{m_A m_B m_C}{m_A + m_B + m_C}\right)^{\frac{1}{2}} \tag{23}$$

Matching the solutions $\psi_\lambda(R_\lambda, r_\lambda, \gamma_\lambda)$ and $\psi_\nu(R_\nu, r_\nu, \gamma_\nu)$ and their normal derivatives along the line $r_\nu = Br_\lambda$, one obtains the desired $S^{v_\nu}_{\bar{\ell}v_\lambda}(\gamma_\lambda, B)$ matrix elements [13].

3. RESULTS

Results for four different systems are presented in this section with the emphasis on total, vibrational state-to-state integral and differential cross sections as well as on rate constants. Rotational state-to-state cross sections will not be discussed as the RIOSA final rotational distributions are, in general, incorrect. The only exception is the rotation distribution obtained in the light-heavy light mass combinations [14], which is present in another chapter.

Except for comparisons with experiment, the RIOSA results will be mainly compared with QCT results which are believed to be correct, at least for total integral cross sections and perhaps also for total differential cross sections. The RIOSA results will also be compared with any EQM or CS results which are available.

3.1. The H + H_2 System [7,15,16]

The RIOSA study on H + H_2 was carried out on the Porter-Karplus potential energy surface [17]. The results for total cross sections from the ground state and the first excited state are shown in figs, (2) and (3a). The branching ratios Γ defined as:

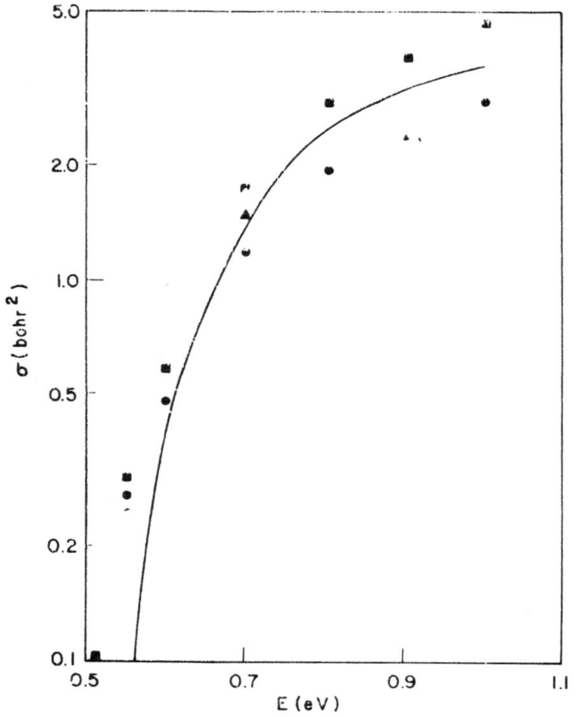

Figure 2. Integral total cross sections for the reaction: H + $H_2(v_i=0)$ → H_2 + H as a function of total energy.
—— QCT results (Ref. 18); ▲ EQM results (Ref. 20)
● ℓ-in results; ■ ℓ-av results.

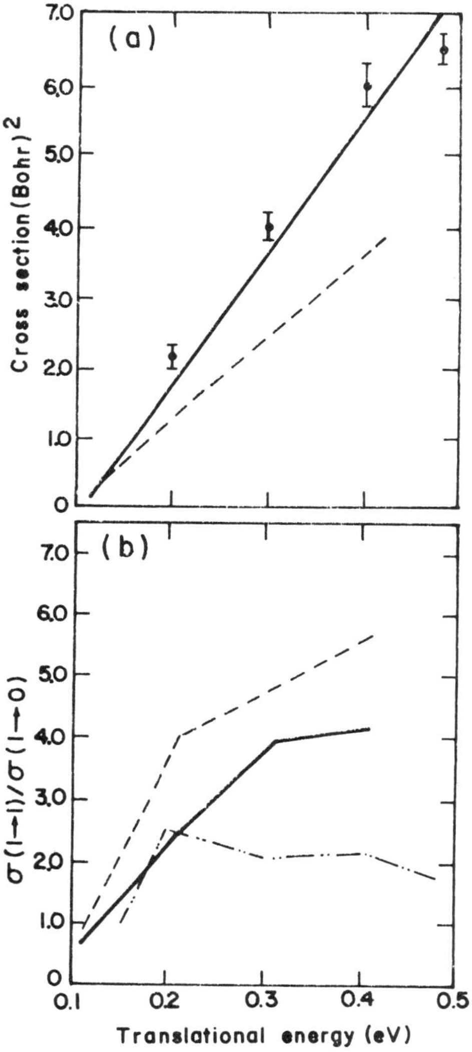

Figure 3. Results for the reaction $H + H_2(v_i=1) \rightarrow H_2 + H$ as a function of translational energy (a) Total integral cross sections:
I QCT results (ref. 19); —— ℓ-av results; - - - - ℓ-in results
(b) Branching ratios $\Gamma(1,0) = \sigma(1 \rightarrow 1)/\sigma(1 \rightarrow 0)$
-..-..-..- QCT results (Ref. 19); —— ℓ-av results; - - - ℓ-in results.

$$\Gamma = \sigma(1 \to 1)/\sigma(1 \to 0) \tag{24}$$

are presented in fig. (3b). In all three figures ℓ-in. ℓ-av and QCT results [18,19] are shown. In fig. (2) EQM results [20] are also presented. For $v_i = 0$, the quality of the fit with both QCT and the EQM results is reasonable. In fact, for intermediate energies, both are located between the ℓ-in and ℓ-av results. The deviations are ∿ 30%. A better fit is shown to exist in fig. (3a). Here the ℓ-av results, which are always considered to be the more accurate, overlap the QCT results [19] very nicely. For the branching ratio the fit with the QCT [19] results is good only for a short energy range for low translational energies. For higher energies the RIOSA branching ratios are much larger - the ℓ-av results are twice as large and the ℓ-in three times.

Comparisons between RIOSA and QCT differential cross sections [15] for $v_i = 1$ and $E_{tot} = 1.2$ eV are presented in fig. (4). Four kinds of

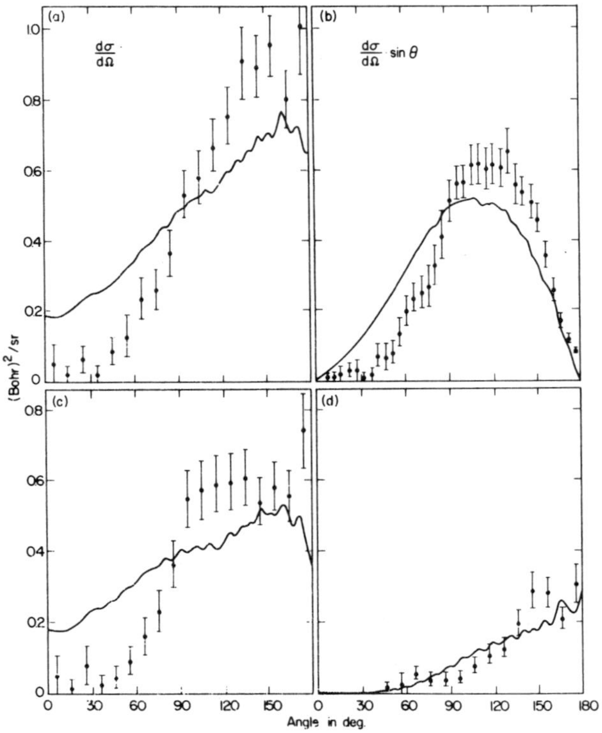

Figure 4. Differential cross sections for $H + H_2(v_i{=}1) \to H_2 + H$ for the total energy $E_{tot} = 1.2$ eV. (a) Total differential cross sections. (b) Total differential cross sections multiplied by sinθ (c) $v_i = 1 \to v_f = 1$ differential cross sections. (d) $v_i = 1 \to v_f = 0$ differential cross sections. ⫿ QCT results (Ref. 15) —— ℓ-av results.

results are shown: $d\sigma/d\omega$ for 1 → all; 1 → 1 and 1 → 0 as well as $\left(\sin\theta \frac{d\sigma}{d\theta}\right)$ for 1 → all. In general the QCT distribution is always more backwards than that of the RIOSA. This feature, as will be seen in what follows, is typical for all cases.

Rate constants for $v_i=0$ are shown in fig. 5. Here the present RIOSA results are compared with EQM [20] and QCT results [18]. The RIOSA rates are about 3 times higher than those of the EQM and about 6 times higher than the QCT rates. The fact that the RIOSA cross sections for the H + H_2 system in the low energy region (around and below the classical threshold) are much larger than the EQM results is the reason for this discrepancy.

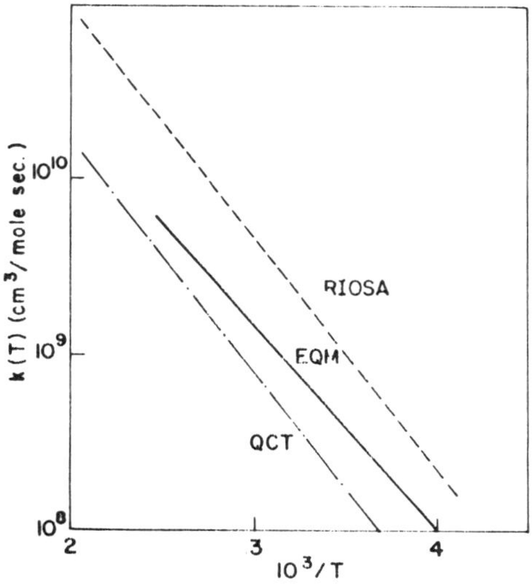

Figure 5. Rate constants as a function of $10^3/T$ for the reaction: $H_2(v_i=0, j_i=0)$ + H → H + H_2. —— EQM results (Ref. 20) -.-.- QCT results (Ref. 17) --- ℓ-av results

3.2. The D + H_2 System [9]

The D + H_2 system was studied employing the LSTH surface [21]. Total cross sections for $v_i = 0,1$ and the branching ratio (see eq. (24)) are shown as a function of translational energy in fig. 6. The results for the cross sections are compared with QCT]22] results and a good fit was obtained for both, i.e. $v_i = 0,1$.

The branching ratios are found to be strongly dependent on energy (like in the H + H_2 case) and for $E_{tran} = 0.4$ eV the value of Γ is ~ 4. This large branching ratio was recently also confirmed experimentally [23].

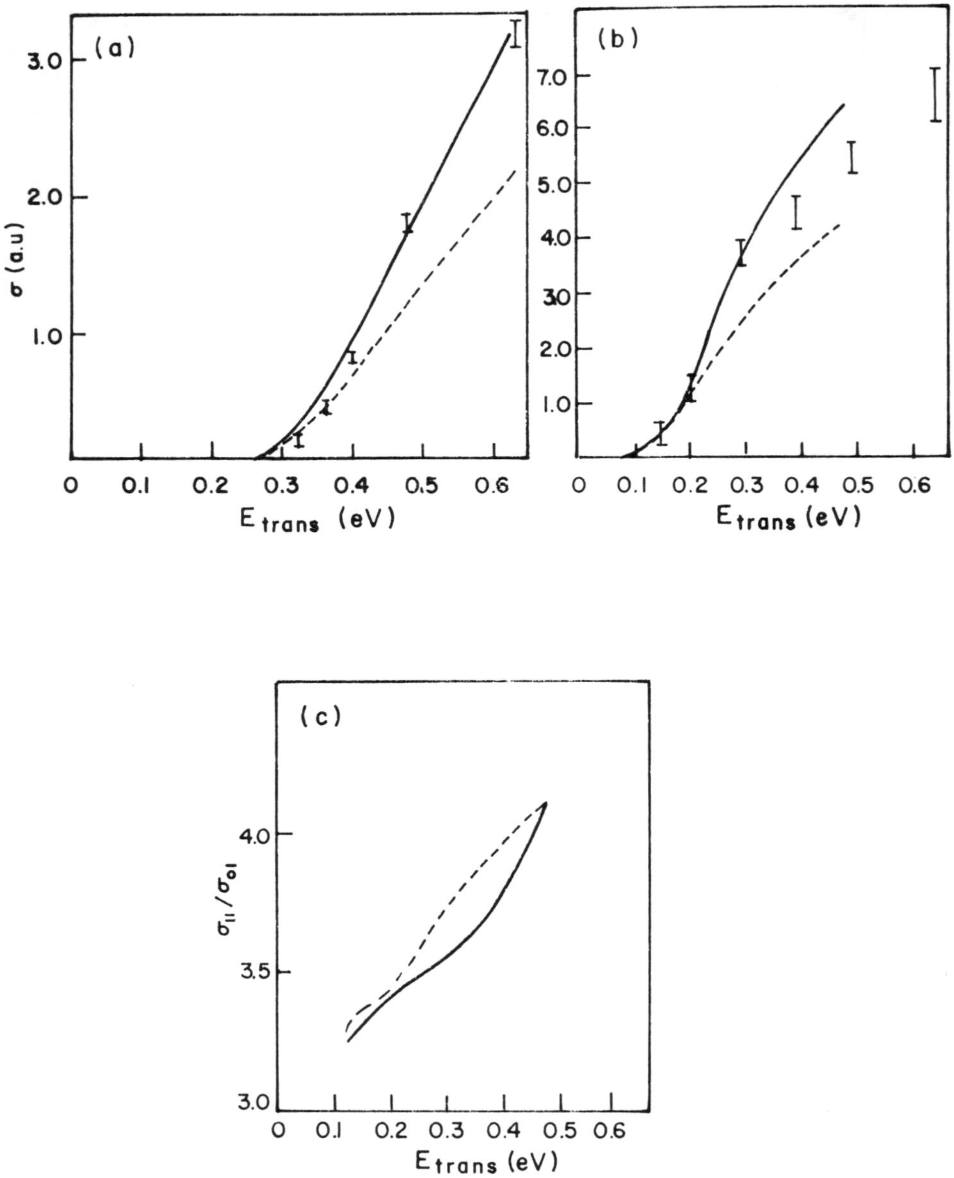

Figure 6. Results for the reactions $D + H_2(v_i) \rightarrow DH + H$ as a function of translational energy. (a) Total integral cross sections for v_i=0 (b) Total integral cross sections for v_i=1 (c) Branching ratios: $\Gamma = \sigma(1 \rightarrow 1)/\sigma(1 \rightarrow 0)$ ⌶ QCT results (Ref. 22) —— ℓ-av results --- ℓ-in results.

Differential cross sections are shown in fig. 7. Results for four different energy values and for the ground and first excited states are presented. A comparison with QCT results [22] was found to be feasible only at one energy value (E_{tr} = 0.48 eV) and for v_i = 0. For

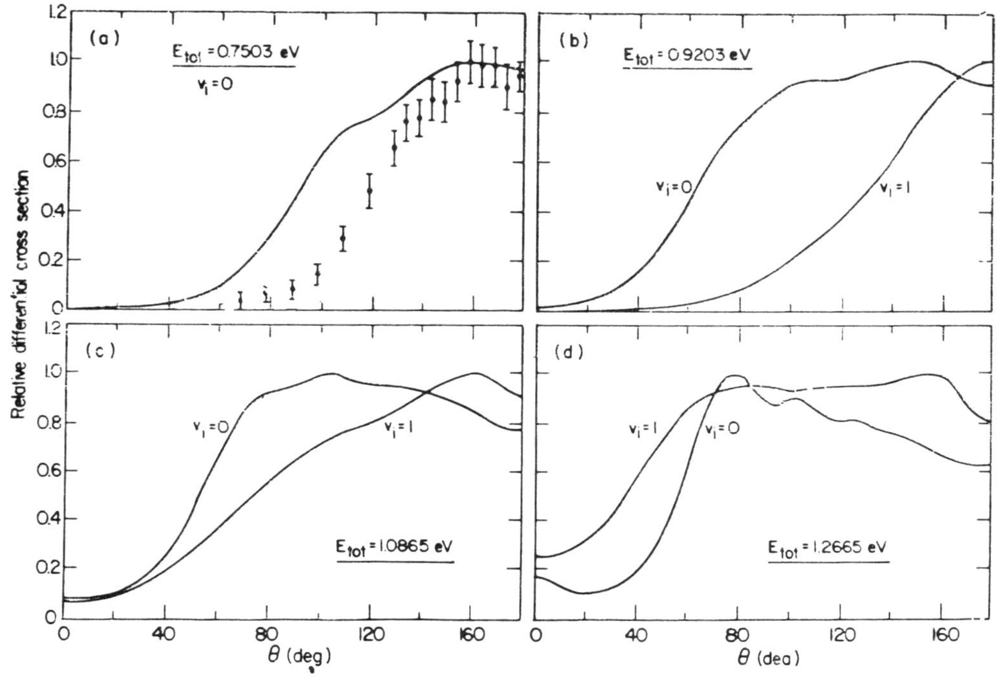

Figure 7. Differential cross sections for the reactions $D + H_2(v_i) \rightarrow DH + H$. —— ℓ-av results ⫯ QCT results (Ref. 22).

this case both curves are backwards, peaking, but the RIOSA curve is broader. In each of the other figures 7(b) - 7(d) two curves are shown, i.e. for v_i = 0,1. At the lower energy the v_i = 0 curve is much broader, but as the energy increases the v_i = 1 curve starts to swell, whereas the v_i = 0 changes only slightly. This makes the v_i = 1 curve broader than that of v_i = 0 for the higher energy. Another feature is the sidewards shift of the v_i = 0 curve as the energy increases. This tendency is less apparent for v_i = 1. So far no explicit experimental center of mass angular distributions were published. But there is strong evidence that for translational energy in the range $1.0 < E_{tr} < 2.0$ eV sidewards peaking is encountered [23]. From private communications we know that for this energy range no sidewards peaking is obtained in QCT calculations. Thus it seems that at least qualitatively the RIOSA, in contrast to the QCT, yields the correct angular behavior.

Rate constants for $v_i = 0$ and $v_i = 1$ as well as the branching ratio K defined as:

$$K = k(1 \rightarrow 1)/k(1 \rightarrow 0)$$

are given in fig. 8. In addition to the RIOSA and the QCT results we also show the experimental results [24,28]. In order to produce rate constants that can be compared with experiment, the dependence of the cross section on the initial rotational (and vibrational) states must be known. According to the RIOSA the total cross sections depend only on the translational energy and are in such a case, independent of the initial rotational state [7]. Mayne et. al. [24] published results employing QCT that show that as long as j is not larger than 4 this assumption is valid. A reasonably good fit among the RIOSA, QCT [22] and experimental results [24,25] is shown to exist for $v_i = 0$. For $v_i = 1$, the RIOSA results are almost twice as large as the QCT results but about 3-5 times smaller than the experimental ones [27-29]. Here it is important to mention that although the deviation between the RIOSA and the experimental results is not small both have the same slope which means that the activation energy as predicted by the RIOSA treatment and found in the experiment are almost identical.

3.3. The F + H_2 System [11,30]:

The study of the F + H_2 system was carried out on the Muckerman V surface [31]. Total cross sections as a function of energy are shown in fig. 9. Five different results are presented; in addition to the ℓ-in, and the ℓ-av we also show CS results due to Redmon and Wyatt, (RW) [32], QCT results [33,34] and also classical IOSA (CIOSA) results as obtained by Jellinek and Baer [35]. It can be seen that there is a good fit between the ℓ-av and the QCT results and that the ℓ-in and the CIOSA results overlap very nicely. The CS results deviate significantly from both the ℓ-av RIOSA and the QCT results and the reason could be due to some changes made by RW to the Muckerman V potential to improve the convergence.

The branching ratio Γ(3,2) defined as:

$$\Gamma(3,2) = \sigma(0 \rightarrow 3)/\sigma(0 \rightarrow 2)$$

is presented in fig. 10. Four curves are shown: two quantum, i.e. the RIOSA (the ℓ-av and ℓ-in branching ratios are identical) and CS curves and two classical curves, the QCT and CIOSA. Although, sometimes large deviations between the two quantum curves are noticed, their energy dependence is similar and they differ significantly from the two classical ones. The fact that the CIOSA fit the QCT branching ratios so well is again an encouraging result for the IOSA in general. (More on this subject is discussed in the next chapter).

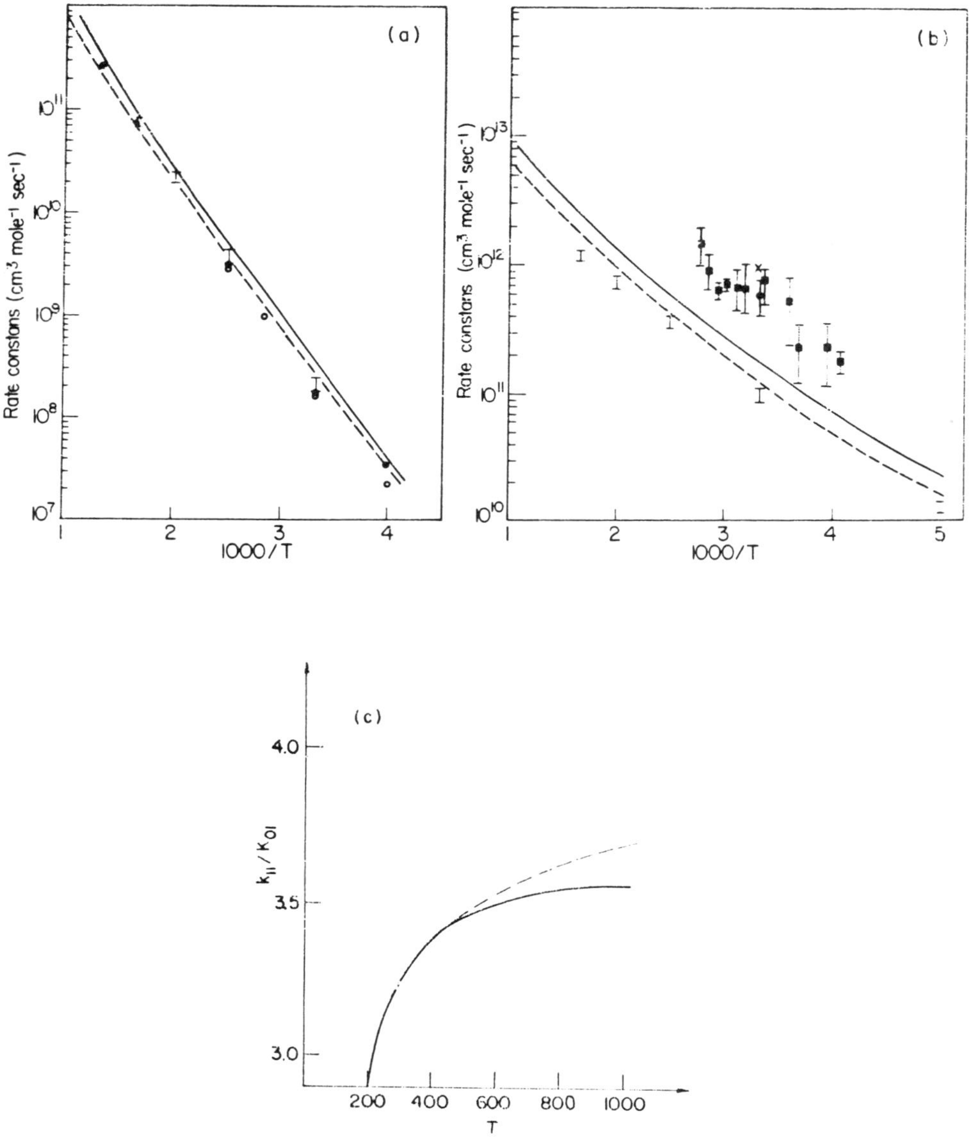

Figure 8. Rate constants for the reactions: $D + H_2(v_i) \rightarrow DH + H$ (a) $v_i = 0$. I QCT results (Ref. 22) —— ℓ-av results,--- ℓ-in results ●o experimental results (Ref. 24,25). (b) $v_i = 1$. I QCT results (Ref. 22) —— ℓ-av results, --- ℓ-in results, ■ Experimental results (Ref. 26) ● Experimental results (Ref. 27) x Experimental result (Ref. 28). (c) Branching ratio $K(1,0) = k(1 \rightarrow 1)/k(1 \rightarrow 0)$ —— ℓ-av results, --- ℓ-in results.

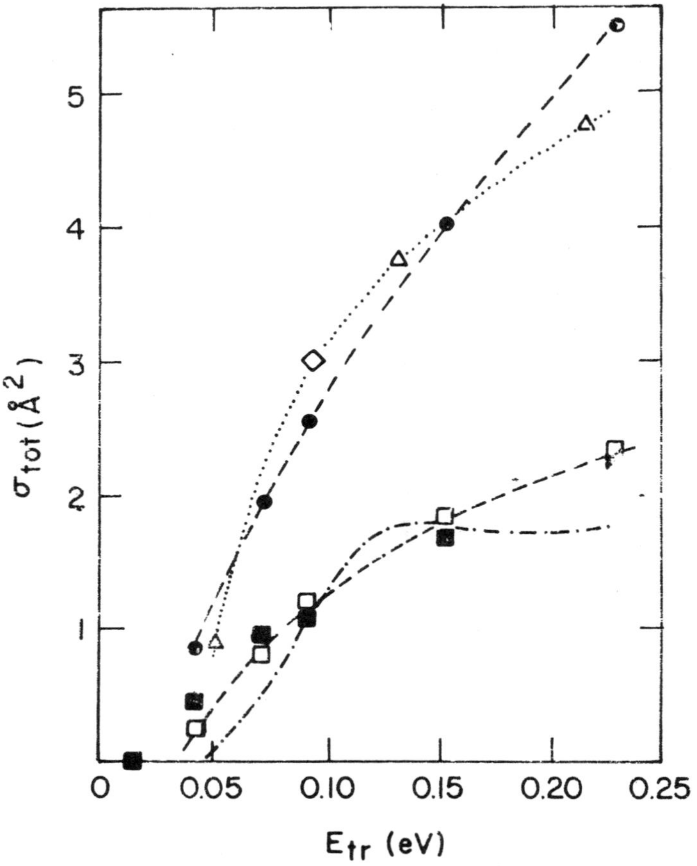

Figure 9. Integral reactive cross sections for the reaction $F + H_2 \rightarrow HF + H$ as a function of translational theory .... QCT results (Ref. 33, 34), -.-. CS results (Ref. 37) ● — ● ℓ-av results, □ — □ ℓ-in results ■ CIOSA results

Differential cross sections for the following reactions

$$F + H_2(v_i = 0) \rightarrow HF(v_f) + H; \quad v_f = 1,2,3$$

are shown in figs. 11-13. The results in figs 11-12 were derived for energy values which are close to those of the experiments [36]. Although the surface is now known to be incorrect, still it was gratifying to find at least one common feature, i.e. the backwards scattering for $v_f = 2$ at the lower energy and the sidewards scattering at higher energy. This, for instance is not the case for

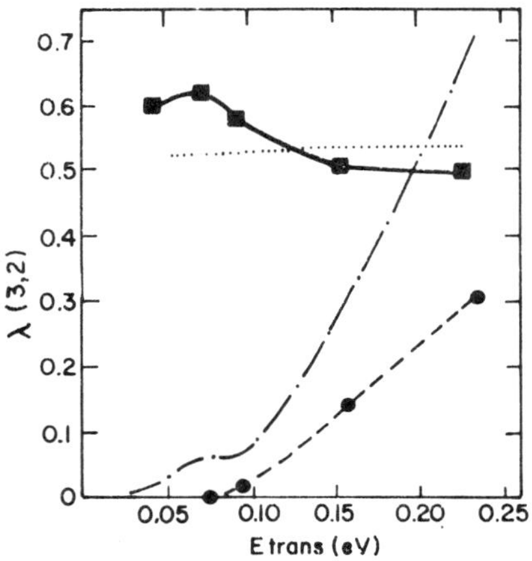

Figure 10. Vibrational branching ratio $\lambda(3,2) = \sigma(v_i = 0 \rightarrow v_f = 3)/\sigma(v_i = 0 \rightarrow v_f = 2)$ as a function of translational energy for the $F + H_2(v_i = 0) \rightarrow HF(v_f) + H$ reaction.
.... QCT results (Ref. 33), -.-.- CS results (Ref. 32)
---- ℓ-av results, ■ CIOSA results.

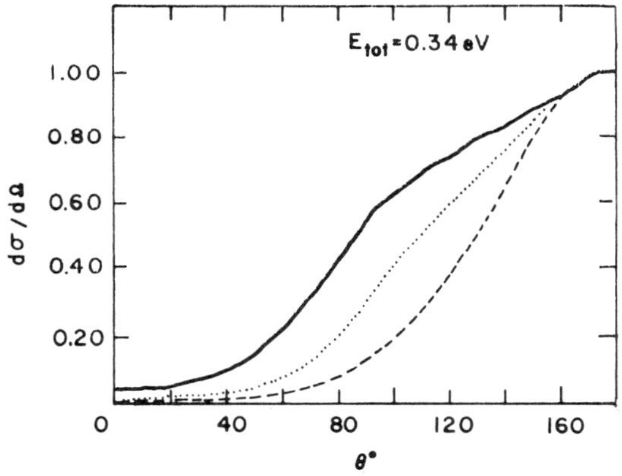

Figure 11. Results for the reaction $F + H_2(v_i) \rightarrow HF(v_f) + H$. Vibrational state resolved differential cross sections for incident translational energy E_{tr} =0.073 eV ≡ 1.68 kcal/mol. The cross sections are normalized to one at $\theta = 180^{\circ}$ (the ratios being 1:2:3 = 0.069 : 1 : 0.008). .... $v_i = 1$, —— $v_i = 2$, --- $v_i = 3$.

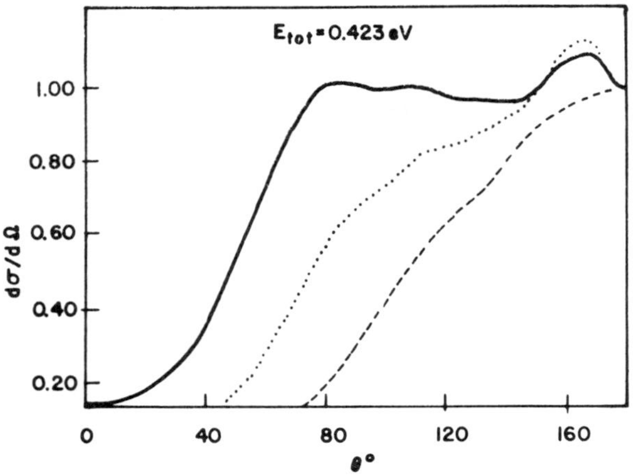

Figure 12. Vibrational state resolved differential cross sections for incident translational energy E_{tr} = 0.156 eV ≡ 3.54 kcal/mol for the $F + H_2(v_i = 0) \rightarrow HF(v_f) + H$ reaction. The cross sections are normalized to one at θ = 180° (the ratio being 1:2:3 = 0.15:1:0.32). .... v_i = 1, —— v_i = 2, --- v_i = 3.

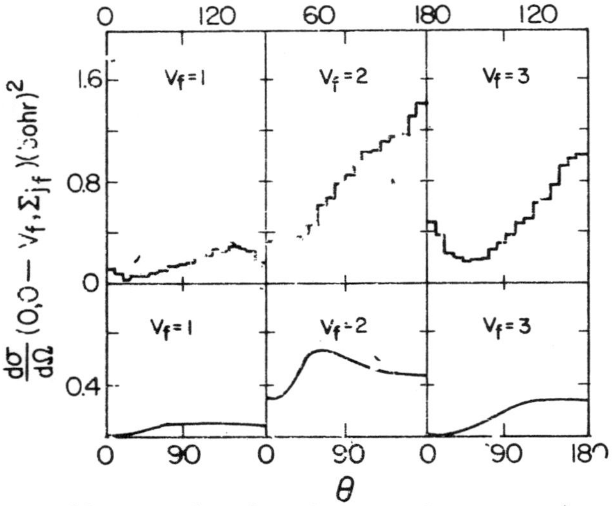

Figure 13. Results for the reaction $F + H_2(v_i = 0) \rightarrow HF(v_f) + H$. A comparison between QCT (Ref. 37) and ℓ-av angular distributions as obtained for E_{tot} = 0.5 eV. Upper panel shows QCT results; lower panel ℓ-av results. Note the sidewards distribution for the ℓ-av v_f = 2 results and the increase in the QCT v_f = 3 curve towards θ ∿ 0.

QCT treatment [37], as can be seen from fig. 13 where the angular distributions due to the two treatments are compared for a high energy.

3.4. The $F + D_2$ System [12]

The $F + D_2$ reaction was studied extensively. However, very few comparisons with other treatments or experiments can be made. Only classical results [31,34,38] are available and the only meaningful comparison can be made with regard to total cross sections. This is presented in fig. 14. As usual the ℓ-av results fit the classical ones very nicely but the ℓ-in are somewhat lower. Since, as mentioned earlier, the Muckerman V surface is wrong, comparison with experiment is not possible.

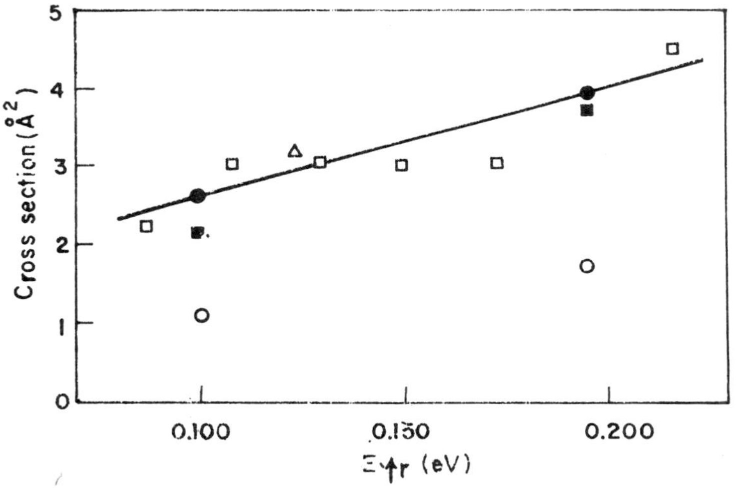

Figure 14. Integral cross sections for the reaction $F + D_2 \rightarrow DF + D$ as a function of translational energy. □ Δ ■ QCT results (Refs. 31, 34, 38) ● ℓ-av results, o ℓ-in results.

4. DISCUSSION

In this work we have concentrated on results which can be compared either with EQM and with results due to less approximate methods or with experiment. In order to have a meaningful comparison with experiment a necessary condition is to have the correct potential energy surface. So far only one such potential energy surface is given, i.e. the H_3 (and its isotopic analogs) surface. This makes the comparison for $D + H_2$, for instance, relevant but rules out any meaningful comparison for the $F + H_2(D_2)$ system.

Our findings may be summarized as follows:

(a) Total Integral Cross Sections

One of the most reliable physical magnitudes that is expected from a numerical treatment is the total cross section. In general the RIOSA ℓ-av total cross sections were found to overlap the corresponding QCT results reasonably well. The only exception is the $H + H_2(v_i = 0)$ case where the deviations are somewhat larger than usual. On the other hand, the RIOSA ℓ-in results, except for the $H + H_2(v_i = 0)$ case, deviate significantly from the QCT results. This particular problem will be discussed further somewhat later.

(b) Vibrational State-to-state integral cross sections

The QCT vibrational state-to-state integral cross sections (VSTSICS) are not expected to be similar to the corresponding QM magnitudes. This belief is based on studies on inelastic and collinear reactive collisions. Thus a fit or a misfit between RIOSA and QCT VSTSICS would not imply whether the RIOSA treatment is relevant or not. The only case from which we can deduce some information on this subject is the $F + H_2(v_i = 0) \rightarrow HF(v_f) + H$ reactions for which CS results exist. The only problem with these CS calculations is that in order to obtain converged results the Muckerman V potential was somewhat changed. The change is expected to affect the total cross section significantly (which was found to be half the size of the corresponding QCT value) and vibrational branching ratio to a lesser extent. As can be seen from fig. 10 the QCT branching ratio $\Gamma(2,3)$ is energy independent, whereas both the CS and the RIOSA yield branching ratios which are strongly energy dependent.

There is also a CS calculation for the reactions $H + H_2(v_i = 1) \rightarrow H_2(v_f) + H$; $v_f = 0,1$ on the Porter-Karplus potential energy surface.[39]. However, due to numerical problems, no fully converged results were obtained. Again, this lack of convergence may affect the absolute values but is expected to yield more reliable values for branching ratios. Here we find that the classical $\Gamma(1,0)$ values are ~ 2 whereas both the CS and RIOSA yield values of ~ 4.

(c) Differential Cross Sections

In comparison with the integral cross sections, a statement regarding the relevance of the RIOSA differential cross section is less conclusive. Except for $H + H_2(v_i = 0)$, EQM differential cross sections are not available and here the fit between EQM and RIOSA results is not satisfactory. Both treatments yield backwards scattering but the EQM curve decreases faster as θ moves away from π. In comparing the RIOSA differential cross sections with QCT results similar deviations were observed; in fact for all cases studied the RIOSA distribution is broader than the classical one. In this respect it should be mentioned that experimental findings support the RIOSA curves and not the QCT ones. The RIOSA, in contrast to the QCT, yields sidewards distribution for higher energies (and for several systems). Angular distributions of this type were found for $F + H_2(D_2)$ (in the same energy range) and there is evidence that this is also the case for $D + H_2$. It could well be that the QCT treatment did not result in sidewards distribution for $F + H_2(D_2)$ due to the fact that the surface is wrong. However, the reliability of the RIOSA treatment would be enhanced if stronger experimental evidence for the sidewards $D + H_2$ angular distribution were available.

(d) Rate Constants

In comparing rate constants one has to be careful because a fit or misfit reflects only on the quality of the lowest portion of energy dependent integral cross sections. Consequently a good fit with QCT results is not necessarily a good sign for a theory because in that energy region the QCT treatment is not always relevant. Nevertheless it was found for $D + H_2(v_i = 0)$ that the QCT results (derived for a correct surface) deviate only slightly from the experimental findings. The RIOSA results (both the ℓ-av and the ℓ-in) were found to be slightly higher than the experimental ones in the low temperature range, but at higher temperatures were very close being the same. A less satisfactory fit was obtained for $v_i = 1$ where the experimental results are about 3-5 times larger than the RIOSA results (the QCT results are two times smaller than the RIOSA results).

A similar situation was encountered in the $F + H_2$ system. The RIOSA results are about 2-3 times larger than the QCT results and about 2-3 times smaller than the experimental ones. However one has to remember that the M5 surface is wrong and so comparison with experiment is probably meaningless.

(e) ℓ-in RIOSA versus ℓ-av RIOSA

When presenting RIOSA integral cross sections both the ℓ-av and the ℓ-in results are always shown. We consistently find that the ℓ-av cross sections compare reasonably well with the QCT results, while the ℓ-in values are always smaller. For vibrational branching ratios no such consistent picture is encountered; whereas for $F + H_2(D_2)$ the results were almost the same slight differences were encountered for $D + H_2(v_i = 1)$ and much larger deviations were obtained for $H + H_2(v_i = 1)$. The question, therefore, is what could be the reason for the large deviations (in the total integral cross sections). There is one source for that discrepancy which is also encountered in the inelastic case and which is due to the different weightings of the various terms in the final summation over ℓ_λ. However from studies of inelastic collisions it is known that ℓ-in and ℓ-av total integral cross sections differ only to a certain extent. Thus there must be another reason which becomes apparent only in exchange collisions. In order to look for that cause we refer to eqs. (12)-(16). The total integral cross sections within the ℓ-in formulation is obtained by employing eq. (16). It is noticed that all what is required is the corresponding primitive γ_λ dependent S-matrix elements. To calculate the total integral cross section within the ℓ-av formulation eqs. (12), (13) and (15) have to be employed. In addition to the primitive S-matrix elements, also the angle of rotation $\Delta(\gamma_\lambda,B)$ (see fig. 1) appears in the expression (this angle is of course missing in the corresponding inelastic expressions). We have calculated, on several occasions, integral (and differential) cross sections employing the ℓ-av expressions, but making $\Delta(\gamma_\lambda,B) = \pi$ (its value at $\gamma_\lambda = 0$). It was found that these modified ℓ-av results are rather close to the ℓ-in results obtained As an example we show in fig. 15 differential cross sections (multiplied by $\sin\theta$) as calculated for $H + H_2$ for the energy $E_{tot} = 0.7$ eV. Four curves are shown, i.e. EQM, ℓ-in, ℓ-av and the modified ℓ-av$(\Delta = \pi)$. It is seen that the ℓ-in and ℓ-av$(\Delta = \pi)$

results are almost identical but that the ℓ-av curve is significantly different and much closer to the EQM curve. Thus, the angle Δ which is eliminated by the algebra within the ℓ-in formulation seems to be essential for obtaining the better results.

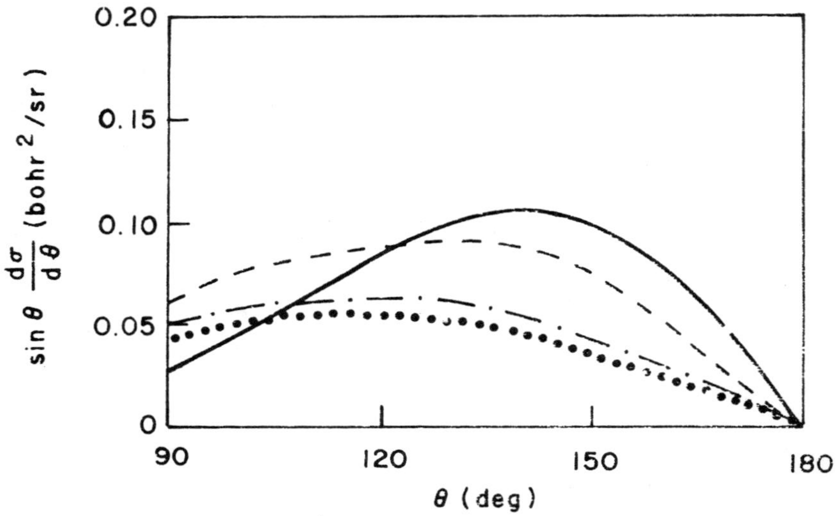

Figure 15. Differential cross sections multiplied by sinθ as obtained for the reaction $H_2(v_i = 0) + H \to H + H_2$ for the total energy $E_{tot} = 0.7$ eV. —— EQM results (Ref. 20), --- ℓ-av results, -.-.- ℓ-in results, ... ℓ-av ($\Delta = \pi$) results.

Support for this finding is given in fig. 9 where QCT, ℓ-in, ℓ-av and CIOSA results are compared. (In that figure the CS results are also given but they, are not relevant for the present discussion.) The CIOSA cross sections are obtained by integrating over the quasi-collinear reactive probabilities - an expression similar to eq. (16). Thus, in this sense, the CIOSA cross sections are ℓ-in type results and as is noticed they are almost identical to the quantum ℓ-in results. This fact supports the previous finding for $H + H_2$ where EQM and QCT cross sections were, for energies above threshold, very similar. Since, as a rule, the ℓ-av cross sections fit the QCT cross sections rather well, the ℓ-av results should be considered to be the more relevant of the two.

There is still a question to be asked: "What is the meaning of the ℓ-av CIOSA?" Knowing the answer to this question will give us a better understanding for the ℓ-av RIOSA in general.

D. J. K. acknowledges the Petroleum Research Fund, administered by the American Chemical Society, for partial support of this research. He also acknowledges partial support under R A Welch Foundation Grant E-608.

REFERENCES

1. Theory of Chemical Reaction Dynamics, edited by M. Baer (Chemical Rubber Company, Boca Raton, 1985).
2. G.C. Schatz and A. Kuppermann, J. Chem. Phys. 62, 2502 (1925); A.B. Elkowitz and R.E. Wyatt, J. Chem. Phys. 62, 2504 (1925); R.B. Walker, E.B. Stechel and J.C. Light, J. Chem Phys. 69, 2922 (1978).
3. A. Kuppermann, G.C. Schatz and M. Baer, J. Chem. Phys. 65, 4596 (1976)
4 (a) V. Khare, D.J. Kouri and M. Baer, J. Chem. Phys. 71, 1188 (1979)
(b) M. Baer, D.J. Kouri and J. Jellinek, J. Chem. Phys. 80, 1431 (1984)
(c) J. Jellinek and D.J. Kouri, in Theory of Chemical Reaction Dynamics Vol. II, edited by M. Baer (Chemical Rubber Company, Boca Raton 1985).
5. M.E. Rose, Elementary Theory of Angular Momentum (Wiley, NY 1952).
6. A.M. Arthurs and A. Dalgarno, Proc. Roy. Soc. London Ser. A. 256, 540 (1960).
7. D.J. Kouri, V.J. Khare and M. Baer, J. Chem. Phys. 75, 1181 (1981).
8. J. Jellinek and M. Baer, J. Chem. Phys. 76, (1982).
9. N. AbuSalbi, D.J. Kouri, Y. Shima and M. Baer, J. Chem. Phys. 82, 2650 (1985).
10. N. AbuSalbi, S-H Kim, D.J. Kouri, M. Baer, Chem. Phys. Lett. 112, 502 (1984).
11. M. Baer, J. Jellinek and D.T. Kouri, J. Chem. Phys. 78, 2962 (1983).
12. N. AbuSalbi, C.L. Shoemaker, D.J. Kouri, J. Jellinek and M. Baer, J. Chem. Phys. 80, 3210 (1984).
13. M. Baer in Theory of Chemical Reaction Dynamics, Vol. I, edited by M. Baer (Chemical Rubber Company, Boca Raton, 1985).
14. D.C. Clary and G. Drolshagen, J. Chem. Phys. 76, 5027 (1982); D.C. Clary, Chem. Phys. 81 379 (1983).
15. M. Baer, H.R. Mayne, V. Khare and D.J. Kouri, Chem. Phys. Lett. 72, 269 (1980).
16. J.M. Bowman and K.T. Lee, J. Chem. Phys. 72, 5071 (1980).
17. R.N. Porter and M. Karplus, J. Chem. Phys. 40, 1105 (1964).
18. M. Karplus, R.N. Porter and R.D. Sharma, J. Chem. Phys. 43, 3259 (1965).
19. H. Mayne, Chem. Phys. Lett. 66, 487 (1967).
20. G.C. Schatz and A. Kuppermann, J. Chem. Phys. 65, 4668 (1976).
21. P. Siegbahn and B. Liu, J. Chem. Phys. 68, 2457 (1978): D.G. Truhlar and C.J. Horowitz, J. Chem. Phys., 68, 2466 (1978); 71, 1514 (1979).
22. H.R. Mayne and J.P. Toennies, J. Chem. Phys. 75, 1794 (1981).
23. J.P. Toennies, Private Communication, 1985.
24. A.A. Westenberg and N. De Haas, J. Chem. Phys. 47, 1393 (1967).
25. D.N. Mitchell and D.J. LeRoy, J. Chem. Phys. 58, 3449 (1973).
26. G.P. Glass and B.K. Chaturverdi, J. Chem. Phys. 77, 3478 (1982).
27. V. Wellhausen, Ph.D. Dissertation, University of Göttingen 1984; see also E. Pollak, N. AbuSalbi and D.J. Kouri, Chem. Phys. Lett. 113, 585 (1985).

28. V.B. Rozenshteyn, Yu. M. Gershenzon, A.V. Ivanov and S.I. Kucheryavii, Chem. Phys. Lett. 105, 423 (1984).
29. C.A. Boonenberg and H.R. Mayne, Chem. Phys. Lett. 108, 67 (1984).
30. Z.H. Zhang, N. AbuSalbi, M. Baer, D.J. Kouri and J. Jellinek, in Resonances in Electron and Molecule Scattering van der Waals Complexes and Reactive Chemical Dynamics, edited by D.G. Truhlar (American Chemical Society, 1984).
31. J.T. Muckerman, in Theoretical Chemistry: Advances and Perspectives, Vol. 6A edited by H. Eyring and D.H. Henderson (Academic Press, NY, 1981) p. 1.
32. M.J. Redmon and R.E. Wyatt, Int. J. Quant. Chem. 11, 343 (1977), Chem. Phys. Lett. 63 209 (1979); see also R.E. Wyatt, J.F. McNutt and M.J. Redmon, Ber. Bunsenges. Phys. Chem. 86 , 437 (1982).
33. J.N.L. Connor, W. Jakubetz, J. Manz and J.C. Whitehead, Chem. Phys. 39 395 (1979).
34. D.F. Feng, E.R. Grant and J.W. Root, J. Chem. Phys. 64, 3450 (1976).
35 J. Jellinek and M. Baer, Chem. Phys. Lett. 82, 162 (1981); J. Chem. Phys. 78, 4494 (1983).
36. D.M. Neumark, A.M. Wodtke, G.N. Robinson, C.C. Hayden and Y.T. Lee, J. Chem. Phys. 82, 3045 (1985).
37. S. Ron, M. Baer and E. Pollak, J. Chem. Phys. 78, 4414 (1983).
38. S. Ron, E. Pollak and M. Baer, J. Chem. Phys. 79, 5204 (1983)
39. G.C. Shatz, Chem. Phys. Lett. 94, 183 (1983).

HYPERSPHERICAL COORDINATE FORMULATION OF THE ELECTRON – HYDROGEN ATOM SCATTERING PROBLEM*

Diane M. Hood and Aron Kuppermann
A. A. Noyes Laboratory of Chemical Physics†
California Institute of Technology
Pasadena, CA 91125 USA

ABSTRACT A formulation is presented for the use of hyperspherical coordinates and local hyperspherical surface functions in the electron-hydrogen atom scattering problem. Some representative numerical results of the application of this formulation are given.

1. INTRODUCTION

The use of hyperspherical coordinates and local surface functions in electron-atom scattering problems[1] and in 3D reactive scattering problems[2] has been suggested for over a decade, but so far converged calculations of differential or integral cross sections of inelastic or reactive processes using this methodology have not been published. The formalism is conceptually simple and in principle very powerful, affording a united treatment of non-reactive and reactive processes for molecule-molecule collisions, and of direct and exchange processes for electron-molecule collisions. It has by now been extensively tested for collinear atom-diatom reactive scattering.[3,4]

One of the difficulties in applying this approach is the accurate and efficient calculation of local hyperspherical surface functions, especially for reactive scattering processes. In the case of the electron-hydrogen atom system, these difficulties are alleviated by the symmetry of the system and the large proton to electron mass ratio. As a result, this is a very convenient system for the application and testing of this methodology. It is also, in some senses, an extreme prototype of light-heavy-light triatomic reactive systems in which the light-light arrangement is either not bound or very closed for energetic reasons.

In Section 2 we define the symmetrized hyperspherical coordinates for the electron-hydrogen atom system and express the hamiltonian in these coordinates. In Section 3 symmetry is discussed. The appropriate symmetry wave functions are introduced in Section 4, the local surface eigenfunctions and energy eigenvalues in Section 5, and the scattering equations and asymptotic analysis in Section 6. Finally, some representative results are given and discussed in Section 7 and a summary of the conclusions is presented in Section 8.

D. C. Clary (ed.), The Theory of Chemical Reaction Dynamics, 193–214.

2. SYMMETRIZED HYPERSPHERICAL COORDINATES

Let the two electrons be designated e_1 and e_2 and the proton p. Let $\mathbf{r}_i$ be the p to e_i vector and $\mathbf{R}_i$ the vector from the center of mass of the p, e_i particle pair to e_j $(i, j = 1, 2,\ i \neq j)$. We introduce the Delves mass-scaled coordinates[5]

$$\begin{aligned} \mathbf{r}'_i &= \left(\frac{\mu_i}{\mu}\right)^{1/2} \mathbf{r}_i \\ \mathbf{R}'_i &= \left(\frac{\nu_i}{\mu}\right)^{1/2} \mathbf{R}_i \end{aligned} \tag{2.1}$$

where μ is the overall reduced mass of the system, μ_i the reduced mass of the p, e_i pair and ν_i the reduced mass of the pe_i, e_j pair. If m and M are the electron and proton masses, respectively, those reduced masses are given by

$$\begin{aligned} \mu &= \left(\frac{m^2 M}{2m + M}\right)^{1/2} \\ \mu_i &= \frac{mM}{m + M} \\ \nu_i &= \frac{m(m + M)}{2m + M} \end{aligned} \tag{2.2}$$

In terms of these mass-scaled coordinates, the hamiltonian for the system, with the motion of its center of mass omitted, is

$$\widehat{H} = -\frac{\hbar^2}{2\mu}(\nabla^2_{\mathbf{r}'_i} + \nabla^2_{\mathbf{R}'_i}) + V_i(r'_i, R'_i, \gamma_i) \tag{2.3}$$

where γ_i is the angle between the $\mathbf{r}'_i$ and $\mathbf{R}'_i$ (or $\mathbf{r}_i$ and $\mathbf{R}_i$) vectors and V_i the interaction potential among the three particles.

The system's hyperradius ρ and hyperangle ω_i are defined by

$$\begin{aligned} \rho &= (\mathbf{r}'^2_i + \mathbf{R}'^2_i)^{\frac{1}{2}} \\ \omega_i &= 2\tan^{-1}\frac{r'_i}{R'_i} \qquad 0 \leq \omega_i \leq \pi \end{aligned} \tag{2.4}$$

The value of ρ is independent of the choice of i whereas that of ω_i is not.[2a,6] The hyperspherical coordinate system i is defined as the set of coordinates formed by ρ, ω_i, the spaced-fixed polar angles θ_i, φ_i of $\mathbf{R}'_i$ (or $\mathbf{R}_i$) and two polar angles which determine the orientation of $\mathbf{r}'_i$ (or $\mathbf{r}_i$). The latter can be chosen either as $\theta_{\mathbf{r}_i}$, $\varphi_{\mathbf{r}_i}$, the space-fixed frame angles, or γ_i, ψ_i, the body-fixed frame angles for which $\mathbf{R}'_i$ is the direction of the corresponding z-axis. Other choices are possible for the four angles which determine the orientations of $\mathbf{r}'_i$ and $\mathbf{R}'_i$, but will not concern us here.

The hyperradius ρ and hyperangle ω_i play a central role in our formulation. In terms of them, the system's kinetic energy operator is given by

$$\widehat{T}_i(\rho, \omega_i, \mathbf{\Omega}_i) = \widehat{T}_\rho(\rho) + \frac{\hat{L}_i^2(\omega_i)}{2\mu\rho^2} + \frac{\hat{j}_i^2(\mathbf{\Omega}_i)}{2\mu\rho^2 \sin^2\frac{\omega_i}{2}} + \frac{\hat{l}_i^2(\mathbf{\Omega}_i)}{2\mu\rho^2\cos^2\frac{\omega_i}{2}} \tag{2.5}$$

where $\hat{j}_i^2$ and $\hat{l}_i^2$ are, respectively, the square of the angular momenta associated with the $\mathbf{r}_i$ and $\mathbf{R}_i$ vectors and depend on the four orientation angles denoted collectively by $\mathbf{\Omega}_i$. The symbols $\widehat{T}_\rho$ and $\hat{L}_i^2$ represent, respectively, the hyperradial kinetic energy and hyperangular angular momentum operators defined by

$$\widehat{T}_\rho = -\frac{\hbar^2}{2\mu}\left(\frac{\partial^2}{\partial\rho^2} + \frac{5}{\rho}\frac{\partial}{\partial\rho}\right) \tag{2.6}$$

and

$$\hat{L}_i^2 = -4\hbar^2\left(\frac{\partial^2}{\partial\omega_i^2} + 2\cot\omega_i\frac{\partial}{\partial\omega_i}\right) \tag{2.7}$$

The grand canonical angular momentum operator $\hat{\Lambda}_i^2$ is defined as

$$\hat{\Lambda}_i^2(\omega_i, \mathbf{\Omega}_i) = \hat{L}_i^2 + \frac{\hat{j}_i^2}{\sin^2\frac{\omega_i}{2}} + \frac{\hat{l}_i^2}{\cos^2\frac{\omega_i}{2}} \tag{2.8}$$

and in terms of it we have

$$\widehat{T}_i(\rho, \omega_i, \mathbf{\Omega}_i) = \widehat{T}_\rho(\rho) + \frac{\hat{\Lambda}_i^2(\omega_i, \mathbf{\Omega}_i)}{2\mu\rho^2} \tag{2.9}$$

So far, these expressions are exact. The potential energy function, in terms of r_1, r_2 and the angle γ between the corresponding vectors, is given by the simple expression

$$V_i(r_1, r_2, \gamma) = -\frac{e^2}{r_1} - \frac{e^2}{r_2} + \frac{e^2}{r_{12}} \tag{2.10}$$

where

$$r_{12} = \left(r_1^2 + r_2^2 - 2r_1r_2\cos\gamma\right)^{\frac{1}{2}} \tag{2.11}$$

When the hyperspherical coordinates ρ, ω_i, γ_i are used, this expression is not as simple; nevertheless, this use does not lead to any significant difficulties. However, major simplifications result if we notice that μ, μ_i and ν_i are all equal to m to order m/M, and to that order $\mathbf{r}'_i$ equals $\mathbf{r}_i$, $\mathbf{R}'_i$ equals $\mathbf{r}_j$, γ_1 and γ_2 equal γ, $\hat{j}_i^2$ equals $\hat{l}_j^2$ and ω_j equals $\pi - \omega_i$. In the limit of vanishing m/M we can rewrite (2.5) and (2.10) respectively as

$$\widehat{T}_i = \widehat{T}_\rho + \frac{\hat{L}_i^2}{2\mu\rho^2} + \frac{\hat{l}_j^2}{2\mu\rho^2\sin^2\frac{\omega_i}{2}} + \frac{\hat{l}_i^2}{2\mu\rho^2\cos^2\frac{\omega_i}{2}} \tag{2.12}$$

and

$$V_i(\rho, \omega_i, \gamma) = -\frac{e^2}{\rho}\left[\frac{1}{\cos\frac{\omega_i}{2}} + \frac{1}{\sin\frac{\omega_i}{2}} - \frac{1}{\sqrt{1-\sin\omega_i\cos\gamma}}\right] \tag{2.13}$$

In the rest of this paper, we will use this infinite proton mass approximation. The inaccuracies it produces are smaller than those resulting from other sources of error in the computations we have performed so far, and if desired can be corrected by either a first order perturbation theory approach or a repetition of the calculations without using the approximation. The last two expressions explicitly display the symmetry of the system, and lead to interesting insights, which justify the slight error they produce.

3. SYMMETRY CONSIDERATIONS

The hamiltonian of the electron-hydrogen atom system is invariant under (a) an exchange of the two electrons, and (b) an inversion of the system through its center of mass (which in the infinite proton mass approximation is the same as an inversion through the proton). The invariance under exchange brings about a quantum number which, for subsequent reasons related to the Pauli principle, is called the spin quantum number S; for wave functions whose orbital part is symmetric/antisymmetric with respect to exchange (and whose spin part is antisymmetric/symmetric) this quantum number is equal to zero/one and corresponds to singlet/triplet states, the exchange parity being therefore $(-1)^S$. The invariance under inversion brings about the quantum number Π which is zero/one for symmetric/antisymmetric states which are said to have even/odd parity.

In addition, if we make a plot of the equipotentials of V in a system of coordinates $OX_iY_iZ_i$ defined by

$$\begin{aligned} X_i &= \rho\sin\omega_i\cos\gamma_i \\ Y_i &= \rho\sin\omega_i\sin\gamma_i \\ Z_i &= \rho\cos\omega_i \end{aligned} \tag{3.1}$$

these equipotentials have a plane of symmetry which, in the infinite proton mass approximation, becomes the OX_iY_i plane. Furthermore, an $i \to j$ coordinate transformation rotates the equipotentials by a fixed angle around the OY_i axis, without changing their shape.[2a] In the infinite proton mass approximation this angle becomes 180°. These properties permit a convenient visualization of the nature of local hyperspherical surface functions when projected onto this space.

A plot of cuts of the equipotential surfaces of V by the OX_iY_i and OX_iZ_i planes is given in Figure 1. For convenience of display, these plots had the range of γ extended from 0 to π to 0 to 2π by adopting the convention $V(\rho, \omega_i, 2\pi-\gamma) = V(\rho, \omega_i, \gamma)$. This reflection of the equipotentials through the OX_iZ_i plane is not used in the actual calculations, since the range of γ_i is only 0 to π. Cuts of equipotentials by planes parallel to OX_iY_i approach circles as $|Z_i|$ increases, since, for a fixed r_i, V becomes independent of γ as r_j increases. This conclusion can also be reached from an analysis of (2.13).

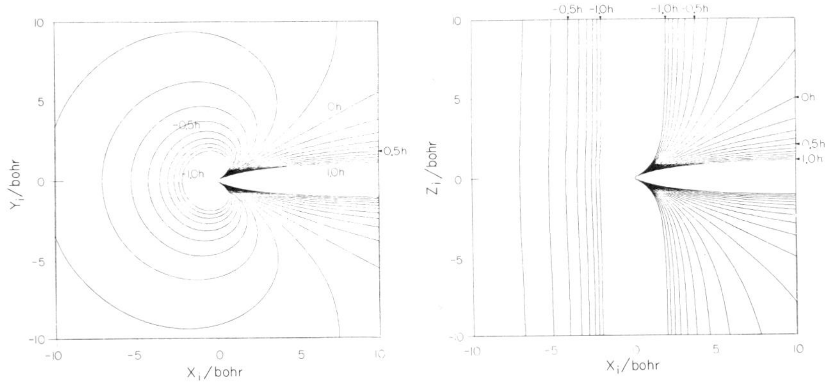

Figure 1. Equipotential surfaces for the electron-hydrogen system in the coordinates defined by Eqs. (2.13). The right (left) part of the figure displays cuts of these equipotentials through the OX_iZ_i (OX_iY_i) planes. The signed numbers on top and right margins of the panels designate the energy of the associated equipotentials in hartree. The energy spacing between neighboring equipotentials is 0.1 hartree.

4. SYMMETRY WAVE FUNCTIONS

For the purpose of calculating wave functions which satisfy the correct physical scattering conditions, it becomes useful to define the symmetry wave functions $\Psi^{JMS\Pi}$ as the solutions of the set of eigenfunction equations

$$\begin{aligned} \hat{H}\Psi^{JMS\Pi} &= E\Psi^{JMS\Pi} \\ \hat{J}^2\Psi^{JMS\Pi} &= J(J+1)\hbar^2\Psi^{JMS\Pi} \\ \hat{J}_z\Psi^{JMS\Pi} &= M\hbar\Psi^{JMS\Pi} \\ \hat{P}_{12}\Psi^{JMS\Pi} &= (-1)^S\Psi^{JMS\Pi} \\ \hat{\mathfrak{I}}\Psi^{JMS\Pi} &= (-1)^\Pi\Psi^{JMS\Pi} \end{aligned} \tag{4.1}$$

In these expressions, the operators appearing on the left hand side are, respectively, the hamiltonian of the system, the square and the laboratory-fixed z-component of its total spacial angular momentum, and the electron exchange and inversion through the center of mass operators. The corresponding eigenvalues are E, $J(J+1)\hbar^2$, $M\hbar$, $(-1)^S$, and $(-1)^\Pi$. For scattering problems E is given *a priori* and is larger than the energy E_0 of an electron plus an infinitely removed hydrogen atom in its ground state. J and M are integers satisfying the usual relations $J > 0$ and $|M| \leq J$, and S and Π have the allowed values discussed in Section 3. The physical scattering wave functions can be expressed as linear combinations of the $\Psi^{JMS\Pi}$.

5. LOCAL SURFACE FUNCTIONS AND ENERGIES

5.1 Definition

We define the local surface functions $\Phi_k^{JMS\Pi}(\omega_i, \mathbf{\Omega}_i; \rho)$ as the simultaneous eigenfunctions of $\hat{J}^2$, $\hat{J}_z$, $\hat{P}_{12}$, $\hat{\mathfrak{I}}$ and the hamiltonian $\overline{H}(\omega_i, \mathbf{\Omega}_i; \rho)$ obtained from $\hat{H}$ by omitting $\hat{T}_\rho(\rho)$. In other words, $\overline{H}$ is the hamiltonian of a particle of mass μ confined to move on a 5-dimensional hypersphere of radius ρ, subject to the potential V. It depends on ρ only parametrically, and is given explicitly by

$$\overline{H}(\omega_i, \mathbf{\Omega}_i; \rho) = \frac{\hat{\Lambda}_i^2}{2\mu\rho^2} + V_i(\rho, \omega_i, \gamma) \tag{5.1}$$

In addition to satisfying the last four of eqs. (4.1), the $\Phi_k^{JMS\Pi}$ satisfy the eigenfunction-eigenvalue equation

$$\overline{H}\Phi_k^{JMS\Pi}(\omega_i, \mathbf{\Omega}_i; \rho) = \varepsilon_k^{JS\Pi}(\rho)\ \Phi_k^{JMS\Pi} \tag{5.2}$$

with the requirement that they be single-valued and continuous. This condition quantizes $\varepsilon_k^{JS\Pi}(\rho)$, making the $\Phi_k^{JMS\Pi}$ constitute an infinite discrete complete set of eigenfunctions of the space spanned by the five angles $\omega_i, \mathbf{\Omega}_i$. The symbol k stands for a set of appropriate quantum numbers which label the functions of this set. For each ρ these functions span regions of configuration space corresponding to the bound arrangement channels $e_1p + e_2$ and $e_2p + e_1$, as well as $e_1 + e_2 + p$ configurations, and are therefore capable of describing the ionization continuum. For large values of ρ, and for $\varepsilon_k^{JS\Pi}(\rho)$ below the ionization energy of the hydrogen atom, the corresponding $\Phi_k^{JMS\Pi}$ are localized mainly in the regions of configuration space corresponding to bound arrangements. The reason for this behavior is the presence of V_i in (5.1). This makes the $\Phi_k^{JMS\Pi}$ constitute a good basis set for describing bound-to-bound scattering processes. By constrast, the K-harmonics or hyperspherical harmonics,[7] which are defined as the eigenfunctions of the grand canonical angular momentum $\hat{\Lambda}_i^2$, are spread out over the entire $\omega_i, \mathbf{\Omega}_i$ angular space, and form a very slowly convergent basis set for expanding bound-to-bound scattering wave functions.[8]

The local surface energy eigenvalues $\varepsilon_k^{JS\Pi}(\rho)$ are degenerate with respect to M and tend to the eigenvalues of an isolated hydrogen atom as ρ increases. Considered as functions of ρ, they are conceptually analogous to the potential energy curves for the electronically adiabatic states of a diatomic molecule, where the independent variable which is "frozen" is the internuclear distance rather than the system's hyperradius. In the large ρ limit, the $\Phi_k^{JMS\Pi}$'s become linear combinations of the degenerate hydrogen atom eigenfunctions having the same eigenvalues $\varepsilon_k^{JS\Pi}(\infty)$.

5.2 Determination of local surface functions and eigenvalues

In order to determine the $\Phi_k^{JMS\Pi}(\omega_i, \mathbf{\Omega}_i; \rho)$ and $\varepsilon_k^{JS\Pi}(\rho)$ it is convenient to choose for $\mathbf{\Omega}_i$ the space-fixed orientation angles θ_i, φ_i, θ_{r_i}, φ_{r_i}, i. e., θ_i, φ_i, θ_j, φ_j. From now on we set $i = 1$, $j = 2$. We now expand the $\Phi_k^{JMS\Pi}$ in the eigenfunctions $\mathcal{Y}_{l_2 l_1}^{JM}$ of $\hat{J}^2$, $\hat{J}_z$, $\hat{l}_1^2$, $\hat{l}_2^2$, according to

$$\Phi_k^{JMS\Pi}(\omega_1, \mathbf{\Omega}_1; \rho) = \sum_{l_2 l_1} \mathcal{Y}_{l_2 l_1}^{JM}(\theta_2, \varphi_2; \theta_1, \varphi_1) f^{JS\Pi}{}_{l_2 l_1}^{k}(\omega_1; \rho) \tag{5.3}$$

where

$$\mathcal{Y}_{l_2 l_1}^{JM}(\theta_2, \varphi_2; \theta_1, \varphi_1) = \sum_{m_1 m_2} C(l_1 l_2 J; m_1 m_2 M) Y_{l_2 m_2}(\theta_2, \varphi_2) Y_{l_1 m_1}(\theta_1, \varphi_1) \tag{5.4}$$

the Y_{lm} and C being, respectively, the usual spherical harmonics and Clebsch-Gordan coefficients. Replacement of (5.3) into (5.2) leads to the following set of coupled equations in the coefficients $f^{JS\Pi}{}_{l_2 l_1}^{k}$:

$$\begin{aligned} &\frac{1}{2\mu\rho^2}\left[\hat{L}_1^2 + \frac{l_2(l_2+1)}{\cos^2\frac{\omega_1}{2}} + \frac{l_1(l_1+1)}{\sin^2\frac{\omega_1}{2}}\right] f^{JS\Pi}{}_{l_2 l_1}^{k}(\omega_1; \rho) \\ &\quad + \sum_{l_2' l_1'} V^{J\Pi}{}_{l_2' l_1'}^{l_2 l_1}(\omega_1; \rho) f^{JS\Pi}{}_{l_2' l_1'}^{k}(\omega_1; \rho) = \varepsilon_k^{JS\Pi}(\rho) f^{JS\Pi}{}_{l_2 l_1}^{k}(\omega_1; \rho) \end{aligned} \tag{5.5}$$

where the $V^{J\Pi}{}_{l_2' l_1'}^{l_2 l_1}$ are the surface potential matrix elements

$$V^{J\Pi}{}_{l_2' l_1'}^{l_2 l_1}(\omega_1; \rho) = \langle \mathcal{Y}_{l_2' l_1'}^{JM} \mid V_1(\rho, \omega_1, \gamma) \mid \mathcal{Y}_{l_2 l_1}^{JM} \rangle \tag{5.6}$$

and can be calculated analytically.

We now define $t_p^{Jl_2 l_1}(\omega_1; \rho)$ as the solutions of the decoupled equations obtained from (5.6) by omitting all coupling terms:

$$\begin{aligned} &\left[\frac{1}{2\mu\rho^2}\left(\hat{L}_1^2 + \frac{l_1(l_1+1)}{\sin^2\frac{\omega_1}{2}} + \frac{l_2(l_2+1)}{\cos^2\frac{\omega_1}{2}}\right) + V^{J\Pi}{}_{l_2 l_1}^{l_2 l_1}(\omega_1; \rho)\right] t_p^{Jl_2 l_1}(\omega_1; \rho) \\ &\quad = \nu_p^{J\Pi l_2 l_1}(\rho)\, t_p^{Jl_2 l_1}(\omega_1; \rho) \end{aligned} \tag{5.7}$$

In this expression p is a symbol which labels the solution being considered. If we further drop the $V^{J\Pi}{}^{l_2 l_1}_{l_2 l_1}(\omega_1;\rho)$ term from this equation its solutions become the analytically known Jacobi polynomials,[5] which when multiplied by $\mathcal{Y}^{JM}_{l_2 l_1}$ yield the hyperspherical harmonics discussed in Section 5.1. Changing from $t_p^{Jl_2 l_1}(\omega_1;\rho)$ to the function $T_p^{Jl_2 l_1}(\omega_1;\rho)$ defined by

$$T_p^{Jl_2 l_1}(\omega_1;\rho) = \sin\omega_1\, t_p^{Jl_2 l_1}(\omega_1;\rho) \tag{5.8}$$

we get the uncoupled equations

$$-\frac{2\hbar^2}{\mu\rho^2}\frac{d^2}{d\omega_1^2}T_p^{Jl_2 l_1}(\omega_1;\rho) + V_{\text{eff}}^{Jl_2 l_1}\,T_p^{Jl_2 l_1}(\omega_1;\rho) = \nu_p^{J\Pi l_2 l_1}(\rho)\;T_p^{Jl_2 l_1}(\omega_1;\rho) \tag{5.9}$$

where $V_{\text{eff}}^{Jl_2 l_1}$ is the effective potential defined as

$$V_{\text{eff}}^{Jl_2 l_1}(\omega_1;\rho) = -\frac{\hbar^2}{2\mu\rho^2}\left[4 - \frac{l_2(l_2+1)}{\cos^2\frac{\omega_1}{2}} - \frac{l_1(l_1+1)}{\sin^2\frac{\omega_1}{2}}\right] + V^{J\Pi}{}^{l_2 l_1}_{l_2 l_1}(\omega_1;\rho) \tag{5.10}$$

and the following boundary condition must be satisfied in order that $t_p^{Jl_2 l_1}(\omega_1;\rho)$ not be divergent at $\omega_1 = 0$ and $\omega_1 = \pi$:

$$T_p^{Jl_2 l_1}(\omega_1 = 0;\rho) = T_p^{Jl_2 l_1}(\omega_1 = \pi;\rho) = 0 \tag{5.11}$$

Eq. (5.9) is easily solved by finite difference or finite element numerical methods.

If l_2 equals l_1 the effective potential $V_{\text{eff}}^{Jl_1 l_1}$ is symmetric with respect to $\omega_1 = \frac{\pi}{2}$. The $T_p^{Jl_1 l_1}$ functions are therefore either symmetric or antisymmetric with respect to $\omega_1 = \frac{\pi}{2}$ (i. e., with respect to electron interchange), and are obtained in separate calculations. As a result

$$\begin{aligned} t_p^{Jl_2 l_1}(\pi-\omega_1;\rho) &= t_p^{Jl_1 l_2}(\omega_1;\rho), \quad \text{for } l_2 \neq l_1 \\ t_p^{Jl_2 l_1}(\pi-\omega_1;\rho) &= \pm\, t_p^{Jl_2 l_1}(\omega_1;\rho) \quad \text{for } l_2 = l_1 \end{aligned} \tag{5.12}$$

The symmetry of $\mathcal{Y}^{JM}_{ll}$ is determined entirely by J, since this function is always even with respect to inversion through the proton (i. e., $(-1)^{\Pi} = +1$). If J is even, the singlet functions T_p^{Jll} will be symmetric with respect to $\omega_1 = \frac{\pi}{2}$ and if J is odd, then it is the triplet basis functions which will have this symmetry.

As a result of this symmetry analysis we conclude that the simple product $\mathcal{Y}^{JM}_{l_2 l_1}(\theta_2,\varphi_2;\theta_1,\varphi_1)\,t_p^{Jl_2 l_1}(\omega_1;\rho)$ does not possess, in general, exchange symmetry.

Consequently, it becomes useful to define, for $l_2 > l_1$, the 5-dimensional angular functions

$$\Theta_{l_2 l_1 p}^{JMS\Pi}(\omega_1, \theta_1, \varphi_1, \theta_2, \varphi_2; \rho) = N_{l_2 l_1}\left[\mathcal{Y}_{l_2 l_1}^{JM}(\theta_2, \varphi_2; \theta_1, \varphi_1)\, t_p^{Jl_2 l_1}(\omega_1; \rho) + (-1)^{\Lambda}\mathcal{Y}_{l_1 l_2}^{JM}(\theta_2, \varphi_2; \theta_1, \varphi_1)\, t_p^{Jl_2 l_1}(\pi - \omega_1; \rho)\right] \tag{5.13}$$

where

$$\Lambda = S + J - \Pi \tag{5.14}$$

and $N_{l_2 l_1}$ is a normalization constant. For $l_1 = l_2 = l$ we define $\Theta_{llp}^{JMS\Pi}$ by

$$\Theta_{llp}^{JMS\Pi}(\omega_1, \theta_1, \varphi_1, \theta_2, \varphi_2; \rho) = \mathcal{Y}_{ll}^{JM}(\theta_2, \varphi_2; \theta_1, \varphi_1)\; t_{p,J+S}^{Jll}(\omega_1; \rho) \tag{5.15}$$

where the additional index $J + S$ on t_p^{Jll} results from the discussion just preceding and following (5.12).

The functions $\Theta_{l_2 l_1 p}^{JMS\Pi}$ have the appropriate symmetry properties under electron exchange and inversion through the proton. For $l_2 \geq l_1$, they are linearly independent and orthonormal, and constitute an appropriate basis set for expanding the surface functions $\Phi_k^{JMS\Pi}$. We note that in defining these $\Theta_{l_2 l_1 p}^{JMS\Pi}$ we have not symmetrized (or antisymmetrized) the $t_p^{Jl_2 l_1}(\omega_1; \rho)$ and $\mathcal{Y}_{l_2 l_1}^{JM}$ functions separately;[9] but only their product. This results in a saving in numerical effort.

If we now expand the $\Phi_k^{JMS\Pi}$ according to

$$\Phi_k^{JMS\Pi}(\omega_1, \mathbf{\Omega}_1; \rho) = \sum_{\substack{l_2 l_1 p \\ l_2 \geq l_1}} a_{l_2 l_1 p}^{JS\Pi k}(\rho)\; \Theta_{l_2 l_1 p}^{JMS\Pi}(\omega_1, \mathbf{\Omega}_1; \rho) \tag{5.16}$$

and replace this expansion in (5.2), we get an eigenvalue-eigenvector set of linear algebraic equations for the unknown coefficients $a_{l_2 l_1 p}^{JS\Pi k}(\rho)$ and eigenvalues $\varepsilon_k^{JS\Pi}(\rho)$ which can be further decoupled using the known symmetry properties of the $V^{J\Pi}{}_{l_2' l_1'}^{l_2 l_1}(\omega_1; \rho)$ defined by (5.6). The matrix of the coefficients of the resulting equations is real and symmetric and results in easily determined real eigenvalues and real, orthonormal eigenvectors.

The expansion in (5.16) converges rapidly, because the basis functions $\Theta_{l_2 l_1 p}^{JMS\Pi}$ already reflect the localization properties introduced by the diagonal matrix elements $V^{J\Pi}{}_{l_2 l_1}^{l_2 l_1}(\omega_1; \rho)$, and the convergence rate increases as ρ increases. This is in sharp contrast with what occurs when using hyperspherical harmonics in the expansion of the surface functions, in which case the rate of convergence decreases as ρ increases,[8] making such an approach impractical for $\rho > 15$ bohr.

6. SCATTERING EQUATIONS AND ASYMPTOTIC ANALYSIS

6.1 Scattering equations

Over a limited range of ρ in the neighborhood of $\bar{\rho}$ let us expand the symmetry wave function $\Psi_{k'}^{JMS\Pi}$ in the local surface function basis set $\Phi_k^{JMS\Pi}(\omega_1, \mathbf{\Omega}_1; \bar{\rho})$ calculated at $\bar{\rho}$, according to

$$\Psi_{k'}^{JMS\Pi} = \rho^{-\frac{5}{2}} \sum_k b^{JS\Pi}{}_k^{k'}(\rho; \bar{\rho}) \Phi_k^{JMS\Pi}(\omega_1, \mathbf{\Omega}_1; \bar{\rho}) \tag{6.1}$$

This is called the locally diabatic representation, and the set of coupled equations which results when replacing it into the first of (4.1), in matrix form, is:

$$\frac{d^2 \mathbf{b}^{JS\Pi}}{d\rho^2} + \mathbf{U}^{JS\Pi} \mathbf{b}^{JS\Pi} = 0 \tag{6.2}$$

where

$$\left[\mathbf{b}^{JS\Pi}(\rho; \bar{\rho})\right]_k^{k'} = \mathbf{b}^{JS\Pi}{}_k^{k'}(\rho; \bar{\rho}) \tag{6.3}$$

$$\mathbf{U}^{JS\Pi} = -\frac{2\mu}{\hbar^2}\left[\mathbf{\Delta V}^{JS\Pi} + \left(\frac{\bar{\rho}}{\rho}\right)^2 \mathbf{e}^{JS\Pi} + \left(\frac{15\hbar^2}{8\mu\rho^2} - E\right)\mathbf{I}\right] \tag{6.4}$$

$$\left[\mathbf{\Delta V}^{JS\Pi}(\rho; \bar{\rho})\right]_k^{k'} = \langle \Phi_{k'}^{JMS\Pi} | V(\rho, \omega_1, \gamma) - \left(\frac{\bar{\rho}}{\rho}\right)^2 V(\bar{\rho}, \omega_1, \gamma) | \Phi_k^{JMS\Pi} \rangle \tag{6.5}$$

and

$$\left[\mathbf{e}^{JS\Pi}(\bar{\rho})\right]_k^{k'} = \varepsilon_k^{JS\Pi}(\bar{\rho})\, \delta_k^{k'} \tag{6.6}$$

We divide the range of ρ of interest, from $\rho_{\min}$ close to zero to $\rho_{\max}$ sufficiently large for the resulting scattering matrix to have become independent of ρ, into intervals sufficiently small for a surface function basis set at a $\bar{\rho}$ in each interval to be appropriate for that interval. We start integrating (6.2) with the initial condition that at $\rho_{\min}$ $\mathbf{b}^{JS\Pi}$ is the null matrix. When the value of ρ separates two such intervals we change surface functions basis sets by imposing the condition that at that value of ρ $\mathbf{b}^{JS\Pi}$ and its derivative with respect to ρ be continuous.

Because of its simplicity, efficiency and adaptability to the hypercube computer architecture being developed at the California Institute of Technology,[10] we have chosen Johnson's logarithmic derivative method[11] to numerically integrate eq. (6.3).

6.2 Asymptotic Analysis

Asymptotically, in arrangement channel i, (=1,2), the scattering wave function should behave as

$$\Psi_{k'}^{JMS\Pi} \underset{R_i \to \infty}{\sim} \sum \frac{1}{R_i} G_{nl_1l_2,i}^{JS\Pi k'}(R_i) \Phi_{nl_1l_2}^{JM\Pi}(r_i, \Omega_i) \tag{6.7}$$

where

$$\Phi_{nl_1l_2}^{JM\Pi}(r_i, \Omega_i) = \mathcal{Y}_{l_2l_1}^{JM}(\Omega_i) R_{nl_1}(r_i) \tag{6.8}$$

and the $R_{nl_1}(r_i)$ is the radial atomic hydrogen wave function. In addition, the functions $G_{nl_1l_2,i}^{JS\Pi k'}(R_i)$ behave asymptotically as

$$G_{nl_1l_2,i}^{JS\Pi k'}(R_i) \underset{R_i \to \infty}{\sim} \frac{1}{v_n^{\frac{1}{2}}} \left[\mathcal{S}_{nl_1l_2,i}(R_i) C_{nl_1l_2,i}^{JS\Pi k'} + \mathcal{C}_{nl_1l_2,i}(R_i) D_{nl_1l_2,i}^{JS\Pi k'} \right] \tag{6.9}$$

where

$$\mathcal{S}_{nl_1l_2,i}(R_i) = \begin{cases} \sin(k_n R_i - l_2\pi/2) & \text{for open channels} \\ \exp(|k_n| R_i) & \text{for closed channels, and} \end{cases} \tag{6.10}$$

$$\mathcal{C}_{nl_1l_2,i}(R_i) = \begin{cases} \cos(k_n R_i - l_2\pi/2) & \text{for open channels} \\ \exp(-|k_n| R_i) & \text{for closed channels.} \end{cases} \tag{6.11}$$

In addition, v_n is the velocity

$$v_n = \hbar |k_n| / \mu \tag{6.12}$$

where k_n is the channel wave number given by

$$k_n = \frac{1}{\hbar} [2\mu(E - E_n)]^{\frac{1}{2}} \tag{6.13}$$

E_n being the energy of a hydrogen atom with principal quantum n. In matrix form (6.9) can be rewritten as

$$\mathbf{G}^{JS\Pi} \underset{R_i \to \infty}{\sim} \mathbf{v}^{-1/2} [\mathcal{S} \mathbf{C}^{JS\Pi} + \mathcal{C} \mathbf{D}^{JS\Pi}] \tag{6.14}$$

where $\mathbf{G}^{JS\Pi}$, $\mathbf{C}^{JS\Pi}$, and $\mathbf{D}^{JS\Pi}$ are square matrices whose rows and columns are spanned by the set of indices $k' \equiv (nl_1l_2, i)$, and $\mathbf{v}$, $\mathcal{S}$, and $\mathcal{C}$ are diagonal matrices of equal order to the previous ones and whose diagonal elements are given by (6.10) through (6.13). All these matrices have rows and columns associated with both arrangement channels ($i = 1, 2$). A reactance matrix $\mathbf{R}^{JS\Pi}$ for partial wave J, spin S, and parity Π is then defined by

$$\mathbf{D}^{JS\Pi} = \mathbf{R}^{JS\Pi} \mathbf{C}^{JS\Pi} \tag{6.15}$$

For computational purposes, a more appropriate form of (6.14), which starts being valid at smaller values of R_i, is

$$\mathbf{G}^{JS\Pi} \underset{R_i \to \infty}{\sim} \left[\mathbf{J} - \mathbf{N} \mathbf{R}^{JS\Pi} \right] \mathbf{C}^{JS\Pi} \tag{6.16}$$

where $\mathbf{J}$ and $\mathbf{N}$ are diagonal matrices whose diagonal elements are, respectively,

$$J_{nl_1l_2} = \frac{1}{v_n^{\frac{1}{2}}} \begin{cases} j_{l_2}(k_n R_i) & \text{for open channels} \\ i_{l_2}(|k_n| R_i) & \text{for closed channels} \end{cases} \tag{6.17}$$

and

$$N_{nl_1l_2} = \frac{1}{v_n^{\frac{1}{2}}} \begin{cases} y_{l_2}(k_n R_i) & \text{for open channels} \\ k_{l_2}(|k_n| R_i) & \text{for closed channels} \end{cases} \tag{6.18}$$

The j_{l_2} and y_{l_2} are regular and irregular spherical Bessel functions,[12] and i_{l_2} and k_{l_2} are modified spherical Bessel functions of the first[12] and third kinds,[13] respectively. Eq. (6.16) reduces to eq. (6.14) in the far asymptotic region. From the open-open sub-block $\mathbf{R}_{oo}^{JS\Pi}$ of $\mathbf{R}^{JS\Pi}$ one can obtain the corresponding sub-block $\mathbf{S}_{oo}^{JS\Pi}$ of the scattering matrix by the standard relation[14]

$$\mathbf{S}_{oo}^{JS\Pi} = \frac{\mathbf{I} + i\mathbf{R}_{oo}^{JS\Pi}}{\mathbf{I} - i\mathbf{R}_{oo}^{JS\Pi}} \tag{6.19}$$

The matrices $\mathbf{R}_{oo}^{JS\Pi}$ and $\mathbf{S}_{oo}^{JS\Pi}$ are real and unitary, respectively, and both are symmetric.[15] The cross sections for all $e+H$ state-to-state processes can be expressed in terms of $\mathbf{S}_{oo}^{JS\Pi}$.

In order to determine $\mathbf{R}_{oo}^{JS\Pi}$ we must put the solution of (6.2), obtained by numerical integration out to $\rho_{\max}$, in the form of (6.16). Since the latter is expressed in terms of the distances R_i (of the isolated electron to the atom) and the former in terms of the hyperradius ρ, we must perform a transformation of variables to a common one, which we will choose to be ρ. This choice is made because we then need only the logarithmic derivative of $\mathbf{b}^{JS\Pi}$ (rather than $\mathbf{b}^{JS\Pi}$ and its derivative with respect to ρ separately) to accomplish this transformation, and to calculate $\mathbf{R}_{oo}^{JS\Pi}$. This approach is particularly suited to the Caltech hypercube architecture.[10]

This transformation of variables is accomplished by using eqs. (2.4), (6.1), and (5.13) through (5.16), as well as

$$\Psi_{k'}^{JMS\Pi} = \sum_{nl_1l_2,i} \mathcal{Y}_{l_2l_1}^{JM}(\theta_i, \varphi_i; \theta_{r_i}, \varphi_{r_i}) \frac{1}{R_i} G_{nl_1l_2,i}^{JS\Pi k'}(R_i) R_{nl_1}(r_i), \tag{6.20}$$

and requiring that the $\Psi_{k'}^{JMS\Pi}$ given by (6.1) and (6.20) be equal, and that the respective ρ-derivatives also be equal. This permits the calculation of the logarithmic derivative of $\mathbf{G}^{JS\Pi}$ from that of $\mathbf{b}^{JS\Pi}$ and of appropriate numerical quadratures of known functions of ω_i. The mathematical details of this constant ρ projection procedure of the numerical solutions of the scattering eq. (6.2) onto the asymptotic arrangement channel will be published elsewhere.[16] Our experience is that it is not only conceptually simple, as just presented, but does not numerical difficulties. Similarly, we will not give here the explicit expressions relating state-to-state cross sections to the $\mathbf{S}_{oo}^{JS\Pi}$ matrix, since these are standard.[17]

7. REPRESENTATIVE RESULTS

Although the emphasis of this paper is on the formalism we used in applying the method of hyperspherical coordinates and local hyperspherical surface functions to the $e + H$ elastic and inelastic scattering problem, we present now a sampling of the results obtained.

The computer used was an FPS164 attached processor with a fast memory of 512 megawords, and a VAX 11/780 host. The appropriate numerical parameters, such as integrator step size, ρ_0, $\rho_{\max}$, frequency of change of surface functions, and number of surface functions used, was adjusted so as to lead to scattering probabilities converged to about 10^{-3} and scattering matrix phases converged to about 10^{-2} radians. Representative computing times and the number of states used are given in Table 1. The surface functions are independent of energy and need be computed only once.

In Figures 2 and 3 we present the surface function energy eigenvalues for some values of J, S, and Π, as a function of ρ, and which converge to the $n = 1$ through 4 energy levels of the H atom at large ρ. Some of these curves present minima, and support bound states, which have been used in the past to model resonances in this system.[18]

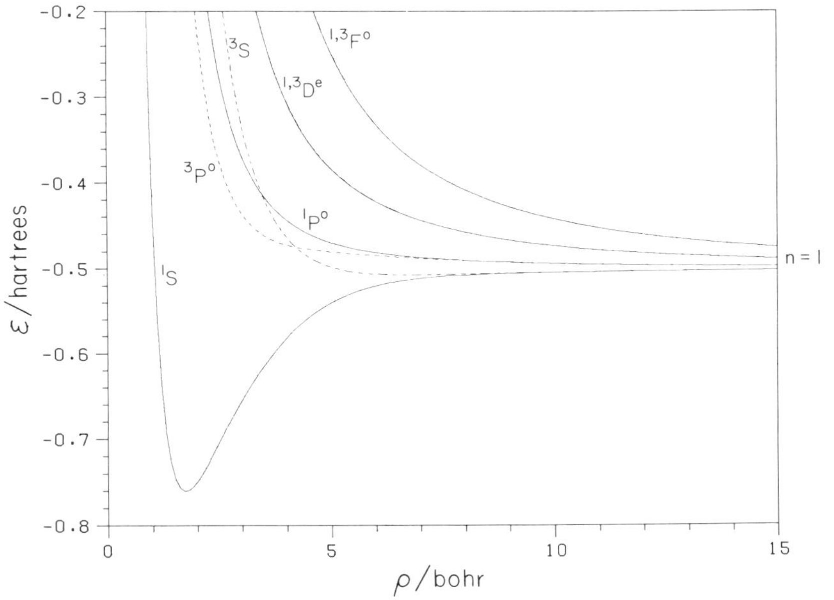

Figure 2. Surface function energies ε, as a function of ρ, for $J = 0$ (S states), 1 (P states), 2 (D states) and 3 (F states), which converge to the $n = 1$ H-atom level as $\rho \to \infty$.

Table I. Computing times.

Partial wave	Number of surface functions	Integration time per energy (sec)	Total time for all surface functions[a] (min)	Time per surface function (sec)
^1S or ^3S	15	20	18	0.36
P^{odd}	25	90	19	0.23
$^1D^{even}$ or $^3D^{even}$	31	130	47	0.36
F^{odd}	34	190	32	0.28

[a] These surface functions were computed at about 200 values of $\overline{\rho}$.

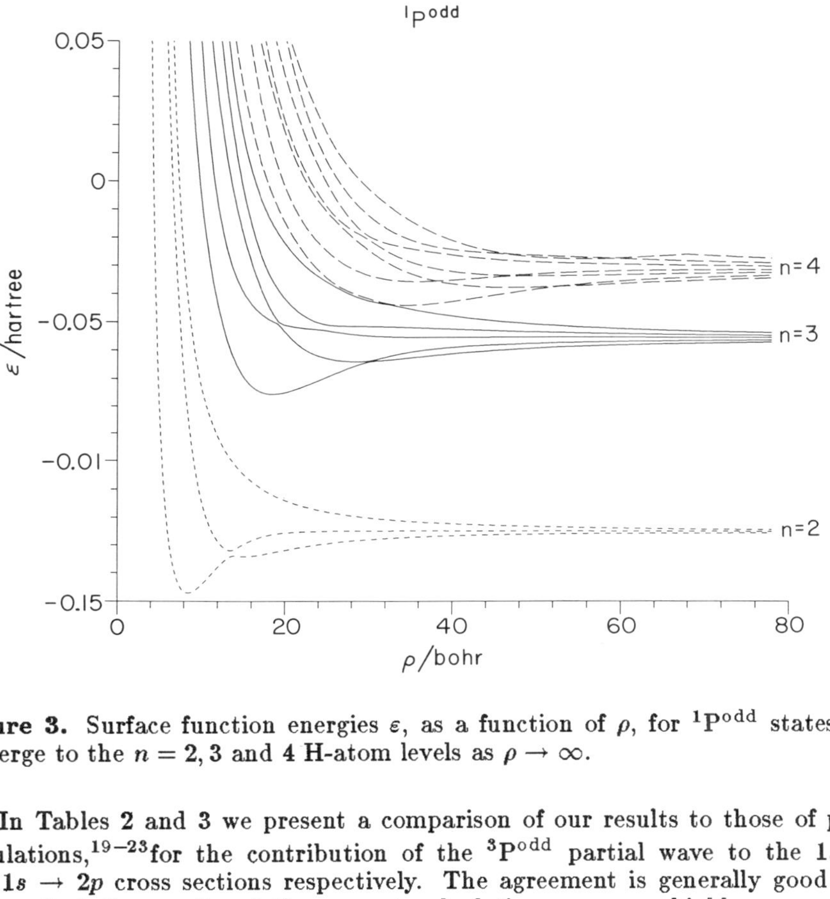

Figure 3. Surface function energies ε, as a function of ρ, for ${}^1\mathrm{P}^{\mathrm{odd}}$ states, which converge to the $n = 2, 3$ and 4 H-atom levels as $\rho \to \infty$.

In Tables 2 and 3 we present a comparison of our results to those of previous calculations,[19–23] for the contribution of the ${}^3\mathrm{P}^{\mathrm{odd}}$ partial wave to the $1s \to 2s$ and $1s \to 2p$ cross sections respectively. The agreement is generally good and we believe that the results of the present calculation are more highly converged and more accurate than the previous ones.

Table II. $^3P^{odd}$ contribution to 1s - 2s cross section.

Kinetic Energy[a]	25 s.f.[b]	BOW[c]	TB[d]	GB[e]	CW/HMM[f]
0.76	0.0384	0.0461	0.0384	0.0429	0.037
0.78	0.0449	0.0567	0.0421	0.0521	0.045
0.81	0.0545	0.0672	0.0503	0.0614	0.053
0.83	0.0584	0.0735	0.0563	0.0668	0.057
0.85	0.0570		0.0596		0.056
0.86	0.0377	0.0516			
0.90	0.0418	0.0582			
0.93	0.0464	0.0517			

[a] Initial electron kinetic energy in rydberg.

[b] Present results for a basis set of 25 surface functions.

[c] Ref. 19.

[d] Ref. 20.

[e] Ref. 21.

[f] Ref. 22. and 23.

Table III. $^3P^{odd}$ contribution to 1s-2p cross section.

Kinetic Energy[a]	25 s.f.[b]	BOW[c]	TB[d]	GB[e]	CW/HMM[f]
0.76	0.0319	0.0478	0.0406	0.0442	0.038
0.78	0.0389	0.0539	0.0456	0.0502	0.041
0.81	0.0446	0.0638	0.0498	0.0584	0.047
0.83	0.0472	0.0674	0.0495	0.0609	0.048
0.85	0.0451		0.0491		0.046
0.86	0.0306	0.0496			
0.90	0.0312	0.0450			
0.93	0.0321	0.0386			

[a] Initial electron kinetic energy in rydberg.

[b] Present results for a basis set of 25 surface functions.

[c] Ref. 19.

[d] Ref. 20.

[e] Ref. 21.

[f] Ref. 22. and 23.

In Fig. 4 the integral cross section for the $1s \rightarrow 2p$ transition is plotted as a function of energy. The arrows on the lower abscissa indicate threshold energies for the $n = 2$, 3, and 4 H-atom levels. The curve is rich in resonances, whose positions agree quite well with those of previous calculations.

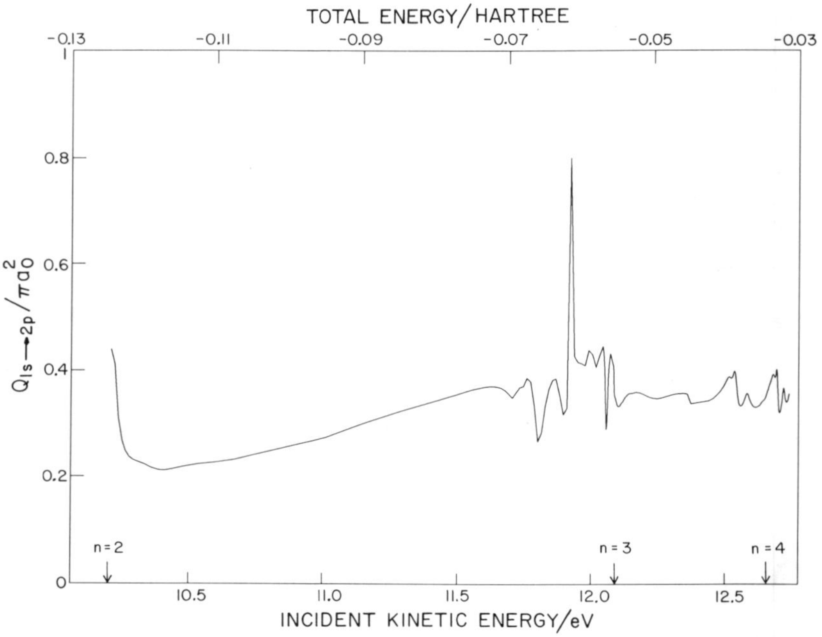

Figure 4. Total cross section $Q_{1s\rightarrow 2p}$ for the $1s \rightarrow 2p$ transition as a function of initial relative electron kinetic energy (lower scale) and total system energy (with respect to the $e + e + p$ configuration).

Fig. 5 displays an Argand diagram for the $(1s1) \rightarrow (2p2)$ element of the scattering matrix for the $^1P^{odd}$ partial wave in the energy region between the $n = 3$ and $n = 4$ thresholds. The counter-clockwise circles indicate the presence of five strong resonances, three of which are very narrow.

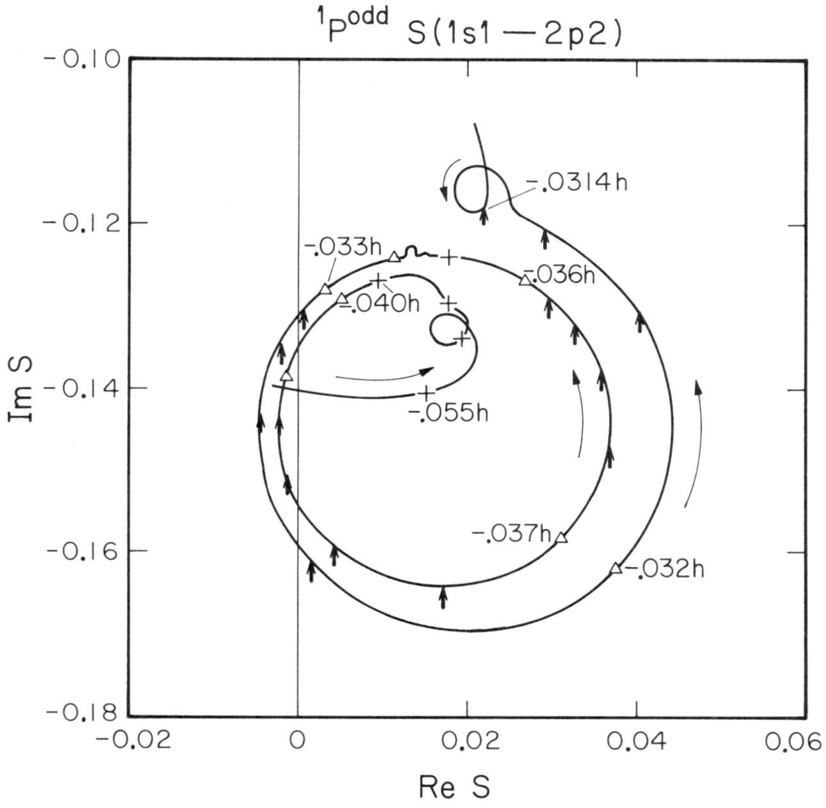

Figure 5. Argand diagram for the $^1P^{odd}$ S-matrix element $(1s1) \rightarrow (2p2)$. The large arrows indicate the direction of increasing energy, from the $n = 3$ to $n = 4$ threshold. The crosses correspond to energies every 0.005 h, the triangles every 0.001 h, and the small arrows every 0.0002 h.

Finally, in Fig. 6 we present the collision lifetime matrix eigenvalues[24] associated with the resonances in the $^1P^{odd}$ partial wave occurring just before the opening of the $n = 4$ channel. As can be seen, some of these resonances live for longer than 10^4 ground state electron orbit times. A detailed description and analysis of these results will be given elsewhere.

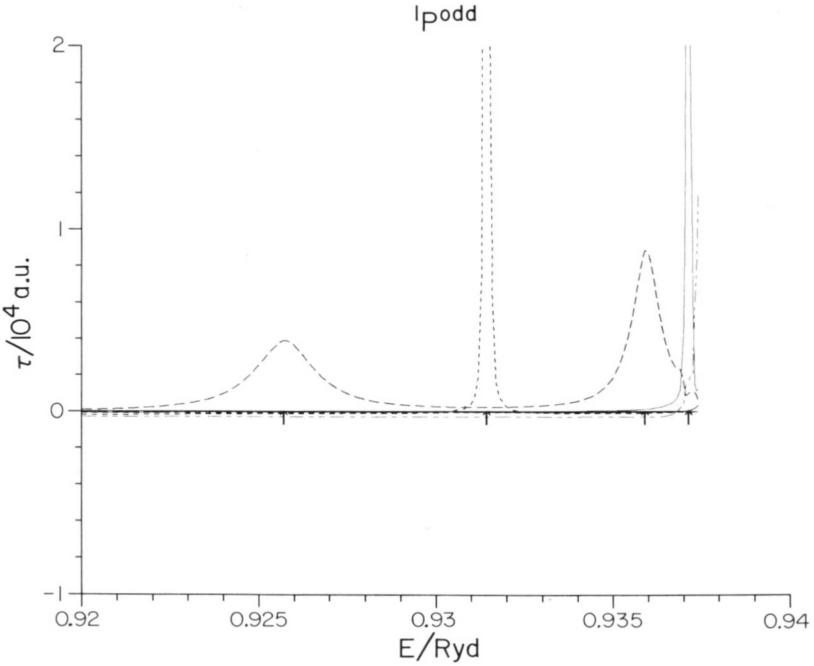

Figure 6. Collision lifetime matrix eigenvalues (in atomic units) for the $^1P^{odd}$ partial wave just below the opening of the $n = 4$ H-atom channel as a function of energy. One atomic unit of time is the classical time it takes an electron in the H-atom ground state to traverse one radian. The ordinates of the off-scale peaks of the dashed and full curves occurring at 0.93145 Ryd and 0.93713 Ryd are 1.3×10^5 and 1.9×10^5 atomic units respectively. The arrows locate the positions of the resonances.

8. SUMMARY AND CONCLUSIONS

We have presented the hyperspherical coordinate formulation for $e + H$ elastic and inelastic scattering using local surface functions and have shown that it is both efficient and accurate. It can in principle be extended to energies above the ionization threshold by including hyperspherical harmonics in the surface function basis set. It also permits a calculation of polarization cross sections. This approach is very promising and should lead to a very complete description of the $e + H$ scattering processes.

REFERENCES

* Research supported in part by the US Department of Energy (Grant No. DE-AS03-83ER13118)
† Contribution Number 7275

1. J. Macek, *J. Phys. B*, **1**, 831 (1968).

2. a) A. Kuppermann, *Chem. Phys. Lett.*, **32**, 374 (1975); b) R. T. Ling and A. Kuppermann, in: *Electronic and Atomic Collisions, Abstracts of Papers of the 9th International Conference on the Physics of Electronic and Atomic Collisions*, Seattle, Washington, 24-30 July 1975, eds. J. S. Risley and R. Geballe (University of Washington Press, Seattle, 1975) pp. 353,354.

3. a) A. Kuppermann, J. A. Kaye, and J. P. Dwyer, *Chem. Phys. Lett.*, **74**, 257 (1980); b) J. A. Kaye and A. Kuppermann, *Chem. Phys. Lett.*, **77**, 573 (1981); c) J. A. Kaye and A. Kuppermann, *Chem. Phys. Lett.*, **78**, 546 (1981).

4. a) G. Hauke, J. Manz, and J. Römelt, *J. Chem. Phys.*, **73**, 5040 (1980);b) J. Römelt, *Chem. Phys. Lett.*, **74**, 263 (1980); c) J. Manz and J. Römelt, *Chem. Phys. Lett.*, **76**, 337 (1980); d) J. Manz and J. Römelt, *Chem. Phys. Lett.*, **77**, 172 (1981); e) J. Manz and J. Römelt, *Chem. Phys. Lett.*, **81**, 179 (1981).

5. L. M. Delves, *Nucl. Phys.*, **9**, 3091 (1959); **20**, 275 (1960).

6. A. Kuppermann, in: *Theoretical Chemistry, Theory of Scattering: Papers in Honor of Henry Eyring*, Vol. 6A, ed. D. Henderson (Academic Press, New York, 1981), pp.79-164.

7. Yu. A. Simonov, *Yad. Fiz.*, **3**, 630 (1966).

8. H. Klar and M. Klar, *Phys. Rev. A*, **17**, 1007 (1978).

9. B. Christensen-Dalsgaard, *J. Chem. Phys.*, **29**, 470 (1984).

10. G. Fox and S. Otto, *Physics Today*, 50, (May 1984).

11. B. R. Johnson, *J. Comp. Phys.*, **13**, 445 (1973); *J. Chem. Phys.*, **67**, 4086 (1977).

12. Handbook of Mathematical Functions, edited by M. Abramowitz and I. A. Stegun (National Bureau of Standards, Washington, D. C., 1964).

13. $k_n(x) = (\pi/2x)^{\frac{1}{2}} K_{n+\frac{1}{2}}(x)$, where K_m is the modified cylindrical Bessel function of the third kind[12].

14. N. S. F. Mott and H. S. W. Massey, The Theory of Atomic Collisions, (Clarendon, Oxford, 1965) 3rd ed., Ch. 14, 15.

15. A. M. Lane and R. G. Thomas, *Rev. Mod. Phys.*, **30**, 257 (1958).

16. D. M. Hood and A. Kuppermann, manuscript in preparation.

17. a) J. M. Blatt and L. C. Biedenharn, *Rev. Mod. Phys.*, **24**, 258 (1952); b) P. G. Burke and K. Smith, *Rev. Mod. Phys.*, **34**, 458 (1962).

18. C. D. Lin, *Phys. Rev. A*, **12**, 493 (1975).

19. P. G. Burke, S. Ormonde, and W. Whitaker, *Proc. Phys. Soc.*, **92**, 319 (1967).

20. A. J. Taylor amd P. G. Burke, *Proc. Phys. Soc.*, **92**, 336 (1967).

21. S. Geltman and P. G. Burke, *J. Phys. B*, **3**, 1062 (1970).

22. J. Callaway and J. W. Wooten, *Phys. Rev. A*, **9**, 1924 (1974).

23. J. Hata, L. A. Morgan, and M. R. C. McDowell, *J. Phys. B*, **13**, 4453 (1980).

24. F. T Smith, *Phys. Rev.*, **118**, 349 (1960).

THE R-MATRIX METHOD

John C. Light
The James Franck Institute and The Department of Chemistry
The University of Chicago
Chicago, Illinois 60637, U.S.A.

ABSTRACT. The R-matrix method is reviewed both in the L^2 and R-matrix propagation implementation. Results are presented for a simple elastic scattering problem showing the improvement in convergence when non-orthogonal L^2 bases are used. Comparably accurate results are obtained with either non-orthogonal sine functions or a distributed Gaussian basis. The implications of these results, together with R-matrix propagation, for reactive scattering are discussed.

1. INTRODUCTION

Although the accurate solution of the appropriate time independent Schrödinger equation, subject to proper boundary conditions, has long been recognized as a rigorous approach to the prediction of chemical reaction dynamics, it is only relatively recently that this has been accomplished for the simplest of chemical reactions, the hydrogen atom-hydrogen molecule exchange reaction [1-3]. Since experimentally far more complex systems are of interest, the main purpose of this Workshop has been to review, discuss, and propose theoretical methods, all approximate to a greater or lesser degree, which can be used for accurate models of real systems to provide reasonably efficient and accurate predictions of their dynamic behavior. Specifically we would like to be able to calculate all experimentally measurable quantities ranging from state-to-state differential cross sections to thermal rate constants given only the accurate (nuclear) potential energy surface(s) for the system in question. Also desirable, of course, would be the ability to predict the course of higher energy collisions involving electronically non-adiabatic transitions, photodissociation via both direct and predissociation routes, and including product distributions, electron-molecule collisions, etc.

Since the exact solution of the Schrödinger equation for most such processes of interest remains far beyond our current capabilities, both in terms of the algorithms for exact solution and the computational resources, both software and hardware, to carry them out, most of this Workshop has focused on the adequacy of various physical approximations

D. C. Clary (ed.), The Theory of Chemical Reaction Dynamics, 215–234.

and "model" approaches to the problem. While these are, of course, currently more useful as predictive tools for complex systems than any "exact" approach (which cannot normally be carried through), the purpose of this article is to present an exact approach, the R-matrix, proposed formally many years ago, which, given improvements in computer resources, theoretical understanding and in development of efficient algorithms may soon yield "exact" solutions to significantly more complex problems in chemical dynamics.

Isolated collision processes are, essentially by definition, governed by short range forces between the interacting particles. In 1947, Wigner and Eisenbud [4,5] proposed that this fact be formally incorporated into quantum scattering theory by dividing the "scattering" coordinate space into two regions, the asymptotic region in which analytic solutions to the Schrödinger equation are known, and the interior, or interaction region, in which all the physical coupling (and mathematical difficulties) occur. Since only asymptotic information is ever observable experimentally, and since the continuity requirements on the wavefunction (solution of the Schrödinger equation) at the interior-exterior boundary determine the asymptotic solutions, Wigner and Eisenbud developed the relations between the interior solutions (evaluated at the boundary) and the desired asymptotic quantity, the S matrix (and thus differential cross sections, etc.). This reaction is the R-matrix --mathematically, the relation between the outward normal gradient of the wavefunction at the boundary, A, and the wavefunction evaluated there:

$$\psi = R\, \underset{\sim}{n}\cdot\underset{\sim}{\nabla}\psi\Big|_{r=A} \tag{I.1}$$

where n is normal to the surface (here trivially $r/|r|$). Since at $r = A$ the wavefunction must have its asymptotic form, namely

$$\psi = IN - O\cdot S \tag{I.2}$$

where IN is an incoming asymptotic solution (e.g., $\frac{e^{-ikr}}{r}$), and O the outgoing solution, the R-matrix is simply related to the S matrix (e.g., phase shift for elastic scattering, δ, with $S = e^{2i\delta}$), by

$$\begin{aligned} \underline{\underline{S}} &= \underline{\underline{W}}^{-1}\,\underline{\underline{W}}^{*} \\ \underline{\underline{W}} &= [\underline{\underline{R}}\,\underline{\underline{O}}' - \underline{\underline{O}}]\underline{\underline{k}}^{-1/2} \end{aligned} \tag{I.3}$$

where the primes denote the outward normal gradient, and k is the wave vector.

In their original formulation, Wigner and Eisenbud used the R-matrix primarily as a formal tool, relating the properties of the solutions of the Schrödinger equation in the internal region in an empirical manner to the resonances observed in nuclear scattering. Although the parameterizations they developed to do this were quite useful, the main purpose of this article is to present the methods developed (and still developing) for the exact evaluation of the R-matrix for more complex systems, and the more general relations of the R-matrix to standard Green's functions. That is, we will present the developments which have taken the original R-matrix theory from a formalism for semi-empirical "explanation" of experimental observations to two quite different "exact" computational approaches: the L^2 basis set expansion in the interior region, and the R-matrix propagation of local analytic solutions throughout the interior region. Although both approaches were developed over a decade ago, some new developments in each area will be reviewed, and a prognosis for applications to more complex problems will be given.

2. L^2 R-Matrix Theory: Basics

In "chemical" scattering theory, we usually wish to solve the non-relativistic Schrödinger equation for the motion of the nucleii on a single adiabatic potential energy surface. We isolate a single coordinate, the scattering coordinate which we will call R, such that as $R \to \infty$ the system is partitioned into non-interacting moieties, i.e., the atoms, molecules, or molecular fragments which are the reactants and products of the collision. In the case of chemical reactions, R may be the hyperradius of a hyperspherical (Delves) [6,7] coordinate system. Thus the Hamiltonian can be partitioned as

$$H = \sum_i h_i + T_R + V(R,\{x\}) \tag{II.1}$$

where the h_i is the Hamiltonian of the isolated fragments (initial and final) and V is the potential. V must go to zero as R goes to infinity (in any chemical channel):

$$V(R,\{x\}) \xrightarrow[R\to\infty]{} 0.$$

The coordinates, $\{x\}$, are the internal coordinates, and T_R is kinetic energy operator containing both the translational kinetic energy operator for R and the overall angular momentum operators.

We assume the asymptotic internal solutions are known

$$h_i\, \phi^i_k = \varepsilon^i_k\, \phi^i_k \tag{II.2}$$

and the two independent translational solutions must be

$$T_R\, \psi_{\ell,k} = -k^2\, \psi_{\ell,k} \qquad (R\to\infty) \tag{II.3a}$$

$$T_R\, \tilde{\psi}_{\ell,k} = -k^2\, \tilde{\psi}_{\ell,k} \qquad (R\to\infty) \tag{II.3b}$$

where ℓ is the total orbital angular momentum and k^2 is the translational energy with respect to a space fixed coordinate system with the origin at the center of mass of the system. ψ and $\tilde{\psi}$ are two independent solutions. Depending on the number and complexity of the fragments, the angular momentum coupling may be more or less complicated [7], but the specific scheme used is not important for this discussion. Here, therefore, we assume that an appropriate scheme has been adopted and that a decomposition of the wavefunction by total angular momentum, J, has been made. We also assume that no external fields are present (thus space is isotropic and our results must be rotationally invariant) so we may choose the space fixed axes such that the projection of the total angular momentum on the space fixed z axis is zero (M = 0).

Given this, the asymptotic wavefunction must be a linear combination of the form

$$\psi_E^{J,M} \underset{R\to\infty}{=} \sum_{\alpha,k} c_{\alpha,k}\, \phi_{k,\alpha}\, \psi_{\ell,k} + \sum_{\alpha,k} d_{\alpha,k}\, \phi_{k,\alpha}\, \psi_{\ell,k} \qquad \text{(II.4)}$$

where the product is over the initial states, ϕ_k, for the appropriate initial internal states in a given chemical channel and the sum (α,k) is over internal states, k, in a given initial channel and over possible outgoing channels, α. The c's and d's are coefficients which are not independent since we will require that all solutions satisfy the boundary condition at the origin:

$$\psi_E^{J,M}(R) \xrightarrow[R\to 0]{} \text{finite (regular b.c.'s).} \qquad \text{(II.5)}$$

We may use the orthogonality of the internal states (we assume the fragments are well enough separated so that the internal states of the different internal channels are orthogonal to each other), take the ψ's to be incoming waves and $\tilde{\psi}$ the outgoing waves to define a column of the S-matrix:

$$\frac{1}{C_{\alpha,n}} \int \phi_{n,\alpha}\, \psi_E^{J,M}\, \pi dx = \delta_{k,n}\, \psi_{\ell,k} + \frac{1}{C_{n,\alpha}} \sum_{\alpha',n'} d^n_{\alpha',n'}\, \tilde{\psi}^n_{\ell',n'} \qquad \text{(II.6)}$$

or

$$S^{J,M}_{n',\alpha',n\alpha} = \frac{1}{C_{n,\alpha}}\, d^n_{\alpha',n'}\ . \qquad \text{(II.7)}$$

(We note the projection has also fixed the initial orbital angular momentum, ℓ, and the index n' refers to internal quantum numbers and ℓ'.)

Recognizing that the individual incoming and outgoing solutions are complex conjugates, we may write (II.6) more simply in matrix form as

$$\underline{\underline{\Psi}}(R) = \underline{\underline{k}}^{-1/2}[\underline{\underline{O}}^* - \underline{\underline{O}}\ \underline{\underline{S}}] \qquad \text{(II.8)}$$

where $\underline{\underline{O}}$ is the diagonal matrix of outgoing solutions, $\tilde{\psi}$. The outward gradient is then

$$\underline{\Psi}' = \underline{k}^{-1/2}[\underline{\underline{O}}^{*\prime} - \underline{\underline{O}}'\underline{\underline{S}}] = -i\ \underline{\underline{k}}^{1/2}\ \underline{\underline{O}}^* - i\ \underline{\underline{k}}^{1/2}\ \underline{\underline{O}}\ \underline{\underline{S}} \qquad (II.9)$$

where $\underline{k}$ is the diagonal matrix of wave vectors, and the $\underline{k}^{-1/2}$ factor norms the solutions to unit incoming flux. Since the R-matrix is defined as the relation between Ψ and Ψ', we have

$$\underline{\Psi} = \underline{\underline{R}}\ \underline{\Psi}' \qquad (II.10)$$

or, relating (II.8-10),

$$\underline{\underline{S}} = [(\underline{\underline{I}} - i\ \underline{\underline{R}}\ \underline{\underline{k}})\underline{\underline{O}}\ \underline{\underline{k}}^{-1/2}]^{-1}\ [(\underline{\underline{I}} + i\ \underline{\underline{R}}\ \underline{\underline{k}})\underline{\underline{O}}^*\ \underline{\underline{k}}^{-1/2}] \qquad (II.11a)$$

$$= \underline{\underline{W}}^{-1}\ \underline{\underline{W}}^*. \qquad (II.11b)$$

Note that the relations (II.10) and (II.11) hold throughout the asymptotic region, the radial dependence of the $\underline{\underline{R}}$ matrix just cancelling that of the outward solution in the expression for the S-matrix. Since the $\underline{\underline{k}}$ and $\underline{\underline{O}}$ matrices are diagonal in the internal state quantum numbers, the boundaries on which $\underline{\underline{R}}$ and $\underline{\underline{O}}$ are evaluated may be different for each internal state. Having established the relation of the R-matrix to the S matrix, we now turn to the much more difficult task of evaluating the R-matrix.

In order to evaluate the R-matrix, we may determine the operator for which it is the matrix representation, and seek to evaluate it directly. Writing the full Schrödinger equation as

$$\{-\frac{d^2}{dR^2} + \sum_i h_i + V + L^2 - E\}\ \Psi = 0 \qquad (II.12)$$

where L^2 is the angular momentum operator, and we have scaled all quantities by $(2\mu/\hbar^2)$. We may now add the Bloch [8] operator to each side:

$$L_0 = \sum_n |n\rangle\ \delta(R-A_n)\ \frac{\partial}{\partial R}\ \langle n| \qquad (II.13)$$

to obtain

$$\{H - E + L_0\}\ \Psi = L_0\Psi. \qquad (II.14)$$

The operator $(H - E + L_0)$ is now easily seen to be Hermitian on the range $0 \leqslant R \leqslant A_n$ for each internal state $|n\rangle$ provided the translational functions satisfy the b.c.'s as $R \to 0$ (i.e., $G\Psi(R,x) \xrightarrow[R\to 0]{} 0$). This permits the formal solution of (II.14)

$$\Psi = \{H - E + L_0\}^{-1}\ L_0\Psi \qquad (II.15)$$

in terms of the Greens function

$$G_{E,L_0} = (H - E + L_0)^{-1}. \qquad (II.16)$$

Projecting Eq. (II.15) on the left with the Bloch operator (putting it into the internal basis representation) and integrating over R, we find the R-matrix of Eq. (II.10) is just the internal state matrix representation of the Green's function evaluated at the channel boundaries:

$$R_{n,n'} = \langle n|\ (H - E - L_0)^{-1}\ |n'\rangle. \qquad \text{(II.17)}$$

The standard L^2 means of evaluating (II.17) is now clear. Choose a multi-dimensional basis satisfying zero derivative b.c.'s for each channel at the channel radii, A_n; evaluate the Hamiltonian matrix in this basis; diagonalize to find the "internal" eigenvalues, λ_i, and eigenvectors, u_i; and project the eigenvectors onto the asymptotic internal states at A_n:

$$\gamma_{in} = \langle u_i|n) \Big|_{R = A_n} . \qquad \text{(II.18)}$$

Then we can evaluate $R_{nn'}$ as

$$R_{nn'} = \sum_i \frac{\gamma^*_{in'}\ \gamma_{in}}{\lambda_i - E} . \qquad \text{(II.19)}$$

The above is difficult to carry out for arbitrary A_n's unless the asymptotic translational and internal bases are used (which will not be very efficient). Thus normally all A_n's are set to the same value, A, and a direct product basis satisfying zero derivative b.c.'s at A is used. As the basis used for expansion in the internal region becomes complete, the R-matrix becomes exact and the approximation to the S matrix converges to the exact result as well. This desired result is, however, difficult to achieve in practice using, as indicated above, an orthonormal basis in R, $(g_i(R))$, satisfying

$$\frac{\partial g_i(R)}{\partial R} \Bigg|_{R = A} = 0 \qquad \text{(II.20a)}$$

$$\int_0^A J(R)d\Omega\ g_i(R) g_j(R) = \delta_{ij} \qquad \text{(II.20b)}$$

(J(R) is the appropriate Jacobian factor).

The reason for this can be viewed in two ways. First, since the Bloch operator b.c.'s effectively constrain the problem to a "box" of size A, the "translational component" of the eigenvalues will increase eventually with the square of the number of translational basis functions, m, and the series converges only as $1/m^2$. The second view is that the true wavefunctions will not have zero derivative at $R = A$, thus requiring many basis functions to permit ψ' to be adequately represented, in a mean square sense, by the basis.

The first view has been addressed in two ways. Buttle, in 1967 [9], proposed that a complete basis could be used, albeit with an approximate Hamiltonian. The so-called Buttle correction does help a great deal (see Section 3) and is constructed as follows. Suppose H^o is a Hamiltonian for which the exact R-matrix is known analytically and in terms of an infinite sum over basis functions. Projecting these basis functions onto our asymptotic internal functions at $R = A$, we find

$$\underline{\underline{R}}^{o,\infty}_{nn'} = \sum_{i=1}^{\infty} \frac{\gamma^o_{in}\,\gamma^o_{in'}}{\lambda^o_i - E} \text{ (analytic)}. \tag{II.21}$$

A Buttle correction (for lack of completeness) is determined by adding R^o to R and subtracting the contribution of basis functions already used, i.e.,

$$\underline{\underline{R}}_{\text{Buttle}} = \underline{\underline{R}} + \underline{\underline{R}}^{o,\infty} - \underline{\underline{R}}^{o,m} \tag{II.22}$$

where $R^{o,m}$ is the R-matrix of H^o using the same basis as for R, and $R^{o,\infty}$ is the analytic R matrix for H^o. Although this helps considerably, it is difficult to implement, particularly for reactive collisions for which a suitable H^o with analytic solutions is difficult to find and evaluate.

Variational corrections to the R-matrix have also been proposed [10,11] and evaluated for elastic and simple inelastic scattering examples. Although very effective in improving convergence, these required the evaluation of new perturbation integrals at each energy and a considerable increase in the complexity of the calculation.

Finally, we take note of the advantages and disadvantages of the standard L^2 R-matrix approach. It is obvious from (II.19) that once the diagonalization of the Hamiltonian in the internal basis and the projection of these eigenfunctions onto the asymptotic states has been accomplished yielding the λ_i's and γ_{in}'s, the evaluation of the R-matrix at a series of energies is very efficient. It requires essentially N^2m operations where N is the number of asymptotically open internal states and m is the number of L^2 basis functions used to diagonalize H. It is also obvious that the R-matrix "box" radius, A, should be as small as possible consistent with the imposition of known (asymptotic) boundary conditions. This increases the spacings of the eigenvalues, λ_i, and hastens convergence with or without the Buttle correction.

However, it has normally been found that 6-15 translational functions per internal state are required for convergence, even with the above corrections. The time required to diagonalize the Hamiltonian matrix in the L^2 basis is thus some 200 to 3000 times as long (an N^3 process) as to diagonalize, locally, the $H(R_i)$ matrix in the internal basis as is required for propagation techniques. Thus only if the product of the number of steps required in propagation techniques, N_s, times the number of energies required, N_E, is of the order of 10^3 will the L^2 approach be worth considering. Given the non-uniform convergence of the uncorrected R-matrix as well, the L^2 approach has been used only rarely for reactive scattering (collinear only [12-14]).

In the next two sections, therefore, we discuss alternatives to the "standard" R-matrix theory described above. In the next section, results and rationalizations for the use of a non-orthogonal L^2 basis are presented, together with a brief review of applications to reaction scattering. In Section 4 we will then present the propagation equations for the piecewise analytic evaluation of the R-matrix.

2. L^2 R-Matrix With Non-Orthogonal Bases

As indicated in the last section, serious problems exist with the standard L^2 R-matrix with respect to slow convergence of the phase shifts (or S-matrix) as the number of translational basis functions is increased. This has been amply demonstrated in a number of papers (e.g., Refs. 8-12). In all of these cases an orthogonal translational basis is used which satisfies fixed log derivative boundary conditions at the R-matrix boundary, $R = A$. As indicated above, the Buttle correction [9] can be added to the R-matrix to account, in an approximate fashion, for the members of the complete basis not included in the explicit R-matrix evaluation. An additional variational correction [10,11] was proposed to improve the results further. Although these procedures help a great deal, they are both "expensive," at least in terms of programming effort for complex systems.

The root of this problem appears to be that the true wavefunction matches the log derivative b.c.'s of the chosen translational basis only at an isolated set of energies. At these energies (near the λ_i of the L^2 expansion of the Green's function for a basis satisfying the Bloch operator b.c.'s, $L_0\emptyset = 0$) the R-matrix results are very accurate for a given basis size, even without the Buttle or variational correction. However, between these energies, the results are very poor.

One other generalization of the standard procedure which was suggested to improve the results was to replace the "zero derivative" Bloch operator of Eq. (II.13) by one which specified a different log derivative b.c., i.e.,

$$L_b \equiv \sum_n |n\rangle\, \delta(R-A_n)\left(\frac{\partial}{\partial R} - b\right) \langle n| . \tag{III.1}$$

In this case one may show [10,15] that, with the Green's function defined by

$$G_{E,L_b} = (H + L_b - E)^{-1} \tag{III.2}$$

and an R^b matrix defined by

$$\underline{\underline{R}}^b_{n,n'} = \langle n|\, G_{E,L_b}\, |n'\rangle \tag{III.3}$$

that the R-matrix is given by

$$\underline{\underline{R}} = [\underline{\underline{I}} - \underline{\underline{R}}^b\, \underline{\underline{L}}_b]^{-1}\, \underline{\underline{R}}^b \tag{III.4}$$

where $\underline{\underline{L}}_b$ is the (diagonal) matrix

$$L_{n,n'} = \delta_{n,n'} \left(A_n \frac{O'_n(A_n)}{O_n(A_n)} - b\right). \qquad \text{(III.5)}$$

In 1972, however, Mori [15] showed that if a fixed (finite) basis is used, i.e., a basis which is independent of b, then the $\underline{\underline{R}}$ and $\underline{\underline{S}}$ matrices defined above are independent of the value of b used. If, of course, the finite basis is changed to satisfy the log derivative b.c.'s at A_n, then the results do depend on b. However, no regular procedure for improving the results by varying the basis and b has ever been established (to my knowledge). Implicit in Mori's work, however, is the idea that one should use basis "functions which are defined in a wider region than the internal region" [15].

In hindsight it is surprising that this was not recognized as the key to accelerated convergence of the R-matrix until the very recent work of Bocchetta and Gerratt [16], although the technique was used several times [12,13] earlier. Schneider [17] explicitly recognized the utility of using functions which did not obey specified b.c.'s at A for each of matrix element evaluation in electron-molecule collisions, and used a floating Gaussian basis and a buttle correction to obtain excellent results with uniform convergence (in E) with a modest basis set. It was, however, only shown in the very recent article of Bocchetta and Gerratt [16] that the Buttle correction itself was unnecessary if Mori's prescription is followed, i.e., if a non-orthogonal (on $0 \leq R \leq A$) basis which does not satisfy the log derivative b.c.'s at A is used.

The introduction of such a basis does, of course, impose a penalty in that one must explicitly orthogonalize the basis. Thus if $\underline{\underline{M}}$ is the matrix of overlap integrals the basis in the "R-matrix box" ($0 \leq \Omega \leq A$), then the Green's function at E is obtained by solving the generalized eigenvalue problem

$$(\underline{\underline{H}} - \lambda \underline{\underline{M}} + \underline{\underline{L}})\ \underline{c} = 0. \qquad \text{(III.6)}$$

Two common approaches would be to Schmidt orthogonalize the translational basis before construction and diagonalization of the $\underline{\underline{H}}$ and $\underline{\underline{L}}$ matrices (as used in Ref. 16), or to use the symmetrically orthogonalized basis using $\underline{\underline{M}}^{-1/2}$ (as used in Section 3). A general description of the latter procedure, applied to R-matrix propagation, was given by Schmalz, Stechel, and Light [18].

In order to verify the ideas presented by Bocchetta and Gerratt [16] and above, we have done some elastic scattering calculations on a simple exponential repulsive potential using both a non-orthogonal sine basis (as in Ref. 16) and a basis of distributed Gaussian functions. The Hamiltonian was that used in Ref. 11:

$$H = -\frac{1}{2m}\frac{d^2}{dR^2} + 0.25\ e^{-2R}$$

$$m = 2400. \qquad \text{(III.7)}$$

Shown in Tables II-IV are results for N = 10, 15, and 20 non-orthogonal basis functions and results for an orthogonal sin basis with and without the Buttle correction. The non-orthogonal bases are:

Sin basis:

$$\phi_n = \sin\left(\frac{n\pi R}{L+\delta r_o} - \frac{n\pi r_c}{L+\delta r_o}\right) \qquad L = A - r_c$$
$$A = 3.75$$

with the non-orthogonality controlled by the value of δr_o; and

Gaussian basis:

$$\phi_n = \left(\frac{\pi}{2A_n c^2}\right)^{1/4} \exp\left(-A_n c^2 (R-R_n)^2\right).$$

Here, R is chosen "semi-classically" according to the prescription given by Hamilton and Light [19].

The specific values of the basis set parameters are given in Table I. Note that the integrals over $0 < R < A$ were performed exactly--the potential is simple enough that the transformation approach recommended by Schneider [17] was not required. The same results (on a finer energy scale) are plotted in Figures 1-3. The arrows on the energy axis show the values of the eigenvalues of $H + L_0$; the bracketed numbers their indices.

As can be seen from the tables and figures, both non-orthogonal bases, the sine functions and the distributed Gaussian basis are much more accurate than the orthogonal sine basis without the Buttle correction. Although the Buttle correction improves the "orthogonal" results considerably, they are still in general less accurate than the non-orthogonal basis results.

The purpose of including the distributed Gaussian basis was two-fold. First, it shows that the improvement obtained by relaxing the b.c.'s on the basis is relatively general, i.e., it is not limited to the use of sine functions for a translational basis. Second, Hamilton and Light [19] have shown that the distributed Gaussian basis is efficient and accurate for multi-dimensional bound state problems. Thus, with a more careful choice of parameters ($\underset{\sim}{R}_n$, A_n, and c) and a larger basis, the original R-matrix method for reactive scattering pioneered almost 15 years ago by Crawford [12] and Der, et al. [13], may well become the method of choice for accurate results on simple systems. Two other advantages of the distributed Gaussian basis for such complex problems are the simplicity of matrix element evaluation [19] and the sparse nature of $\underset{\sim}{H}$ and $\underset{\sim}{M}$, particularly for large multi-dimensional systems.

Finally, I would like to suggest that the use of the name "non-orthogonal bases" is somewhat of a misnomer. The bases are, in fact, orthogonalized before the evaluation of the Green's function, and the key to their success lies not in their non-orthogonality but in the fact that they are not eigenfunctions of a Bloch operator, i.e., they do not satisfy a fixed b.c. at $R = A$. This permits a more accurate representation of $\psi/\psi'\big|_{R=A}$. Since the easy means of constructing orthogonal bases

TABLE I. Basis Set Parameters

A. Orthogonal sine: $u_i(r) \,\alpha\, \sin\left((i - \frac{1}{2})\pi\ r/A\right)$ $A = 3.75$

B. Non-orthogonal sine: $u_i(r) \,\alpha\, \sin\left((i - \frac{1}{2})\pi\ (r - r_c)/(A - r_c + \delta r_o)\right)$

N	δr_o	r_c
10	0.3	0.9
15	0.3	0.7
20	0.3	0.6

C. Distributed Gaussians ($N = 10$, $c = 0.5$, $E_{MAX} = 0.06$)

R_i	A_i	R_i	A_i
0.913972	1.335353	2.699403	2.390719
1.346657	1.592159	3.021256	2.413319
1.706485	2.059219	3.343117	2.413219
2.043522	2.240886	3.664983	2.413185
2.374507	2.324606	3.986848	2.413185

D. Distributed Gaussians ($N = 15$, $c = 0.5$, $E_{MAX} = 0.08$)

R_i	A_i	R_i	A_i
0.719194	2.471731	2.625778	5.424926
1.037225	3.034651	2.840458	5.465281
1.293238	4.142084	3.053531	5.506475
1.528574	4.670089	3.266608	5.506339
1.755979	4.972958	3.479687	5.506250
1.977002	5.191772	3.692768	5.506217
2.194855	5.306315	3.905848	5.506217
2.41116	5.385194		

E. Distributed Gaussians ($N = 20$, $c = 0.5$, $E_{MAX} = 0.10$)

R_i	A_i	R_i	A_i
0.581592	3.743869	2.410220	9.415539
0.840002	4.641693	2.573165	9.414919
1.045745	6.461937	2.736126	9.511025
1.233388	7.489598	2.897419	9.609485
1.411147	8.135684	3.058715	9.609250
1.583980	8.614677	3.220013	9.511031
1.751854	8.869729	3.382970	9.510914
1.919753	9.045892	3.544269	9.608865
2.084340	9.322482	3.705570	9.608828
2.247270	9.416398	3.866870	9.608828

TABLE II. Phase Shifts for N = 10. Columns A, distributed Gaussian basis; B, non-orthogonal sines; C, orthogonal sines; D, orthogonal sines with Buttle correction; E, Exact.

E(a.u.)	A	B	C	D	E
0.010	-12.5064	-12.5114	-12.9929	-12.5538	-12.5031
0.012	-13.3290	-13.0797	-13.7299	-13.2204	-13.0773
0.014	-13.9090	-13.5642	-15.3125	-14.2411	-13.5545
0.016	-14.4944	-14.0410	-14.9352	-12.5865	-13.9582
0.018	-15.8736	-14.3706	-16.7463	-12.5416	-14.3040
0.020	-17.3840	-14.8681	-18.5265	-12.5481	-14.6031

in a finite region is to fix the b.c.'s, the easy way to generate a basis which does not satisfy any fixed b.c.'s at R = A is to make it non-orthogonal. The non-orthogonality itself is, however, a nuisance in evaluation, and a simple method to generate a flexible orthogonal basis without fixed b.c.'s would probably be useful.

TABLE III. Phase Shifts for N = 15. Columns A, distributed Gaussian basis; B, non-orthogonal sines; C, orthogonal sines; D, orthogonal sines with Buttle correction; E, Exact.

E(a.u.)	A	B	C	D	E
0.010	-12.5016	-12.5020	-12.6735	-12.5039	-12.5031
0.012	-13.0772	-13.0776	-13.4248	-13.0835	-13.0773
0.014	-13.5528	-13.5533	-13.5557	-13.5531	-13.5545
0.016	-13.9584	-13.9591	-14.4828	-13.9703	-13.9582
0.018	-14.3022	-14.3031	-14.3153	-14.3028	-14.3040
0.020	-14.6039	-14.6050	-15.2400	-14.6182	-14.6031
0.022	-14.8626	-14.8631	-14.8745	-14.8640	-14.8637
0.024	-15.1042	-15.0961	-15.8098	-15.1332	-15.0919
0.026	-15.3112	-15.2924	-15.2989	-15.2914	-15.2925
0.028	-15.5179	-15.4776	-16.1896	-15.6504	-15.4693
0.030	-15.7221	-15.6270	-17.0605	-16.0395	-15.6254
0.032	-15.9249	-15.7848	-16.1828	-15.5786	-15.7632
0.034	-15.9229	-15.8918	-17.3950	-15.6349	-15.8848
0.036	-16.2972	-16.1020	-18.6684	-15.6489	-15.9921
0.038	-17.0929	-16.3376	-19.8146	-15.4250	-16.0864

TABLE IV. Phase Shifts for N = 20. Columns A, distributed Gaussian basis; B, non-orthogonal sines; C, orthogonal sines; D, orthogonal sines with Buttle correction; E, Exact.

E(a.u.)	A	B	C	D	E
0.010	-12.5016	-12.5017	-12.6152	-12.5023	-12.5031
0.012	-13.0774	-13.0776	-13.3316	-13.0791	-13.0773
0.014	-13.5530	-13.5531	-13.5549	-13.5531	-13.5545
0.016	-13.9583	-13.9585	-14.3203	-13.9609	-13.9582
0.018	-14.3029	-14.3030	-14.3114	-14.3030	-14.3040
0.020	-14.6031	-14.6033	-15.0240	-14.6065	-14.6031
0.022	-14.8627	-14.8629	-14.8698	-14.8630	-14.8637
0.024	-15.0919	-15.0921	-15.5514	-15.0963	-15.0919
0.026	-15.2915	-15.2919	-15.2962	-15.2920	-15.2925
0.028	-15.4691	-15.4696	-15.8929	-15.4745	-15.4693
0.030	-15.6243	-15.6251	-15.7699	-15.6264	-15.6254
0.032	-15.7624	-15.7635	-16.0353	-15.7674	-15.7632
0.034	-15.8841	-15.8853	-16.4377	-15.8912	-15.8848
0.036	-15.9907	-15.9930	-16.0503	-15.9926	-15.9921
0.038	-16.0880	-16.0883	-16.7161	-16.0998	-16.0864

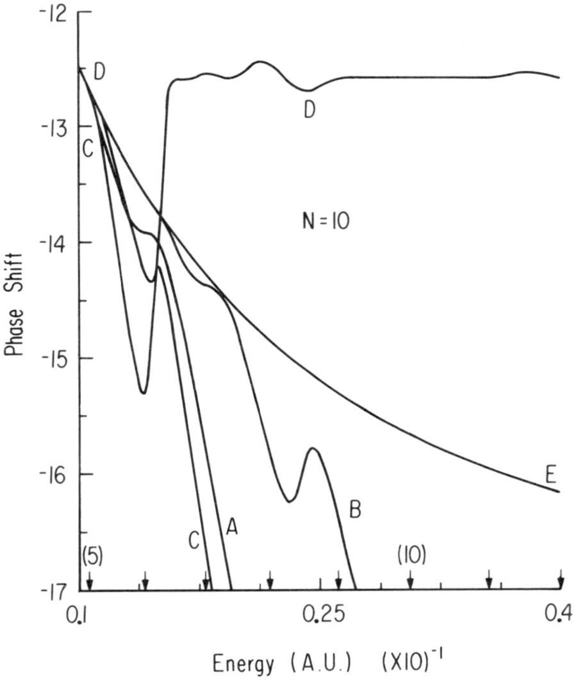

Figure 1. Plot for Table II.

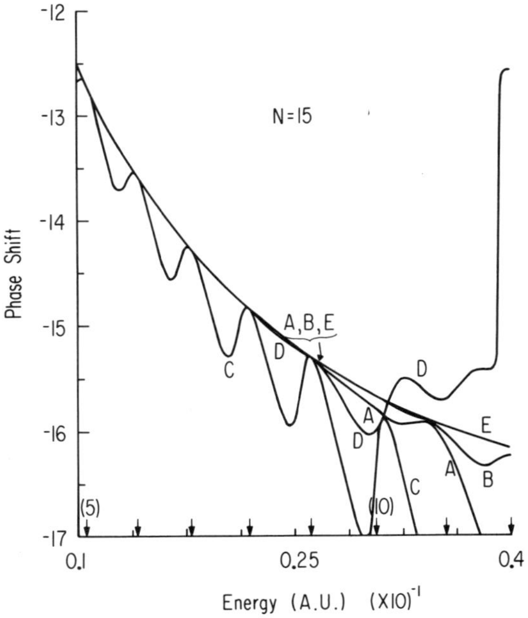

Figure 2. Plot for Table III.

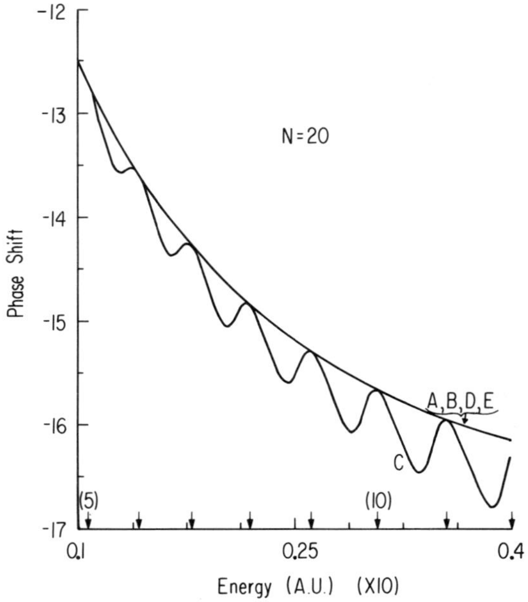

Figure 3. Plot for Table IV.

4. R-Matrix Propagation

It is well established [20,5] and reasonable that the number of translational functions required for convergence of an L^2 expansion for the R-matrix depends on the potential, the energies desired, and the size of the "R-matrix box," i.e., A. As the size of the box is reduced for an N function basis, the eigenvalues of $\underline{\underline{H}}+\underline{\underline{L}}$ are increased and spread out, and the sum in Eq. (II.19) will converge more rapidly (for $E \lesssim \lambda_{N/2}$). Also it is obvious that the eigenvalues, λ_i, near E must be given relatively accurate, and a smaller A (for fixed N) usually helps in this as well. However, since at A one has only the L^2 approximation to the Green's function (on $0 \leqslant R \leqslant A$), the connection between this "local" R-matrix and the S-matrix must be made. For A in the asymptotic region, this relation is given in Eq. (II.11), but we want A to be as small as possible for efficiency. We therefore consider below two methods of connecting the R-matrix to the S-matrix using a small "R-matrix box," i.e., small value of A. The first is a simple WKB (adiabatic) connection, and the second the more accurate R-matrix propagation [20,21].

We see from Eq. (II.10) that the R-matrix relates the exact wavefunction and derivative at the R-matrix boundary, A. If A is not near a classical turning point and is outside the channel coupling region, then with a high degree of accuracy we can take the wave function to be of the WKB form. For initial state i, it is:

$$\psi_i = \frac{c_i^*}{\sqrt{k_i(R)}} \exp\{-i \int_A^R k_i(R)\,dR\}\, \phi_i(x) - $$

$$- \sum_j S_{ji}\, \phi_j(x) \frac{c_j}{\sqrt{k_j(R)}} \exp\{+i \int_A^R k_i(R)\,dR\} \qquad \text{(IV.1)}$$

where the ϕ_i are the (perhaps adiabatically deformed) internal states and $k_i(R)$ is the local wave vector in state i:

$$k_i = [\frac{2\mu}{\hbar^2} (E - V_{ii}(R))]^{1/2}. \qquad \text{(IV.2)}$$

In order for the phases to be correct we require ψ_i to have the proper $e^{\pm ikR}$ form as $R \to \infty$, i.e.,

$$c_i^* \exp\{-i \int_A^R k_i(R)\,dR\} \underset{R\to\infty}{=} e^{-ik_iR}. \qquad \text{(IV.3)}$$

If we define the phase shift for state i from $R = A$ to ∞ by

$$\delta_i(A) = \int_A^\infty [k_i(R) - k_i]\,dR \qquad \text{(IV.4)}$$

we see that the c_i are just the phase factors

$$c_i = e^{-i\delta_i(A) + ik_iA} . \qquad (IV.5)$$

Using (IV.1) in (II.10) we see that

$$\underset{=}{S} = [\underset{=}{c}\,\underset{=}{k}^{-1/2} - i\,\underset{=}{R}\,\underset{=}{c}^*\,\underset{=}{k}^{1/2} + \underset{=}{R}\,\underset{=}{c}^*\,\underset{=}{k}'/2\underset{=}{k}^{3/2}]^{-1}$$
$$\times [\underset{=}{c}\,\underset{=}{k}^{-1/2} + i\,\underset{=}{R}\,\underset{=}{c}^*\,\underset{=}{k}^{1/2} + \underset{=}{R}\,\underset{=}{c}^*\,\underset{=}{k}'/2\underset{=}{k}^{3/2}] \qquad (IV.6)$$

where $\underset{=}{c}$ and $\underset{=}{k}$ are diagonal matrices given by (IV.2) and (IV.5) evaluated at $R = A$. ((IV.6) was, in fact, used in the numerical results presented in the last section.) Although this WKB analysis is accurate where V'/k^2 is small and where there is no channel coupling, R-matrix propagation may be used quite generally to relate an R-matrix on surface A to an R-matrix on a different surface, $A' > A$.

Since R-matrix propagation has been reviewed several times [22,23] and is widely used, it will not be presented in great detail here. The basic idea of the R-matrix propagation is that over an interval of the scattering coordinate say (R_i, R_{i+1}), in which the potential does not vary much, analytic solutions for the translational functions at arbitrary E and satisfying fixed b.c.'s at R_i, R_{i+1} can be found. These are just the trigonometric or hyperbolic functions (for constant potentials) or Airy functions (for linear potentials). The R-matrix propagation equations [14,20] are then the algorithm for the sequential construction of the R-matrix over an entire region $A \leqslant R \leqslant A'$ from the "sector" R matrices, $\underset{=}{r}_i$.

If we assume that $(\underset{=}{H} - E\underset{=}{I})$ in an internal basis at $R_i' = (R_i + R_{i+1})/2$ has been diagonalized by a transformation T_i to yield the internal eigenfunctions $\theta_n(x, R_i')$ and eigenvalues $\lambda_n^2(R_i')$, then the translational functions in the sector satisfying zero derivative b.c.'s at R_i and R_{i+1} are

$$\begin{aligned} \phi_n^L &= \cos|\lambda_n|(R - R_i) & \lambda_n^2 &\leqslant 0 \\ &= (\coth|\lambda_n|(R - R_i)) & \lambda_n^2 &> 0 \\ \phi_n^R &= \cos|\lambda_n|(R - R_{i+1}) & \lambda_n^2 &\leqslant 0 \\ &= (\coth|\lambda_n|(R - R_{i+1})) & \lambda_n^2 &> 0 \end{aligned} \qquad (IV.7)$$

and the Green's function corresponding to zero derivative b.c.'s is

$$\underset{=}{G}_i(R,R') = \begin{cases} \underset{=}{\phi}^L(R)\,(\underset{=}{W}^T)^{-1}\,(\underset{=}{\phi}^R(R'))^T & R < R' \\ \underset{=}{\phi}^R(R)\,(\underset{=}{W})^{-1}\,(\underset{=}{\phi}^L(R'))^T & R' < R \end{cases} \qquad (IV.8)$$

where $\underline{\underline{W}}$ is the matrix Wronskian of the solutions:

$$\underline{\underline{W}} = (\underline{\underline{\phi}}^{L})^{T} \underline{\underline{\phi}}^{R'} - \underline{\underline{\phi}}^{L'^{T}} \underline{\underline{\phi}}^{R} . \qquad \text{(IV.9)}$$

For these b.c.'s in the sector, we then have the R-matrix type relation for the scattering function at the sector surfaces in the locally diagonal basis:

$$\begin{pmatrix} \underline{F}(R_i) \\ \underline{F}(R_{i+1}) \end{pmatrix} = \begin{pmatrix} \underline{\underline{G}}_i(R_i,R_i) & \underline{\underline{G}}_i(R_i,R_{i+1}) \\ \underline{\underline{G}}_i(R_{i+1},R_i) & \underline{\underline{G}}_i(R_{i+1},R_{i+1}) \end{pmatrix} \begin{pmatrix} \underline{F}'(R_i) \\ -\underline{F}'(R_{i+1}) \end{pmatrix} \qquad \text{(IV.10)}$$

where $\underline{F}$ is the vector of coefficients, $f_n(R)$, of the wavefunction in sector i

$$\psi(x,R) = \sum_n f_n(R)\, \theta_n(x;\, R_i') \qquad R_i \leqslant R \leqslant R_{i+1} \qquad \text{(IV.11)}$$

In the diagonal representation the sector Green's functions are diagonal and are the negative of the sector R-matrices defined in Ref. 21:

$$-\underline{\underline{G}}_{nn}(R_i,R_i) = -\underline{\underline{G}}_{nn}(R_{i+1},R_{i+1})$$

$$= (\underline{\underline{r}}_1^{(i)})_{nn} = (\underline{\underline{r}}_4^{(i)})_{nn} = \begin{cases} |\lambda_n|^{-1}\coth|\lambda_n\Delta| & \lambda_n^2 > 0 \\ -|\lambda_n|^{-1}\cot|\lambda_n\Delta| & \lambda_n^2 \leqslant 0 \end{cases}$$

$$\text{(IV.12)}$$

$$-\underline{\underline{G}}_{nn}(R_i,R_{i+1}) = -\underline{\underline{G}}_{nn}(R_{i+1},R_i)$$

$$= (\underline{\underline{r}}_2^{(i)})_{nn} = (\underline{\underline{r}}_3^{(i)})_{nn} = \begin{cases} |\lambda_n|^{-1}\operatorname{csch}|\lambda_n\Delta| & \lambda_n^2 > 0 \\ -|\lambda_n|^{-1}\csc|\lambda_n\Delta| & \lambda_n^2 \leqslant 0 \end{cases}$$

The R-matrix propagation equations are obtained by assuming the R-matrix is known at R_i (in the basis of section i-1)

$$\underline{F}^{(i-1)}(R_i) = \underline{\underline{R}}^{(i-1)}(R_i)\, \underline{F}^{(i-1)'}(R_i) . \qquad \text{(IV.13)}$$

This is then combined with the transformation from the diagonal basis at R_{i-1}' to the diagonal basis at R_i'

$$\underline{\underline{Q}}_{i-1,i} \equiv \underline{\underline{T}}_{i-1}^{T}\, \underline{\underline{T}}_i \qquad \text{(IV.14)}$$

and the sector R-matrix for sector i

$$\begin{pmatrix} F^{(i)}(R_i) \\ F^{(i)}(R_{i+1}) \end{pmatrix} = \begin{pmatrix} r_1^{(i)} & r_2^{(i)} \\ r_3^{(i)} & r_4^{(i)} \end{pmatrix} \begin{pmatrix} -F^{(i)'}(R_i) \\ F'(R_{i+1}) \end{pmatrix} \tag{IV.15}$$

to yield the R-matrix at R_{i+1} in the basis of sector i:

$$R^{(i)}(R_{i+1}) = r_4^{(i)} - r_3^{(i)} Z^{(i)} r_2^{(i)} \tag{IV.16a}$$

$$Z^{(i)} \equiv [r_1^{(i)} + Q^T_{i-1,i} R^{(i-1)} Q_{i-1,i}]^{-1}. \tag{IV.16b}$$

The advantage of using the R-matrix propagation, of course, lies in the stability of the algorithm and, for slowly varying potentials, in the large step size possible. Although the propagation must be repeated at each energy, the creation of the Hamiltonian matrix, and its diagonalization, must be done only once. The eigenvalues and the Q matrices can be saved for each step and used for other energies.

One apparent limitation of using the R-matrix propagation is that only (inverse) log derivative information is carried along--the wavefunction is only determined asymptotically with the imposition of b.c.'s. This makes the accumulation of overlap integrals for photodissociation, for example, somewhat more complicated. At this Workshop, however, R. B. Walker [24] showed that the log derivative matrix can be propagated using the information evaluated in the R-matrix propagation, with a very similar algorithm. This permits the efficient propagation of overlap information in a relatively simple fashion analogous to that used by Kulander, et al. for collinear "reactive" photodissociation [25] and by Heather, et al. for 3-D triatomic photodissociation [26].

5. Summary

Although the L^2 R-matrix method for scattering was proposed many years ago, its implementation has been quite restricted because of the slow convergence of the "unadorned" R-matrix to numerically adequate results. For inelastic scattering the use of a simple zero order Hamiltonian with an analytic solution to generate the Buttle correction makes it a viable option for some systems. In particular the increase in memory and speed of large computers make large basis calculations feasible.

For reactive scattering, however, the Buttle correction cannot be applied easily since there exists no reasonable zero order Hamiltonian with two or more chemical channels for which analytic solutions are known. Although a Buttle type correction may be determined in a basis, it adds substantially to the computational effort required.

The results presented in Ref. 16 by Bocchetta and Gerratt and in this article [27] using non-orthogonal bases for L^2 R-matrix calculations are therefore most interesting for their implications for reactive scattering. Since convergence using these bases appears to be as good as that with a Buttle correction (using an orthogonal basis) and appears to be very uniform with N and E, it is likely that one may dispense with the Buttle correction entirely. For reactive scattering R-matrix calculations this is, indeed, an important development.

As currently implemented the L^2 R-matrix calculations (with non-orthogonal functions) appear to give adequate accuracy up to an energy of about 1/2 to 2/3 of the highest "R-matrix box" eigenvalue (for 1-dimension). This is, of course, about the level at which the accuracy of the eigenvalues begins to degrade as well. Thus we may expect that in higher dimensions the level of accuracy of the scattering calculation will be determined by the level of accuracy of the eigenvalues at the corresponding scattering energy, a quantity which is relatively easy to determine.

The size of the "R-matrix box" is, of course, a prime determinant of the eigenvalues of the Hamiltonian plus Bloch operator, and, as shown quite definitively by Schneider and Walker [28], is directly related to accuracy for a fixed number of functions. Thus an appropriate strategy [16] for reactive scattering would be to use a non-orthogonal L^2 basis in the strong interaction region (where V, V', V", etc., are large), with the "box" as small as possible. Outside this region, where the potentials do not vary too fast, the R-matrix propagation method becomes very efficient. Thus a hybrid L^2-propagation R-matrix method would seem to be the approach of choice for complex systems such as chemical reactions.

Acknowledgment: I am grateful to Robert Whitnall who did the calculations for the numerical results presented in Section 3. This research was supported by the National Science Foundation under Grants CHE-8203453 and CHE-8505001.

References

1. G. C. Schatz and A. Kuppermann, J. Chem. Phys. 65, 4642 (1976); 65, 4668 (1976); 62, 2502 (1975).

2. A. B. Elkowitz and R. E. Wyatt, J. Chem. Phys. 63, 702 (1975).

3. R. B. Walker, E. B. Stechel, and J. C. Light, J. Chem. Phys. 69, 2922 (1978); CECAM Report on Workshop on Reactive Scattering, ed. C. Moser, Orsay, France, pp. 1-49.

4. E. P. Wigner and L. Eisenbud, Phys. Rev. 72, 29 (1947).

5. For an extensive review, see A. M. Lane and R. G. Thomas, Rev. Mod. Phys. 30, 257 (1958).

6. L. M. Delves, Nucl. Phys. 9, 391 (1959); 20, 275 (1960)

7. A. Kuppermann, Theor. Chem.: Advances and Perspectives, 6A, 79 (1981).

8. C. Bloch, Nucl. Phys. 4, 503 (1957).

9. P. J. A. Buttle, Phys. Rev. 160, 719 (1967).

10. E. J. Heller, Chem. Phys. Lett. 23, 102 (1973).

11. D. J. Zvijac, E. J. Heller, and J. C. Light, J. Phys. B8, 1016 (1975).

12. O. H. Crawford, J. Chem. Phys. 55, 2563 (1971).

13. R. Der. O. Gebhart, and R. Haberlandt, Chem. Phys. Lett. 27, 107 (1974).

14. D. J. Zvijac and J. C. Light, Chem. Phys. 12, 237 (1976).

15. A. Mori, Phys. Rev. C5, 1795 (1972).

16. C. J. Bocchetta and J. Gerratt, J. Chem. Phys. 82, 1351 (1985).

17. B. Schneider, Chem. Phys. Lett. 31, 237 (1975).

18. E. B. Stechel, T. G. Schmalz, and J. C. Light, J. Chem. Phys. 70, 5640 (1978); T. G. Schmalz, E. B. Stechel, and J. C. Light, J. Chem. Phys. 70, 5660 (1979).

19. I. P. Hamilton and J. C. Light, submitted to J. Chem. Phys.

20. J. C. Light and R. B. Walker, J. Chem. Phys. 65, 4272 (1976).

21. E. B. Stechel, R. B. Walker, and J. C. Light, J. Chem. Phys. 69, 3518 (1978).

22. J. C. Light, R. B. Walker, E. B. Stechel, and T. G. Schmalz, Comp. Phys. Commun. 17, 89 (1979).

23. J. V. Lill, T. G. Schmalz, and J. C. Light, J. Chem. Phys. 78, 4456 (1983).

24. R. B. Walker, Private Communication.

25. K. C. Kulander and J. C. Light, J. Chem. Phys. 73, 4337 (1980).

26. R. Heather and J. C. Light, J. Chem. Phys. 78, 5513 (1983); 79, 147 (1983).

27. The calculations were done by R. Whitnall.

28. B. I. Schneider and R. B. Walker, J. Chem. Phys. 70, 2466 (1979).

THE TIME DEPENDENT WAVEPACKET METHOD : APPLICATION TO COLLISION INDUCED DISSOCIATION PROCESSES

C.Leforestier
Laboratoire de Chimie Théorique
Université de Paris-Sud
91405 ORSAY (FRANCE)

ABSTRACT. After reviewing the time dependent wavepacket method as applied to collision induced dissociation processes,we report accurate quantum results for reactive and non reactive collinear A+BC systems. Both systems display a vibrational enhancement effect in the low energy region. While the non reactive systems exhibit a vibrational inhibition effect at higher energies,a more complex behavior is observed in the reactive case. Below the classical dissociation threshold,the non reactive systems display tunnelling tails which decrease with the initial vibrational excitation of the diatomic molecule. The reactive system displays important quantum effects at energies well above the classical dissociation threshold.

1. INTRODUCTION

Collision Induced Dissociation (CID) calculations constitute a challenging problem in the field of reaction dynamics :The full 3D treatment of the simplest system,A+BC → A+B+C,is out of scope for the present time unless we resort to the classical approximation. Only a few 1D calculations (i.e. collinear or perpendicular) have appeared in the literature so far[1-15],most of them being concerned with non reactive systems modeling an H_2+Rg type collision. Three different methods have mainly been used for these calculations,namely

i) the time dependent wavepacket method[1-4]
ii) the Close Coupling method,based on spherical coordinates[10-11]
iii) the semi-classical (Classical S-matrix) method[13-15].

The rotational degrees of freedom (i.e. 2D or 3D calculations) are known to be important for a correct description of molecular energy transfer since the early works of Kelley and Wolfsberg[16] and Bergeron and Chapuisat[17]. This puts an important restriction on the validity of the dissociation mechanism as obtained from 1D calculations. Nevertheless such quantum calculations are the only way to estimate the adequacy of classical mechanics for the study of dissociation processes. Also, as will be discussed in the conclusion,these collinear calculations open the road to more realistic developments.

D. C. Clary (ed.), The Theory of Chemical Reaction Dynamics, 235–246.

In section 2, we review the time dependent wavepacket method, which is now well codified since the work of Mc Cullough and Wyatt[18]. Section 3 is devoted to a comparison of classical and quantum results for reactive and non reactive collinear calculations. Finally we discuss the possibility of extending these calculations to more realistic cases, such as approximate 3D quantum treatments.

2. THE TIME DEPENDENT WAVEPACKET METHOD

The collinear A+BC dissociative collision can be treated in a straightforward manner, using the Time Dependent WavePacket (TDWP) method. The reason is that the dissociative continuum of the BC molecule is handled automatically within the space discretization scheme of the grid. As the basic method has already been described in detail elsewhere[18], it will be only outlined here, emphasizing the technical points and some new features which lead to a significant reduction in computation time.

2.1. Initial Conditions

The method consists in computing the time evolution of a wavepacket ψ, which represents initially an atom A impinging on a molecule BC in a given vibrational state v :

$$\psi(x,y,t=0) = F(x,x_0).u_v(y) \tag{1}$$

where the (x,y) coordinates, which diagonalize the kinetic energy operator, can be either the usual skewed coordinates[19] for a collinear collision or the (r,R) coordinates for a perpendicular one[15].

Most of the calculations have used a Gaussian k-distribution centered around the mean value k_0:

$$F(x,x_0) = \{\hbar/2\pi\}^{1/2} \int_{-\infty}^{+\infty} \phi(k,k_0).\exp\{-ik(x-x_0)\}dk$$

$$= \{2\pi\delta^2\}^{-1/4} \exp\{-(x-x_0)^2/4\delta^2 - ik_0x\} \tag{2}$$

$$\phi(k,k_0) = \{2\delta^2/\pi\}^{1/2} \exp\{-2\delta^2(k-k_0)^2\} \tag{3}$$

As is well known, this initial distribution in momentum space allows to extract from the final wavepacket results for a whole range of collision energies $\hbar^2k^2/2\mu_{A,BC}$ centered around the mean collision energy $\hbar^2k_0^2/2\mu_{A,BC}$. The reliability of the results as a function of the distance $|k-k_0|$ will be discussed in section 2.3.

Other translational distribution functions F can be defined: e.g. Mazur and Rubin have used the Fourier transform of a Boltzmann energy distribution[20]. The use of such distributions leads to quantities which are more directly related to experimental conditions.

2.2 Propagation of the Wavepacket

The time evolution of the initial wavepacket (eq.1) is given by the evolution equation :

$$\psi(t) = \exp\{-iH(t-t_0)/\hbar\}.\psi(t_0) \tag{4}$$

$$H = -\hbar^2/2\mu.\{\partial^2/\partial x^2 + \partial^2/\partial y^2\} + V(x,y).$$

To treat the space dependence of this partial differential equation one defines a grid $\{x_0,x_0+\Delta x,\ldots,x_0+M.\Delta x\}x\{y_0,y_0+\Delta y,\ldots,y_0+N.\Delta y\}$ and requires that the solution satisfies the above equation (4) at each node. To evaluate the second derivatives at node (x_p,y_q),one can use the finite difference scheme :

$$\partial^2\psi/\partial x^2\Big|_{x_p,y_q} = \Delta x^{-2}\{D_0^{(d)}\psi_{pq} + D_1^{(d)}(\psi_{p+1,q} + \psi_{p-1,q}) + \ldots + D_d^{(d)}(\psi_{p+d,q} + \psi_{p-d,q})\} \tag{5}$$

where d is the finite difference order ($d\geqslant 1$) and the $D_i^{(d)}$ are the corresponding coefficients. In Table 1,we compare several finite difference schemes using different orders for evaluating the second derivative of cos18x as a function of the mesh size Δx. This function roughly models the actual space dependence of the wavepackets to be propagated. The total CPU time being proportional to $n^2.(2d+1)$,it can be seen from Table 1 that high order schemes ($d\geqslant 3$) can lead to a significant reduction in computation time for a same accuracy.

n ($=\lambda/\Delta x$)	d=1	d=3	d=5	d=7	d=9
4	2(-1)	2(-2)	3(-3)	5(-4)	1(-4)
8	5(-2)	4(-4)	4(-6)	6(-8)	<1(-9)
12	2(-2)	4(-5)	8(-8)	<1(-9)	<1(-9)
16	1(-2)	6(-6)	6(-9)	<1(-9)	<1(-9)

Table 1 : Relative error on the evaluation of $d^2\cos 18x/dx^2$ as a function of
i) the number n of grid points per wavelength λ
ii) the finite difference order d

For each variable r,the discretization scheme is equivalent to using a Dirac delta functions basis set $\{|r_i\rangle, i=1,\ldots,N\}$; from the "closure" relation $1 = \sum_i |r_i\rangle\langle r_i|$,one can write

$$|\psi\rangle = \sum_i |r_i\rangle\langle r_i|\psi\rangle = \sum_i R_i\,|r_i\rangle \tag{6}$$

Such a delta function basis set is well suited for local operators,such as V which is diagonal in this basis set :

$$V|\psi\rangle = \Sigma\, V|r_i\rangle\langle r_i|\psi\rangle = \Sigma\, V(r_i)\,R_i\,|r_i\rangle\,. \tag{7}$$

In order to handle in a similar way the non local operator $\partial^2/\partial r^2$,one can use the pseudospectral method[21]. In this method,one defines another basis set $\{|\phi_n\rangle, n=1,\ldots,N\}$ on which one expands the function $|\psi\rangle$:

$$|\psi\rangle = \sum_n c_n \, |\phi_n\rangle \; .$$

The expansion coefficients $\underline{R}$ and $\underline{c}$ transform according to the relation

$$\underline{R} = \underset{\sim}{M}.\underline{c} \qquad \Longleftrightarrow \qquad \underline{c} = \underset{\sim}{M}^{-1}.\underline{R} \tag{8}$$

where the $\underset{\sim}{M}$ matrix elements are defined as

$$M_{in} = \langle r_i | \phi_n \rangle = \phi_n(r_i) \; . \tag{9}$$

One can then define the ϕ's as the eigenfunctions of the non local operator under consideration;e.g. in the case of the $\partial^2/\partial r^2$ operator, these will be the plane wave functions :

$$\partial^2/\partial r^2 \; |\phi_n\rangle = - \, k_n^2 \; |\phi_n\rangle \; . \tag{10}$$

The effect of the $\partial^2/\partial r^2$ operator upon the wavefunction $|\psi\rangle$ is computed in the eigenfunctions representation $\{|\phi_n\rangle\}$ according to the scheme :

$$\nabla^2|\psi\rangle \; : \; \begin{vmatrix} R_1 \\ \cdot \\ \cdot \\ \cdot \\ R_n \end{vmatrix} \xrightarrow{\underset{\sim}{M}^{-1}.\underline{R}} \begin{vmatrix} c_1 \\ \cdot \\ \cdot \\ \cdot \\ c_n \end{vmatrix} \xrightarrow{x(-k_i^2)} \begin{vmatrix} c'_1 \\ \cdot \\ \cdot \\ \cdot \\ c'_n \end{vmatrix} \xrightarrow{\underset{\sim}{M}.\underline{c}'} \begin{vmatrix} R'_1 \\ \cdot \\ \cdot \\ \cdot \\ R'_n \end{vmatrix} \tag{11}$$

This method has recently been used by Kosloff and Kosloff[22] in conjunction with a Fast Fourier Transform (FFT) to switch back and forth between the $\{|r_i\rangle\}$ and $\{|\phi_n\rangle\}$ basis sets: The FFT operations number is varying as $N.\log N$ instead of N^2 for the $M.\underline{R}$ matrix product. As pointed out by these authors,the main reason to use this method is that it can achieve a far better accuracy than the finite difference method. In table 2,we report such a comparison for the evaluation of $d^2\sin 18x/dx^2$, using sine functions for the $\{\phi_n\}$ basis set.

$\lambda/\Delta x$	pseudo-spectral	Finite Difference d=3	d=5	d=7	d=9
2.6	3(-5)	1(-1)	6(-2)	3(-2)	2(-2)

Table 2 : Relative error on the evaluation of $d^2\sin 18x/dx^2$ as given by the pseudospectral method (using 64 sine functions) and several finite difference schemes.

When using such a pseudospectral method,one must be cautious with the accuracy of the method near the edges of the grid. The results reported in Table 2 correspond to an evaluation at a point located near the middle of the grid. Table 3 displays the accuracy on the evaluation of $d^2\{\sin 18x/\{1+\exp(-\alpha x+A)\}\}/dx^2$ as a function of the grid point location. The α and A parameters have been chosen such that the sine function $\sin 18x$ is reduced by about three orders of magnitude at the first node, as compared to its value near the middle of the grid. From Table 3 we can see that the grid must be extended slightly beyond the classical turning point in order that a good accuracy is reached in the physical region of interest.

Grid point location	2	6	10	14	18	22	26
Relative error	2(-1)	7(-2)	1(-2)	2(-3)	2(-3)	4(-4)	4(-5)

Table 3 : Relative error on the evaluation of $d^2\{\sin 18x/\{1+\exp(-\alpha x+A)\}\}/dx^2$ (see text) as a function of the grid point location. The pseudospectral basis set consists of 64 sine functions.

Two different explicit schemes have been shown to be particularly efficient in order to propagate the wavepacket in time:

i) The Richardson scheme[23,24] which corresponds to a third order expansion of the evolution operator (eq.(4))

$$\psi(t+\Delta t) = \psi(t-\Delta t) - 2i\Delta t/\hbar . H\psi(t) + O(\Delta t^3) \quad . \tag{12}$$

ii) An expansion of the evolution operator in a Chebychev series[25]

$$\exp\{-iH\Delta t\} = \sum_{n=0}^{N} a_n \Phi_n(-iH\Delta t) \tag{13}$$

where the a's are the expansion coefficients and the Φ's the complex Chebychev polynomials. This scheme is specially adapted to long propagation times because the error decreases exponentially once N is large enough. The error in propagation can thus be kept lower than the error arising from the spatial discretization scheme.

2.3. Final Analysis

When the reaction is complete, one can extract the state-to-state probabilities by projecting the final wavepacket $\psi(x,y,T)$ onto the asymptotical eigenstates. For example the inelastic probabilities $P^I_{v\to v'}(k)$ at the collision energy $E_c = \hbar^2k^2/2\mu_{A,BC}$ are computed from the formula

$$P^I_{v\to v'}(k) = k/k_{v'} . \{4\pi^2|\Phi(k,k_0)|^2\}^{-1} \left| \iint dxdy \; u_{v'}(y) . e^{-ik_{v'}x} . \psi(x,y,T) \right|^2 \tag{14}$$

where $\Phi(k,k_0)$ is the distribution function of the incoming wavepacket in momentum space (see eq.(3)). A similar formula holds for the reactive probabilities $P^R_{v\to v''}$. The dissociation probability is computed as

$$P^D_v(k) = 1 - \sum_{v'} P^I_{v\to v'}(k) - \sum_{v''} P^R_{v\to v''}(k) \quad . \tag{15}$$

In order to estimate the accuracy of the calculations, several tests can be performed[3-4] :

i) The first one consists in checking if the inelastic and reactive probabilities sum up to 1 below the dissociation energy. Table 4 displays the results obtained for the reactive-dissociative H + HD system[4], as a function of the distance $|k-k_0|$ (see also figure 1).

$|k-k_0|$ (a.u.)	$\Phi(k,k_0)$	$P^I + P^R$
$k_1-k_0=3.3$	.05	.995
$k_2-k_0=5.6$	.002	.975

Table 4 : Sum of the inelastic and reactive probabilities below the dissociation energy as a function of the distance $|k-k_0|$. A value δ of 0.25 has been used in eq.(3) giving the amplitude $\Phi(k,k_0)$.

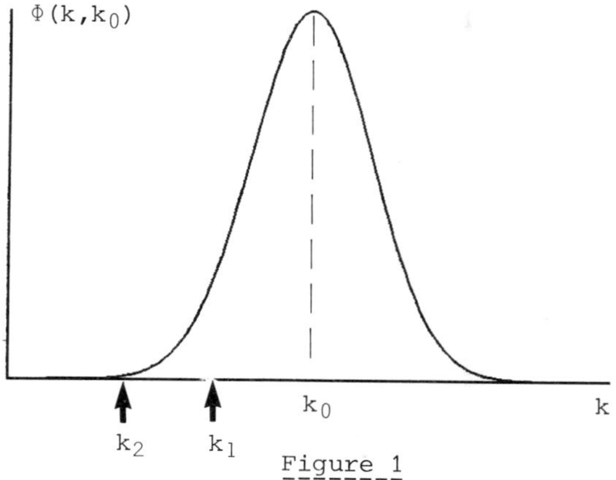

Figure 1

ii) As a second test, one can compare the probabilities arising from two distinct wavepackets with overlapping energy distributions. In Table 5 we report the state-to-state $P^I_{0\to6}$ and dissociation P^D_0 probabilities computed from two wavepackets for an inelastic H + H_2(0) system[3].

E_T ($\hbar\omega$)	WP I (k_0=26.3) $P^I_{0\to6}$	P^D_0	WP II (k_0=30.4) $P^I_{0\to6}$	P^D_0
	2.11(-2)	.001	2.13(-2)	.000
13	5.69(-2)	.003	5.67(-2)	.005
15	9.53(-2)	.019	9.54(-2)	.019
17	1.18(-1)	.065	1.18(-1)	.064
19	1.19(-1)	.153	1.20(-1)	.149
21	1.03(-1)	.278	1.04(-1)	.272
23	8.00(-2)	.423	8.12(-2)	.414

Table 5 : Comparison of the inelastic $P^I_{0\to6}$ and dissociation P^D_0 probabilities arising from two distinct wavepackets as a function of the total energy. The calculations correspond to an inelastic H + H_2 model system.

iii) Finally one can check if the microreversibility principle is satisfied by comparing the state-to-state probabilities $P_{v\to v'}(E_T)$ and $P_{v'\to v}(E_T)$ computed from wavepackets originating respectively in the v and v' vibrational states. In table 6 are reported the inelastic probabilities $P^I_{0\to1}$ and $P^I_{1\to0}$ for an inelastic H+DH model system[3], calculated in the common energy range of the two wavepackets.

E_T ($\hbar\omega$)	$P_{0\to1}$	$P_{1\to0}$
11	2.31(-1)	2.30(-1)
15	7.78(-2)	7.77(-2)
19	2.07(-2)	2.07(-2)
23	4.91(-3)	4.94(-3)
27	1.11(-3)	1.10(-3)
31	2.64(-4)	2.71(-4)

Table 6 : Comparison of the direct $P_{0\to1}$ and reverse $P_{1\to0}$ for different total energies. These calculations correspond to a non reactive H+DH model system.

3. RESULTS

In this section we will discuss separately the results obtained for non reactive systems from those obtained for reactive ones,as they display very different behaviors.

3.1. Non Reactive Systems

The results reported[2,3,14] correspond to Rg-H_2 model systems,but where the mass of the rare gas atom Rg has been set to 1 a.u.m. Amongst these systems is included the H-H_2 one using a L.E.P.S. potential energy surface,as studied by Kulander et al.[2],because it displays no reactive scattering for the collision energies considered. All these calculations lead to the similar trends exemplified on figure 2,namely:

i) The dynamical threshold is much higher than the energetic one when the diatomic molecule is initially in its ground vibrational state. The high threshold energies observed decrease rapidly however with initial excitation of the diatomic molecule.

ii) There is a strong vibrational enhancement effect at low energies and a vibrational inhibition one at high energies. This feature has been discussed by Hunt and Child[26] using a classical S-matrix phase space approach. Briefly the underlaying mechanism can be explained as follows: In a non reactive A+BC(v_0) $\to$ A+BC(v) collision,the more excited the initial state v_0 the broader will be the corresponding final distribution over vibrational states v. Henceforth the final distribution associated to an initially excited v_0 state will hit the dissociation limit first. When increasing the total energy,the v_0=0 state will eventually lead to a nearly complete dissociation. For this same total energy,the broader final distribution associated to the initially excited v_0 state will still display some components on the highest vibrational states,and thus results in a vibrational inhibition effect.

iii) By comparison with quasi-classical results,the quantum tails for dissociation are larger when the diatomic molecule is initially in a low vibrational state. These quantum tails have been found more important for the (1,2,1) mass combination as compared to the (1,1,1) one. Above the threshold region,there is a reasonable agreement between the quasi-classical and quantum results,except for the oscillations which are not reproduced classically. Such oscillations have already appeared in the CID study of a truncated harmonic oscillator by Johnson and

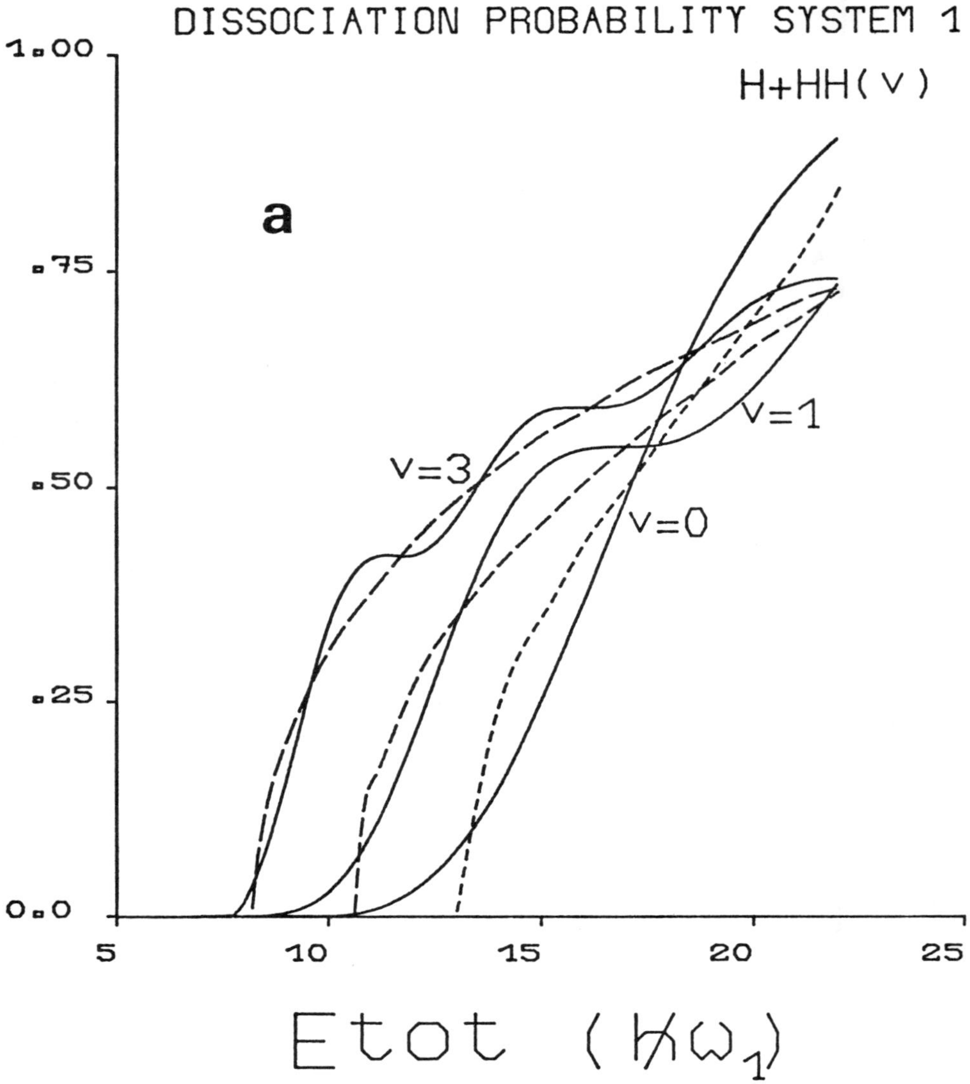

Figure 2 : Comparison of quantum (___) and classical (- -) dissociation probabilities for a model H + H_2(v) inelastic system as a function of the total energy. The energy is expressed in units of the fundamental frequency ω of the diatomic molecule. The dissociation energy corresponds to 6ℏω.

Roberts[13] and in the dissociation of a forced Morse oscillator by Hunt and Sridharan[27]. In each case the number of bumps in the dissociation curve was found to be equal to the initial vibrational number of the oscillator.

3.2. Reactive Systems

To date only three reaction-dissociation calculations have been reported by Manz and Römelt[10],Kaye and Kuppermann[11] and Leforestier[4,28].The former two used the hyperspherical coordinates method to study model $X-X_2$ systems,bearing respectively one and two bound states asymptotically. The latter one,whose results are presented on figure 3,will be discussed below; it corresponds to a model H-HD system,HD and H_2 bearing respectively 7 and 6 bound states.

Figure 3 enlightens two distinct features from the inelastic case (see figure 2 for a comparison) which are:

i) The near coincidence of the energetic and dynamic thresholds for dissociation. This feature has not been observed for inelastic systems. Even on a reactive surface,when no reactive scattering occurs at the dissociation energy,no coincidence between the two thresholds is observed[2]. The condition for the near coincidence of energetic and dynamic thresholds is the presence of concomitant reactive scattering at the dissociation energy. It is related to the fact that energy transfer is more efficient in reactive collisions.

ii) Small quantum tails at the dissociation threshold but important quantum effects at higher energies. Just above threshold and up to twice the dissociation energy,there is a very good agreement between the classical and quantum results. But unlike the inelastic case,one can note large discrepancies at higher total energies. These discrepancies which appear as sharp peaks in the classical dissociation probability curves,are due to anti-threshold effects[29,30]: These peaks result from the inelastic probability curve falling off. One can see however that the higher the initial vibrational state of the diatomic molecule,the better is the agreement between the classical and quantum curves.

4. DISCUSSION

The Time Dependent WavePacket (TDWP) method has been greatly improved in the last few years,essentially from the work of Kosloff et al.[22,25]. This method leads to dissociation probabilities converged within 1% or better,as shown in this paper. Besides giving a time description of the collision process,the TDWP method handles only one column of the S-matrix at the same time: This feature is particularly interesting for dissociation processes. Also the extra work which results from adding the time variable to the Schrödinger equation,leads to state-to-state probabilities for a full range of collision energies.

The quantum calculations reported so far help to determine the validity of a classical description of CID. In the inelastic case,one should expect quantum effects at threshold,this effect having been found more important for the (1,2,1) mass combination as compared to

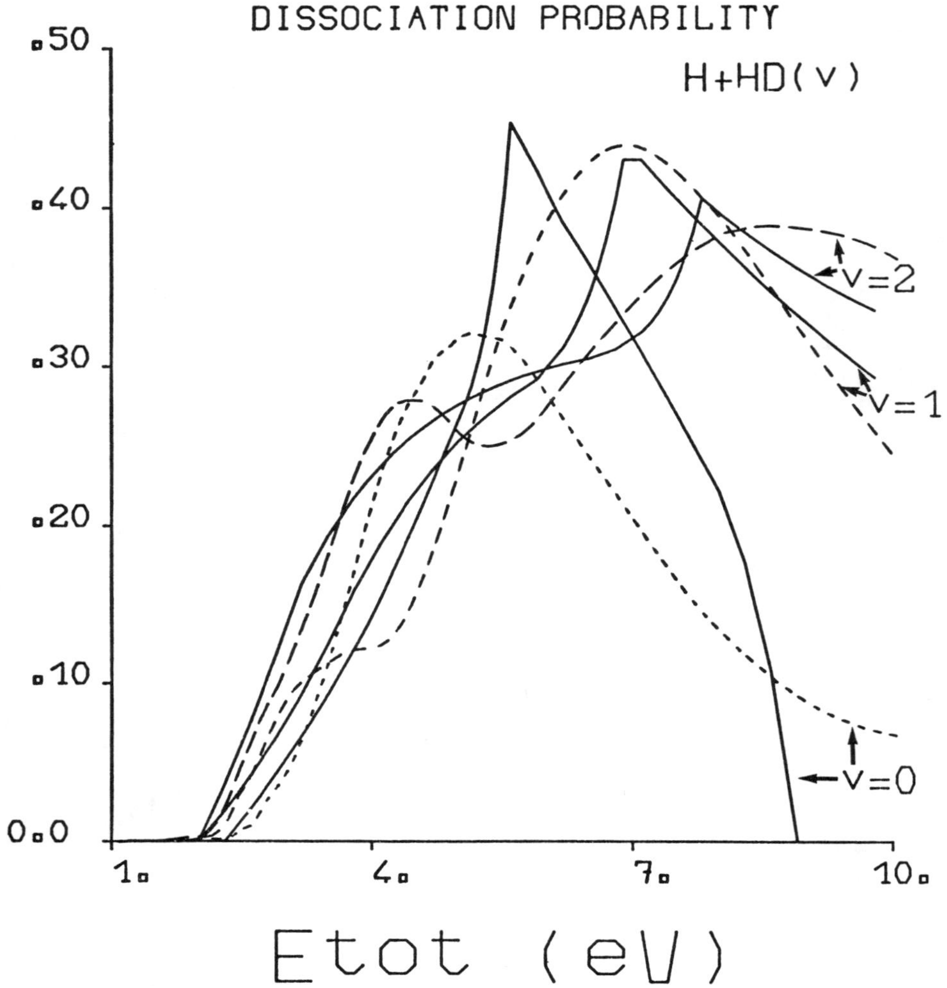

Figure 3 : Comparison of quantum (- -) and classical (___) dissociation probabilities for a model H+HD(v) reactive system as a function of the total energy. The dissociation energy corresponds to 1.77 e.V.

the (1,1,1) one. These threshold effects can produce large errors in the calculation of thermal rate constants. The results obtained at higher energies display a vibrational inhibition effect and show there is a reasonable agreement between the classical and quantum curves. On the converse,the results obtained in the reactive case display small quantum tails in the threshold region for dissociation,but large quantum effects at higher energies due to anti-threshold effects. Such effects could persist in 3D calculations whenever the reaction cross section collapses at some energy. More calculations on reactive-dissociative systems need to be performed in order to get a clearer picture of these effects.

While only exact 1D calculations are feasible for the present time, approximate 3D quantum treatments can be used. The Infinite Order Sudden Approximation (IOSA) appears particularly well suited to study the dissociation of a heavy diatomic molecule collided by a light atom: This approximation considers that the diatomic molecule does not rotate during the collision. Within IOSA[31],the full 3D treatment is replaced by a series of 1D calculations corresponding to:

i) different values of the angle γ between the axis of the diatomic molecule and the direction of the incident atom,

ii) different values of the relative angular momentum ℓ.

The dissociation cross section is then computed from the corresponding dissociation probabilities averaged over γ and ℓ. Such a study is presently undertaken on the $He-Ar_2$ system.

REFERENCES

1. L.W.Ford,D.J.Diestler and A.F.Wagner,J.Chem.Phys.,63(1975),2019.
2. K.C.Kulander,J.Chem.Phys.,69(1978),5064;
   J.C.Gray,G.A.Fraser,D.G.Truhlar and K.C.Kulander,J.Chem.Phys.,73 (1980),5726.
3. C.Leforestier,G.Bergeron and P.Hiberty,Chem.Phys.Letters,84(1981), 385;
   G.Bergeron,P.Hiberty and C.Leforestier,Chem.Phys.,93(1985),253.
4. C.Leforestier,Chem.Phys.,87(1984),241.
5. G.Wolken,J.Chem.Phys.,63(1975),528.
6. E.W.Knapp,D.J.Diestler and Y.W.Liu,Chem.Phys.Letters,49(1977),379.
7. E.W.Knapp and D.J.Diestler,J.Chem.Phys.,67(1977),4969.
8. L.H.Beard and D.A.Micha,J.Chem.Phys.,73(1980),1193.
9. G.D.Barg and A.Askar,Chem.Phys.Letters,76(1980),609.
10. J.Manz and J.Römelt,Chem.Phys.Letters,77(1981),172.
11. J.A.Kaye and A.Kuppermann,Chem.Phys.Letters,78(1981),546.
12. M.I.Haftel and T.K.Lim,Chem.Phys.Letters,89(1982),31.
13. L.L.Johnson and R.E.Roberts,Chem.Phys.Letters,7(1970),480.
14. I.Rusinek and R.E.Roberts,J.Chem.Phys.,65(1976),872;
    ibid.,68(1978),1147.
15. I.Rusinek,J.Chem.Phys.,72(1980),4518.
16. J.D.Kelley and M.Wolfsberg,J.Chem.Phys.,53(1970),2967.
17. G.Bergeron and X.Chapuisat,Chem.Phys.Letters,11(1971),334.
18. E.A.Mc Cullough and R.E.Wyatt,J.Chem.Phys.,54(1971),3578.
19. P.J.Kuntz in:"Modern Theoretical Chemistry",Vol.2,ed.W.H.Miller, Plenum Press,New York,1976.
20. J.Mazur and R.J.Rubin,J.Chem.Phys.,31(1959),1395.
21. S.A.Orszag,Studies Applied Math.,51(1972),253;ibid.,50(1971),293;
    S.A.Orszag,Phys.Rev.Letters,26(1971),1100;
    D.Gottlieb and S.Orszag,Numerical analysis of spectral methods, theory and application(SIAM,Philadelphia,1977)
22. D.Kosloff and R.Kosloff,J.Comp.Phys.,52(1983),35.
23. D.M.Young and R.T.Gregory,"A survey of numerical mathematics", Addison-Wesley,Reading,1973.
24. A.Askar and A.S.Cakmak,J.Chem.Phys.,68(1978),2794;
    R.J.Rubin,J.Chem.Phys.,70(1979),4811.
25. H.Tal-Ezer and R.Kosloff,J.Chem.Phys.,81(1984),3967.
26. P.M.Hunt and M.S.Child,J.Phys.Chem.,86(1982),1116.
27. P.M.Hunt and S.Sridharan,J.Chem.Phys.,77(1982),4022.
28. C.Leforestier,to be published.
29. S.C.Leasure and J.M.Bowman,Chem.Phys.Letters,39(1976),462.
30. G.Bergeron and C.Leforestier,Chem.Phys.Letters,71(1980),519.
31. G.A.Parker and R.T Pack,J.Chem.Phys.,68(1978),1585.

THE DISTORTED WAVE THEORY OF CHEMICAL REACTIONS

J. N. L. Connor
Department of Chemistry
University of Manchester
Manchester M13 9PL
England

ABSTRACT. A review is given of recent applications of the distorted wave (DW) method to the theory of chemical reactions. A brief account of the following topics is included: the formal DW theory of reactions, static and adiabatic methods for choosing the distortion potentials, and the removal of the 3 Euler angles from the 6 dimensional DW integral. Applications of various DW theories to the $H+H_2$, $H+F_2$, $O(^3P)+H_2$, $O(^3P)+C(CH_3)_4$, $O(^3P)+HC(CH_3)_3$, $He+H_2^+$, $F+H_2$ and Cl+HCl chemical reactions and isotopic variations are discussed.

1. INTRODUCTION

The distorted-wave Born-approximation (DWBA) is a quantum technique for calculating differential cross sections, integral cross sections and product state distributions for three dimensional (3D) chemical reactions of the type

$$A+BC(v,j,m_j) \rightarrow AB(v',j',m_{j'})+C,$$

where v is a vibrational quantum number, j is a rotational quantum number with m_j the projection quantum number. The DWBA is not a single theory, but is a generic name covering an infinite number of possible approximations.

The simplest DW theories assume that the single most important collisional event in the entrance channel, A+BC, and in the exit channel, AB+C, is elastic scattering. The reaction is then treated as a perturbation on the elastic scattering. Distortion potentials are introduced to describe the elastic scattering in the entrance and exit channels. This approximation of considering reactive scattering as a perturbation between elastic scattering states should intuitively be valid for many chemical reactions, since elastic integral cross sections are typically several orders of magnitude larger than reactive cross sections.

The DWBA has a number of important properties:

(a) The conservation of energy, linear momentum and angular momentum is treated exactly. In contrast, some other approximate theories of

D. C. Clary (ed.), The Theory of Chemical Reaction Dynamics, 247–283.

reactive scattering violate these conservation laws. For example, in the energy-sudden (ES) approximation, which is a part of the infinite-order-sudden (IOS) technique, the rotational energy levels of BC are assumed to be degenerate and energy conservation is lost.

(b) The quantum states of the reactant and product molecules are correctly described. This is not the case in the usual quasiclassical trajectory method, for example, where the "boxing" of the trajectories to form "quantized" vibrational-rotational states can sometimes be a severe approximation.

(c) The exact hamiltonian for the collision is used. In some other theories, approximate hamiltonians are employed. For example, in the centrifugal-sudden or coupled-states (CS) approximation, which also forms part of the reactive IOS method, certain off-diagonal coupling terms in the exact body-fixed hamiltonian operator are ignored.

(d) The mathematical formalism of the DWBA is general and is not restricted to a particular reaction, or class of reactions. This does not imply, of course, that the numerical effort in applying the DWBA is the same for all reactions, or that the error in the DW method is approximately constant for different collision systems. It will be argued in Section 2 that the most favourable chemical reactions for the DWBA are those whose potential energy surfaces have large barriers between reactants and products.

(e) The DWBA provides all the information about a reactive collision that would be obtained from an exact quantum treatment. Suppose, for example, the differential cross section $d\sigma/d\Omega$ for a transition $(v, j, m_j) \rightarrow (v', j', m_{j'})$ is required, then the DWBA could calculate this very detailed quantity.

(f) The transition amplitudes are calculated for each transition $(v, j, m_j) \rightarrow (v', j', m_{j'})$ one at a time and not simultaneously. This means that only those transitions of interest need be included in the computations. The DWBA does not involve complicated matchings of wavefunctions in different arrangement channels, and the manipulation of large matrices is avoided.

(g) The DW method can be improved in a systematic manner to yield (in principle) the exact quantum result. As already mentioned, the DWBA neglects inelastic non-reactive collisions and treats the reaction as a perturbation between elastic scattering states in the different arrangement channels. Including the inelastic non-reactive scattering yields the coupled-channels distorted-wave (CCDW) method, which involves solving sets of coupled differential equations for the inelastic collisions; the reaction is still treated as a perturbation however. Approximate methods for solving the coupled equations can also be used, as in the centrifugal-sudden distorted-wave (CSDW) method. Removing the perturbative treatment of the reaction in the CCDW approximation yields the coupled-reaction-channels (CRC) method.

The DWBA, CCDW and CRC methods are established techniques in the theory of direct nuclear reactions (see, for example, the books by Austern [3], Glendenning [43] and Satchler [72]). In the theory of chemical reactions, the first large scale DWBA calculation was reported in 1967 by Karplus and Tang [51] for the $H+H_2$ reaction. However, it was only in 1974 that Choi and Tang [15] demonstrated a computationally effective method for handling the 6D distorted-wave integral that occurs in the theory (again for $H+H_2$). The first extensive DWBA calculations for a reaction (other than $H+H_2$ and its isotopic variants) that could be meaningfully compared with experiment, were those of Clary and the writer [21,26] in 1979 for the $H+F_2$ reaction. In 1984, Schatz et al. [75] reported the first 3D CCDW calculation for $H+H_2$, as did Choi et al. [20] using a more approximate CCDW theory. No accurate 3D CRC calculation has yet been carried out.

In the remainder of this review, I shall outline the relevant theory in Section 2, whilst a survey of DWBA and CCDW calculations and results will be given in Section 3. Conclusions are in Section 4.

2. OUTLINE OF THE THEORY

The formalism of the DWBA as applied to chemical reactions has four main steps:

(a) derivation of formal expressions, both exact and approximate, for the differential cross section $d\sigma/d\Omega$ for a transition $(v,j,m_j) \rightarrow (v',j',m_{j'})$,
(b) choosing distortion potentials for the elastic scattering in the entrance and exit channels,
(c) removal of the 3 Euler angles describing the orientation of the ABC triangle in space. This operation allows the 6D distorted wave integral to be reduced to a sum of 3D integrals,
(d) numerical implementation of the expressions obtained in steps (a)-(c).

A full derivation of all the equations in the DWBA is a lengthy, albeit straightforward, exercise. In this section, I shall just summarize some of the key equations and ideas. A detailed, self- contained, exposition of many theoretical aspects of the DWBA has recently been written by Tang [103].

2.1 Formal Scattering Theory

The formal DW theory of rearrangement collisions is a well established topic in scattering theory. Detailed derivations can be found in books concerned with collision theory (for example, Messiah [59], Rodberg and Thaler [71]) or the theory of direct nuclear reactions (for example, Austern [3], Glendenning [43], Satchler [72]), see also Choi et al. [19].

The differential cross section for a transition $(v,j,m_j) \rightarrow (v',j',m_{j'})$ can be written in terms of a T matrix element as follows

$$\frac{d\sigma}{d\Omega}_{v',j',m_{j'}\leftarrow v,j,m_j} = \frac{\mu_{A,BC}\mu_{C,AB}}{(2\pi\hbar^2)^2}\frac{k_{v',j'}}{k_{v,j}}\left|T_{v',j',m_{j'}\leftarrow v,j,m_j}\right|^2, \qquad (1)$$

where $\mu_{A,BC}$ and $\mu_{C,AB}$ are the reduced masses of the (A,BC) and (C,AB) systems respectively and $k_{v,j}$ and $k_{v',j'}$ are the wavenumbers of the initial and final collision systems respectively. To avoid a cumbersome notation, the entrance and exit channels will be labelled by

α and β respectively, so that Eq. (1) becomes

$$\frac{d\sigma_{\beta\alpha}}{d\Omega} = \frac{\mu_\alpha\mu_\beta}{(2\pi\hbar^2)^2}\frac{k_\beta}{k_\alpha}|T_{\beta\alpha}|^2. \tag{2}$$

The hamiltonian for the collision can be written in terms of either the entrance or exit channel coordinates

$$H = H_\alpha + V_\alpha = H_\beta + V_\beta. \tag{3}$$

In Eq. (3), H_α and H_β are the hamiltonians for the non-interacting systems. For example, if $\mathbf{R}_\gamma$ with $\gamma = \alpha$ or β is the vector for the atom relative to the centre-of-mass of the molecule and $\mathbf{r}_\gamma$ is the vector for the internal motion of the molecule, then in the entrance channel

$$H_\alpha \equiv H_\alpha(\mathbf{R}_\alpha,\mathbf{r}_\alpha) = -\frac{\hbar^2}{2\mu_\alpha}\nabla^2_{\mathbf{R}_\alpha} - \frac{\hbar^2}{2\mu'_\alpha}\nabla^2_{\mathbf{r}_\alpha} + V_{BC}(r_\alpha), \tag{4}$$

where μ'_α is the reduced mass of BC and $V_{BC}(r_\alpha)$ is the potential energy for the isolated BC molecule. A similar expression can be written down for the exit channel. The term V_γ in Eq. (3) is the interaction potential: it is that part of the total potential energy surface that tends to zero as $R_\gamma \to \infty$.

In terms of the quantities introduced in Eqs. (2)-(4), an exact expression for the T matrix element is

$$T_{\beta\alpha} = \langle \Phi_\beta|V_\beta|\Psi_\alpha^{(+)}\rangle. \tag{5}$$

In this equation, Φ_β is an eigenfunction of H_β, that is,

$$(H_\beta - E)\Phi_\beta = 0, \tag{6}$$

and consists of a vibrational-rotational wavefunction for AB multiplied by a plane wave for the relative motion of C and AB. The quantity $\Psi_\alpha^{(+)}$ is the exact wavefunction for the system, with the superscript indicating an outgoing boundary condition. It satisfies

$$(H_\alpha + V_\alpha - E)\Psi_\alpha^{(+)} = 0. \tag{7}$$

Another exact expression for $T_{\beta\alpha}$ is

$$T_{\beta\alpha} = \langle \Psi_\beta^{(-)}|V_\alpha|\Phi_\alpha\rangle, \tag{8}$$

where

$$(H_\alpha - E)\Phi_\alpha = 0, \tag{9}$$

and

$$(H_\beta + V_\beta - E)\Psi_\beta^{(-)} = 0. \tag{10}$$

$\Psi_\beta^{(-)}$ is the exact wavefunction with an incoming boundary condition.

The Born approximation (BA) is obtained by replacing $\Psi_\alpha^{(+)} \to \phi_\alpha$ in the post form (5), or by replacing $\Psi_\beta^{(-)} \to \phi_\beta$ in the prior form (8), giving

$$T_{\beta\alpha}(BA) = \langle \phi_\beta | V_\beta | \phi_\alpha \rangle = \langle \phi_\beta | V_\alpha | \phi_\alpha \rangle. \qquad (11)$$

The post and prior forms of the BA are identical as indicated in Eq. (11).

The BA essentially treats the interaction potential as a first order perturbation on the motion of the non-interacting reactant and product collision systems. This approximation is not expected to be reliable for chemical reactions at low collision energies. Consider, for example, a reaction with a large barrier between reactants and products, such as $H+H_2$. The distortion of the incident plane wave by the barrier is clearly an important physical process. This suggests a better description of the reactive scattering might be obtained if the distortion of the relative motion in the initial and final channels is taken into account. This is the aim of the DWBA.

In the DW method, the interaction potential is written as the sum of two terms

$$V_\alpha = V_\alpha{}^0 + V_\alpha', \qquad (12)$$

$$V_\beta = V_\beta{}^0 + V_\beta', \qquad (13)$$

where the distortion potential $V_\gamma{}^0$ gives rise only to non-reactive scattering, and V_γ' is the term that induces reaction. The scattering from $V_\gamma{}^0$ is assumed to represent a soluble problem, so that the distorted wavefunctions, $\chi_\gamma{}^{(\pm)}$, which satisfy

$$(H_\alpha + V_\alpha{}^0)\chi_\alpha{}^{(+)} = E\chi_\alpha{}^{(+)}, \qquad (14)$$

$$(H_\beta + V_\beta{}^0)\chi_\beta{}^{(-)} = E\chi_\beta{}^{(-)}, \qquad (15)$$

can be calculated. An exact expression for the T matrix element is now given (in the post form) by

$$T_{\beta\alpha} = \langle \chi_\beta{}^{(-)} | V_\beta' | \Psi_\alpha{}^{(+)} \rangle, \qquad (16)$$

or equivalently in the prior form by

$$T_{\beta\alpha} = \langle \Psi_\beta{}^{(-)} | V_\alpha' | \chi_\alpha{}^{(+)} \rangle. \qquad (17)$$

The DWBA is obtained by making the approximation $\Psi_\alpha{}^{(+)} \to \chi_\alpha{}^{(+)}$ in Eq. (16) or $\Psi_\beta{}^{(-)} \to \chi_\beta{}^{(-)}$ in Eq. (17):

$$T_{\beta\alpha}(DWBA) = \langle \chi_\beta{}^{(-)} | V_\beta' | \chi_\alpha{}^{(+)} \rangle, \qquad (18)$$

$$= \langle \chi_\beta{}^{(-)} | V_\alpha' | \chi_\alpha{}^{(+)} \rangle. \qquad (19)$$

The post and prior forms are equal (for a proof see, for example, Glendenning [43] or Satchler [72]), although if further analytical or numerical approximations are introduced, this property may no longer hold. As a check on Eqs. (16) - (19), note that if $V_\gamma^0 \equiv 0$, so that $V'_\gamma \equiv V_\gamma$, then Eqs. (5), (8) and (11) are regained. An alternative form for $T_{\beta\alpha}$(DWBA) follows from Eqs. (14) and (19):

$$T_{\beta\alpha}(\mathrm{DWBA}) = \langle \chi_\beta^{(-)} | H - E | \chi_\alpha^{(+)} \rangle, \tag{20}$$

in which V'_γ does not explicitly appear.

Equations (16) and (17) for $T_{\beta\alpha}$ are exact results. In particular it does not matter how V_γ is partitioned into $V_\gamma^0 + V'_\gamma$ provided only that V_γ^0 gives rise to elastic or inelastic non-reactive scattering (otherwise, another term has to be added to the right hand sides of Eqs. (16) and (17)).

The expressions (16) and (17) also involve the exact wavefunction. Clearly knowing Ψ_γ is equivalent to exactly solving the original Schrödinger equation (7) or (10) for the 3D reactive collision. The importance of these exact expressions for $T_{\beta\alpha}$ is that they allow systematic approximations to be made for Ψ_γ, which are often amenable to physical interpretation. Note that Ψ_γ need only be approximated in those regions of configuration space for which $T_{\beta\alpha}$ is non-zero.

When Ψ_γ is approximated by χ_γ, as in the DWBA, it is clear that $T_{\beta\alpha}$ now depends on the choice of distortion potentials V_α^0 and V_β^0. In fact, since there are an infinite number of ways of partitioning V_γ into V_γ^0 + V'_γ, the term DWBA covers an infinity of possible approximations.

The distortion potentials used in the DWBA are chosen so that they only give rise to elastic scattering in the incident and final channels. (They are discussed in more detail in Section 2.2). Thus $\chi_\beta^{(-)}$ and $\chi_\alpha^{(+)}$ each correspond to a single internal state. The DWBA approximation $\Psi_\alpha^{(+)} \rightarrow \chi_\alpha^{(+)}$ in Eq. (16) is evidently a severe one and it is useful to summarise the assumptions behind the DWBA:

(a) The transfer of the atom B takes place directly from the initial to the final state of the system. Inelastic couplings in the entrance and exit channels are therefore ignored, as are resonance effects and reactions involving long-lived complexes. However, since the starting equation (16) is exact, it is possible that future researches will lead to improved approximations for $\Psi_\alpha^{(+)}$ that can handle these effects (see also the CCDW below).

(b) The distorted wavefunctions $\chi_\alpha^{(+)}$ and $\chi_\beta^{(-)}$ are assumed to correctly describe the elastic scattering in the entrance and exit channels. In principle, this could be checked by comparison with experimental data. It is unlikely that the distortion potentials currently being used (see Section 2.2) do in fact correctly reproduce the non-reactive elastic scattering. Furthermore, different distorted wavefunctions may have the same asymptotic phase shifts, and so give rise to the same elastic scattering cross sections, yet differ in the inner region that is important for the overlap integral (18). In fact, it may be that the distorted wavefunctions contributing to the DW overlap integral should not be constrained to reproduce the non-reactive elastic scattering data.

(c) The reaction is assumed to be sufficiently weak that it can be treated as a perturbation between the elastic scattering states. Note that it is probably not correct to say that the reaction is being treated to first order in the DWBA, since the choice of distortion potentials may include higher order effects in a phenomenological manner.

Assumptions (a) and (b) of the DWBA can be partially removed by going to the CCDW approximation. In this method, the distortion potential V_γ^0 is chosen so that it allows inelastic as well as elastic non-reactive collisions. Formally we can write

$$\chi_\gamma = \sum_{\gamma'} \chi_{\gamma'\gamma} \tag{21}$$

where the $\chi_{\gamma'\gamma}$ satisfy a finite set of coupled differential equations in the γ arrangement channel. Using Eq. (20), the expression for the T matrix element becomes

$$T_{\beta\alpha}(\mathrm{CCDW}) = \sum_{\beta'} \sum_{\alpha'} < \chi_{\beta'\beta}^{(-)} | H - E | \chi_{\alpha'\alpha}^{(+)} >. \tag{22}$$

Although sets of coupled equations need to be solved for the inelastic scattering in the entrance and exit channels, note that the $T_{\beta\alpha}$ are still calculated one at a time. The CCDW is clearly much more involved numerically than the simpler DWBA. The numerical effort can be reduced somewhat by using approximate methods to solve the coupled equations; for example, the CS approximation, giving rise to the CSDW method.

2.2 Distortion Potentials

In the DWBA, distortion potentials are introduced to describe the elastic scattering in the initial and final channels. There are many ways of choosing these distortion potentials. One way (the static approximation) is to assume that the reactant or product molecule is unperturbed by the incoming or receding atom respectively. Another way (the adiabatic approximation) assumes the molecule adiabatically adjusts (wholly or partially) to the presence of the atom.

In the static method, the distorted wavefunction is written in the form

$$\chi_{v,j,m_j}^{(\pm)}(\mathbf{R}_\gamma, \mathbf{r}_\gamma) = F_{v,j}^{(\pm)}(R_\gamma)\eta_{v,j}(r_\gamma)Y_{j,m_j}(\hat{\mathbf{r}}_\gamma)/R_\gamma r_\gamma, \tag{23}$$

where the first term on the rhs of Eq. (23) is the translational wavefunction which describes the elastic scattering and the remaining terms are the vibrational-rotational wavefunction of the unperturbed molecule. Using these static wavefunctions for the reactant and product molecules give rise to the static-static distorted wave (SSDW) method.

For many potential energy surfaces with barriers, the dominant term in a Legendre polynomial expansion of the interaction potential is the leading term, that is, the spherically averaged potential:

$$V^S(R_\gamma, r_\gamma) = \frac{1}{2}\int_0^\pi V_\gamma(R_\gamma, r_\gamma, \cos\theta_\gamma)\sin\theta_\gamma d\theta_\gamma, \tag{24}$$

where θ_γ is the angle between the vectors $\mathbf{R}_\gamma$ and $\mathbf{r}_\gamma$. Averaging this potential over the molecular wavefunction defines the static distortion potential for the elastic scattering

$$V^{0}_{v,j}(R_\gamma) = \int_0^\infty \eta^2_{v,j}(r_\gamma)V^{s}(R_\gamma,r_\gamma)dr_\gamma. \tag{25}$$

It is evident that a different static distortion potential is obtained for each value of v and j.

Now for some highly exoergic reactions such as T + F_2, there are over 1000 product TF vibrational-rotational states open. To avoid having to calculate a different distortion potential for each j, the usual rigid-rotor approximation can be made. In this case, we can write

$$\chi^{(\pm)}_{v,j,m_j}(\mathbf{R}_\gamma,\mathbf{r}_\gamma) = F^{(\pm)}_{v,j}(R_\gamma)\eta_v(r_\gamma)Y_{j,m_j}(\hat{\mathbf{r}}_\gamma)/R_\gamma r_\gamma, \tag{26}$$

with the static distortion potential defined by

$$V^{0}_{v}(R_\gamma) = \int_0^\infty \eta^2_{v}(r_\gamma)V^{s}(R_\gamma,r_\gamma)dr_\gamma, \tag{27}$$

which is only dependent on the vibrational quantum number v.

In Eq. (27), it is still necessary to perform a quadrature over the vibrational wavefunction. This can be avoided if r_γ is replaced by its expectation value $< r_\gamma >_v$ for a given vibrational state.

$$V^{0}_{v}(R_\gamma) = V^{s}(R_\gamma,r_\gamma =< r_\gamma >_v). \tag{28}$$

An even simpler approximation is to replace r_γ by $r_{\gamma e}$, the equilibrium distance of the molecule

$$V^{0}_{e}(R_\gamma) = V^{s}(R_\gamma,r_\gamma = r_{\gamma e}). \tag{29}$$

This distortion potential no longer depends on v.

A different kind of static distortion potential has been used for the $H+D_2 \rightarrow HD+D$ reaction [10,31]. In this reaction, the centre of mass and centre of charge of the HD molecule do not coincide, with the result that the interaction potential in the exit channel is highly anisotropic. This suggests that a physically more reasonable distortion potential is obtained by using the preferred configuration of the atoms (collinear in this case) rather than performing a spherical average. The collinear distortion potentials are defined by

$$V^{0}_{v}(R_\gamma) = \int_0^\infty \eta^2_{v}(r_\gamma)V_\gamma(R_\gamma,r_\gamma,\cos\theta_\gamma = -1)dr_\gamma, \tag{30}$$

or more simply by

$$V^{0}_{v}(R_\gamma) = V_\gamma(R_\gamma,r_\gamma =< r_\gamma >_v,\cos\theta_\gamma = -1). \tag{31}$$

The spherically-averaged and collinear distortion potentials can be considered as two limiting cases. Thus, the spherically-averaged potential averages over all possible configurations of the atom with respect to the molecule, leading to a potential that will be too repulsive in general. The collinear distortion potential, on the other hand, overemphasizes the collinear configuration, resulting in a potential that is too attractive. In fact, a better distortion potential would be complex valued and energy dependent, in addition to depending on the internal state of the molecule. No applications to chemical reactions have yet been made with such a distortion potential.

In contrast to the static DW theories just discussed, the adiabatic theories allow the molecule to partially or wholly adjust to the presence of the atom. For reactions with large barriers, this should be a reasonable approximation at low collision energies, since the effect of the barrier is to slow down the approaching reactants, allowing the molecule time to adjust to the incoming atom.

In the vibrationally-adiabatic-distorted-wave (VADW) method, the vibrational wavefunction of the molecule adjusts adiabatically to the presence of the atom, but the rotational wavefunction is held static. This method was introduced by Clary and the writer [21] to provide a practical DW technique for reactions with large numbers of product rotational states, such as $H + F_2$. The distorted wavefunctions take the form

$$\chi^{(\pm)}_{v,j,m_j}(\mathbf{R}_\gamma,\mathbf{r}_\gamma) = F^{(\pm)}_{v,j}(R_\gamma)\eta_v(R_\gamma,r_\gamma)Y_{j,m_j}(\hat{\mathbf{r}}_\gamma)/R_\gamma r_\gamma, \qquad (32)$$

where the adiabatic vibrational wavefunction satisfies

$$\left[\frac{-\hbar^2}{2\mu_\gamma'}\frac{\partial^2}{\partial r_\gamma{}^2} + V^s(R_\gamma,r_\gamma)\right]\eta_v(R_\gamma,r_\gamma) = \epsilon_v(R_\gamma)\eta_v(R_\gamma,r_\gamma). \qquad (33)$$

The change in the vibrational eigenvalue as a function of R_γ then provides the distortion potential for the elastic scattering

$$V^D_v(R_\gamma) = \epsilon_v(R_\gamma) - \epsilon_v(R_\gamma = \infty). \qquad (34)$$

A more sophisticated approximation is to allow both the vibrational and rotational wavefunction of the molecule to adiabatically adjust to the incoming atom. This method was introduced by Karplus and Tang [51] for the reactant channel wavefunction $\chi_\alpha{}^{(+)}$. We can formally write

$$\chi^{(+)}_\alpha(\mathbf{R}_\alpha,\mathbf{r}_\alpha) = F^{(+)}_\alpha(R_\alpha)\eta_\alpha(\mathbf{R}_\alpha,\mathbf{r}_\alpha)/R_\alpha r_\alpha, \qquad (35)$$

where $\eta_\alpha(\mathbf{R}_\alpha,\mathbf{r}_\alpha)$ is the adiabatic molecular wavefunction. For the rotational ground state, $\eta_\alpha(\mathbf{R}_\alpha,\mathbf{r}_\alpha)$ can be factored into a perturbed vibrational part, $\eta_v(R_\alpha,r_\alpha)$, and a perturbed rotational part, $\Phi_v(R_\alpha,\cos\Theta_\alpha)$. The calculation of $\Phi_v(R_\alpha,\cos\Theta_\alpha)$ is simplified by approximating the interaction potential by the first two non-zero terms of its Legendre polynomial expansion. It is then found that $\Phi_v(R_\alpha,\cos\Theta_\alpha)$ satisfies an equation for a hindered rotor. This method forms part of the rotationally-adiabatic-distorted-wave (RADW) approximation (also called the ADW method). For rotationally excited states, the method is similar but more complicated (Choi et al. [18]).

In the RADW approximation, the adiabatic molecular wavefunction is assumed to be a product of a vibrational part and a rotational part. This approximation evidently neglects the coupling between vibration and rotation. Including it leads to the converged adiabatic distorted wave (CADW) method, also known as the adiabatic T matrix method (Sun et al. [93]). In this technique, it is necessary to solve a double well problem [92] in order to determine the molecular wavefunction.

In this subsection, four different ways of choosing the distorted wavefunctions have been discussed: static, VA, RA and CA. It is clear that the difficulty of calculating these wavefunctions increases in the order: static < VA < RA < CA. In particular, the RA and CA wavefunctions are considerably more difficult to compute than the static or VA ones, because of the requirement to treat the rotational motion adiabatically. Also, the treatment of excited rotational states is much more complicated than for the ground rotational state. However, we shall see in Section 3 that for $H+H_2$ improving the quality of the distorted wavefunctions results in more accurate cross sections.

It should also be noted that for each method, additional approximations may be introduced to make the method easier to apply. This means it is necessary to examine papers in the literature carefully, in order to find out precisely how the distortion potential is defined.

2.3 Evaluation of the Distorted Wave Integral

It was shown in Section 2.1 that simple formal expressions can be derived for the T matrix element. However, the simplicity of these expressions is deceptive. Consider, for example, the distorted wave integral (18). The integrand is over the vectors $\mathbf{R}_\gamma$ and $\mathbf{r}_\gamma$. Thus the DW integral is actually 6 dimensional. The numerical evaluation of a 6D integral, with an oscillatory integrand, is clearly a difficult problem, which would severely limit the usefulness of the DWBA as a practical technique.

However for an A+BC reaction, only 3 coordinates are necessary to describe the internal motion of the ABC system; the remaining 3 coordinates describe the rotation of the system as a whole, and correspond to the fact that the total angular momentum of the system is conserved. Thus by changing coordinates in the DW integral (18), it can be reduced from a 6D integral to (a sum of) 3D integrals. However, this reduction can be carried out in more than one way.

The use of a body-fixed reference frame together with total angular momentum conservation was discussed by Miller [64], whilst Brodsky and Levich [13] employed a space-fixed frame. However, the first demonstration of a numerically practical procedure was by Choi and Tang [15] in 1974. In their method, the projections of the molecular angular momenta of the entrance and exit channels are referred to the same space fixed axis, which is taken to be the direction of the incident wave vector $\mathbf{k}_{v,j}$. Most calculations in the literature are based on the Choi-Tang formulation.

A different procedure for reducing the DW integral has been described by Suck [79]. This is based on the transfer of rotational angular momentum $\mathbf{j}_{\beta\alpha}$ from the reactant molecule to the product molecule. If $\mathbf{j}_\gamma$ is the rotational angular momentum vector and $\boldsymbol{\ell}_\gamma$ is the orbital angular momentum vector, then since $\mathbf{j}_\alpha+\boldsymbol{\ell}_\alpha = \mathbf{j}_\beta+\boldsymbol{\ell}_\beta = \mathbf{J}$ = constant vector, it follows that

$$\mathbf{j}_{\beta\alpha} = \mathbf{j}_\beta-\mathbf{j}_\alpha = \boldsymbol{\ell}_\alpha-\boldsymbol{\ell}_\beta. \tag{36}$$

This method has some advantages compared to the total angular momentum representation [79]. It is widely used in the theory of direct nuclear reactions (Glendenning [43], Satchler [72], see also Levine [54]).

Depending on the representation used, the reduction of the 6D DW integral is a lengthy exercise in angular momentum algebra. Details will not be presented here. Instead the equation for the T matrix element for a transition $(v, j, m_j) \rightarrow (v', j', m_{j'})$ will be given for the simplest case of the SSDW theory using the distorted wavefunctions (26) in the Choi-Tang approach:

$$T_{\beta\alpha} = 2\pi\sum_{\ell'} (2\ell'+1)\left[\frac{(\ell'-|m_j-m_{j'}|)!}{(\ell'+|m_j-m_{j'}|)!}\right]^{1/2} P_{\ell'}^{|m_j-m_{j'}|}(\cos\Theta)A_{\beta\alpha},$$

where

$$A_{\beta\alpha} \equiv A_{\ell'}(v,j,m_j \rightarrow v',j',m_{j'}) = \epsilon(m_j-m_{j'})[(2j+1)(2j'+1)]^{1/2}$$

$$\times \sum_J \sum_\ell i^{\ell-\ell'} \exp[i(\delta_{\alpha\ell}+\delta_{\beta\ell'})] \langle \ell 0 j m_j | J m_j \rangle \langle \ell' m_j - m_{j'} j' m_{j'} | J m_j \rangle$$

$$\times \frac{(2\ell+1)}{2J+1} \sum_M \sum_{m'} \epsilon(M)\epsilon(-m')\epsilon(m'-M) \langle \ell m' j M-m' | JM \rangle \langle \ell' 0 j' M | JM \rangle$$

$$\times \left[\frac{(\ell-|m'|)!(j-|M-m'|)!(j'-|M|)!}{(\ell+|m'|)!(j+|M-m'|)!(j'+|M|)!}\right]^{1/2}$$

$$\times \int_0^\infty dr_\beta \int_0^\infty dR_\beta \int_0^\pi d\Theta_\beta \Big[\sin\Theta_\beta r_\alpha r_\beta \eta_v(r_\alpha)\eta_{v'}(r_\beta) L_{vj\ell}(R_\alpha) L_{v'j'\ell'}(R_\beta)$$

$$\times V_\beta'(r_\beta, R_\beta, \cos\Theta_\beta) P_\ell^{|m'|}(\cos\Theta'') P_j^{|M-m'|}(\cos\Theta') P_{j'}^{|M|}(\cos\Theta_\beta)\Big]. \quad (37)$$

In this expression, $\epsilon(m)$ is a phase factor defined by $\epsilon(m) = (-1)^m$ for $m \geq 0$, $\epsilon(m) = 0$ for $m < 0$, $L_{vj\ell}(R_\alpha)$ is a translational radial wavefunction with asymptotic phase shift $\delta_{\alpha\ell}$ and the angles Θ, Θ' and Θ'' are defined by $\cos\Theta = \hat{k}_{v,j} \cdot \hat{k}_{v',j'}$, $\cos\Theta' = \hat{r}_\alpha \cdot \hat{R}_\beta$ and $\cos\Theta'' = \hat{R}_\alpha \cdot \hat{R}_\beta$. The variables r_α, R_α can be expressed in terms of the integration variables using standard identities for a triangle. When $j = m_j = 0$ or $j' = m_{j'} = 0$, Eq. (37) can be simplified.

The numerical evaluation of Eq. (37) is still a complicated task, with the most difficult part being the accurate calculation of the 3D integrals. In practice, it is found that the region around the classical turning points for the translational radial wavefunctions makes a major contribution to the integrand. However, as the collision energy increases, the integrand becomes more oscillatory, making it more difficult, or impossible, to evaluate the 3D integrals accurately.

Equation (37) is for the simple case of the SSDW theory with the distorted wavefunctions (26). The reduction of the 6D integral for the other DW theories is similar, but more complicated.

3. APPLICATIONS

The DW theory of reactive molecular collisons is just one of many theoretical approaches that have been applied to simple three-body reactions of the A+BC type. For general reviews of the theory of chemical reactions with many references to the literature, see Connor [28,29], Micha [63], Wyatt [109], Walker and Light [108], Schatz [74], Baer [7], Basilevsky and Ryaboy [8] and other Chapters in this book.

The Born and DW formalisms outlined in Section 2.1 were the basis of many early theories of chemical reactions, see for example, Golden and Peiser [44], Golden [45,46], Micha [60–62], Suplinskas and Ross [95], Pirkle and McGee [67,68], Gelb and Suplinskas [39], Levich et al. [53], Brodskii et al. [12] and Eu et al. [38]. However, in these early papers, it was necessary to introduce numerous simplifying approximations of uncertain validity in order to arrive at a tractable theory. This early work will not be reviewed here. More recent approximate theories based on the DW formalism, in particular Born and Franck–Condon theories of chemical reactions, are discussed in the general reviews mentioned above.

Another topic that will not be discussed is the DW theory of coplanar (Walker and Wyatt [105,106]) and collinear reactions (Walker and Wyatt [104], Gilbert and George [42], Madden [57], Babamov et al. [4–6] and Lopez et al. [55]).

Tables I and II summarize all DWBA calculations known to the writer which attempt an accurate numerical evaluation of the DW integral (18). The Tables also include a few unpublished calculations which should shortly appear in the literature. Table I is concerned with $H+H_2$ and its isotopes, whilst all other reactions are given in Table II.

The first and second columns of the Tables give the reaction and potential energy surface used. Standard abbreviations are employed for the names of the potential surfaces. Thus, PK = Porter–Karplus potential surface No 2 for $H+H_2$, LSTH = Liu–Siegbahn–Truhlar–Horowitz potential surface for $H+H_2$, YLL = Yates–Lester–Liu potential surface for $H+H_2$, LEPS = extended London–Eyring–Polanyi–Sato potential surface and DIM = diatomics-in-molecules potential surface.

The third and fourth columns of the Tables give information on the distorted wavefunctions $\chi_\beta^{(-)}$ and $\chi_\alpha^{(+)}$, and in particular the approximation used for the vibration–rotation wavefunction of the molecule – see Eqs. (23), (26), (32) and (35). The abbreviations used are : S = static or A = adiabatic approximation for the individual vibrational or rotational wavefunction, and CA = converged-adiabatic approximation, when the coupling between vibration and rotation is included. There have also been a few calculations which use a semi- adiabatic wavefunction for χ_γ. In these calculations, the vibrational or rotational wavefunction is held static, but the adiabatic potential is used in solving for the translational wavefunction $F_\gamma^{(\pm)}(\mathbf{R}_\gamma)$. This approximation is denoted S^a in the Tables.

The fifth column gives an abbreviation for the DW method employed. In the SSDW method, the vibration and rotational wavefunctions of the reactant and product molecules are held static. The VADW theory treats the vibrational wavefunctions of the reactant and product molecules adiabatically, but keeps the rotational wavefunctions static. In the RADW

Table I. DWBA calculations for the $H+H_2$ reaction and its isotopic variations. The abbreviations used in the Table are explained in the text.

reaction	potential surface	$\chi_\beta^{(-)}$ vib.	$\chi_\beta^{(-)}$ rot.	$\chi_\alpha^{(+)}$ vib.	$\chi_\alpha^{(+)}$ rot.	method	references	comments
$H+H_2$	PK	S	S	S	S	SSDW	[15,23,51,96,97]	$j=0 \to j'=0$ at $E_{tr} = 0.5$ eV
	PK	S	S	S	S	SSDW	[20,89]	$j=0 \to j'=1$ at 9 energies
	PK	S^a	S	S^a	S	SSDW	[25]	$j=0 \to j'\leq 5$ at $E_{tr} = 0.5$ eV
	PK	A	S	A	S	VADW	[23,25]	$v\leq 2, j=0 \to v'\leq 2, j', m_{j'}$
	PK	A	S	A	S	VADW	[24]	$j\leq 2 \to j'\leq 4$ at $E_{tot} = 0.6$ eV
	PK	S	S	A	A	RADW	[15,51,96,97]	$j=0 \to j'=0$ at 3 energies
	PK	S	S	A	A	RADW	[14,15]	$j=0 \to j'=1$ at $E_{tr} = 0.5$ eV
	PK	S	S	A	A	RADW	[17,100]	$j=0 \to j'\leq 4$ at 6 energies
	PK	S	S	A	A	RADW	[25]	$j=0 \to j'\leq 7$ at 12 energies
$H+H_2$	LSTH	S	S	S	S	SSDW	[93]	$\sigma(0,0)$
	LSTH	A	S	A	S	VADW	[25]	$\sigma(0,0)$ at 5 energies
	LSTH	S	S	A	A	RADW	[25]	$\sigma(0,0)$ at 5 energies
	LSTH	S	S	CA		CADW	[93]	$j=0 \to j'\leq 6$ at 5 energies
$D+H_2$	PK	S	S	A	A	RADW	[99,100]	$j=0 \to j'\leq 6$ at 8 energies
	PK	S	S	A	A	RADW	[18]	$j\leq 1 \to j'=1$ at $E_{tr} = 0.48$ eV
	PK	S	S	A	A	RADW	[25]	$j=0 \to j'\leq 8$ at 8 energies
	PK	S	S	A	A	RADW	[90]	$v=1, j=0 \to v'\leq 1$
$D+H_2$	YLL	S	S	A	A	RADW	[16,100]	$j=0 \to j'\leq 6$ at 3 energies
$D+H_2$	LSTH	A	S	A	S	VADW	[25]	$j=0 \to j'\leq 8$ at $E_{tr} = 0.4$ eV
	LSTH	S	S	A	A	RADW	[110]	$j=0 \to j'\leq 5$ at 7 energies
	LSTH	S	S	A	A	RADW	[25]	$j=0 \to j'\leq 8$ at 7 energies
	LSTH	S	S	A	A	RADW	[90]	$v\leq 1, j=0 \to v'\leq 1$
	LSTH	S	S	A	A	RADW	[91]	$j=1 \to j'\leq 5$ at 4 energies
	LSTH	S	S	CA		CADW	[94]	$j=0 \to j'\leq 6$ at 5 energies
$H+D_2$	LSTH	S	S	S	S	SSDW	[30,86]	$j=0 \to v'\leq 2, j'$ at 2 energies
	LSTH	S	S	S	S	SSDW-C	[10,31]	$j=0 \to v'\leq 2, j'$ at 4 energies
	LSTH	A	S	A	S	VADW	[30]	$j=0 \to v'\leq 1, j'\leq 7$ at $E_{tr} = 0.55$ eV
$H+T_2$	?	S	S	A	A	RADW	[101]	conference abstract

method, the vibrational and rotational wavefunctions of the reactant molecule are both treated adiabatically, but the product molecule is held static. The CADW method is similar except that the CA approximation is used for the reactant molecule (the product molecule is still kept static). There can be several variations of each technique, depending on the precise definition of the distortion potentials used in the calculations, as has been discussed in Section 2.2 for the SSDW method.

The final column in the Tables gives some indication of the scope of each calculation. Typically, the quantities calculated are the state-to-state energy-dependent differential and integral cross sections for a transition $(v,j \rightarrow v',j')$ where

$$\sigma(v,j \rightarrow v',j') = (2j+1)^{-1} \sum_{m_j} \sum_{m_{j'}} \sigma(v,j,m_j \rightarrow v',j',m_{j'})$$

Note that many of the earlier calculations only considered $j=0$ and $j'=0$, because the DW theory is then much simpler (cf. Eq. (37)). An $m_{j'}$ in the Tables indicates that information on transitions involving the projection quantum number is also reported. The vibrational quantum numbers can be assumed to be $v=0$ and $v'=0$ unless stated otherwise.

3.1 DWBA Calculations for the H+H_2 Reaction and Isotopic Variations

Table I shows that many DW calculations have been reported for the canonical model of H+H_2 on the PK surface (for the concept of a canonical model, see Connor [28]). The DW results can be compared with the detailed exact quantum calculations of Schatz and Kuppermann [73]. Similarly, for H+H_2 on the LSTH surface, the DW results can be compared with the exact quantum integral cross sections $\sigma(v=0,j=0)$ reported by Walker et al. [107], where

$$\sigma(v,j) = \sum_{v'} \sum_{j'} \sigma(v,j \rightarrow v',j').$$

The general conclusion from these comparisons for H+H_2 is that the different DW theories give qualitatively similar results for relative quantities such as state-to-state differential cross sections and rotational product distributions. However, the absolute values of cross sections are different in the various DW theories, the order being: SSDW $<$ VADW $<$ RADW $<$ CADW $\lesssim$ exact. This order illustrates that as systematically better approximations (based on physical understanding) are made to the exact wavefunction $\Psi_\alpha^{(+)}$ in the T matrix element (16), the magnitudes of the DW cross sections approach the exact result. It should be emphasized, however, that these conclusions have only been tested at low energies, where the accurate quantum results have been calculated, and for the LSTH surface only limited exact results are available [107]. Next some of the DW results will be considered in more detail.

The first attempt at an exact evaluation of the DW integral was made in 1967 by Karplus and Tang [51] for H+H_2 on PK. They also introduced the SSDW and RADW approximations. Karplus and Tang [51] and Tang and Karplus [96,97] reduced the 6D DW integral to a 5D one, which they evaluated numerically for the $j=0 \rightarrow j'=0$ transition by the SSDW method at one energy and by the RADW technique at three energies. They also performed more extensive calculations with the help of a collinear approximation [51, 97], including an application to the $D+H_2$ reaction (Tang [98]). Most of this early work has been reviewed by Karplus [52].

The method developed by Choi and Tang [15] in 1974 for reducing the 6D DW integral to a sum of 3D integrals significantly extended the usefulness of the DWBA as a practical technique. Using this method, they reported in 1976 a large scale RADW calculation for $H+H_2$ on PK for many transitions, $j=0 \rightarrow j'\leq 4$, at 6 energies [17]. (This calculation also revised some of their earlier results [14, 15]). Comparison with the exact quantum cross sections [73] showed good agreement for the magnitude of the $j=0 \rightarrow j'=1$ cross section. However, the agreement for the product rotational distributions was less good. For example, the RADW distribution [17] always peaked at $j'=2$ whereas the quantum distribution peaks at $j'=1$. Nevertheless, it appeared that the RADW method gave good absolute values; this was assumed to be true in the period 1976–80.

The accuracy of the RADW method for $H+H_2$ on PK was reinvestigated in 1981 (Clary and Connor [25]). This study showed that the RADW cross sections are always smaller than the exact ones by factors of 2 to 8; the same result was obtained for the LSTH surface [25]. The RADW rotational distributions [25] agreed better with the exact quantum results than did the earlier calculations of Choi and Tang [17]. The inaccuracies in the earlier work [14, 15, 17] were shown to arise from errors in the elastic scattering adiabatic distortion potential [25]. Figure 1 compares with the exact results for $H+H_2$ on PK, the RADW differential cross section for $j=0 \rightarrow j'=1$ at a translational energy of E_{tr} = 0.327 eV, whilst Figure 2 compares the rotational distributions at the same energy (Clary and Connor [25]). There is good agreement between the exact and RADW results.

The simpler SSDW and VADW methods have also been applied to $H+H_2$ on PK (Clary and Connor [23–25], Suck Salk and Lutrus [89], Choi et al. [20]). Relative distributions are generally similar to those from the RADW method, but the absolute values of the cross sections are smaller in the order: SSDW < VADW < RADW. The close agreement of the VADW and RADW distributions is illustrated in Figures 1 and 2 (Clary and Connor [23, 25]). Even very detailed quantities such as the $m_{j'}$ dependence of differential cross sections, are quite well predicted by the VADW method. Figure 3 compares differential cross sections for $j=0$, $m_j=0 \rightarrow j'=3$, $|m_{j'}|\leq 3$ with the exact quantum results for $H+H_2$ on PK (Clary and Connor [23]). The helicity representation is used in this figure in which the axis of quantization of both the incoming and outgoing rotational states is chosen to coincide with the direction of the incident and final wave vectors respectively.

All the DW calculations discussed above have been for the reactant molecule in its ground state, with $v=0$, $j=0$. The first DW calculations for a rotationally or vibrationally excited reactant molecule were carried out by the VADW method (Clary and Connor [23, 24]). Figure 4 compares with the exact results [73], the VADW rotational product distributions for $j=0, 1, 2$ for $H+H_2(j)$ on PK [24]. There is close agreement between the two calculations. The effect of vibrational excitation [23] for the $H_2(v=2, j=0)$ molecule is illustrated in Figure 5 (also for PK), which plots $\sigma(v=2, j=0 \rightarrow v'\leq 2)$ against translational energy where

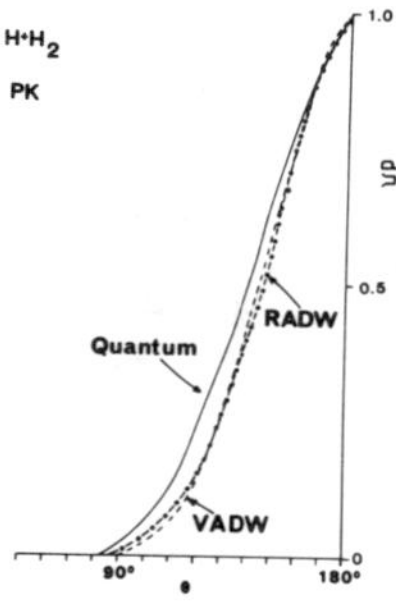

Figure 1 Accurate quantum, RADW and VADW differential cross sections $d\sigma/d\Omega(0,0 \rightarrow 0,1)$ for $H+H_2$ on PK at E_{tr} = 0.327 eV normalized at 180°.

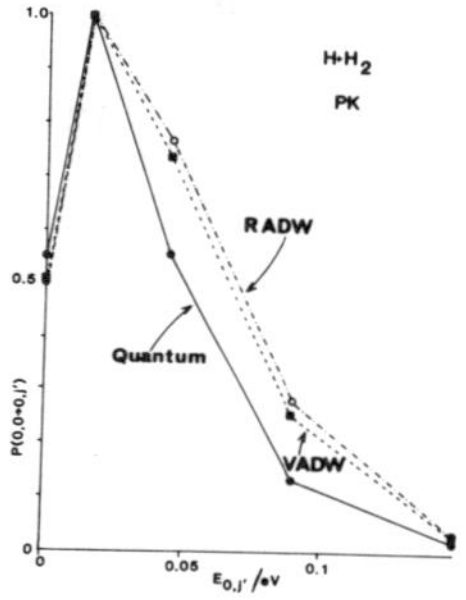

Figure 2 Accurate quantum, RADW and VADW rotational product distributions $P(0,0 \rightarrow 0,j')$ for $H+H_2$ on PK at E_{tr} = 0.327 eV plotted against rotational energy.

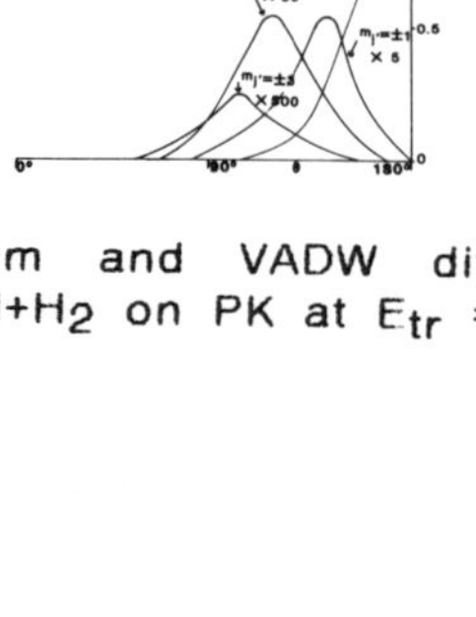

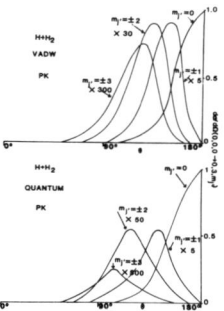

Figure 3 Accurate quantum and VADW differential cross sections $d\sigma/d\Omega(0,0,0 \rightarrow 0,3,m_{j'})$ for $H+H_2$ on PK at E_{tr} = 0.327 eV normalized at 180°.

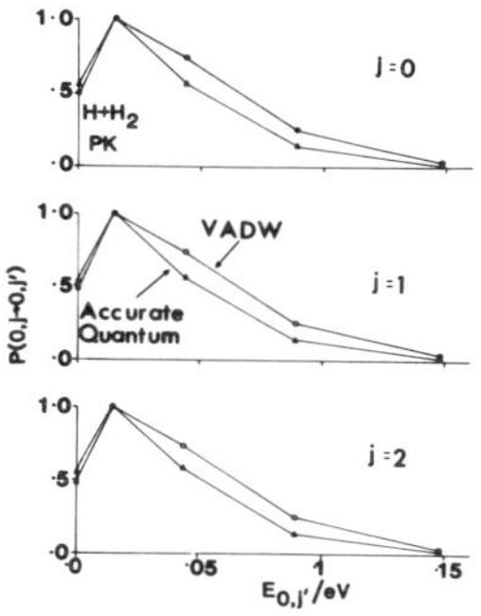

Figure 4 Accurate quantum and VADW rotational product distributions P(O, j → 0, j') for $H+H_2$ or PK at E_{tot} = 0.60 eV plotted against rotational energy.

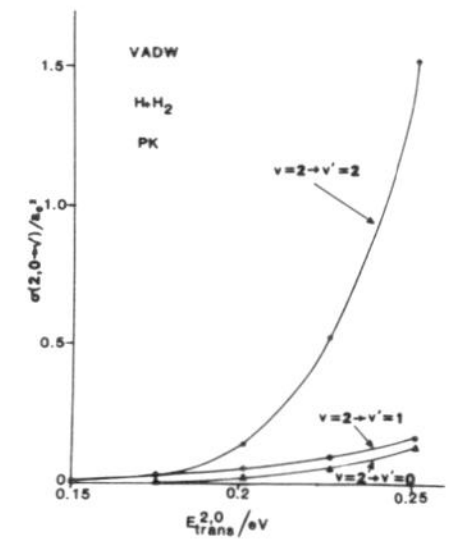

Figure 5 VADW cross sections σ(v=2,0 → v'≤2) for $H+H_2$ on PK plotted against translational energy.

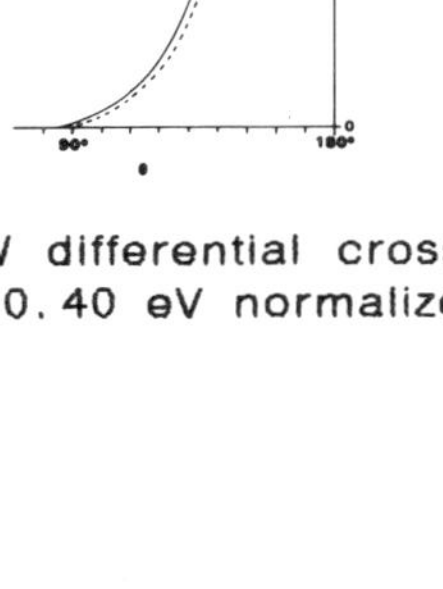

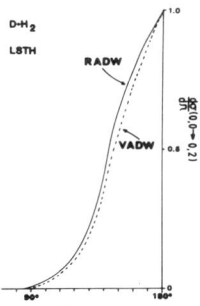

Figure 6 RADW and VADW differential cross sections dσ/dΩ(0,0 → 0,2) for $D+H_2$ on LSTH at E_{tr} = 0.40 eV normalized at 180°.

$$\sigma(v,j \rightarrow v') = \Sigma_{j'} \sigma(v,j \rightarrow v',j').$$

The $H+H_2(v=0,j=0)$ reaction on the LSTH potential surface has also been investigated by the DWBA (Clary and Connor [25], Sun et al. [93]), with results similar to these discussed above for the PK surface. An interesting finding is that the $\sigma(0,0)$ cross sections calculated by the CADW method are larger than the RADW ones. They also agree well with the exact quantum $\sigma(0,0)$ cross sections at low energies, although at the highest energy of $E_{tr} = 0.327$ eV, the CADW cross section is smaller than the exact one by about 30%.

The VADW, RADW and CADW methods have also been applied to the $D+H_2$ reaction on a variety of potential surfaces (see Table I). Molecular beam and kinetic data is also available for this reaction. As for the $H+H_2$ reaction, the various DW approximations give similar relative quantities, but different absolute values for the cross sections in the order: VADW < RADW < CADW. Figure 6 shows the VADW and RADW differential cross section for the $j=0 \rightarrow j'=2$ transition at $E_{tr} = 0.4$ eV on the LSTH surface, whilst Figure 7 illustrates the rotational distributions for the same system (Clary and Connor [25]). These figures clearly demonstrate the close agreement of the VADW and RADW theories for this case. Using the CADW method for $D+H_2(v=0,j=0)$ on LSTH, and the RADW technique for $D+H_2(v=0,j=1)$, Sun et al. [94] have reported good agreement with all available experimental data for this reaction.

There has recently been considerable theoretical interest in the $H+D_2$ reaction. Two important experiments have measured vibrational-rotational product distributions under nearly single collision conditions [41,58]. The experiments are performed at $E_{tr} = 1.3$ eV, with a smaller product contribution coming from collisions with $E_{tr} = 0.55$ eV. It is difficult to carry out a DW calculation using adiabatic distorted wavefunctions at a collision energy as high as 1.3 eV. This is because it is necessary to calculate the adiabatic eigenvalues at sufficiently small values of R_γ that the numerical evaluation of the elastic radial wavefunction can be started in the classically forbidden region. It becomes progressively more difficult to do this as E_{tr} increases.

There have been two SSDW calculations for $H+D_2$ on LSTH at 1.3 eV (Connor and Southall [30], Suck Salk et al. [86]) with results agreeing closely with each other. However, comparison with the experimental data shows that the SSDW product distribution are, in general, less rotationally excited than the experimental ones. An example is illustrated in Figure 8, where the SSDW rotational distribution for $v'=0$ (Connor and Southall [30]) is compared with a linear surprisal fit to the experimental data of Gerrity and Valentini [41]. Also shown is an extrapolation made by Marinero et al. [58] to $v'=0$ from their experimental $v'=1$ and $v'=2$ data, using a linear surprisal assumption. Figure 8 shows that all three distributions are in close agreement for $j' \leq 4$, but there are differences for $j'>4$.

In the usual SSDW method, the spherical average of the interaction potential is used to define the distortion potentials (see Eq. (24)). Bowers et al. [10] have suggested that for systems such as $H+D_2$, which possess a strongly anisotropic interaction in the exit channel, a better distortion potential is obtained using the preferred collinear configuration of the atoms. This will be called the SSDW-C method, and has been

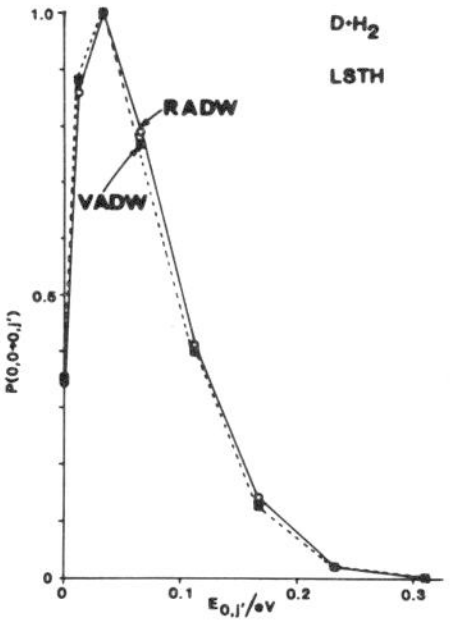

Figure 7 RADW and VADW rotational product distributions for $D+H_2$ on LSTH at E_{tr} = 0.40 eV plotted against rotational energy.

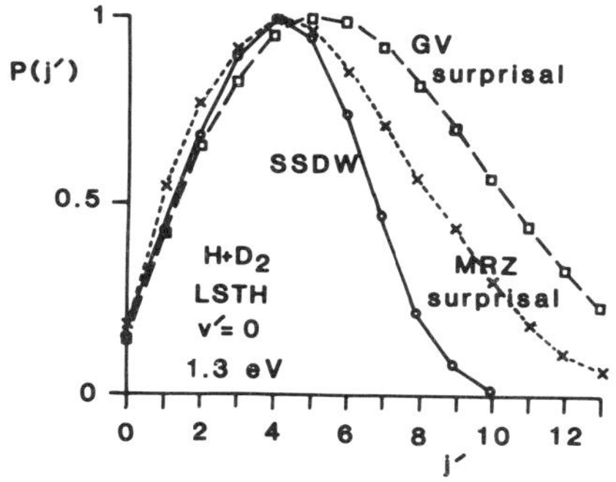

Figure 8 Rotational product distributions into v′=0 for $H+D_2$ at E_{tr} = 1.3 eV.

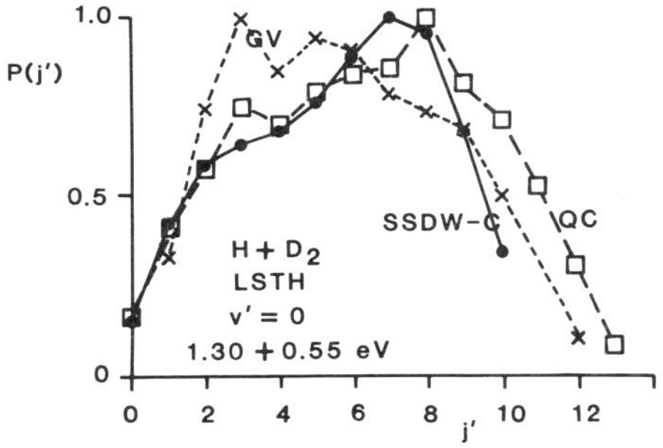

Figure 9 Rotational product distributions into v′=0 for $H+D_2$ from collisions with E_{tr} = 0.55 and 1.3 eV.

briefly discussed in Section 2.2. Using the distortion potential defined by Eq. (31) for $H+D_2$ on LSTH at 1.3 eV results in more product rotational excitation than does the standard SSDW technique (Bowers et al. [10]).

The more general collinear distortion potential (30) has also been applied to $H+D_2$ on LSTH at 0.55 eV and 1.3 eV (Connor and Southall [31]). Figure 9 compares the SSDW-C rotational distribution for v′=0 with the experimental results of Gerrity and Valentini [41], and with the results of quasiclassical (QC) trajectory calculations (Blais and Truhalar [9]). Because the absolute values of the DW cross sections are inaccurate, to allow for a meaningful comparison with experiment, the ratio of the SSDW-C total cross sections at 0.55 eV and 1.3 eV have been scaled to the quasiclassical ratio. Figure 9 shows that all three results are in quite good agreement with each other. The shoulder at low j′ arises from collisions with E_{tr} = 0.55 eV.

3.2 DWBA Calculations for Other Reactions

Table II reports DWBA calculations for reactions containing a heavier atom than H, D or T. Since $H+H_2$ is an isoergic reaction, the number of open product states is relatively small. In contrast, most of the reactions in Table II possess a large number of product vibrational-rotational states. For example, the highly exoergic $H+F_2$ reaction at E_{tr} = 0.106 eV has 373 open HF vibrational-rotational states, whilst TF has 1023 open product states. In addition, it is found experimentally that product rotational states with $j'\neq 0$ are the most populated in many cases. This implies that DW calculations which only compute j′=0 cannot meaningfully be compared with experiment in these cases.

The VADW method was specifically designed for reactions like $H+F_2$ which possess a large number of open product states (Clary and Connor [21,26]). This method allows the reactant and product vibrational wavefunctions to distort adiabatically, but keeps the rotational wavefunctions static. An adiabatic treatment of the rotational degrees of freedom, although presumably more accurate, would be very cumbersome for reactions like $H+F_2$, requiring the diagonalization of very large matrices. In addition, it is straightforward to apply the VADW method to reactions with rotationally excited reactants, whereas it is much more difficult to do the same in a RA or CA treatment. In fact, there are no DW calculations in Table II using the RADW or CADW techniques - only the VADW and SSDW methods have been employed.

The $H+F_2$ reaction was the first one, beyond $H+H_2$ and its isotopes, for which it was meaningful to compare large scale DW calculations with experiment (Clary and Connor [21]). This reaction has the advantage that a reasonable LEPS potential surface is available (Jonathan et al. [49]) and the product distributions are insensitive to variations in j and E_{tr} for the thermal energy range. This last property allows a fixed energy calculation for $H+F_2$(v=0, j=0) to be meaningfully compared with thermal infrared chemiluminescence data (Polanyi and Sloan [69], Brandt and Polanyi [11]).

Figure 10 shows the VADW distortion potential (34) for elastic scattering in the entrance channel $H+F_2$(v=0), whilst Figure 11 shows the corresponding distortion potential for the exit channel HF(v′=6)+F. The distortion potential $V_0^D(R_\alpha)$ for the entrance channel has a small barrier,

Table II. DWBA calculations for reactions other than $H+H_2$ and its isotopic variations.

reaction	potential surface	$\chi_\beta^{(-)}$		$\chi_\alpha^{(+)}$		method	references	comments
		vib.	rot.	vib.	rot.			
$H+F_2$	LEPS	A	S	A	S	VADW	[21,26]	$j'=0 \rightarrow v'\leq 11, j', m_{j'}$ at 4 energies
		A	S	S^a	S	VSDW	[21]	$j'=0 \rightarrow v'\leq 11, j'$ at 2 energies
		S^a	S	A	S	SVDW	[21]	$j'=0 \rightarrow v'\leq 11, j'$ at 2 energies
$Mu+F_2$	LEPS	A	S	A	S	VADW	[26]	$j'=0 \rightarrow v'\leq 3, j', m_{j'}$ at $E_{tr} = 0.106$ eV
$D+F_2$	LEPS	A	S	A	S	VADW	[26]	$j'=0 \rightarrow v'\leq 16, j', m_{j'}$ at $E_{tr} = 0.106$ eV
$T+F_2$	LEPS	A	S	A	S	VADW	[26]	$j'=0 \rightarrow v'\leq 19, j', m_{j'}$ at $E_{tr} = 0.106$ eV
$O(^3P)+H_2$	ab initio	A	S	A	S	VADW	[22]	$v\leq 2, j\leq 2 \rightarrow v'\leq 2, j', m_{j'}$ at 12 energies
$O(^3P)+H_2$	LEPS	A	S	A	S	VADW	[22]	$v\leq 2, j\leq 2 \rightarrow v'\leq 2, j', m_{j'}$ at 12 energies
$O(^3P)+C(CH_3)_4$	LEPS	A	S	A	S	VADW	[27]	$j=0 \rightarrow j'\leq 6$ at $E_{tr} = 0.23$ eV
$O(^3P)+HC(CH_3)_3$	LEPS	A	S	A	S	VADW	[27]	$j=0 \rightarrow v'=1, j'\leq 6$ at $E_{tr} = 0.23$ eV
$He+H_2^+$	DIM	S	S	S	S	SSDW	[111]	$v\leq 1$ $j=0 \rightarrow v', j'=0$
	DIM	S	S	S	S	SSDW	[87]	$v=2, j=1 \rightarrow j'=1$ at $E_{tr} = 0.35$ eV
	DIM	?	?	?	?	?	[102]	$v\leq 2$, conference abstract
$F+H_2$	LEPS	S	S^a	S	S^a	SSDW	[78]	$j=0 \rightarrow v'\leq 3, j'$ at 5 energies
	LEPS	S	S	S	S	SSDW	[35,80-83,88]	$j=0 \rightarrow v'\leq 3, j'=0$
	LEPS	S	S	S	S	SSDW	[34,36,37,85,88]	$j=0 \rightarrow v'\leq 3, j'$

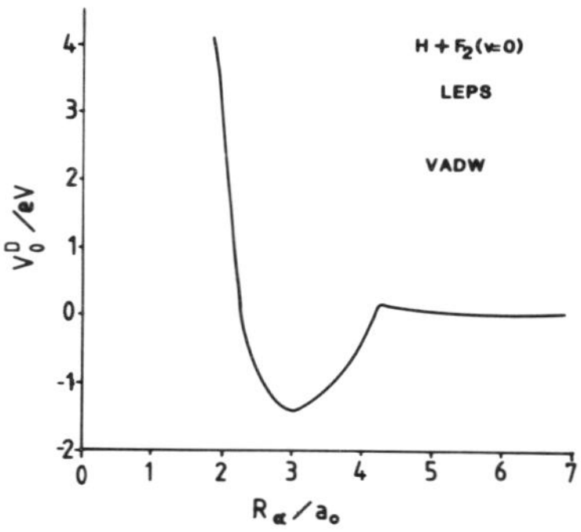

Figure 10 VADW distortion potential for elastic scattering for $H+F_2(v=0)$ on a LEPS surface.

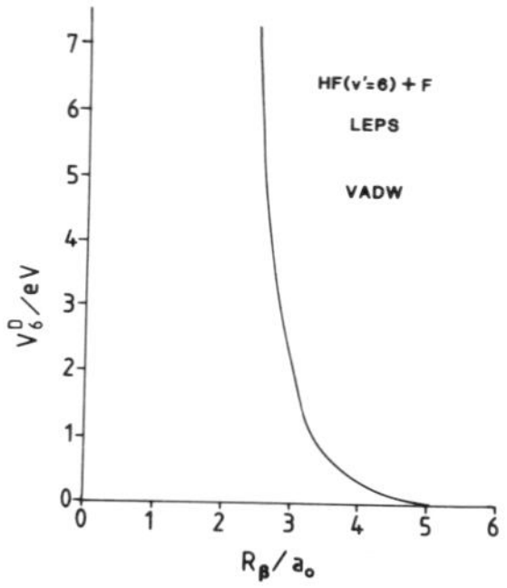

Figure 11 VADW distortion potential for elastic scattering for $HF(v'=6)+F$ on a LEPS surface.

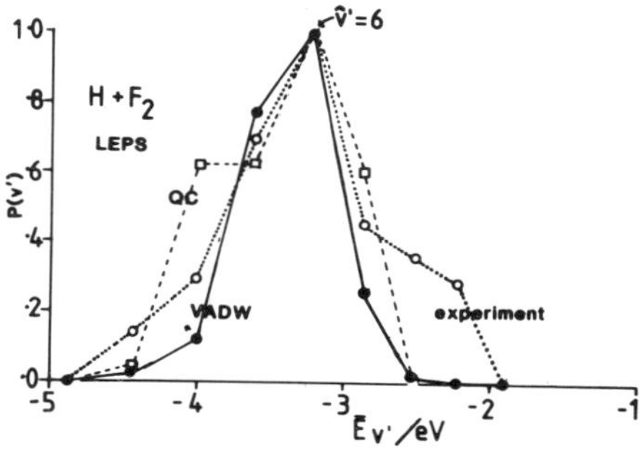

Figure 12 VADW vibrational product distribution for $H+F_2(v=0, j=0)$ on a LEPS surface at E_{tr} = 0. 106 eV plotted against vibrational energy. The quasiclassical (QC) and experimental results are for thermal reactants at T = 300 K.

followed by a well, and finally a strongly repulsive region. This shape is a consequence of the "attractive" nature of the potential energy surface, which has an early saddle point. The distortion potential $V_6^D(R_\beta)$ for the exit channel has an exponential shape and is very similar for all values of v'.

Figure 12 compares the VADW vibrational distribution for $H+F_2$(v=0, j=0) at E_{tr} = 0.106 eV with thermal quasiclassical trajectory results at 300 K (from Figure 1 of Polanyi et al. [70]) and with the thermal chemiluminescence experimental data (from Table V and Fig. 9 of Brandt and Polanyi [11]). There is good agreement between all three distributions, with the most populated vibrational state being v'=6.

Figure 13 shows a similar comparison for the $D+F_2$ reaction. The VADW distribution is for $D+F_2$(v=0, j=0) at E_{tr} = 0.106 eV and the quasiclassical results are those computed by Jonathan et al. [50]. The experimental data is the very recent fast flow measurement made by Dzelzkalns and Kaufman [33]. Note that the VADW calculations were carried out before the experimental measurement was made. Again, all three distributions agree well, with the most populated state being v'=9. However, one defect of all the theoretical distributions in Figures 12 and 13 is that they are narrower than the experimental ones.

Semi-adiabatic calculations have also been carried out (Clary and Connor [21]) for $H+F_2$(v=0, j=0) at E_{tr} = 0.106 eV (see Table II). The calculation using the VA approximation for reactants, and the S^a approximation for the product vibrations, gives a vibrational distribution that agrees closely with the full VADW calculation. However, agreement is poor when the S^a is used for reactants and the VA for products [21]. This interesting result is probably related to the position of the barrier on the potential surface. An accurate description of the reaction dynamics is particularly important around the barrier region, and for $H+F_2$, the barrier is located early in the entrance valley.

Figure 14 shows the VADW rotational distributions for $H+F_2$ for the four most populated vibrational states, namely v'=4, 5, 6, 7. Also illustrated are the thermal experimental chemiluminescence rotational distributions (Polanyi and Sloan [69]). The agreement between the VADW and experimental results is seen to be good. Note that $H+F_2$ is a light+heavy-heavy atom reaction, with the consequence that the amount of energy released into rotational degrees of freedom is very small. The VADW rotational distributions predicted for $D+F_2$(v=0, j=0) at E_{tr} = 0.106 eV are shown in Figure 15. There are no experimental measurements to compare with for this system. Vibrational and rotational distributions for $Mu+F_2$ and $T+F_2$ together with other properties are reported in [26].

The VADW method also gives cross sections $\sigma(0,0,0 \rightarrow v',j',m_{j'})$ for rotational product states with different $m_{j'}$ values. In every case, it has been found for $X+F_2$ (where X = Mu, H, D or T) that

$$\sigma(0,0,0 \rightarrow v',j',|m_{j'}|) > \sigma(0,0,0 \rightarrow v',j',|m_{j'}+1|).$$

This relation also holds for the $H+H_2$ reaction (see Figure 3), as well as for the reactions of $O(^3P)$ discussed below. It is related to the fact that the potential energy surfaces of all these reactions favour the collinear configuration.

The $O(^3P)+H_2(v \leq 2, j \leq 2)$ reaction on two potential surfaces has also

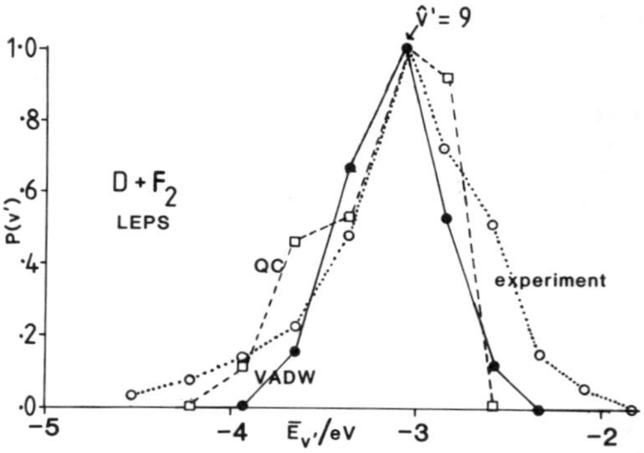

Figure 13 VADW vibrational product distribution for $D+F_2(v=0, j=0)$ on a LEPS surface at E_{tr} = 0.106 eV plotted against vibrational energy. The quasiclassical (QC) and experimental results are for thermal reactants.

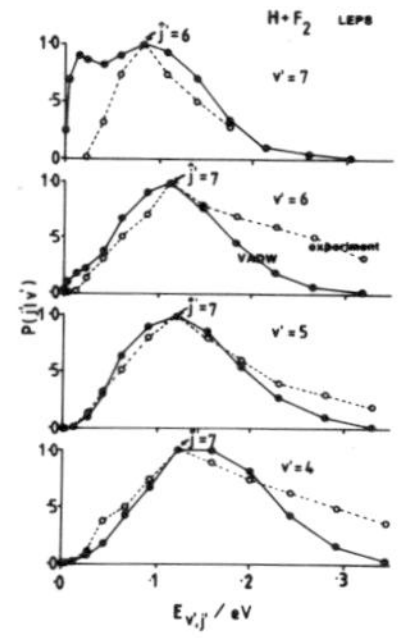

Figure 14 VADW rotational product distributions for $H+F_2(v=0, j=0)$ on a LEPS surface at E_{tr} = 0.106 eV plotted against rotational energy. The experimental results are for thermal reactants at T = 300 K.

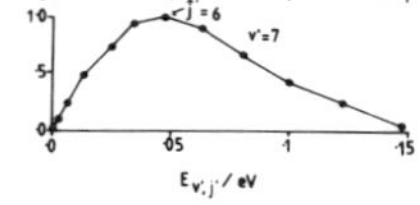

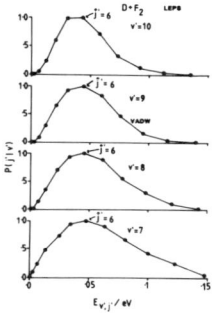

Figure 15 VADW rotational product distributions for $D+F_2(v=0, j=0)$ on a LEPS surface at E_{tr} = 0.106 eV plotted against rotational energy.

been studied by the VADW method (Clary and Connor [22]). One potential surface is a LEPS form due to Johnson and Winter [48], whilst the other is an ab initio one, fitted by a function containing 56 adjustable parameters (Schinke and Lester [77]). Both surfaces have large barriers of 0.54 eV and 0.60 eV respectively.

The VADW method predicts that the vibrational distributions obey the adiabaticity rule: $v \rightarrow v'=v$ transitions are favoured over $v \rightarrow v' \neq v$ transitions. This finding is similar to the $H+H_2(v \leq 2)$ case illustrated in Figure 5.

Figure 16 shows the differential cross sections $d\sigma(0,0)/d\Omega$ for both surfaces at two translational energies. On both surfaces, the peak moves away from the backward direction as E_{tr} increases. The rotational distributions for $v=0$, $j=0 \rightarrow v'=0$, $j' \leq 11$ are illustrated in Figure 17. Note that as E_{tr} becomes larger, the rotational excitation of the product OH molecule rapidly increases. Again, the behaviour for both surfaces is similar. Note also that the energies in Figure 17 are close to the 3D quasiclassical thresholds. In this threshold energy range, it is particularly difficult to compute converged rotational distributions by quasiclassical techniques.

Although the $O(^3P)+H_2$ reaction has been intensively studied by the VADW method, as well as by other theoretical techniques (see in addition Table III), no state-to-state experimental measurements have yet been made for it. However, several molecular beam experiments of the reaction of $O(^3P)$ with saturated hydrocarbons HR, using laser induced fluorescence detection of the product OH molecule, have been reported (Andresen and Luntz [2], Dutton et al. [32]). These experiments have found very low rotational excitation of the OH molecule, with primary, secondary and tertiary hydrocarbons yielding nearly identical rotational distributions. Evidently the reactions are all direct, with a strong preference for a collinear transition state.

Figure 18 shows the VADW rotational distribution for $O(^3P)$ reacting with the tertiary hydrocarbon $HC(CH_3)_3$ ≡ isobutane into $v'=1$ at E_{tr} = 22.2 kJ mol^{-1} (Clary et al. [27]). The VADW result is compared with experimental measurements (Andresen and Luntz [2]) and with quasiclassical trajectory computations (Luntz and Andresen [56]). The same LEPS potential surface has been used in the VADW and quasiclassical calculations. There is good qualitative agreement between the calculated and experimental rotational distributions. Note that the theoretical and experimental rotational distributions in Figure 18 have been plotted as a function of the (theoretical and experimental) rotational energy of OH, rather than using the rotational quantum number (Clary et al. [27]).

Table II shows there have been many SSDW calculations for the $F+H_2$ reaction using the Muckerman 5 LEPS potential surface. The first DW calculation for this reaction was a semi-adiabatic one by Shan et al. [78]. Suck (Suck Salk) and coworkers have used the transferred angular momentum formalism (Suck [79], see Eq. (36)) to make detailed SSDW investigations of many aspects of the $F+H_2$ reaction. They have studied topics such as: the preferred geometric configuration, the contribution of different partial waves to the differential cross section, the role of angular momentum transfer, wavenumber matching in the entrance and exit channels, and the post-prior equivalence (see Eqs. (18) and (19)).

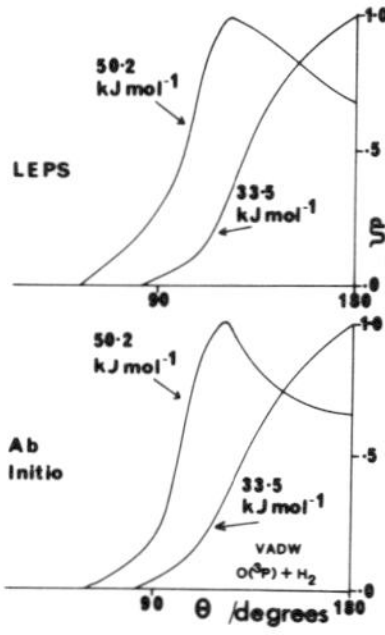

Figure 16 VADW differential cross sections $d\sigma(0,0)/d\Omega$ for $O(^3P)+H_2$ at two translational energies normalized at 180°.

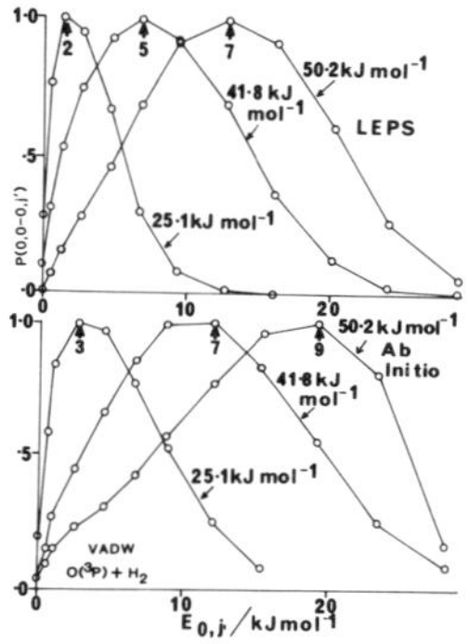

Figure 17 VADW rotational product distributions $P(0,0 \rightarrow 0,j')$ at three translational energies for $O(^3P)+H_2$ plotted against rotational energy.

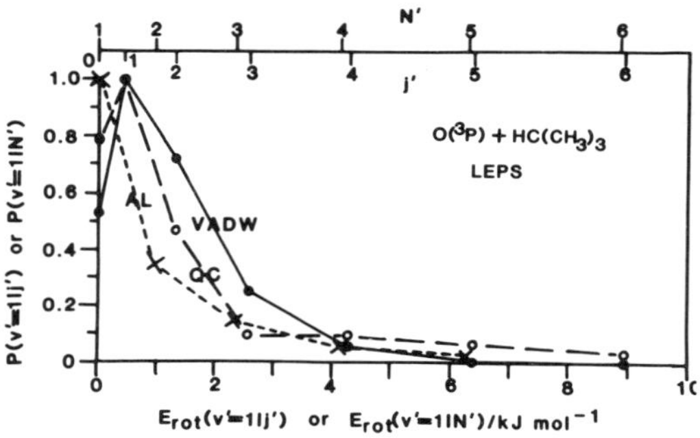

Figure 18 VADW, quasiclassical (QC) and experimental (AL) rotational product distributions of OH(v'=1) for the $O(^3P)+HC(CH_3)_3$ reaction at E_{tr} = 22.2 kJ mol^{-1} plotted against rotational energy.

Many of their earlier calculations were restricted to $j'=0$ (Suck [80, 81], Suck and Emmons [82, 83, 88], Emmons and Suck [35]) but more recently results for $j'\neq0$ have been reported (Emmons and Suck [34, 36], Suck Salk et al. [85], Emmons et al. [37], Suck and Emmons [88]).

However, it is unlikely that the SSDW method using the simple distortion potentials discussed in Section 2.2. can accurately describe the dynamics of the $F+H_2$ reaction on the Muckerman 5 surface. Approximate 3D quantum calculations for $F+H_2$ indicate that the resonances obtained in exact collinear quantum reaction probabilities are not completely averaged out on going to 3D. As emphasized in Section 2, the DWBA theories currently in use should work best for direct reactions and not reactions involving resonances. In addition, the recent detailed experimental molecular beam data of Neumark et al. [65] indicates that resonances are important for $F+H_2$, although the Muckerman 5 surface is not adequate to explain them. Furthermore, since the product distributions are sensitive to the value of the translational energy [65], it follows that it is probably not meaningful to compare a fixed energy calculation for $F+H_2(v=0, j=0)$ with thermal experimental results, such as infra-red chemiluminescence measurements. In summary, it is likely that techniques more elaborate than the simple DWBA method will be needed for reactions with dynamics as complicated as $F+H_2$.

Table II shows that the SSDW method has also been applied to the $He+H_2^+$ reaction, but only preliminary results have been reported (Zuhrt et al. [111] Tang et al. [102], Suck Salk and Emmons [87]). For example, Zuhrt et al. [111] only considered transitions into $j'=0$.

3.3 Coupled-channel Distorted-wave Calculations

All the DWBA calculations listed in Tables I and II assume that elastic scattering is the only non-reactive process in the initial and final arrangement channels. However, this is an inaccurate description of the non-reactive scattering, because inelastic, as well elastic, collisions can occur. In particular, rotational transitions occur readily even at low collision energies.

Including the inelastic scattering in the incident and exit channels leads to the CCDW approximation, which was introduced in Section 2.1. Table III lists DW calculations which approximate $\chi_\alpha^{(+)}$ and $\chi_\beta^{(-)}$ (or both) by a finite CC expansion (see also Suck [84]). Also included are calculations which use the CS approximation to solve the inelastic coupled equations.

Schatz et al. [75] have applied the CCDW method to $H+H_2$ on PK. Comparison with the exact quantum results [73], shows that the CCDW technique is accurate for energies where the total reaction probability for each partial wave is less than about 0.1. At higher energies, the CC expansions do not converge as additional closed channels are added. A similar result has been obtained in collinear $H+H_2$ calculations [47].

Since the reaction is still treated as a perturbation in the CCDW method, it is to be expected that this method will eventually break down as E_{tot} increases (like the simpler DWBA theories). Nevertheless, for many reactions with high barriers, the perturbation assumption should be valid at thermal energies, where the dynamics is dominated by tunnelling

Table III. DW calculations which include inelastic scattering in the entrance or exit channels

reaction	potential surface	$\chi_\beta^{(-)}$	$\chi_\alpha^{(+)}$	method	reference	comments
$H+H_2$	PK	CC	CC	CCDW	[75]	$j \leqslant 3 \to j' \leqslant 3$ at $E_{tot} = 0.5$ eV
	PK	CS	CS	CSDW	[75]	$j \leqslant 3 \to j' \leqslant 3$ at 3 energies
	PK	S S	CC	SCDW	[20]	$v \leqslant 1, j=0 \to j' \leqslant 5$ at 4 energies
$O(^3P)+H_2$	LEPS	CS	CS	CSDW	[76]	$j \leqslant 6 \to j' \leqslant 9$ at 6 energies
$O(^3P)+HD$	LEPS	CS	CS	CSDW	[76]	$j \leqslant 6 \to j' \leqslant 9$ at 4 energies
$O(^3P)+DH$	LEPS	CS	CS	CSDW	[76]	$j \leqslant 6 \to j' \leqslant 9$ at 4 energies
$O(^3P)+D_2$	LEPS	CS	CS	CSDW	[76]	$j \leqslant 6 \to j' \leqslant 9$ at 4 energies
Cl+HCl	LEPS	CS	CS	CSDW	[1]	work in progress

through the barrier. Schatz et al. [75] also found the simpler CSDW method gave results in good agreement with the exact quantum values. Related work suggests that the errors in the CS approximation should be less than 30%.

Choi et al. [20] have extended the DWBA to include inelastic non-reactive scattering. However in their formalism, the CC method is only applied to $\chi_\alpha^{(+)}$; the other distorted wavefunction $\chi_\beta^{(-)}$ is treated by the SS method, as in the RADW and CADW approximations. Comparison with the exact quantum results for $H+H_2$ on PK, shows the resulting cross sections are too small. Evidently, for an isoergic reaction like $H+H_2$, it is important to include the inelastic couplings in both the entrance and exit channels, if accurate absolute cross sections are to be calculated.

Choi et al. [20] also studied how the integral cross section for the $j=0 \rightarrow j'=1$ transition in $H+H_2$, changed when the full vibrational-rotational basis set was replaced by a pure vibrational one, and a pure rotational one for v=0. They found that the cross section using the pure vibrational basis set is only slightly larger than that from the SSDW method. The cross section using the pure rotational basis set is an order of magnitude larger, but is still smaller than that from the full vibrational-rotational basis set.

The CSDW method has been used by Schatz [76] in an extensive study of the $O(^3P)+H_2$,HD,DH,D_2 reactions. For the $O(^3P)+H_2$ reaction, the rotational product distributions agree well with those calculated by the VADW method [22], see Figure 17. The CSDW rate coefficients are larger than quasiclassical ones at low temperatures owing to tunnelling effects through the high barrier on the surface. This study, as well as the one for $H+H_2$ [75], indicates that the CSDW method should be an effective technique for reactions with high barriers. In particular, it should work well for energies close to the reaction threshold, where the quasiclassical trajectory method is least accurate, but which is the energy region which contributes most to thermal rate coefficients. Another reaction with a large barrier is the heavy + light-heavy atom system Cl+HCl. Preliminary results indicate that the CSDW method can be used successfully for this type of reaction as well (Amaee et al. [1]).

4. CONCLUDING REMARKS

Many quantum mechanical theories have been proposed in the literature for 3D chemical reactions. Only a few theories have been shown to be viable in practice. Even fewer have been applied to reactions other than $H+H_2$ and its isotopes.

The DW theory is a practical method for studying the details of direct 3D chemical reactions. The DW approach encompasses an infinite number of possible approximations. Those discussed in this review include the SSDW, VADW, RADW, CADW, CSDW and CCDW methods. All these methods treat the reaction as a perturbative transition between non-reactive scattering states in the initial and final arrangement channels. Note that the calculations reported so far have used the usual Jacobi coordinates. It would be interesting to see if more accurate theories would result if natural collision coordinates or hyperspherical coordinates were used instead.

For the $H+H_2$ reaction, all DW methods give similar results at low collision energies for relative quantities such as differential cross sections. In such a situation, the simpler theories can be used, i.e., the SSDW and VADW methods. The VADW technique has been shown to be viable for reactions with as many as 1000 product vibrational-rotational states. More research on better methods for choosing the distortion potentials in these simple DW theories would be valuable.

As the approximations used in the DW theories are systematically improved, the absolute values of the cross sections approach the exact results (for $H+H_2$ at least). If true in general, this indicates that the CSDW and CCDW methods should be reliable for the threshold region of reactions with large barriers, where the dynamics is dominated by tunnelling.

In earlier reviews of reactive scattering, the DW approach was described as "inadequate for reactive collisions" (George and Ross [40], p. 272) or "merely a qualitative description" (Nikitin and Zülicke [66], p. 97). The research described in this review shows that these conclusions are wrong.

ACKNOWLEDGEMENTS

I am very grateful to B. Amaee (Manchester), D.C. Clary (Cambridge) and W.J.E. Southall (Manchester) for many discussions on the distorted wave method. I thank Ms. D. Turner for her rapid and accurate typing of the manuscript. I also thank G.C. Schatz (Evanston), S.H. Suck Salk (Rolla) and K.T. Tang (Tacoma) for sending me preprints of their recent work. Support of this research by the Science and Engineering Research Council (U.K.) and NATO is gratefully acknowledged.

REFERENCES

1. B. Amaee, J.N.L. Connor and G.C. Schatz, work in progress (1985).
2. P. Andresen and A.C. Luntz, *The chemical dynamics of the reactions of O(^{3}P) with saturated hydrocarbons. I. Experiment*, J. Chem. Phys., **72**, 5842-50 (1980).
3. N. Austern, *Direct Nuclear Reaction Theories* (Wiley-Interscience, New York, 1970).
4. V.K. Babamov, V. Lopez and R.A. Marcus, *Dynamics of hydrogen atom and proton transfer reactions. Simplified analytic two-state formulae*, Chem. Phys. Lett., **101**, 507-11 (1983).
5. V.K. Babamov, V. Lopez and R.A. Marcus, *Dynamics of hydrogen atom and proton transfer reactions. Nearly degenerate asymmetric case*, J. Chem. Phys., **78**, 5621-8 (1983); erratum *ibid.*, **81**, 4182 (1984).
6. V.K. Babamov, V. Lopez and R.A. Marcus, *An exponentiated DWBA formula for H-atom transfers. Extensions to lower barrier potentials and to higher energies*, J. Chem. Phys., **80**, 1812-6 (1984); erratum *ibid.*, **81**, 4181 (1984).

7. M. Baer, *A review of quantum-mechanical approximate treatments of three-body reactive systems*, Adv. Chem. Phys. **49**, 191-309 (1982).
8. M. V. Basilevsky and V. M. Ryaboy, *Quantum Dynamics of linear triatomic reactions*, Adv. Quant. Chem., **15**, 1-83 (1982).
9. N. C. Blais and D. G. Truhlar, *Calculated product-state distributions for the reaction $H+D_2 \rightarrow HD+D$ at relative translational energies 0.55 and 1.30 eV*, Chem. Phys. Lett., **102**, 120-5 (1983).
10. M. S. Bowers, B. H. Choi, R. T. Poe and K. T. Tang, *Quantum mechanical determination of product state distributions in the $H+D_2 \rightarrow HD+D$ reaction*, Chem. Phys. Lett., **116**, 239-44 (1985).
11. D. Brandt and J. C. Polanyi, *Energy distributions among reaction products. XI. $H+ClF \rightarrow HF+Cl$, $HCl+F$*, Chem. Phys., **35**, 23-34 (1978).
12. A. M. Brodskii, V. G. Levich and V. V. Tolmachev, *Wave-mechanical theory for the cross sections of gas-phase substitution reactions* II. *Overlap integral and general properties and energetic and angular dependences of the cross section*, Khim. Vys. Energ., **4**, 195-201 (1970); English translation: High Energy Chem., **4**, 171-6 (1970).
13. A. M. Brodsky and V. G. Levich, *Theory of the simplest substitution reactions*, J. Chem. Phys., **58**, 3065-81 (1973).
14. B. H. Choi and K. T. Tang, *Adiabatic distorted wave calculation of $H+H_2$ reactive scattering*, J. Chem. Phys., **61**, 2462-4 (1974).
15. B. H. Choi and K. T. Tang, *Theory of distorted-wave Born approximation for reactive scattering of an atom and a diatomic molecule*, J. Chem. Phys., **61**, 5147-57 (1974).
16. B. H. Choi and K. T. Tang, *Three-dimensional quantum mechanical studies of $D+H_2 \rightarrow HD+H$ reactive scattering. II*, J. Chem. Phys., **63**, 2854-60 (1975).
17. B. H. Choi and K. T. Tang, *Three-dimensional quantum mechanical studies of the $H+H_2$ reactive scattering*, J. Chem. Phys., **65**, 5161-80 (1976).
18. B. H. Choi, R. T. Poe, J. C. Sun and K. T. Tang, *Reactive scattering of rotationally excited target molecules with adiabatic theory*, J. Chem. Phys., **73**, 4381-9 (1980).
19. B. H. Choi, R. T. Poe, J. C. Sun, K. T. Tang and Y. Y. Yung, *Transition matrix theory of molecular reactive scattering*, J. Chem. Phys., **74**, 5686-93 (1981).
20. B. H. Choi, R. T. Poe and K. T. Tang, *Coupled channel distorted wave method of atom-molecule reactive scattering: Application to para to ortho hydrogen molecule conversion*, J. Chem. Phys., **81**, 4979-90 (1984).
21. D. C. Clary and J. N. L. Connor, *Application of the vibrationally adiabatic and static distorted wave Born approximations to the reaction $H+F_2(v=0,j=0) \rightarrow HF(v',j')+F$*, Chem. Phys. Lett., **66**, 493-7 (1979).
22. D. C. Clary and J. N. L. Connor, *The $O(^3P)+H_2(v\leq 2,j,m_j) \rightarrow OH(v'\leq 2,j',m_{j'})+H$ reaction. A vibrationally adiabatic distorted wave study using a LEPS and fitted ab initio potential energy surface*, Mol. Phys., **41**, 689-702 (1980).
23. D. C. Clary and J. N. L. Connor, *Distorted-wave calculations for the three dimensional chemical reaction $H+H_2(v\leq 2,j=0) \rightarrow H_2(v'\leq 2,j',m_{j'})+H$*, Chem. Phys., **48**, 175-81 (1980).

24. D.C. Clary and J.N.L. Connor, *Vibrationally adiabatic distorted wave calculation for the rotationally excited reaction* $H+H_2(v=0,j) \rightarrow H_2(v'=0,j')+H$, J. Chem. Phys., **74**, 6991-3 (1981).
25. D.C. Clary and J.N.L. Connor, *Comparison of the rotationally adiabatic and vibrationally adiabatic distorted wave methods for the* $H+H_2(v=0,j=0) \rightarrow H_2(v'=0,j')+H$ *and* $D+H_2(v=0,j=0) \rightarrow DH(v'=0,j')+H$ *chemical reactions*, Mol. Phys., **43**, 621-39 (1981).
26. D.C. Clary and J.N.L. Connor, *The vibrationally adiabatic distorted wave method for direct chemical reactions: Application to* $X+F_2(v=0,j=0) \rightarrow XF(v',j',m_{j'})+F$ $(X=Mu,H,D,T)$, J. Chem. Phys., **75**, 3329-39 (1981).
27. D.C. Clary, J.N.L. Connor and W.J.E. Southall, *Reactions of* $O(^3P)$ *with saturated hydrocarbons: Vibrationally adiabatic distorted wave calculations of product rotational distributions for two triatomic model reactions*, J. Chem. Phys., (in the press).
28. J.N.L. Connor, *Theory of molecular collisions and reactive scattering*, Ann. Rep. Chem. Soc., **70A**, 5-30 (1973).
29. J.N.L. Connor, *Reactive molecular collision calculations*, Comput. Phys. Commun. **17**, 117-43 (1979).
30. J.N.L. Connor and W.J.E. Southall, *The reaction* $H+D_2 \rightarrow HD+D$: *Distorted wave calculations at* $E_{trans}(v=0,j=0)$ = 0.55 *and* 1.3 *eV*, Chem. Phys. Lett., **108**, 527-31 (1984).
31. J.N.L. Connor and W.J.E. Southall, *The reaction* $H+D_2 \rightarrow HD+D$ *at 0.55, 0.98, 1.10 and 1.30 eV: A distorted wave study*, unpublished manuscript (1985).
32. N.J. Dutton, I.W. Fletcher and J.C. Whitehead, *Laser-induced fluorescence determination of the internal state distributions of* $OH(X^2\Pi)$ *produced in molecular beam reactions of* $O(^3P)$ *with some cyclic hydrocarbons*, Mol. Phys., **52**, 475-83 (1984).
33. L.S. Dzelzkalns and F. Kaufman, *Vibrational relaxation of highly excited diatomics. VI.* $DF(9\leq v\leq 12)+N_2$, *CO*, CO_2 *and* N_2O *and* $HF(v=5-7)+CO$, J. Chem. Phys., **80**, 6114-21 (1984).
34. R.W. Emmons and S.H. Suck, *Distorted-wave Born-approximation study of angular distributions for state-to-state rearrangement collisions: Role of orbital angular momentum*, Phys. Rev., **25A**, 178-86 (1982).
35. R.W. Emmons and S.H. Suck, *Equivalence between the prior- and post-interaction forms in the distorted-wave Born-approximation transition amplitude*, Phys. Rev., **25A**, 2385-7 (1982).
36. R.W. Emmons and S.H. Suck, *State-to-state and state-to-all states reactive scattering angular distributions:* $F+H_2 \rightarrow HF+H$, Phys. Rev., **27A**, 1803-11 (1983).
37. R.W. Emmons, C.R. Klein and S.H. Suck Salk, *Variation of direct-process contribution with collision energy in reactive scattering*, Phys. Rev., **29A**, 1131-4 (1984).
38. B.C. Eu, J.H. Huntington and J. Ross, *Direct interaction theory of reactive molecular collisions:* $K+Br_2$ *system*, Can. J. Phys., **49**, 966-70 (1971).
39. A. Gelb and R.J. Suplinskas, *Influence of the distorted wave approximation in calculations of chemical reaction cross sections;* Ar^++HD, J. Chem. Phys., **53**, 2249-57 (1970).

40. T.F. George and J. Ross, *Quantum dynamical theory of molecular collisions*, Ann. Rev. Phys. Chem., **24**, 263-300 (1973).
41. D.P. Gerrity and J.J. Valentini, *Experimental study of the dynamics of the $H+D_2 \rightarrow HD+D$ reaction at collision energies of 0.55 and 1.30 eV*, J. Chem. Phys., **81**, 1298-313 (1984).
42. R.G. Gilbert and T.F. George, *On the distorted wave approximation for chemical reactions*, Chem. Phys. Lett., **20**, 187-92 (1973).
43. N.K. Glendenning, *Direct Nuclear Reactions* (Academic Press, New York, 1983).
44. S. Golden and A.M. Peiser, *The quantum mechanics of chemical kinetics of homogeneous gas phase reactions II. Approximations for displacement reaction between an atom and a diatomic molecule*, J. Chem. Phys., **17**, 630-43 (1949); erratum *ibid.*, **17**, 842 (1949).
45. S. Golden, *Note on the quantum-mechanical calculation of reaction rates*, J. Chem. Phys., **21**, 2071-2 (1953).
46. S. Golden, *Adequacy of the Born approximation in the calculation of chemical reaction rates: A reply to Yasumori and Sato*, J. Chem. Phys., **22**, 1938-9 (1954).
47. L.M. Hubbard, S-h. Shi and W.H. Miller, *Multichannel distorted wave Born appriximation for reactive scattering*, J. Chem. Phys., **78**, 2381-7 (1983)
48. B.R. Johnson and N.W. Winter, *Classical trajectory study of the effect of vibrational energy on the reaction of molecular hydrogen with atomic oxygen*, J. Chem. Phys., **66**, 4116-20 (1977).
49. N. Jonathan, S. Okuda and D. Timlin, *Initial vibrational energy distributions determined by infra-red chemiluminescence III. Experimental results and classical trajectory calculations for the $H+F_2$ system*, Mol. Phys., **24**, 1143-64 (1972); erratum *ibid.*, **25**, 496 (1973).
50. N.B.H. Jonathan, J.P. Liddy, P.V. Sellers and A.J. Stace, *Initial vibrational energy distributions determined by infrared chemiluminescence: the D/F_2 system*, Mol. Phys., **39**, 615-27 (1980).
51. M. Karplus and K.T. Tang, *Quantum-mechanical study of $H+H_2$ reactive scattering*, Disc. Faraday Soc., **44**, 56-67 (1967).
52. M. Karplus, *Special results of theory: Distorted waves*, in *Molecular Beams and Reaction Kinetics*, Proc. of the Int. School of Physics "Enrico Fermi", Course 44, edited by Ch. Schlier (Academic, New York, 1970) pp. 407-26.
53. V.G. Levich, A.M. Brodskii and V.V. Tolmachev, *Wave theory for the cross sections of gas-phase substitution reactions I. Derivation of the equations for the cross sections*, Khim. Vys. Energ., **4**, 101-7 (1970); English translation: High Energy Chem., **4**, 87-92 (1970).
54. R.D. Levine, *Simplistic analysis of reactive scattering II. Initial and final distortions*, Israel J. Chem., **8**, 13-28 (1970).
55. V. Lopez, V.K. Babamov and R.A. Marcus, *A simple DWBA ("Franck-Condon") treatment of H-atom transfers between two heavy particles*, J. Chem. Phys., **81**, 3962-6 (1984).
56. A.C. Luntz and P. Andresen, *The chemical dynamics of the reactions of $O(^3P)$ with saturated hydrocarbons. II. Theoretical model*, J. Chem. Phys., **72**, 5851-6 (1980).

57. P.A. Madden, *The exponential approximation for collinear reactive scattering*, Mol. Phys., **29**, 381-8 (1975).
58. E.E. Marinero, C.T. Rettner and R.N. Zare, *$H+D_2$ reaction dynamics. Determination of the product state distributions at a collision energy of 1.3 eV*, J. Chem. Phys., **80**, 4142-56 (1984).
59. A. Messiah, *Quantum Mechanics*, translated from the French by J. Potter (North-Holland, Amsterdam, 1970) Vol II, Chap. XIX.
60. D.A. Micha, *A quantum mechanical model for simple molecular reactions*, Ark. Fys. **30**, 411-23 (1965).
61. D.A. Micha, *The exchange reaction of H and H_2*, Ark. Fys., **30**, 425-36 (1965).
62. D.A. Micha, *Angular distribution of products of hydrogen atom-hydrogen molecule reactions*, Ark. Fys., **30**, 437-47 (1965).
63. D.A. Micha, *Quantum theory of reactive molecular collisions*, Adv. Chem. Phys., **30**, 7-75 (1975).
64. W.H. Miller, *Distorted-wave theory for collisions of an atom and a diatomic molecule*, J. Chem. Phys., **49**, 2373-81 (1968).
65. D.M. Neumark, A.M. Wodtke, G.N. Robinson, C.C. Hayden and Y.T. Lee, *Molecular beam studies of the $F+H_2$ reaction*, J. Chem. Phys., **82**, 3045-66 (1985).
66. E.E. Nikitin and L. Zülicke, *Selected Topics of the Theory of Chemical Elementary Processes*, Lecture Notes in Chemistry, No. 8 (Springer, Berlin, 1978).
67. J.C. Pirkle, Jr., and H.A. McGee, Jr., *Perturbed Morse oscillator approximation in reactive collisions. I. An attractive potential*, J. Chem. Phys., **49**, 3532-40 (1968).
68. J.C. Pirkle, Jr., and H.A. McGee, Jr., *Perturbed Morse oscillator approximation in reactive collisions. II. A repulsive potential*, J. Chem. Phys., **49**, 4504-8 (1968).
69. J.C. Polanyi and J.J. Sloan, *Energy distribution among reaction products. VII. $H+F_2$*, J. Chem. Phys., **57**, 4988-98 (1972).
70. J.C. Polanyi, J.L. Schreiber and J.J. Sloan, *Distribution of reaction products (theory), XI. $H+F_2$*, Chem. Phys., **9**, 403-21 (1975).
71. L.S. Rodberg and R.M. Thaler, *Introduction to the Quantum Theory of Scattering*, (Academic, New York, 1967).
72. G.R. Satchler, *Direct Nuclear Reactions* (Clarendon Press, Oxford 1983).
73. G.C. Schatz and A. Kuppermann, *Quantum mechanical reactive scattering for three-dimensional atom plus diatom systems. II. Accurate cross sections for $H+H_2$*, J. Chem. Phys., **65**, 4668-92 (1976).
74. G.C. Schatz, *Overview of reactive scattering*, in *Potential Energy Surfaces and Dynamics Calculations for Chemical Reactions and Molecular Energy Transfer*, edited by D.G. Truhlar (Plenum, New York, 1981) pp. 287-310.
75. G.C. Schatz, L.M. Hubbard, P.S. Dardi and W.H. Miller, *Coupled channel distorted wave calculations for the three-dimensional $H+H_2$ reaction*, J. Chem. Phys., **81**, 231-40 (1984).
76. G.C. Schatz, *A coupled states distorted wave study of the $O(^3P)+H_2(D_2,HD,DH)$ reaction*, J. Chem. Phys., (in the press).

77. R. Schinke and W.A. Lester Jr., *Trajectory study of* $O+H_2$ *reactions on fitted ab initio surfaces. I. triplet case*, J. Chem. Phys., **70**, 4893-902 (1979); erratum *ibid.*, **72**, 6821, (1980).
78. Y. Shan, B.H. Choi, R.T. Poe and K.T. Tang, *Three-dimensional quantum mechanical study of the* $F+H_2$ *reactive scattering*, Chem. Phys. Lett., **57**, 379-84 (1978).
79. S.H. Suck, *Theory of atom-diatom reactive scattering based on the distorted-wave Born approximation*, Phys. Rev., **15A**, 1893-9 (1977); erratum *ibid.*, **24A**, 2865 (1981).
80. S.H. Suck, *A DWBA study of angular distributions for the state-to-state reactive scattering angular process of* $F+H_2 \rightarrow HF+H$, Chem. Phys. Lett., **77**, 390-3 (1981).
81. S.H. Suck, *The kernel of DWBA transition amplitude in atom-diatom reactive scattering*, Int. J. Quant. Chem., **19**, 441-50 (1981).
82. S.H. Suck and R.W. Emmons, *Effect of partial wave interference on angular distributions and sideways scattering in rearrangement collisions*, Chem. Phys. Lett., **79**, 93-6 (1981).
83. S.H. Suck and R.W. Emmons, *Two-body rearrangement collision of atom-diatomic-molecule system: Role of wave-number matching*, Phys. Rev., **24A**, 129-37 (1981).
84. S.H. Suck, *Theory of atom-diatom rearrangement collisions based on the coupled-channel Born approximation*, Phys. Rev., **27A**, 187-98 (1983).
85. S.H. Suck Salk, R.W. Emmons and C.R. Klein, *Role of angular momentum match in state-to-state reactive scattering and product rotational state distributions*, Phys. Rev., **29A**, 1135-9 (1984).
86. S.H. Suck Salk, C.R. Klein and C.K. Lutrus, *DWBA predicted relative product state distribution for* $H+D_2 \rightarrow HD+D$, Chem. Phys. Lett., **110**, 112-4 (1984).
87. S.H. Suck Salk and R.W. Emmons, *Preferential angular momentum transfer in state-to-state reactive scattering*, Phys. Rev., **29A**, 2906-8 (1984).
88. S.H. Suck and R.W. Emmons, *Importance of relative angular momentum coherence in state-to-state rearrangement collisions (reactive scattering)*, The 9th Symposium of Korean Science and Technology, 3-6 July 1984, Seoul, Korea, 1, 46-9 (1984).
89. S.H. Suck Salk and C.K. Lutrus, *Comparison between approximate (perturbation) and exact (close-coupling) three-dimensional quantal methods in reactive scattering*, J. Chem. Phys., (in the press).
90. J.C. Sun, B.H. Choi, R.T. Poe and K.T. Tang, *Quantum theory of* $D+H_2$ *rearrangement collision: Effects of vibrational excitation*, Phys. Rev. Lett., **44**, 1211-4 (1980).
91. J.C. Sun, B.H. Choi, R.T. Poe and K.T. Tang, *Three-dimensional quantum mechanical studies of* $D+H_2 \rightarrow HD+H$ *reactive scattering. IV. Cross sections and rate constants with rotationally excited target molecules*, J. Chem. Phys., **73**, 6095-107 (1980).
92. J.C. Sun, B.H. Choi, R.T. Poe and K.T. Tang, *Three dimensional effects on the linear adiabatic molecular wavefunctions in the* $H+H_2$ *system*, Chem. Phys. Lett., **82**, 255-9 (1981).
93. J.C. Sun, B.H. Choi, R.T. Poe and K.T. Tang, *Adiabatic T matrix theory for three dimensional reactive scattering: Application to the* (H, H_2) *system*, J. Chem. Phys., **78**, 4523-32 (1983).

94. J.C. Sun, B.H. Choi, R.T. Poe and K.T. Tang, *Three dimensional quantum mechanical studies of $D+H_2 \rightarrow DH+H$ reactive scattering. V. Cross sections and rate constants from the adiabatic T matrix theory*, J. Chem. Phys., **79**, 5376-85 (1983).
95. R.J. Suplinskas and J. Ross, *Perturbed-stationary-state calculation of collisions in a reactive system*, J. Chem. Phys., **47**, 321-30 (1967).
96. K.T. Tang and M. Karplus, *Quantum Theory of (H,H_2) Scattering: Two-body potential and elastic scattering*, J. Chem. Phys., **49**, 1676-92 (1968).
97. K.T. Tang and M. Karplus, *Quantum theory of (H,H_2) Scattering: Approximate treatments of reactive scattering*, Phys. Rev., **4A**, 1844-58 (1971).
98. K.T. Tang, *Quantum cross sections of $D+H_2 \rightarrow HD+H$ reaction*, J. Chem. Phys., **57**, 1808-9 (1972).
99. K.T. Tang and B.H. Choi, *Three-dimensional quantum mechanical studies of $D+H_2 \rightarrow HD+H$ reactive scattering*, J. Chem. Phys., **62**, 3642-58 (1975).
100. K.T. Tang and B.H. Choi, *Three-dimensional quantum mechanical studies of $H+H_2$ and $D+H_2$ reactive scatterings*, in *Electronic and Atomic Collisions*, Abstracts of Papers of the IXth International Conference on the Physics of Electronic and Atomic collisions, edited by J.S. Risley and R. Geballe (Univ. of Washington Press, Seattle, 1975) Vol. 1, pp. 367-8.
101. K.T. Tang and J.R. Grover, *Reconciliation of crossed-beam results on the hydrogen exchange reaction*, in International Conference on the Physics of Electronic and Atomic Collisions. 10th, Abstracts of Papers, edited by Commissariat à l'Energie Atomique, Paris, (North-Holland, Amsterdam, 1977) Vol. 1, p. 26.
102. K.T. Tang, Y.Y. Yung, B.H. Choi and R.T. Poe, *Three-dimensional quantum mechanical studies of $He+H_2^+ \rightarrow HeH^++H$ reactive scattering*, in *Electronic and Atomic Collisions*, XIth International Conference on the Physics of Electronic and Atomic Collisions, 29 August - 4 September 1979, Kyoto International Conference Hall, Kyoto, Japan, Abstracts of Contributed Papers, edited by K. Takayanagi and N. Oda (North-Holland, Amsterda, 1979) pp. 894-5.
103. K.T. Tang, *Approximate treatments of reactive scattering: the T matrix approach*, in *Theory of Chemical Reaction Dynamics*, edited by M. Baer (CRC Press, Boca Raton, 1985) Vol. II.
104. R.B. Walker and R.E. Wyatt, *DWBA study of the collinear $H+H_2$ reaction*, Chem. Phys. Lett., **16**, 52-6 (1972).
105. R.B. Walker and R.E. Wyatt, *Three-dimensional reaction cross sections from planar scattering data*, Mol. Phys., **28**, 101-11 (1974).
106. R.B. Walker and R.E. Wyatt, *Investigation of the planar $H+H_2$ reaction near threshold*, J. Chem. Phys., **61**, 4839-47 (1974).
107. R.B. Walker, E.B. Stechel and J.C. Light, *Accurate H_3 dynamics on an accurate H_3 potential surface*, J. Chem. Phys., **69**, 2922-3 (1978).
108. R.B. Walker and J.C. Light, *Reactive molecular collisions*, Ann. Rev. Phys. Chem., **31**, 401-33 (1980).

109. R. E. Wyatt, *Reactive scattering cross sections II: Approximate quantal treatments*, in *Atom-Molecule Collision Theory. A Guide for the Experimentalist*, edited by R. B. Bernstein (Plenum, New York, 1979) pp. 477–503.

110. Y. Y. Yung, B. H. Choi and K. T. Tang, *Three dimensional quantum mechanical studies of* $D+H_2 \rightarrow HD+H$ *reactive scattering. III. On the ab initio potential energy surface*, J. Chem. Phys., **72**, 621–9 (1980).

111. Ch. Zuhrt, F. Schneider and L. Zülicke, *The distorted-wave Born approximation applied to chemically reactive systems. Endoergic exchange processes* $H_2^+(He,H)HeH^+$, Chem. Phys. Lett., **43**, 571–5 (1976).

THE REPRESENTATION AND USE OF POTENTIAL ENERGY SURFACES IN THE WIDE VICINITY OF A REACTION PATH FOR DYNAMICS CALCULATIONS ON POLYATOMIC REACTIONS

Donald G. Truhlar, Franklin B. Brown, and Rozeanne Steckler
Department of Chemistry
University of Minnesota
Minneapolis, Minnesota 55455

and Alan D. Isaacson
Department of Chemistry
Miami University
Oxford, Ohio 45056

ABSTRACT. We consider three aspects of potential energy surface representations for dynamics calculations on polyatomic systems, with special emphasis on generalized transition state theory and tunneling calculations. (i) We present methods for calculating the vibrational energies of generalized transition states from either a cartesian or internal-coordinate force field and including the effect of mode-mode couplings on the rate constant by perturbation theory and the Pitzer-Gwinn approximation. (ii) We discuss practical aspects in the use of *ab initio* gradient-based electronic structure calculations for the calculation of cartesian force fields for a set of stationary points on the potential energy surface or for a sequence of generalized transition states. (iii) We discuss recent progress on the development of global analytic representations for potential energy surfaces of polyatomic reactions. Such global representations can be used to generate either cartesian or internal-coordinate force fields for generalized transition states, and they can also be used to compute the potential energy surface far from the minimum energy path as may be required for tunneling calculations in some cases.

1. INTRODUCTION

The calculation of reaction rates is generally carried out in two steps. In the first step one calculates or models the potential energy surface,[1,2] PES (or surfaces; however in the present report we limit our attention to electronically adiabatic reactions for which only a single surface is involved). In the second step one calculates dynamical quantities, using the PES as given.[1,3] It is becoming increasingly clear, however, that these two steps should not be performed independently. In the first place, the dynamics calculations are expected to be more sensitive to some features of the PES than to others, and it would be desirable (in the practical case where the PES is not equally accurate

D. C. Clary (ed.), The Theory of Chemical Reaction Dynamics, 285–329.

for all possible geometries) to expend the greatest fraction of the theoretical effort on those features of the PES that are expected to have the greatest effect on the dynamical results of interest.[4] In the second place, when the PES is based on *ab initio* electronic structure calculations it is not practical economically to perform calculations for all possible geometries of the reacting molecule or molecules. The disparity between the number of calculations needed to map a reasonable grid of all relevant geometrical parameters and the number of affordable calculations grows rapidly with the number N of atoms involved. To span each internal coordinate with only 10 points already requires 10^{3N-6} geometries to be considered, which is unaffordable for $N \geq 4$. Of course an accurate PES is not needed at all possible geometries, and thus, when using the *ab initio* approach, we want to calculate--and fit or represent --the potential only where it is really important.

One approach to circumvent the problems mentioned above is to combine a local representation of the PES in the vicinity of the reaction path with dynamics calculations based on localized bottlenecks and localized semiclassical tunneling paths. A large amount of experience has been gained with these methods[5-10] for reactions with only a few atoms. In our group we have made extensive tests of the reliability of such methods by comparing the results to those from accurate quantal dynamics calculations for simple systems and to experiment. On the basis of these tests we can conclude that methods based on localized dynamical bottlenecks and localized semiclassical tunneling paths are capable of accurate predictions of thermal and some state-selected reaction rates, kinetic isotope effects, and threshold energies for overall reaction[9-22] and sometimes for reaction into specific product vibrational states,[19,23,24] as well as predictions of resonance energies and lifetimes and branching ratios for decay[25-29] and spectroscopic tunneling splittings.[30] Reaction path methods have also been applied to treat energy transfer in nonreactive processes.[31-33] Methods based on an expansion in reaction-path coordinates about the minimum-energy path are sometimes called reaction-path Hamiltonian (RPH) methods,[8,27,28,34-36] although the idea is older than the name.[37-39] The emphasis in much RPH research is on a correct formulation of the kinetic energy in reaction-path coordinates.[34,35,40] In the present paper, however, we wish to emphasize the representation of the potentials, especially for systems with four or more atoms. We also wish to emphasize that in many cases the PES must be known in regions beyond those where it can be predicted by a quadratic or other expansion about the minimum-energy path, e.g., even in regions where reaction-path coordinates are not unique.[15,18-22] These wider regions are still localized though and can be identified with reasonable confidence so that we do not need a complete global representation of the PES.

There are two reasons why one needs to go beyond a quadratic expansion about the minimum-energy path. The first is anharmonicity, which may be especially important for low-frequency modes, and which is essential for even a qualitatively correct treatment of bifurcating reaction paths. It is also very important for quantitative calculations of low-temperature rate constants to include anharmonicity of high-frequency modes. The second is tunneling in systems with intermediate and large curvature of the reaction path. In such systems the best semiclassical tunneling paths may be very far from the reaction path. In such regions reaction-path coordinates can become multivalued, making the RPH kinetic energy operator invalid. To treat such regions we transform locally to mass-scaled cartesians, which are valid everywhere. For systems with large reaction-path curvature [for short, we call these large-curvature (LC) systems], the region over which we require the potential may be envisioned as a multidimensional tube surrounding the minimum-energy path and including extra wide regions for possible tunneling paths on the concave sides of elbows. It is convenient to call this extended region around the reaction path the reaction swath, and we arrive at the reaction-swath potential (RSP) as an intermediate construct, or level of required knowledge, between the RPH on one hand and the global PES on the other.

Two quantities which play a primary role in the use of an RPH for dynamics calculations are the generalized free energy of activation curves $\Delta G_T^{GT,0}(s)$ and the vibrationally adiabatic potential curves $V_a^g(n_i,s)$. The former are used for variational transition state theory (VTST) calculations of thermal reaction rates with classical reaction-coordinate motion, and the latter are used for calculating overall and state-specific threshold energies, tunneling probabilities, and the properties of resonance states. Section 2 reviews the basic definitions of these quantities and also reviews the independent-normal-mode (INM) approximation that provides the simplest way to actually calculate $\Delta G_T^{GT,0}(s)$ and $V_a^g(n_i,s)$ for polyatomic systems. The INM approximation may be implemented harmonically or anharmonically but it includes only principal force constants† in normal coordinates. Then we discuss a better method to treat anharmonicity by first modelling the potential energy in curvilinear internal coordinates and then transforming it to a normal coordinate representation. The motivations for this approach are presented, and the practical procedures necessary for calculations are outlined in detail.

Section 2 also includes a brief review of the large-curvature and least-action tunneling approximations with emphasis on delineating the regions of the PES required for such calculations.

Sections 3 and 4 are concerned with the representation of the potential energy information that is needed as input for the calculations of Section 2. In particular, to carry out the calculations of Section 2 we must be able to generate the PES at any point near the reaction path for small-curvature (SC) systems and at any point in the reaction swath, as defined above, for LC systems.

Section 3 discusses *ab initio* calculations of the RPH by so-called "gradient methods", which are algorithms for the direct calculation of

---

†We use the convention, from spectroscopy, that "principal" force constants describe the potential within a single normal or internal-coordinate mode whereas "interaction" force constants describe mode-mode coupling. This convention avoids confusion with "diagonal" and "off-diagonal" matrix elements in a perturbation theory treatment of anharmonicity.

PES derivatives. Gradient techniques are better suited to calculating expansions about the reaction path than to calculating the full RSP, and we will limit this initial discussion to the RPH. In a calculation based on *ab initio* gradient methods, the potential may conveniently be represented in terms of a finite number of force constant matrices, each corresponding to an expansion about a different point on or near the minimum-energy path. This obviates the need for choosing specific functional forms, but it raises a number of new questions about computational economics and step sizes. These will be discussed and simulated gradient calculations based on global PES's will be presented to demonstrate some practical difficulties.

Section 4 discusses methods of representing the PES, or at least the full RSP, in terms of globally defined functional forms. This section begins with a review of methods for fitting atom-diatom PES's and methods developed previously for representing polyatomic PES's. Then we discuss a new approach. The most important elements in the new approach are that the globally defined functional form is required to be accurate only in the reaction swath, and it is flexibilized in this swath by making globally significant potential parameters explicit functions of selected coordinates. Illustrative examples and possible pitfalls are also included.

2. DYNAMICAL CALCULATIONS

2.1. Variational Transition State Theory, Vibrationally Adiabatic Potential Curves, and Tunneling

In canonical variational transition state theory (CVTST or, for short, CVT) the rate constant for a temperature T is calculated in three steps. First one calculates the hybrid generalized transition state theory (GTST or, for short, GT) rate constant $k^{GT}(T,S^{GT})$ as a function of the location S^{GT} of the generalized transition state.[9,41-44] The word "hybrid" here refers to the fact that in this calculation the reaction coordinate is treated classically but all other degrees of freedom are quantized, and the word "generalized" refers to the fact that the generalized transition state is not required to pass through the saddle point as in conventional[45] transition state theory. In the second step one minimizes $k^{GT}(T,S^{GT})$ with respect to S^{GT} yielding the hybrid CVT rate constant[9,41-44]

$$k^{CVT}(T) = \min_{S^{GT}} k^{GT}(T,S^{GT}) \quad . \tag{1}$$

The words "canonical" and "variational" in CVT refer to the fact that in this step the dividing surface is variationally optimized for the canonical ensemble specified by T. In the third step one multiplies $k^{CVT}(T)$ by a transmission coefficient $\kappa(T)$ to account for quantal effects on the reaction coordinate, yielding the final estimated rate constant:[9,42-44]

$$k^{CVT/\kappa}(T) = \kappa(T)\, k^{CVT}(T) \; . \tag{2}$$

In principle the generalized transition state can be any hypersurface in phase space except for the constraint that it must divide reactants from products.[46] However, in practice it would be difficult to calculate $k^{GT}(T,S^{GT})$ for arbitrary dividing surfaces as well as to perform the variational step of eq. (1) for all possible surfaces. To make VTST practical one must define a subset of all possible dividing surfaces for which these steps are realizable and yet which is capable of yielding the required accuracy. It is our contention, based on extensive computational experience, that such a subset is provided by a one-parameter sequence of surfaces perpendicular to a physically chosen reference path that leads from reactants to products. In most cases this reference path is chosen to be the minimum-energy path through mass-scaled or mass-weighted cartesians.[38,47–50†] Here we call this path the MEP; sometimes it is called the intrinsic reaction coordinate (IRC). The distance along the MEP from a reference point (which is usually defined as the highest saddle point if there is one) is called the reaction coordinate s, and the dividing surfaces are parametrized by the value of s at which they intersect the MEP. In the vicinity of their intersection with the MEP, the dividing surfaces are taken to be hyperplanes in the mass-scaled cartesian space that are perpendicular to the MEP; more globally they are bent if necessary to insure that they separate reactants from products.

The hybrid GTST rate constant for the generalized transition state at s is[42–44]

$$k^{GT}(T,s) = (\tilde{k}T/h)K^{\ddagger,0}\exp[-\Delta G_T^{GT,0}(s)/\tilde{k}T] \tag{3}$$

where $\tilde{k}$ is Boltzmann's constant, h is Planck's constant, $K^{\ddagger,0}$ is unity for unimolecular reactions and the reciprocal of the standard-state concentration for bimolecular reactions, and $\Delta G_T^{GT,0}(s)$ is the generalized standard-state free energy of activation. The subscript on $\Delta G_T^{GT,0}(s)$ denotes the temperature. The generalized free energy of activation is expressed as[51]

$$\Delta G_T^{GT,0}(s) = -\tilde{k}T \ln [K_{eq}^{GT}(T,s)/K^{\ddagger,0}] \tag{4}$$

where $K_{eq}^{GT}(T,s)$ is a quasiequilibrium constant for forming generalized

---

†In mass-scaled cartesians the three cartesian coordinates of atom A are scaled by $(m_A/\mu)^{1/2}$ where m_A is the mass of A and μ is an arbitrary convenient mass. In mass-weighted cartesians the coordinates of atom A are weighted by $m_A^{1/2}$. Mass-scaled coordinates have units of length; mass-weighted coordinates have units of mass$^{1/2}$ length, and are usually given as $u^{1/2}$ Å or $u^{1/2}$ a_0, where 1 u = 1 universal (^{12}C) atomic mass unit = 1822.887 m_e, and 1 a_0 = 1 bohr = 0.5291771 Å.

transition states from reactants. For a unimolecular reaction

$$K_{eq}^{GT}(T,s) = \frac{Q^{GT}(T,s)}{Q^{R}(T)} \exp[-V_{MEP}(s)/\tilde{k}T] \tag{5}$$

and for a bimolecular one

$$K_{eq}^{GT}(T,s) = \frac{Q^{GT}(T,s)}{\Phi^{R}(T)} \exp[-V_{MEP}(s)/\tilde{k}T] \tag{6}$$

where $Q^{GT}(T,s)$ and $Q^{R}(T)$ are the partition functions for the generalized transition state and reactants, respectively, $\Phi^{R}(T)$ is $Q^{R}(T)$ per unit volume, and $V_{MEP}(s)$ is the Born-Oppenheimer potential at the point where the generalized transition state intersects the MEP.

In eqs. (5) and (6) the zero of energy for $V_{MEP}(s)$, $Q^{R}(T)$, and $\Phi^{R}(T)$ is at the equilibrium geometry of reactants, and the zero of energy for $Q^{GT}(T,s)$ is $V_{MEP}(s)$. It is also very popular to define partition functions with respect to the local zero point energy. For this purpose we define

$$V_{a}^{G}(s) = V_{MEP}(s) + \varepsilon^{G}(s) \tag{7}$$

and

$$\Delta V_{a}^{G}(s) = V_{a}^{G}(s) - \varepsilon^{RG} \tag{8}$$

where $\varepsilon^{G}(s)$ is the zero point energy at s, and ε^{RG} is the zero point energy of reactants. [$\Delta V_{a}^{G}(s)$ is called the ground-state adiabatic potential curve.] Then eqs. (5) and (6) become

$$K_{eq}^{GT}(T,s) = \frac{\tilde{Q}^{GT}(T,s)}{\tilde{Q}^{R}(T) \text{ or } \tilde{\Phi}^{R}(T)} \exp[-\Delta V_{a}^{G}(s)/\tilde{k}T] \tag{9}$$

where, for example,

$$\tilde{Q}^{GT}(T,s) = Q^{GT}(T,s) \exp[\varepsilon^{G}(s)/\tilde{k}T] \tag{10a}$$

$$= \sum_{\alpha} d_{\alpha} \exp\{-[\varepsilon(\alpha,s) - \varepsilon^{G}(s)]/\tilde{k}T\} \ . \tag{10b}$$

In eq. (10b), $\varepsilon(\alpha,s)$ is the energy of level α of the generalized transition state and d_{α} is the degeneracy of level α.

In calculating $Q^{GT}(T,s)$, we fix the system in a hypersurface orthogonal to the reaction path at a fixed value of s. This is equivalent to an adiabatic approximation with all generalized-transition-state modes treated as adjusting adiabatically to changes in s. The usual approximation is to further assume that $\varepsilon(\alpha,s)$ is a sum of electronic, vibrational, and rotational energies. The vibrational modes may be additionally decoupled from each other by the harmonic adiabatic approximation but are coupled nonadiabatically in the kinetic energy operator and anharmonically by the potential. The kinetic coupling element for modes k and k' is denoted by $\mathcal{B}_{kk'}(s)$ and it arises from the twisting of the normal modes around the MEP and into each other. Each element $\mathcal{B}_{kk'}(s)$ is the scalar product of the generalized normal mode vector for mode k' and the derivative of the generalized normal mode vector for mode k with respect to the reaction coordinate at s. Since the MEP is curved, there are also coupling elements between the generalized normal mode motions and motion along the reaction coordinate. These elements are called $\mathcal{B}_{kF}(s)$, where F indexes the reaction coordinate and there are F-1 generalized normal modes (F = 3N-5 or 3N-6 for N-atom generalized transition states that are linear or nonlinear, respectively). Each term $\mathcal{B}_{kF}(s)$ can be written in terms of the scalar product of the generalized normal mode vector of vibration k and the derivative of the gradient (representing motion along the reaction path) with respect to the reaction coordinate at s.

So far we have outlined the CVT formalism for calculating thermal reaction rates. In our CVT calculations we have neglected the nonadiabatic coupling elements, in which case CVT calculations require only the potential energy along the MEP and the energy levels for a sequence of generalized transition states. Transmission coefficients in the small-curvature semiclassical adiabatic (SCSA) approximation[10,52,53] depend on these same quantities plus the $\mathcal{B}_{kF}(s)$ curvature elements. Several other interesting reaction attributes may also be calculated from $V_{MEP}(s)$, $\varepsilon(\alpha,s)$, and $\mathcal{B}_{kF}(s)$. For example, for interpretative purposes we are often interested in the location of the variationally optimized dividing surface; this is called the canonical variational transition state, and it is located at the maximum of $\Delta G_T^{GT,0}(s)$. The overall translational threshold energy in the absence of tunneling is given in the VTST approximation by the maximum of $\Delta V_a^G(s)$. Threshold energies for reactant or product molecules with a specific vibrational quantum number n for some high-frequency mode are sometimes given by the maxima of

$$V_a^g(n,s) = V_{MEP}(s) + \varepsilon^g(n_i,s) \tag{11}$$

where $\varepsilon^g(n_i,s)$ is the energy of the level of the generalized transition state that has quantum number n_i for the mode, i, that correlates to the specific reactant or product mode and quantum number 0 for all other modes. (The superscript g denotes the system is in the ground state for all modes whose quantum numbers are not explicitly specified whereas G denotes ground state for all modes.) Resonance energies are sometimes given by the energy levels of $V_a^g(s)$, and this quantity also sometimes serves as an effective potential for tunneling. Resonance decay probabilities depend on the $\mathcal{B}_{kF}(s)$ as well.

When the canonical variational transition state is strongly dependent on temperature, a more consistent theory is provided by improved canonical variational transition state theory (ICVTST or, for short,

ICVT).[44] As for the quantities discussed in the last paragraph, the ICVT approximation to the thermal rate constant may be calculated from $V_{MEP}(s)$ and the set of $\varepsilon(\alpha,s)$.

We mentioned in Sect. 1 that for LC systems the semiclassical tunneling paths may pass through regions where the RPH breaks down. We originally proposed two somewhat complicated schemes for calculating the tunneling probabilities in LC systems,[20,21] but later work[22] showed that almost identical results could be obtained with a simpler prescription for the tunneling paths. The approximation incorporating this simpler prescription is called the large-curvature approximation, version 3, or LC3.[54] In this approximation the semiclassical tunneling paths are straight lines through mass-scaled coordinates from an adiabatic translational turning point on the MEP in the entrance channel to a translational turning point on the MEP in the exit channel. For ground-state reactants or products the adiabatic translational turning points are defined as the points where an energy parameter $\tilde{E}$ equals $V_a^G(s)$; for excited states they are computed from $V_a^g(n_i,s)$ or its generalization. The energy parameter takes on all values from ε^{RG} to the total energy. The region of coordinate space between the LC3 path at the lowest total energy for which tunneling must be considered and the region where a quadratic expansion about the MEP is valid is included in the reaction swath; clearly the swath becomes wider when lower-energy tunneling processes must be considered.

For intermediate reaction-path curvature, one may use either the SCSA or LC3 approximation, but even more accurate results are obtained by a least-action (LA) method.[22,54] In the LA method, the tunneling paths are linear interpolations between the MEP and the LC3 paths. Thus this method does not require knowing the potential over a wider swath than is necessary for the LC3 method.

Babamov and Marcus[55] have proposed tunneling models in which the tunneling paths correspond to a fixed hyper-radius, where the hyper-radius is the distance from the origin in mass-scaled hyperspherical coordinates. These require a knowledge of the potential over about the same swath as required for LC3 calculations.

2.2. Independent-Normal-Mode Approximation

It should be clear from Sect. 2.1 that the generalized-transition-state energy levels $\varepsilon(\alpha,s)$ play a central role in VTST and related theories. Usually one writes $\varepsilon(\alpha,s)$ as a sum of electronic, vibrational, and rotational energies, in which case the partition functions become products of electronic, vibrational, and rotational factors. The electronic problem is often well approximated by assuming that reaction occurs with appreciable probability only on a single potential energy surface, and the rotational problem is usually treated accurately enough by simple classical approximations. The vibrational energies, $\varepsilon_{vib}(n_1,\ldots,n_{F-1},s)$, where n_m is a vibrational quantum number, and vibrational partition functions, $Q_{vib}^{GT}(T,s)$, however, are not obtained as straightforwardly, at least if one desires high accuracy.

In many respects the vibrations of generalized transition states are like those of ordinary molecules, and thus the generalized-transition-state vibrational partition function $Q^{GT}_{vib}(T,s)$ may be calculated by many statistical methods developed for ordinary molecules (see Ref. 54 and references therein). One important distinction between a generalized transition state and an ordinary molecule, though, is that the former, being a hypersurface orthogonal to the reaction path, is missing one vibrational degree of freedom, which corresponds to the reaction-coordinate motion. To account for this we must calculate $\varepsilon_{vib}(n_1,\ldots,n_{F-1},s)$ and $Q^{GT}_{vib}(T,s)$ in the (F-1)-dimensional subspace that is orthogonal not only to overall translations and rotations, as for a real molecule, but also to the reaction path. To accomplish the dimensionality reduction we use the projection operator method of Miller et al.[34] In this method the harmonic frequencies and corresponding generalized normal modes are determined by diagonalizing the projected force constant matrix $\underset{\approx}{F}^{P}(s)$. This matrix is related to the force constant matrix $\underset{\approx}{F}(s)$, defined as the matrix of second derivatives of the potential energy with respect to mass-scaled cartesians, by[34,35]

$$\underset{\approx}{F}^{P}(s) = [\underset{\approx}{1} - \underset{\approx}{P}(s)]\underset{\approx}{F}(s)[\underset{\approx}{1} - \underset{\approx}{P}(s)] \tag{12}$$

where $\underset{\approx}{1}$ is the unit matrix and $\underset{\approx}{P}(s)$ is the projector which projects onto the mode directions corresponding to the three overall translations, the two or three rotations, and the motion along the reaction path. Thus, diagonalizing $\underset{\approx}{F}^{P}$ will yield 6 or 7 zero eigenvalues corresponding to the projected motions and 3N-6 or 3N-7 generalized normal mode frequencies which correspond to the vibrations orthogonal to the reaction path.

The simplest approach to treating the bound vibrational motions, in terms of both the computational effort and the amount of information required about the PES, is the harmonic approximation, under which the vibrational energy levels are given by

$$\varepsilon_{vib}(n_1,n_2,\ldots,n_{F-1},s) = \sum_{m=1}^{F-1} (n_m + \tfrac{1}{2})hc\bar{\nu}_m(s) \tag{13}$$

where n_m and $\bar{\nu}_m(s)$ are the vibrational quantum number and frequency in cm^{-1} for mode m, respectively, c is the speed of light in cm per unit time, and the energy is measured from the bottom of the vibrational well. The vibrational partition function,

$$Q^{GT}_{vib}(T,s) = \sum_{n_1,n_2,\ldots,n_{F-1}} \exp[-\beta\varepsilon_{vib}(n_1,n_2,\ldots,n_{F-1},s)] \tag{14}$$

where $\beta \equiv (\tilde{k}T)^{-1}$, is thus separable in the harmonic approximation and equals

$$Q^{GT}_{vib}(T,s) = \prod_{m=1}^{F-1} Q^{GT}_{vib,m}(T,s) \tag{15}$$

where the vibrational partition function for mode m is given by

$$Q^{GT}_{vib,m}(T,s) = \sum_{n_m} \exp[-\beta\varepsilon_{vib,m}(n_m,s)] \tag{16}$$

with $\varepsilon_{vib,m}(n_m,s) = (n_m + \frac{1}{2})hc\bar{\nu}_m(s)$ in the harmonic approximation. The summation in eq. (16) should be terminated with the last term for which $\varepsilon_{vib,m}(n_m,s)$ is less than $D-V_{MEP}(s)$, where D is the lowest dissociation energy of the system.[51,56] However, assuming that the contributions from energy levels above $D-V_{MEP}(s)$ are negligible for the temperature being considered and extending the summation in eq. (16) over all harmonic levels, it can be summed analytically to yield

$$Q^{GT}_{vib,m}(T,s) = \exp[-hc\bar{\nu}_m(s)\beta/2]\{1 - \exp[-hc\bar{\nu}_m(s)\beta]\}^{-1} \ . \tag{17}$$

Since, in general, the vibrational degrees of freedom are anharmonic, substantial errors can be obtained in CVT rate constants computed under the harmonic approximation.[43,57-59] As an example of the effect of including anharmonicity in the calculation of quantal CVT rate constants, we consider the reaction $OH + H_2 \rightarrow H_2O + H$, which has been studied[59] using the analytic PES obtained by Schatz and Elgersma[60] by a fit to the <u>ab initio</u> calculations of Walch and Dunning.[61] For this reaction, the CVT/SCSAG rate constants obtained with the harmonic approximation for the bound vibrational motions were found to overestimate the best anharmonic results [obtained within the independent-normal-mode (INM) framework described below] by factors of 2.27 at 298 K and 1.32 at 2400 K. [Note: The G at the end of SCSAG or other tunneling method abbreviations denotes that $\kappa(T)$ is based on ground-state tunneling probabilities.]

One practical approach to the inclusion of vibrational anharmonicity is to neglect the mode-mode coupling of the normal modes and to employ an approximate anharmonic potential curve to describe the motion along each generalized normal mode of the reacting system independently. This is called the INM method.[54,59] In this approach, the vibrational energy is just the sum of the vibrational energies within each mode,

$$\varepsilon_{vib}(n_1,n_2,\ldots,n_{F-1},s) = \sum_{n=1}^{F-1} \varepsilon_{vib,m}(n_m,s) \tag{18}$$

so that eqs. (15) and (16) are still valid. In order to compute the approximate anharmonic vibrational energy levels of mode m, we must consider the potential energy along this mode, i.e., along the generalized normal coordinate $Q_m(s)$. This coordinate can be expressed as a linear combination of mass-scaled cartesian displacements $\Delta\underset{\sim}{x}$ from the bottom of the vibrational well,

$$Q_m(s) = \Delta\underset{\sim}{x} \cdot \underset{\sim}{L}_m(s) \tag{19}$$

where $\underset{\sim}{L}_m(s)$ is a column of the unitary matrix $\underset{\approx}{L}(s)$ that diagonalizes the projected matrix $\underset{\approx}{F}^P(s)$ of eq. (12):[34,59,62]

$$[\underset{\approx}{L}(s)]^T \underset{\approx}{F}^P(s) \underset{\approx}{L}(s) = \underset{\approx}{\Lambda}(s) \tag{20}$$

where a superscript T denotes a transpose. The nonzero eigenvalues $k_{mm}(s)$ of the diagonal matrix $\underset{\approx}{\Lambda}(s)$ are the principal normal-coordinate quadratic force constants, which are related to the normal-mode frequencies (in cm^{-1}) by

$$\bar{\nu}_m(s) = [k_{mm}(s)/\mu]^{\frac{1}{2}}/2\pi c \tag{21}$$

where μ is defined in Sect. 2.1. The potential energy along mode m can be expressed as

$$V_m[Q_m(s),s] = \tfrac{1}{2}k_{mm}(s)[Q_m(s)]^2 + k_{mmm}(s)[Q_m(s)]^3 + + k_{mmmm}(s)[Q_m(s)]^4 + \ldots \tag{22}$$

where $k_{mmm}(s)$, $k_{mmmm}(s)$, etc. are higher-order principal normal-coordinate force constants, and are related to the third, fourth, etc. directional derivatives of the potential energy along the normal-mode direction $\underset{\sim}{L}_m(s)$. While formally correct, eq. (22) is not directly useful for the present discussion because even if sufficient information about the potential energy in the region of the bottom of the well is available for the calculation of the higher-order force constants in eq. (22), for a general polyatomic system with a relatively large number of vibrational modes there is no practical way to use these force constants to obtain accurately the large number of energy levels required by eq. (16). For this purpose, in modes possessing cubic anharmonicity [i.e., $k_{mmm}(s) \neq 0$] it is useful to replace the general potential of eq. (22) by a Morse function,

$$V_{M,m}[Q_m(s),s] = D_e(s)\{\exp[-\beta_{M,m}(s)Q_m(s)] - 1\}^2 \tag{23}$$

where the dissociation energy $D_e(s) = D - V_{MEP}(s)$ and where the range parameter $\beta_{M,m}(s)$ is chosen so that the Morse potential has the correct quadratic force constant at its minimum:

$$\beta_{M,m}(s) = [k_{mm}(s)/2D_e(s)]^{\frac{1}{2}} . \tag{24}$$

We refer to this method of choosing $D_e(s)$ and $\beta_{M,m}(s)$ as the Morse I approximation.[42,51,59] The energy levels of this potential are given by

$$\varepsilon_{vib,m}(n_m,s) = hc\bar{\nu}_m(s)(n_m + \tfrac{1}{2})[1 - x_{M,m}(s)(n_m + \tfrac{1}{2})] \tag{25}$$

where $\bar{\nu}_m(s)$ is the harmonic frequency of eq. (21) and $x_{M,m}(s)$ is the unitless Morse anharmonicity constant given by

$$x_{M,m}(s) = hc\bar{\nu}_m(s)/4D_e(s) \ . \qquad (26)$$

This approach has been shown to provide satisfactory treatments for the bound stretching motion in collinear atom-diatom collisions[10,42,44] as well as for the four vibrational modes of the $OH + H_2$ system[59] that possess cubic anharmonicity. It should also be pointed out that in many cases the results obtained using the Morse I approximation agree well with those obtained by fitting the Morse function to the true quadratic and cubic force constants of the potential. The Morse I approximation appears to be suitable for general application to vibrational modes possessing cubic anharmonicity, and it has the advantage that it does not require derivatives of the potential higher than second.

Some vibrational modes that, due to symmetry, have no cubic anharmonicity [i.e., $k_{mmm}(s)=0$] cannot be described well by the Morse model. Examples of such modes include bends of linear systems, out-of-plane bends of planar systems, and certain stretching motions (such as the asymmetric stretch in the water molecule). In cases where $k_{mmmm}(s)$ is known, either from differentiating the actual PES or from fitting the potential along the mode to some simple functional form, such modes can be treated by a quadratic-quartic model, which has been shown to provide satisfactory results in atom-diatom systems[43,44,57,58] and for the out-of-plane bending motion in the $OH + H_2$ system.[59] In this approach, the potential of eq. (22) is truncated after the quartic term and the energy levels for the resulting quadratic-quartic potential are approximated accurately by an analytic procedure obtained by a perturbation-variation method discussed elsewhere.[43,57,63]

Although the INM approach allows for the inclusion of anharmonic effects within each individual mode in a practical and relatively accurate manner, it ignores the couplings between the modes, which have been shown to be quite important for obtaining accurate vibrational partition functions in the H_2O and SO_2 molecules[64] and which are probably also important in describing the bound vibrational motions of a reacting system along the reaction path. Mode-mode couplings are considered in the next subsection.

2.3. Mode-Mode Couplings and Vibrational Energy Calculations for an Internal-Coordinate Force Field

The majority of VTST calculations performed to date have been for atom-diatom collisions.[11] For that kind of collision, reasonably accurate calculations of the vibrational energy levels are possible without excessive labor. For example, for a collinear minimum-energy path the vibrations orthogonal to the path consist of one stretch and a twofold degenerate bend. Use of a curvilinear bend coordinate[43,44,57,65] reduces the bend-stretch coupling, and principal anharmonicity can be included accurately in the bend by the harmonic-quartic approximation described above or by the WKB approximation. The stretch can also be treated accurately by the WKB approximation.[15] It is also possible to estimate the effect of bend-rotational coupling,[57] and in particularly interesting cases one could realistically do even better. For polyatomic systems, the effort to systematically improve the quantal or semiclassical calculation of the multidimensional bound vibrational energy levels rapidly becomes impractical as the number of atoms increases, especially in the context of canonical VTST calculations, for which a knowledge of a large number of energy levels is required at each location of the dividing surface along the reaction path. Furthermore, the quantal or semiclassical calculation of vibrational energy levels requires detailed information about the PES, while for many polyatomic systems the available information may consist only of a set of geometries, energies, gradients, and quadratic force constants (frequencies) along the reaction path. Strategies different from those used for atom-diatom collisions are thus clearly required for treating the bound vibrational motions in polyatomic reacting systems. One possible element of commonality, however, would be to use curvilinear internal coordinates to reduce mode-mode coupling.

One possible way to include mode-mode couplings in normal coordinates is by perturbation theory. The perturbation-theory expressions for the energies of a polyatomic system are usually given in terms of dimensionless normal coordinates, $\{q_m(s),\ m=1,2,\ldots,F-1\}$. These are related to the mass-scaled normal-coordinates $\{Q_m(s),\ m=1,2,\ldots,F-1\}$ of eq. (19) by

$$q_m(s) = 2\pi[c\mu\bar{\nu}_m(s)/h]^{\frac{1}{2}}Q_m(s) \ . \tag{27}$$

In these coordinates the vibrational potential energy can be evaluated in cm^{-1} by

$$\begin{aligned}[V(q_1,q_2,\ldots,q_{F-1},s) - V_{MEP}(s)]/hc &= \tfrac{1}{2}\sum_i \bar{\nu}_i(s)[q_i(s)]^2 + \\ &+ \sum_{i\le j\le k} \bar{k}_{ijk}(s)q_i(s)q_j(s)q_k(s) + \\ &+ \sum_{i\le j\le k\le \ell} \bar{k}_{ijk\ell}(s)q_i(s)q_j(s)q_k(s)q_\ell(s)\end{aligned} \tag{28}$$

where $\bar{k}_{ijk}(s)$ and $\bar{k}_{ijk\ell}(s)$ are the cubic and quartic dimensionless normal coordinate force constants (in cm^{-1}), respectively, which are related to the appropriate third and fourth derivatives, respectively, of the potential energy with respect to the dimensionless normal coordinates. We also define $\bar{k}_{211} = \bar{k}_{112}$, etc. Although force constants with $i > j$ do not appear in eq. (28), they *do* appear below in eq. (30). In eq. (28) we have followed the usual practice of truncating the Taylor series expansion of the potential energy at quartic terms. If the cubic and quartic force constants in eq. (28) are known, they can be related to the vibrational energy via perturbation theory. A standard procedure is to treat cubic terms to second order and quartic terms to first order. This yields:[66]

$$\varepsilon_{vib}(n_1,n_2,\ldots,n_{F-1},s)/hc = \sum_i \bar{\nu}_i(s)(n_i+\tfrac{1}{2}) + \sum_{i,j} x_{ij}(s)(n_i+\tfrac{1}{2})(n_j+\tfrac{1}{2}) \tag{29}$$

where, omitting the dependencies on s to simplify the expressions,

$$x_{ii} = \tfrac{1}{4}\left[6\bar{k}_{iiii} - \frac{15\bar{k}^2_{iii}}{\bar{\nu}_i} - \sum_{j\neq i}\left(\frac{\bar{k}^2_{iij}}{\bar{\nu}_j}\right)\frac{8\bar{\nu}^2_i - 3\bar{\nu}^2_j}{4\bar{\nu}^2_i - \bar{\nu}^2_j}\right] \tag{30}$$

and

$$x_{ij} = \tfrac{1}{2}\left\{\bar{k}_{iijj} - \frac{6\bar{k}_{iii}\bar{k}_{ijj}}{\bar{\nu}_i} - \frac{4\bar{k}^2_{iij}\bar{\nu}_i}{4\bar{\nu}^2_i - \bar{\nu}^2_j} - \sum_{k\neq i,j}\frac{\bar{k}_{iik}\bar{k}_{kjj}}{\bar{\nu}_k} - \tfrac{1}{2}\sum_{k\neq i,j}\bar{k}^2_{ijk}\bar{\nu}_k\left[\frac{\bar{\nu}^2_k - \bar{\nu}^2_i - \bar{\nu}^2_j}{(\bar{\nu}_i+\bar{\nu}_j+\bar{\nu}_k)(\bar{\nu}_i+\bar{\nu}_j-\bar{\nu}_k)(\bar{\nu}_i-\bar{\nu}_j+\bar{\nu}_k)(\bar{\nu}_i-\bar{\nu}_j-\bar{\nu}_k)}\right]\right\}. \tag{31}$$

Equations (29)-(31) are for the case of nondegenerate vibrations; the modifications in these equations for degenerate vibrations may be found elsewhere.[66-68] For the discussion below we emphasize that eqs. (29)-(31) are based on a knowledge of the cubic and some of the quartic dimensionless normal coordinate force constants.

As discussed above, the neglect of the normal coordinate interaction force constants often causes a great loss of accuracy. However, for a moderate-sized polyatomic reacting system, the direct calculation of the large number of normal-coordinate interaction force constants at each location of the dividing surface along the reaction path is not only impractical, but also requires more information about the potential energy surface than is usually available. It may be useful in such cases to consider the representation of the potential energy surface in terms of more physically meaningful curvilinear internal coordinates s_a (e.g., bond stretches and bond-angle bends). If we use 3N-6 internal coordinates for an N-atom reactant molecule we may write its potential as:

$$V^R = \sum_{a\leq b} K_{ab}s_a s_b + \sum_{a\leq b\leq c} K_{abc}s_a s_b s_c + \sum_{a\leq b\leq c\leq d} K_{abcd}s_a s_b s_c s_d\,. \tag{32}$$

In the present treatment we will pay special attention to the case where the s_a are "valence coordinates", which consist of bond stretches, bond-angle bends, out-of-plane bends, and bond torsions.[69] (In the more general case one could also include interpair distances for nonbonded atoms.) If the cubic and quartic interaction force constants are neglected in such a representation, far less loss of accuracy occurs than when they are neglected in the normal-coordinate representation. This has been explicitly demonstrated in a recent study of the vibrational partition functions for the H_2O and SO_2 molecules.[64] As an example of the differences between the normal coordinate and curvilinear internal coordinate representations of the potential energy embodied in eqs. (28) and (32), respectively, consider the bending and stretching motions in the CO_2 molecule. For geometries near linear, Pariseau et al.[67] showed that, up to the energy corresponding to about ten bending vibrational quanta, the minimum in the potential energy along a C-O bond stretching coordinate, as the bending angle is varied, describes a nearly circular path with a radius equal to the C-O equilibrium bond distance. Thus, in the internal-coordinate representation of the potential energy, the effects of bending and stretching motions are nearly separable (i.e., the interaction internal-coordinate force constants involving the bending and stretching internal coordinates are quite small), while in normal coordinates, which are linear combinations of mass-scaled or mass-weighted cartesians, a circular bending path can result only by substantial bend-stretch couplings of the uncoupled straight-line motions of the nuclei.

A further advantage of representing the potential energy in the internal coordinates is that if the principal anharmonic internal coordinate force constants K_{aaa} and K_{aaaa} cannot be calculated directly from the available information, they can often be predicted sufficiently accurately by modelling the potential energy along a particular curvilinear internal coordinate direction by a simple functional form.[64] For example, bond stretches can be modelled in terms of the quadratic force constant K_{aa} and the dissociation energy D_e by the Morse I approximation described above, and linear A-B-C bending motions can be modelled in terms of the AC diatomic Morse parameters D_e^{AC}, β_M^{AC}, and r_e^{AC} by the anti-Morse bend approximation:[70]

$$V_{AM}(\Phi) = V_{AM}^{AC}[r_{AC}(\Phi)] - V_{AM}^{AC}[r_{AC}(\Phi=\pi)] , \tag{33}$$

where Φ is the bond angle,

$$V_{AM}^{AC}(r_{AC}) = (\gamma D_e^{AC}/2)\{2 \exp[-\beta_M^{AC}(r_{AC} - r_e^{AC}) + \\ + \exp[-2\beta_M^{AC}(r_{AC} - r_e^{AC})]\} , \tag{34}$$

γ is adjusted to reproduce $K_{\Phi\Phi}$, and $K_{\Phi\Phi\Phi\Phi}$ is obtained by differentiation. Sometimes even quadratic force constants can be estimated;[57,58,70,71] for example, for Cl-H-H generalized transition states, satisfactory results have been obtained with a value of 0.5 for γ.

Neglecting all of the higher-order (cubic and quartic) cross terms in eq. (32) yields the harmonic-general-plus-anharmonic valence force field (HG/AVFF):†

$$V^R = \sum_{a \leq b} K_{ab} s_a s_b + \sum_a (K_{aaa} s_a^3 + K_{aaaa} s_a^4) . \tag{35}$$

By repeated application of the chain rule for derivatives, this potential energy can be transformed through quartic terms to the representation in dimensionless normal coordinates of eq. (28) in the standard way.[66,67,72-74] Since the internal coordinates are curvilinear while the normal coordinates are not, this transformation is necessarily nonlinear.

In general the potential may be written as a function of 3N-6 internal coordinates. In some cases, e.g., CH_4, there are more than 3N-6 valence coordinates.[74] One may always delete sufficient coordinates from the list to obtain an independent set.[75] In some cases, however, either to take advantage of symmetry or obtain or use transferable and physically meaningful force constants, it is convenient to use more than 3N-6 internal coordinates. In such a case one or more coordinates are redundant. Another possibility is that a redundancy condition is satisfied only for a restricted range of geometries, including the reference geometry of the force field; this is sometimes called a constraint.[76,77] If redundant coordinates are retained or there is a constraint, linear force constants need not be zero [i.e., terms of the form $K_a s_a$ may appear in eq. (35)].[75,76] Both linear and higher-order force constants become nonunique when redundant coordinates are used.[62a,75]

To use the HG/AVFF as described above, we must first determine the quadratic force field from the available information about the PES. Then the anharmonic terms can be modelled or calculated directly in internal coordinates. If the internal coordinates are independent there is a unique transformation from the normal-coordinate force field to the internal-coordinate one, and in particular a harmonic general force field may be calculated from the normal-coordinate one; alternatively the harmonic general force field may be calculated uniquely from any global PES by the chain rule. If there are redundancies, then these procedures do not yield unique force constants. In such cases one should model the force field directly in internal coordinates or introduce subsidiary conditions on the force constants.

In order to apply the HG/AVFF model to polyatomic generalized transition states, we must reference the bond stretches and bend coordinates to an arbitrary point on the MEP, making them functions of s. We then obtain

---

†A valence force field includes only valence coordinates and principal force constants; a general force field also includes interactions. Thus the harmonic valence terms are the principal ones in the first sum in eq. (35), the harmonic general field consists of the whole first sum, and the anharmonic valence terms comprise the second sum.

$$V = \sum_a K_a(s)s_a(s) + \sum_{a\leq b} K_{ab}(s)s_a(s)s_b(s) + \\ + \sum_a \{K_{aaa}(s)[s_a(s)]^3 + K_{aaaa}(s)[s_a(s)]^4\} \quad (36)$$

where we have included the linear force constants $K_a(s)$ because the first derivatives do not vanish at a general location on the MEP, even for independent, unconstrained internal coordinates. The extension of eq. (36) to include internal-coordinate anharmonic mode-mode couplings is straightforward and simply consists of adding terms like $K_{abc}(s)s_a(s)s_b(s)s_c(s)$ to eq. (36).

The transformation from the internal coordinate force constants $K_a(s)$, $K_{ab}(s)$, etc. to the dimensionless normal coordinate force constants $\bar{k}_i(s)$, $\bar{k}_{ij}(s)$, etc. can be accomplished through a set of several steps. For this purpose it is convenient to employ a dual notation for the atomic cartesian coordinates. Let $i = A_i\gamma_i$ be an index such that A_i can denote any of the atoms A,B,C,..., and γ_i can be x, y, or z. Then the unscaled atomic cartesian coordinates are denoted X_i such that X_1, X_2, and X_3 denote the x, y, and z coordinates of the first atom, X_4, X_5, and X_6 denote the x, y, and z coordinates of the second atom, etc. We then define difference cartesians X_{ij} for $i\neq j$ and $\gamma_i=\gamma_j$ as

$$X_{ij} = X_{A_i\gamma_i A_j\gamma_j} = X_i - X_j \ . \quad (37)$$

These quantities are not needed for i=j or $\gamma_i\neq\gamma_j$.

For the first step of the transformation, we express the internal coordinates in terms of the difference cartesians. The length of the A-B bond is thus given by

$$r_{AB} = \Big(\sum_{\gamma=x,y,z} X^2_{A\gamma B\gamma} \Big)^{\frac{1}{2}} \quad (38)$$

while the angle A-B-C can be expressed as

$$\Phi_{ABC} = \cos^{-1}\Big[\Big(\sum_{\gamma=x,y,z} X_{B\gamma C\gamma}X_{B\gamma A\gamma}\Big)/r_{BA}r_{BC}\Big] \ . \quad (39)$$

Corresponding expressions for the other two types of internal coordinates (out-of-plane bending and torsional angles) are given elsewhere.[78] The difference cartesian force constants k''_{ij}, $k''_{ijk\ell}$, $k''_{ijk\ell mn}$,..., are the derivatives of V with respect to the difference cartesians X_{ij}; X_{ij} and $X_{k\ell}$; X_{ij}, $X_{k\ell}$, and X_{mn};.... These are related to the internal coordinate force constants by the nonlinear transformations:[79]

$$k''_{ij} = \sum_a b^a_{ij} K_a \quad (40)$$

$$k''_{ijk\ell} = \sum_a b^a_{ijk\ell} K_a + \sum_{a\leq b} b^a_{ij} b^b_{k\ell} K_{ab} \quad (41)$$

$$k''_{ijk\ell mn} = \sum_a b^a_{ijk\ell mn} K_a + \sum_{a \le b} (b^a_{k\ell mn} b^b_{ij} + b^a_{ijmn} b^b_{k\ell} + b^a_{ijk\ell} b^b_{mn}) K_{ab} + \sum_{a \le b \le c} b^a_{ij} b^b_{k\ell} b^c_{mn} K_{abc} \quad (42)$$

and

$$\begin{aligned} k''_{ijk\ell mnop} = {} & \sum_a b^a_{ijk\ell mnop} K_a + \sum_{a \le b} (b^a_{ijk\ell} b^b_{mnop} + b^a_{ijmn} b^b_{k\ell op} + \\ & + b^a_{ijop} b^b_{k\ell mn} + b^a_{k\ell mnop} b^b_{ij} + b^a_{ijmnop} b^b_{k\ell} + b^a_{ijk\ell op} b^b_{mn} + \\ & + b^a_{ijk\ell mn} b^b_{op}) K_{ab} + \sum_{a \le b \le c} (b^a_{ijk\ell} b^b_{mn} b^c_{op} + \\ & + b^a_{ijmn} b^b_{k\ell} b^c_{op} + b^a_{ijop} b^b_{k\ell} b^c_{mn} + b^a_{k\ell mn} b^b_{ij} b^c_{op} + \\ & + b^a_{k\ell op} b^b_{ij} b^c_{mn} + b^a_{mnop} b^b_{ij} b^c_{k\ell}) K_{abc} + \\ & + \sum_{a \le b \le c \le d} b^a_{ij} b^b_{k\ell} b^c_{mn} b^d_{op} K_{abcd} \end{aligned} \quad (43)$$

where we have omitted the argument s on all coefficients, and the new coefficients, b^a_{ij}, $b^a_{ijk\ell}$, etc. are the partial derivatives of internal coordinate a with respect to difference cartesian X_{ij}, difference cartesians X_{ij} and $X_{k\ell}$, etc. These coefficients may be expressed analytically for each type of internal coordinate.[67,74,78] Some examples are given in the Appendix.

In the second step of the transformation, the difference cartesian force constants $k''_{ij}(s)$, $k''_{ijk\ell}(s)$, etc. are converted to mass-scaled cartesian force constants $k'_i(s)$, $k'_{ij}(s)$, etc. From the definition of the difference cartesians given in eq. (37), it is easily seen that this transformation is linear and that the transformation coefficients are given by (with x_i a mass-scaled cartesian):

$$t^i_{op} = \frac{\partial X_{op}}{\partial x_i} = \left(\frac{\mu}{m_{A_o}}\right)^{\frac{1}{2}} \delta_{oi} - \left(\frac{\mu}{m_{A_p}}\right)^{\frac{1}{2}} \delta_{pi} \quad (44)$$

where m_{A_p} is the mass of atom A_p and μ is defined in Sect. 2. This transformation may then be written explicity as

$$k'_i(s) = \sum_{m<n} k''_{mn}(s) t^i_{mn} \quad (45)$$

$$k'_{ij}(s) = \sum_{m<n\le o<p} k''_{mnop}(s) t^i_{mn} t^j_{op} \quad (46)$$

$$k'_{ijk}(s) = \sum_{m<n\leq o<p\leq q<r} k''_{mnopqr}(s)t^{i}_{mn}t^{j}_{op}t^{k}_{qr} \tag{47}$$

and

$$k'_{ijk\ell}(s) = \sum_{m<n\leq o<p\leq q<r\leq s<t} k''_{mnopqrst}(s)t^{i}_{mn}t^{j}_{op}t^{k}_{qr}t^{\ell}_{st} \ . \tag{48}$$

The array $k'_i(s)$ is just grad V in mass-scaled cartesians. In the third step of the force constant transformation, we project out of $k'_{ij}(s)$ the contributions from motion along the reaction path and from overall rotations and translations,[34,59] as discussed above, and diagonalize the projected matrix. The nonzero eigenvalues $k_{mm}(s)$ provide the generalized normal-mode frequencies $\bar{\nu}_m(s)$ via eq. (21), and the associated eigenvectors $L_m(s)$ [the columns of the matrix $\underset{\approx}{L}(s)$ of eq. (20)] yield the generalized normal modes. The dimensionless normal coordinates are then given by

$$q_m(s) = \underset{\sim}{\Delta x} \cdot \underset{\sim m}{\ell}(s) \tag{49}$$

where

$$\underset{\sim m}{\ell}(s) = 2\pi[c\mu\bar{\nu}_m(s)/h]^{\frac{1}{2}}\underset{\sim m}{L}(s) \ . \tag{50}$$

The fourth and final step of the transformation of the force constants from difference cartesians to dimensionless normal coordinates, i.e., the transformation from mass-scaled cartesian coordinates to dimensionless normal coordinates, is thus linear and is given for the cubic and quartic force constants appearing in eqs. (30) and (31) by

$$\bar{k}_{ijk}(s) = \sum_{m\leq n\leq r} k'_{mnr}(s)\ell_{im}(s)\ell_{jn}(s)\ell_{kr}(s) \tag{51}$$

and

$$\bar{k}_{ijk\ell}(s) = \sum_{m\leq n\leq r\leq u} k'_{mnru}(s)\ell_{im}(s)\ell_{jn}(s)\ell_{kn}(s)\ell_{\ell u}(s) \ . \tag{52}$$

Having obtained the dimensionless normal coordinate force constants from those in internal coordinates, eqs. (29)-(31) can be used to obtain the perturbation theory approximation to the vibrational energy levels.

In many cases the energy levels predicted by perturbation theory will be sufficiently accurate and the above scheme will be completely satisfactory. This will be true especially for calculating $V_a^G(s)$ or $\Delta G_T^{GT,0}(s)$ at low T. However for moderately large anharmonicity or higher-energy levels, the accuracy of perturbation theory becomes worse.[64,68,80] An alternative procedure for estimating the vibrational partition function in such cases is the Pitzer-Gwinn method.[64,81,82] This method is based on the fact that the ratio $\tilde{Q}_{vib}(T,s)/\tilde{Q}^H_{vib}(T,s)$ of the anharmonic to harmonic quantal partition functions, with the zero of energy located at the zero-point level (indicated by the tilde), is given correctly by the corresponding ratio $Q_{vib,C}(T,s)/Q^H_{vib,C}(T,s)$ of classical (C) partition functions in both the low- and high-temperature limits. The approximation is to assume that this relationship holds at all temperatures. The quantal anharmonic partition function $\tilde{Q}_{vib}(T,s)$ is thus approximated as

$$\tilde{Q}_{vib}(T,s) \cong \tilde{Q}^{H}_{vib}(T,s)[Q_{vib,C}(T,s)/Q^{H}_{vib,C}(T,s)] . \tag{53}$$

This approach has already been shown to provide accurate results for the vibrational partition functions of the bound molecules H_2O and SO_2,[64] and eq. (53) should be equally applicable for generalized transition states. The harmonic partition functions are given by[81]

$$Q^{H}_{vib,C}(T,s) = \prod_{i=1}^{F-1} \tilde{k}T/[hc\bar{\nu}_i(s)] \tag{54}$$

and

$$\tilde{Q}^{H}_{vib}(T,s) = \prod_{i=1}^{F-1} \{1 - \exp[-hc\bar{\nu}_i(s)\beta]\}^{-1} . \tag{55}$$

The classical anharmonic partition function for the potential of eq. (28) is[83]

$$Q_{vib,C}(T,s) = (2\pi\mu\tilde{k}T/h^2)^{(F-1)/2} \int dQ_1(s)\ldots dQ_{F-1}(s) \times \exp\{-\beta[V - V_{MEP}(s)]\} \tag{56}$$

where the integrations are over the range $(-\infty,+\infty)$ in mass-scaled normal coordinates defined by

$$Q_m(s) = [h/c\mu\bar{\nu}_m(s)]^{\frac{1}{2}} q_m(s)/2\pi . \tag{57}$$

Contributions from energies greater than $D-V_{MEP}(s)$ should be excluded from the integrand. For small systems the integration can be performed conveniently by Gauss-Hermite quadrature formulas, while for larger systems Monte Carlo numerical integration[84-86] may be more efficient. The anharmonic partition function $Q_{vib}(T,s)$, with the zero of energy located at the bottom of the well, can then be obtained by combining $\tilde{Q}_{vib}(T,s)$ with the zero point energy calculated from eq. (29) by:

$$Q_{vib}(T,s) = \tilde{Q}_{vib}(T,s)\exp[-\beta\varepsilon_{vib}(0,0,\ldots,0,s)] . \tag{58}$$

A possible pitfall in the approach discussed above is that the effective potential energy surface of eq. (28) may not provide an accurate representation of the true potential energy surface in a large enough region about the bottom of the vibrational well, i.e., in the region about the bottom of the well where the integrand of eq. (56) is significant. For example, large cubic or large negative quartic force constants can lead to large, anomalous wells in the effective potential energy that cause great difficulty in the convergence of the numerical integration for $Q_{vib,C}(T,s)$. In such cases, one may need to resort to different choices for the models assumed for the potential energy along the internal-coordinate directions, or to calculating explicitly the third and fourth derivatives of the potential energy with respect to the internal coordinates, or to global fitting techniques such as discussed in Sect. 4. In such cases it may be advantageous to use eq. (56) but with V in internal coordinates. Thus, rather than transform the internal-coordinate expression for V to normal coordinates through quartic terms before evaluating eq. (56), one may numerically calculate the internal coordinates and the untransformed V for each point in the quadrature grid as eq. (56) is evaluated. This has the advantage that globally meaningful untruncated potential approximations (like the Morse I approximation[42,51]) may be employed in internal coordinates to ensure a well behaved potential for eq. (56), and the truncated normal-coordinate expression need be used only to estimate the zero point energy.

Since we have emphasized second-order perturbation theory in this section, it might be useful to point out that the effect of the $\mathcal{B}_{kk'}(s)$ nonadiabatic coupling elements could also be included by second-order perturbation theory, using a procedure analogous to that of Barton and Howard.[87]

We have concentrated on vibrations in this section and have not considered hindered rotations, Coriolis coupling, or related complications. These kinds of complications will be at least as important for generalized transition states as for bound molecules, and these complications will have to be addressed in future work.

Another approach to including mode-mode coupling in the hybrid rate constant is to perform a multidimensional, nonseparable quantal Monte Carlo calculation. Voter[88] has recently given a convenient formulation that could be applied to $\Delta G_T^{GT,0}(s)$. It includes quantal effects by a Fourier expansion of Feynmann path integrals[89] and allows for importance sampling as required for Monte Carlo calculations on processes with high activation energy. This formulation would be expected to be particularly convenient if the transmission coefficient is close to unity and hence need not be evaluated. When the transmission coefficient is to be calculated also, one requires $V_a^G(s)$. This can be calculated from the zero-temperature limit of $\Delta G_T^{GT,0}(s)$ or by calculating the zero point energies from a reaction-path Hamiltonian. The use of different methods for $\Delta G_T^{GT,0}(s)$ at $T \neq 0$ and $V_a^G(s)$ may be justified by (i) the sensitivity of low-temperature results to very small energy errors, which are hard to make completely negligible in a Monte Carlo calculation, and (ii) the increasing importance of anharmonicity, and hence nonseparability of the normal modes, as T increases.

3. DETERMINATION OF THE REACTION-PATH HAMILTONIAN FROM AB INITIO CALCULATIONS

One of the most important advances achieved in ab initio electronic structure theory in the last 15 or so years has been the capability of determining analytic gradients of energies computed from many types of wavefunctions.[90-100] The gradient of the energy is the vector of partial derivatives of the energy with respect to each of the cartesian coordinates of the molecule or an equivalent set of internal coordinates.[91] Gradients are extremely valuable in locating and characterizing stationary points on multidimensional PES's.[101-104] There are two advantages of using analytic gradients rather than multiple evaluations of the energy followed by numerical differentiation. First, the analytic gradients are more reliable because there are no artifacts caused by a poor choice of step size. Second, the use of analytic gradients is computationally much more efficient. Using a crude approximation to numerically determine the gradient for a nonlinear polyatomic molecule with N atoms and 3N-6 degrees of freedom requires 3N-5 energy calculations while only one calculation, which is 2 to 5 times longer than one energy calculation, is required for determining analytic gradients.[91] Thus the use of analytic gradients becomes more efficient computationally as the number of atoms N increases. The use of analytic gradients makes the calculation of stationary points (reactant, product, and saddle point geometries) computationally feasible for a large number of reacting systems with 3 or more atoms.

With the availability of analytic gradients, numerical second derivative matrices have been determined and employed in more efficient algorithms for locating stationary points.[101-106] Also, for many types of wavefunctions analytic second derivatives are computationally feasible to determine.[93,97,98,107-111] One gains the same advantages over numerical differentiation using analytic second derivatives as already discussed for using analytic gradients. Furthermore, analytic second derivatives do not contain artifacts from a non-optimal distribution of points used in the numerical determination of the second-derivative matrix. Finally, it should be noted that even though ab initio analytic second derivatives are not currently coded for all types of correlated wavefunctions, the general formulas for determining analytic second derivatives for most types of wavefunctions have been derived[112,113] and we can look forward to their computer implementation. (Even higher derivatives are realizable with the recent calculation of analytic third derivatives for an SCF H_2O calculation.[114]) In summary, state-of-the-art ab initio techniques can provide analytic gradients for most types of wavefunctions and analytic second-derivative matrices for many types of wavefunctions for systems with several ($2 < N < 10$) atoms.

Given these techniques, the determination of a useful RPH for a chemical reaction can proceed at several levels of approximation. The first step in determining the RPH is locating the stationary points that correspond to the reactant, product, and saddle point geometries of the reaction system. The vibrational frequencies and normal mode directions as well as the imaginary frequency corresponding to the reaction coordinate can be determined by diagonalizing the second derivative matrix for the system, and cubic force fields can be obtained by numerical first derivatives of second derivative matrices. In Subsect. 3.1 we discuss determining the RPH from this first level of information; namely, the potential energy and quadratic or cubic force fields at the reactants, products, and saddle point geometries. At each of these geometries the gradient is zero. To calculate the potential along the reaction path (MEP) requires following the path of steepest descent in mass-scaled or mass-weighted coordinates from the saddle point to both the reactant and product geometries. Computationally, this requires taking an initial step off the saddle point in both the product and reactant directions along the direction of the normal mode with the imaginary frequency and then following the path of steepest descent along the direction of the negative gradient. At each point along the reaction path, the frequencies of the vibrations that are orthogonal to the reaction path can be determined by eqs. (20) and (21). Thus, using state-of-the-art computer hardware and ab initio methods, the determination of the reaction path and RPH is feasible for many systems of chemical interest. In Subsect. 3.2 we will discuss some significant practical problems to consider when using ab initio methods to determine an RPH.

3.1. Simple Interpolatory Methods for Reaction Path Calculations

Although RPH's based upon ab initio calculations of the reaction path and the vibrational frequencies perpendicular to the reaction path have been determined for several polyatomic systems (see, e.g., Refs. 7, 115-119), there are many polyatomic reactions for which a set of high quality ab initio calculations along the reaction path is currently not computationally feasible. However, for many such systems it is feasible to optimize the geometries and determine the vibrational frequencies at the set of stationary points along the reaction path, including any saddle point geometries. In such cases one possible way of constructing an RPH is by interpolating the geometry, the potential energy, and the vibrational frequencies along the reaction path by using information either pertaining solely to the stationary points or based on a small number of points including the stationary points.[120] This approach has been applied successfully to collinear A + BC reactions.[120] For the high-barrier reactions $H + H_2 \rightarrow H_2 + H$ and $Cl + HD \rightarrow HCl + D$, the potential energy and the real vibrational frequency in the vicinity of the saddle point were fit to quadratic functions in s. For the reactions $F + H_2 \rightarrow HF + H$ and $I + H_2 \rightarrow HI + H$, which have small intrinsic barrier heights, an asymmetric Eckart barrier[121,122] and a gaussian form were used to model the barrier while an exponential form similar to that used by Quack and Troe[123,124] for triatomic dissociation reactions was used to represent the vibrational frequency. Reaction probabilities based upon classical microcanonical variational transition state theory[51] were determined and compared for both the interpolated RPH and the exact RPH of the global potential energy surfaces[125-128] for each of these reactions. The results from this investigation indicate that the interpolation method works reasonably well for estimating the reaction probability, even though it fails to predict the position of the generalized transition state dividing surface with high accuracy. In particular, for the systems studied, the error in the reaction probabilities is at most 18% while the predicted deviation of the variational transition state from the saddle point differs as much as 260% from the predicted position using the noninterpolated RPH. Nonetheless, it is very encouraging that reasonable reaction probabilities could be obtained using this interpolation scheme based upon a minimal amount of information concerning the reaction path. Notice that because three points were used in the vicinity of the saddle point, the method is equivalent to using numerical differentiation of the saddle point quadratic force field. If anharmonicity is neglected, one can perform conventional transition state theory calculations based on quadratic force fields for reactants and products, but with a generalization of the method of Ref. 120, one can perform approximate VTST calculations if one simply adds the cubic or cubic and quartic force constants at the saddle point.

An even simpler method, but one which will often be much less reliable, is to interpolate based only on quadratic force fields at the stationary points. Recently, Carrington *et al.*[129] have approximated the RPH for the isomerization of vinylidene to acetylene using an interpolation method based solely on a set of *ab initio* energies and force constants at four stationary points along the reaction path (vinylidene, acetylene, and two symmetrically equivalent transition states). The potential energy along the reaction path was interpolated by an even sixth-order polynomial, and a quadratic form was used for interpolating the vibrational frequencies of the vibrational modes perpendicular to the reaction path. These authors also included the $\mathcal{B}_{kF}(s)$ curvature components and the $\mathcal{B}_{kk'}(s)$ nonadiabatic coupling coefficients in their RPH, and they fit these to a quadratic form.

Other workers[130] have employed gaussian-type functions for fitting the curvature as a function of the reaction coordinate. In many A + BC reactions, curvature is not a simple function of s and the reaction probabilities are very sensitive to it. Attempts to approximate the curvature in such cases led to large quantitative errors;[131] however, Carrington *et al.*[129] used their RPH to calculate the lifetime of vinylidene and these lifetime calculations were not as sensitive to reaction-path curvature as are most of the tunneling probabilities that have been studied in our group. Thus Ref. 129 provides an instructive example of obtaining an interpolated RPH from a minimal amount of *ab initio* data.

In this subsection we have presented two examples of calculations where the RPH required for dynamical calculations has been based upon an interpolation of the potential energy, vibrational frequencies, and curvature and nonadiabatic coupling coefficients from a knowledge of PES properties at a small number of points along the reaction path. We pointed out, however, that if attention is restricted only to stationary points, one will generally obtain an accurate approximation to the dependence of the vibrational frequencies on the reaction coordinate in the vicinity of the saddle point only if the input data includes at least some of the third and fourth, as well as the second, derivatives at the saddle point.[120] Simple interpolatory schemes may be the only feasible method for investigating systems with many atoms and degrees of freedom. Nonetheless, further work comparing and refining these schemes is needed before the results can be considered reliable.

3.2. Steepest-Descent-Path Calculations for Constructing RPH's

In this subsection we discuss several practical considerations that arise when constructing RPH's for polyatomic systems from a set of *ab initio* calculations of the energy, gradient, and force constant matrix at a series of points along a reaction coordinate. To illustrate some of these considerations, we will use the $CH_3 + H_2 \rightarrow CH_4 + H$ reaction and the inversion "reaction" of NH_3 as examples. For the $CH_3 + H_2$ reaction, two different kinds of empirical global PES's have been proposed in the literature[132-134] (and are discussed in Sect. 4), and several sets of *ab initio* calculations have also been carried out.[117,135-147] For the NH_3 inversion, Wolfsberg and coworkers[148] have proposed an empirical PES based in part on an accurate anharmonic force field. We have used Raff's global PES for the $CH_3 + H_2$ reaction and Wolfsberg's for the ammonia inversion to construct RPH's. We used the former RPH for VTST and semiclassical tunneling calculations of the abstraction rate constant, and we used the latter to calculate the splitting of the two lowest-energy vibrational levels caused by tunneling through the low (5.2 kcal/mol) inversion barrier.

First, we consider the problem of following the gradient to determine the MEP and the functions $V_{MEP}(s)$ and $V_a^G(s)$. As mentioned above, this requires following the path of steepest descent by taking steps in the direction of the negative gradient. However, when one uses a finite step size, one "zigzags" back and forth across the true MEP. Thus a compromise must be reached between using a small enough step size that the true MEP is followed very closely, and using a large enough step size that the number of *ab initio* calculations is affordable.

For *ab initio* calculations of RPH's in the literature, it has not always been stated what step size was used, but practical considerations have apparently dictated the use of fairly large steps. We will now discuss a few examples where step sizes are given in the literature, in each case expressing the step size in mass-weighted cartesians rather than mass-scaled cartesians because the mass-weighted choice makes it easier to compare different systems. Gray *et al.*[116] used a step size of 0.19 $u^{\frac{1}{2}}a_0$ for studying the HNC → HCN isomerization, and they did find oscillations of the computed MEP; however, they were able to smooth these oscillations by "hand" since only a small portion of the reaction path was required for determining tunneling probabilities in their application. Similarly, Schmidt *et al.*[149] stated that they typically used steps of 0.15 $u^{\frac{1}{2}}a_0$ in their study of the rotational barrier in silaethylene. In this study, these authors used the method of Ishida *et al.*[150] for stabilizing the oscillations in the calculated MEP. In particular, they performed an energy minimization along the bisector of the angle formed by the normalized negative gradients from two adjacent calculated points to return to the true MEP. In another example, for the isomerization $CH_3O \rightarrow CH_2OH$, Colwell and Handy[118,151] used a step size of 0.05 $u^{\frac{1}{2}}a_0$ near the saddle point and followed the gradient with a step of 0.1 $u^{\frac{1}{2}}a_0$ when $V_{MEP}(s)$ was about 40% below the value of the $V_{MEP}(s)$ at s=0.

Experience in our group in studying the dynamics of many A + BC type reactions[10,13,15,16,18-28,42,44,51-53,70,122,131] has indicated that small step sizes are necessary, especially when the oscillations are not stabilized. For production runs on A + BC reactions we have used step sizes in the range 4×10^{-5} to 2×10^{-3} $u^{\frac{1}{2}}a_0$. To investigate this point further for the present discussion, we have graphically examined the convergence properties of the calculated MEP and $V_{MEP}(s)$ and $V_a^G(s)$ curves with respect to the step size for the reactions $H + H_2$, $OH + H_2$, and $CH_3 + H_2$. In these studies we used the global PES's of Truhlar and Horowitz,[152] Schatz and Elgersma,[60] and Raff,[132] respectively; the gradients were analytically evaluated; and the force constants were determined by converged numerical differentiation. The harmonic oscillator approximation was used in treating the vibrations that are perpendicular to the reaction path. The MEP's were computed using step sizes ranging from 7×10^{-2} to 1.3×10^{-4} $u^{\frac{1}{2}}a_0$. Figures 1 and 2 illustrate the resulting computed $V_{MEP}(s)$ and $V_a^G(s)$ curves, respectively, for the $CH_3 + H_2$ reaction. These figures show that both $V_{MEP}(s)$ and $V_a^G(s)$ are quite sensitive to the step size, and they are only well converged when the step size is 1×10^{-3} $u^{\frac{1}{2}}a_0$ or less. Severe oscillations occur in the $V_a^G(s)$ curve computed using step sizes greater than 1×10^{-3} $u^{\frac{1}{2}}a_0$, and these would have a large effect on rate constant calculations. Figure 1 also illustrates the cautionary fact that, even though the $V_{MEP}(s)$ curve resulting from a calculation with a large step size may be smooth, it is not necessarily converged. Similar results were obtained for the $H + H_2$ and $OH + H_2$ reactions, for which the MEP's seem reasonably well converged for step sizes of 1×10^{-2} to 1×10^{-3} $u^{\frac{1}{2}}a_0$, respectively. In previous work on the $OH + H_2$ system where more stringent convergence criteria were employed, a step size of 1.3×10^{-4} $u^{\frac{1}{2}}a_0$ was used.[59] Furthermore, it was demonstrated for this system that, when using very small step sizes, the stabilization method of Ishida _et al._[150] actually slowed convergence of the calculated MEP.[59] These results are all consistent in indicating that relatively small step sizes are required for obtaining converged MEP and $V_{MEP}(s)$ and $V_a^G(s)$ curves for reactive systems.

A second point to consider in constructing $V_a^G(s)$ is that if numerical differentiation is used to calculate the force constant matrix, then the results may be sensitive to the distribution of points and the step size used in the difference formulas. We have found that frequencies calculated using the GAMESS[153] codes can vary significantly based upon using 2- or 3-point numerical differentiation formulas and a step size ranging from 0.01 to 0.0001 a_0. Of course, for SCF calculations this problem is eliminated with the use of analytic second derivatives as used by Colwell and Handy.[118,151]

Another kind of difficulty emerges in calculating the MEP and $V_{MEP}(s)$ and $V_a^G(s)$ curves for an isomerization reaction. For such reactions, $V_{MEP}(s)$ and $V_a^G(s)$ have double-well character. We will use the inversion of NH_3 as an example. In this case, because of symmetry, we need to compute the RPH only for one side of the saddle point. We will use the left side for our example. To determine the MEP in the barrier region between the well and the saddle point, one follows the negative gradient from the D_{3h} saddle point structure towards the C_{3v} equilibrium structure. In this case since one is proceeding to the left, s is decremented at each step, starting at zero. To determine the MEP on the other side of the minimum, one starts from points high on the repulsive wall beyond the well and follows the negative gradient in toward the well. While carrying out this step we don't know an absolute origin for s that is consistent with the scale to the right of the minimum so we use a temporary origin at the initial point. The reaction coordinate referred to this temporary origin is called $\hat{s}$. Since we are proceeding to the right, $\hat{s}$ is incremented at every step. One places the initial points higher and higher on the potential until $V_{MEP}(\hat{s})$ and $V_a^G(\hat{s})$ in the

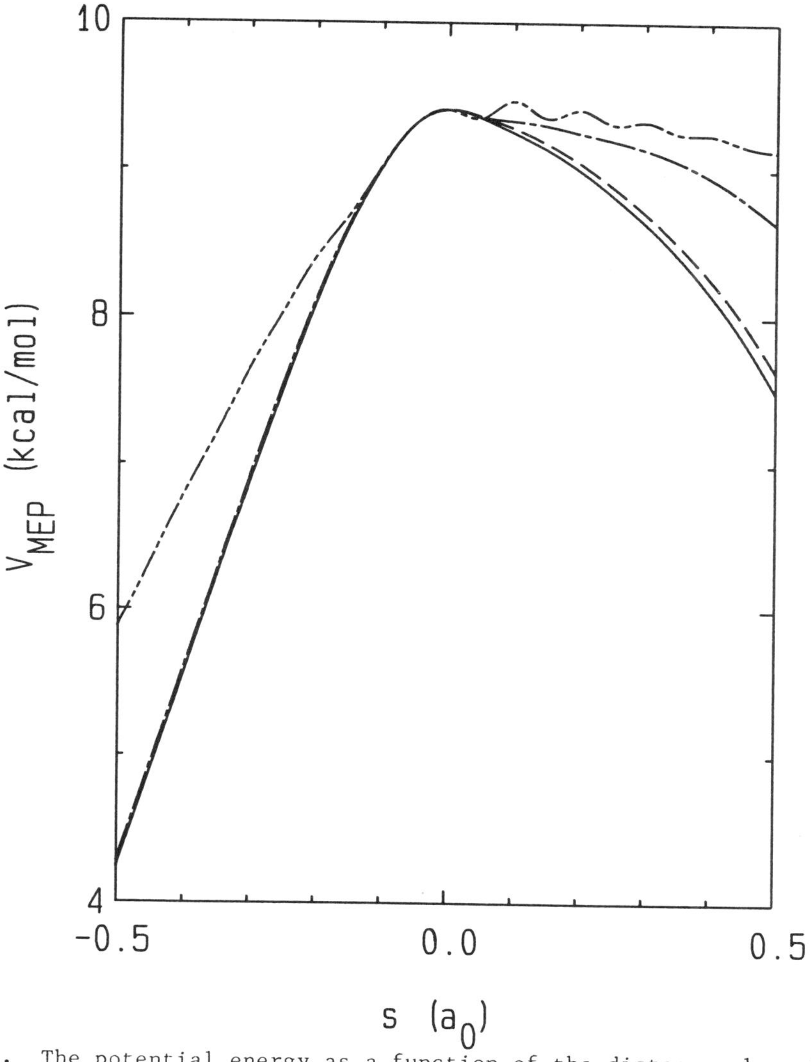

Fig. 1. The potential energy as a function of the distance along the calculated MEP (through mass-scaled coordinates with $\mu = m_{CH_3}m_{H_2}/m_{CH_5}$) for the CH_3 + HH' → CH_4 + H' reaction on the Raff potential energy surface. The zero of V_{MEP} is taken to be at the CH_3 + HH' asymptote. The different curves correspond to using different step sizes in following the path of steepest descent to determine the MEP. The step sizes of the curves are: 0.050 a_0 (—— - -), 0.025 a_0 (—— -), 0.010 a_0 (— — —), and 0.001 a_0 (——). Results for 0.0001 a_0 would be superimposable on those for 0.001 a_0 to within plotting accuracy. To convert these step sizes to mass-weighted coordinate space, multiply by $\mu^{\frac{1}{2}}$.

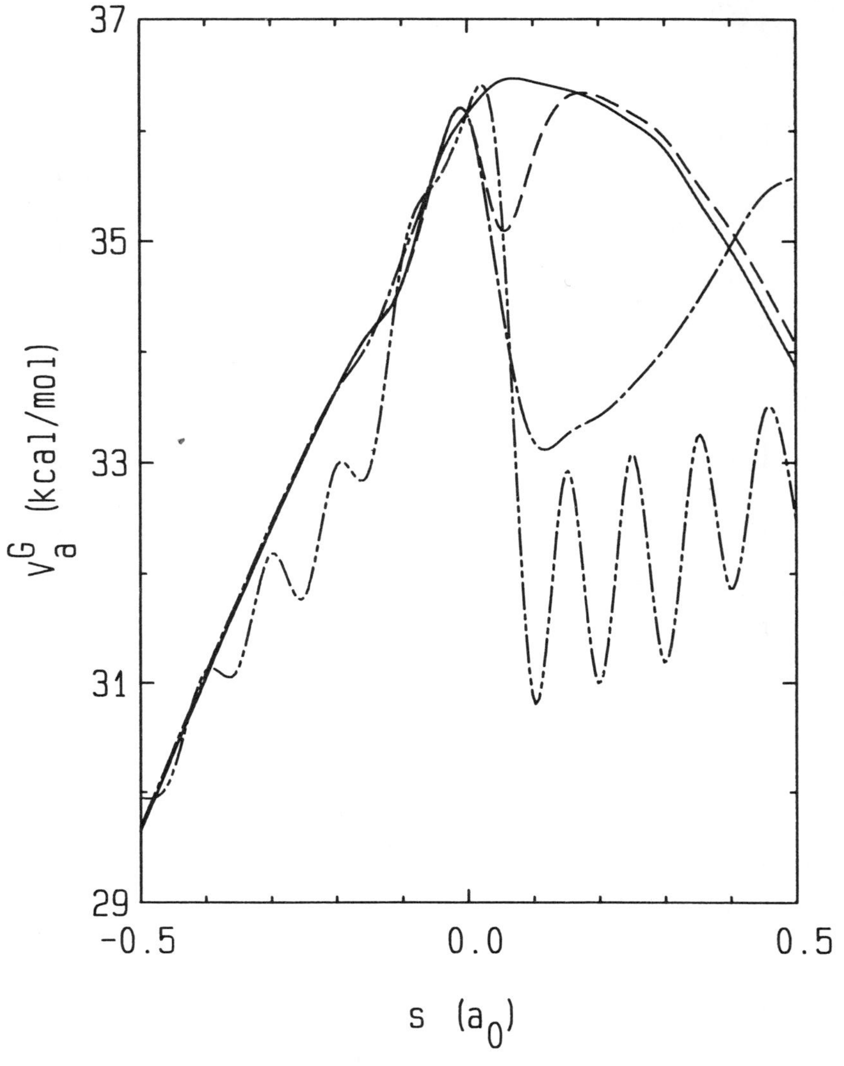

Fig. 2. The ground-state adiabatic potential energy as a function of the distance along the calculated MEP for the CH_3 + HH' → CH_4 + H' reaction on the Raff potential energy surface. The zero of energy is CH_3 + HH' at classical equilibrium; thus the curve tends to the zero point energy of the reactants at $s = -\infty$. The different curves correspond to using different step sizes in following the gradient to determine the MEP. The key for the curves is the same as in Fig. 1, and again the results for 0.001 a_0 and 0.0001 a_0 are superimposable within plotting accuracy.

low-energy region of dynamical interest have converged with respect to the location of the starting point.

To perform the tunneling calculations we require a smooth $V_a^G(s)$ curve for this double-well system. From following the negative gradient starting at the saddle point we determined $V_{MEP}(s)$ and $V_a^G(s)$ for the region to the right of the equilibrium NH_3 structure. Similarly, by following the negative gradient starting from high on the potential and proceeding to the well, we determined $V_{MEP}(\hat{s})$ and $V_a^G(\hat{s})$ to the left of the equilibrium structure. However, a special problem arises in following the negative gradient as one approaches the minimum from either direction, since the gradient approaches zero, and the reaction path calculated with practical-sized steps may show significant zigzagging. This causes the distance along such a path to be artificially longer than the distance along the true MEP and also introduces other errors. Thus we must handle three problems: (i) correct for the elongation of $V_a^G(\hat{s})$ and $V_a^G(s)$ near the minimum, (ii) convert $\hat{s}$ values to s values, and (iii) smoothly join $V_a^G(\hat{s})$ and $V_a^G(s)$ at the minimum to form a continuous $V_a^G(s)$ curve.

To correct for the elongation of the MEP and the resultant errors in $V_a^G(\hat{s})$ and $V_a^G(s)$ we assume that the elongation only occurs when $V_{MEP}(\hat{s})$ or $V_{MEP}(s)$ is less than 0.1 kcal/mol above $V_{MEP}(s)$ at the equilibrium structure, which is taken to be the zero of energy. Therefore we fit $V_{MEP}(\hat{s})$ and $V_{MEP}(s)$ from regions where they lie between 0.1 and 0.2 kcal/mol to the following forms:

$$V_{MEP}(\hat{s}) = \hat{a}(\hat{s} - \hat{s}_0)^2 + \hat{b}(\hat{s} - \hat{s}_0)^3 \tag{59a}$$

and

$$V_{MEP}(s) = a(s - s_0)^2 + b(s - s_0)^3 \ . \tag{59b}$$

In these equations a, $\hat{a}$, b, $\hat{b}$, s_0, and $\hat{s}_0$ are the fitting parameters, of which only the last two will be used. The value obtained for s_0 represents the location along the reaction coordinate of the equilibrium structure of NH_3. Then $\hat{s}$ is converted to s by the following change of origin:

$$s = \hat{s} - \hat{s}_0 + s_0 \ . \tag{60}$$

At this point we have two segments of $V_a^G(s)$ available, one for $s < s_0$ and one for $s > s_0$. We also know $V_a^G(s_0)$ accurately from a standard vibrational analysis at the equilibrium geometry; this was performed prior to any MEP calculations. By connecting the $V_a^G(s)$ functions smoothly together at s_0, consistent with $V_a^G(s_0)$, we will have corrected for the effects of numerical zigzagging. This is accomplished by simultaneously fitting the data from both sides of s_0, using only data where the values of these curves are between 0.1 and 0.2 kcal/mol above $V_a^G(s_0)$. These data are then fit to the following functional form:

$$V_a^G(s) = V_a^G(s_0) + A(s-s_0) + B(s-s_0)^2 + C(s-s_0)^3 \quad . \qquad (61)$$

The final $V_a^G(s)$ curve is a spline fit to a set of values on a grid. When $V_a^G(s)$ at the grid point is greater than $V_a^G(s_0) + 0.2$ kcal/mol, the directly computed value is used as input to the spline routine; but when $V_a^G(s)$ is less than this the input to the spline routine is calculated from eq. (61). The final vibrational energy level splitting was the same when this procedure was repeated with the cubic term missing in eq. (61) so we assume that the order of the polynomial in eq. (61) is sufficiently high to represent the $V_a^G(s)$ curve within 0.2 kcal/mol of its minimum.

With the continuing development of computer hardware and *ab initio* methods and codes, the construction of RPH's for polyatomic systems may become routine. However, at present, care must be exercised in using *ab initio* results to achieve a practical balance between affordability and distortions in the results because of errors in following the gradient with an unconverged step size. One avenue that might provide fertile ground for exploration would be to use interpolation methods like those discussed in Subsect. 3.1 in conjunction with input data at increasing numbers of *ab initio* points along the reaction coordinate. This might provide a reliable and cost effective method for constructing RPH's based upon *ab initio* data.

4. GLOBAL POTENTIALS

Although, as discussed in Sect. 2, there are still significant difficulties in the practical treatment of anharmonicity and mode coupling, VTST and semiclassical tunneling calculations have reached a high enough state of development that in practical applications the accuracy of a calculated thermal rate constant will usually be more limited by the uncertainties in the potential energy surface than by the errors introduced by the approximate treatment of the dynamics. The *ab initio* steepest-descent-path techniques discussed in Sect. 3 provide one promising avenue for supplying the required PES data. In many cases though one will require a more global PES, either because large-curvature tunneling paths must be considered or because semiempirical adjustments are to be considered or both. In attempting either to construct semiempirical surfaces or to fit *ab initio* calculations, we require flexible analytic procedures. The difficulty of reliably representing PES data in an analytic form when it is available or obtainable has been just as serious of a stumbling block in recent years as has been the uncertainty in, and the difficulty of generating, the original data. In this section we will address some of the issues involved in the analytic representation of PES's or RSP's for polyatomic systems.

A promising starting point in the design of polyatomic PES's is to extend in some manner the methods that have been used with success for atom-diatom PES's. This should be an especially promising approach when one only requires an accurate potential in the reaction swath for an atom-transfer reaction where at most two bond lengths differ significantly from their equilibrium values. One example of this approach was developed by Raff[132] and applied to CH_3 + HT ↔ CH_4 + T. This method, which is a polyatomic generalization of the multiparameter LEPS scheme (MLEPS)[154,155] that has been applied so successfully to A + BC reactions, seems to us to have a number of practical advantages. One advantage of this procedure is that the MLEPS scheme can be made very flexible by making the Sato parameters explicit functions of local variables such as internal angles or bond lengths.[156,157] For A + BC this allows one to make localized changes in the PES, and the hope is that one can embed this same flexibility in a polyatomic surface by starting with the MLEPS function for the dependence of the potential on the lengths of bonds that are made and broken. A worthwhile goal for this kind of treatment is to obtain a functional form for which one can refine one area of the surface in order to agree with ab initio calculations or experimental data without changing the rest of the surface.

In Raff's PES for CH_3 + HT, the potential is a sum of four three-center MLEPS potentials, one for each of the C-H-T moieties, and an angle-dependent term to control the change of the methyl moiety from trigonal planar to tetrahedral. By breaking up the potential in this manner, Raff reduced the problem of modelling a six-body interaction to that of modelling several three-body ones.

Table I compares the saddle point characteristics of the Raff surface to those calculated by the ab initio polarization-configuration-interaction (PolCI) method.[144,145] The Raff surface has a lower and earlier, but also thinner barrier. Since, all other things being equal, one expects higher barriers to have higher imaginary frequencies, it

TABLE I

Saddle point characteristics

Quantity	PolCI[a]	Raff[b]
R_{C-H_1,H_2,H_3} (a_0)	2.04	2.07
R_{C-H_4} (a_0)	2.78	3.02
$R_{H_4-H_5}$ (a_0)	1.74	1.48
$V^{\ddagger}$ (kcal/mol)[c]	10.7	9.4
$\bar{\nu}^{\ddagger}$ (cm^{-1})	974i	1478i

[a]Ref. 144 and 145

[b]Ref. 132

[c]relative to $CH_3 + H_2$.

poses an interesting challenge to try to adjust the Raff surface to agree better with the *ab initio* one for all these characteristics.

The basic design of the Raff surface suggests that the methods for localized adjustments mentioned above for atom-diatom MLEPS potentials could be used to correct the discrepancies between the Raff surface and the PolCI calculations. In these methods single parameters in the MLEPS function are replaced by functional forms. The functional forms used should be chosen such that except for a localized region of strong interaction, they go smoothly to the values required to give the surface its correct asymptotic limits and general global form. The "turning off and on" of these localized functional forms is best accomplished through the use of switching functions. When choosing an appropriate switching function it is important to maintain the analyticity of the PES at least through second derivatives and preferably through fourth derivatives. Flexibility can be built in by using adjustable parameters in exponential, hyperbolic, and gaussian functions.

A problem that sometimes occurs in reaction-path Hamiltonians, especially for bend potentials,[118,151,158] is the bifurcation of the reaction path. This occurs when a harmonic frequency becomes imaginary, and for the Raff surface this occurs for bends on both sides of the saddle point. *Ab initio* calculations can be helpful in determining if the bifurcation is an artifact of the form of the analytic potential function or if it is present in the actual system. When the MEP bifurcates it is probably best to base the RPH on a reference path centered on the ridge between two equivalent MEP's.[20,158] This requires extra effort when computing vibrational energy levels since the vibrational potential becomes a double-minimum one, but it probably reduces mode-mode coupling, which (see Sect. 2) is hard to treat accurately.

In making adjustments to the Raff surface we found that while we are indeed able to make localized changes, the changes caused by varying individual Sato parameters are not nearly as independent of each other as was the case[157] with the atom-diatom reaction F + H_2. To raise the saddle point to approximately the height of the PolCI one, all three Sato parameters need to be adjusted simultaneously in order to prevent other local maxima and minima from occurring.

An important consideration in the design of an analytic polyatomic PES is to know which region of the PES is most important for the dynamics. In VTST, an accurate PES is necessary only in the reaction swath. Thus, for the hybrid rate constants it is necessary to have an accurate representation of the potential along the MEP and for small deviations from it in the vicinity of dynamical bottlenecks, and for semiclassical tunneling calculations it is sometimes necessary to know the PES for larger deviations from the MEP along a lengthy segment of it. For VTST/κ calculations it is especially important that the frequencies as functions of s go smoothly and realistically from their reactant values to realistic saddle point values and then to their product values. This is an area where *ab initio* calculations can be very useful.

In Sect. 2 we discussed the practical advantages of valence coordinates for modelling anharmonicity. As discussed there, when the valence coordinates are non-redundant the valence-coordinate force field can be calculated directly from a global PES in any internal coordinate system, e.g., interpair distances, as have been widely employed. Hase and coworkers[159-166] have made great progress, however, in modelling PES's directly in terms of valence coordinates. So far they have concentrated on association, dissociation, and isomerization reactions, but their methods could also be used, for example, in conjunction with MLEPS functions, for some of the interactions in atom-transfer reactions. The basic idea in all of Hase's valence-coordinate surfaces is to write the potential as a sum of Morse functions for all bonds, and harmonic, harmonic-quartic, or Taylor series potentials for all bends and torsions. The parameters in these terms are optimized using least-squares techniques with ab initio or spectroscopic input data.

Cobos and Troe[167] have demonstrated a strong sensitivity of the calculated rate constant for the dissociation of methane to a range parameter that they used to control the decay of the force constants involving the breaking bond to their asymptotic values. The same sensitivity to force constants can occur in atom-transfer reactions. Just as in the CH_4 dissociation, when an atom B is transferred between A and C, one must model the rate of decrease of an A-B-C bend potential as the AB or BC bond is broken. One of the first attempts to do this was by Johnston and Goldfinger.[168] They modelled the bending force constant by attenuating the equilibrium value according to the bond order of AB and BC. Sims and coworkers[169,170] have further tested the validity of this kind of model of the bending force constants for various hydrocarbons. They checked the effect of using both the square root of the product of the bond orders or the product of the bond orders when computing the force constants and then compared to experimental values. It was found that the product of the bond orders is the preferred choice if either of the bonds is undergoing a major change as the case would be during a chemical reaction. This suggests that some method of smoothly varying the bond order from 0 to 1 during the course of a reaction would be useful for modelling a bend potential on a polyatomic PES for VTST/κ calculations. Quack and Troe[123] used this same kind of idea in their statistical adiabatic channel model calculations of triatomic dissociation, for which they modelled the bond order as an exponential function of the deviation of the bond length from equilibrium. The exponential function they used contains an adjustable range parameter. More recently, however, Duchovic et al.,[166] in the course of designing a surface for the dissociation of methane, have performed ab initio calculations that indicate that the bending force constant may decay to zero more like a gaussian than an exponential.

Of course once one decides how to model the variation of a bending force constant with bond distance, this variation is easily incorporated into the PES if it is expressed in valence coordinates. As a final example of the flexibility of MLEPS functions, however, we point out that this can also be done by varying the Sato parameters in Raff's potential for CH_3 + HT. In particular we were able to adjust the C-H-T bend potential for a 40° deviation from collinearity to agree quite well with ab initio bend potentials for a 1.6 a_0 segment of the MEP by making one of the C-T triplet parameters a function of both the C-H-T bend angle and the H-T bond length. Although this procedure is mathematically quite different from the anti-Morse bend potential discussed in

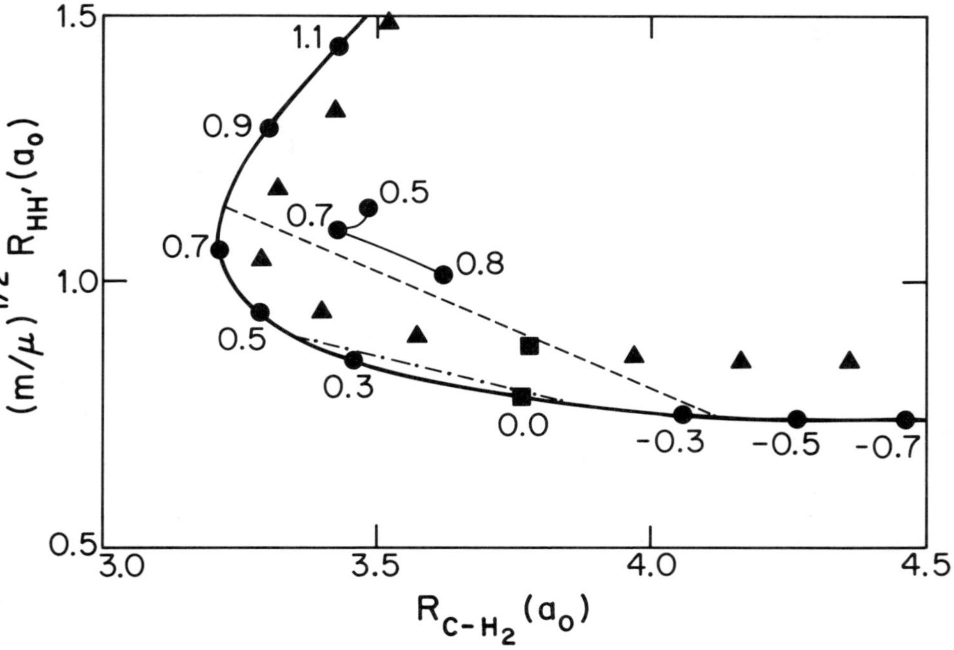

Fig. 3. Minimum-energy path and related quantities for CH_3 + HH' → CH_4 + H'. All quantities for this figure are computed in the harmonic approximation for the vibrations. The abscissa is the distance from C to the center of mass of HH' and the ordinate is the scaled HH' distance, where the scale factor is $(m/\mu)^{\frac{1}{2}}$, $m = m_H/2$, and $\mu = m_{CH_3} m_{H_2} / m_{CH_5}$. The solid curve is the MEP, and the labels on it denote the values of the reaction coordinate s, where s is the distance along the MEP through mass-scaled coordinates with $\mu = m_{CH_3} m_{H_2} / m_{CH_5}$. The chain line and the dashed line are the ground-state LC3 tunneling paths for tunneling from one translational turning point of $V_a^G(s)$ to the other at the most probable tunneling energies (31.52 and 34.59 kcal/mol) at 350 and 250 K, respectively. The triangles denote coordinates of systems with zero amplitude in all their vibrational coordinates except the C-H-H' stretch, which is placed at the classical turning point on the concave side of the MEP. One of the points on the MEP and one of the triangles have been changed to squares to denote s=0. The other triangles are evenly spaced in s with interval $\Delta s = 0.2\ a_0$. The light curve connects the coordinates of systems with $s = 0.5$–$0.8\ a_0$ with zero amplitude in all vibrational coordinates except the C-H-H' stretch, which is placed at the radii of curvature for this mode.

Sect. 2, it is physically quite similar since the repulsive triplet C-T interaction potential in a C-H-T MLEPS function controls the bending potential for a C-H-T bend.

The final practical problem that we wish to emphasize in this chapter is the necessity to consider the whole reaction swath. We will use the PES of Raff for CH_3 + HH' to illustrate how one may estimate the region over which the potential must be known. First of all we calculated the RPH for this system using the harmonic approximation and general methods presented elsewhere.[54] (This RPH is also discussed in Sect. 2.) At every point along the MEP, we calculated the C-H and H-H' bond lengths for the bonds being made and broken, and from them we calculated the Jacobi-like coordinates used as abscissa and ordinate in Fig. 3. The MEP is plotted as a solid curve in Fig. 3, which thereby becomes a two-dimensional internal-coordinate projection of the full-dimensional steepest descent path through mass-scaled coordinates. The distance s measured through the full set of mass-scaled coordinates along the MEP is shown at intervals of 0.2-0.3 a_0. Next, at every 0.2 a_0 along the MEP, we displaced the system through the full set of mass-scaled cartesian coordinates to the classical turning point (on the concave side

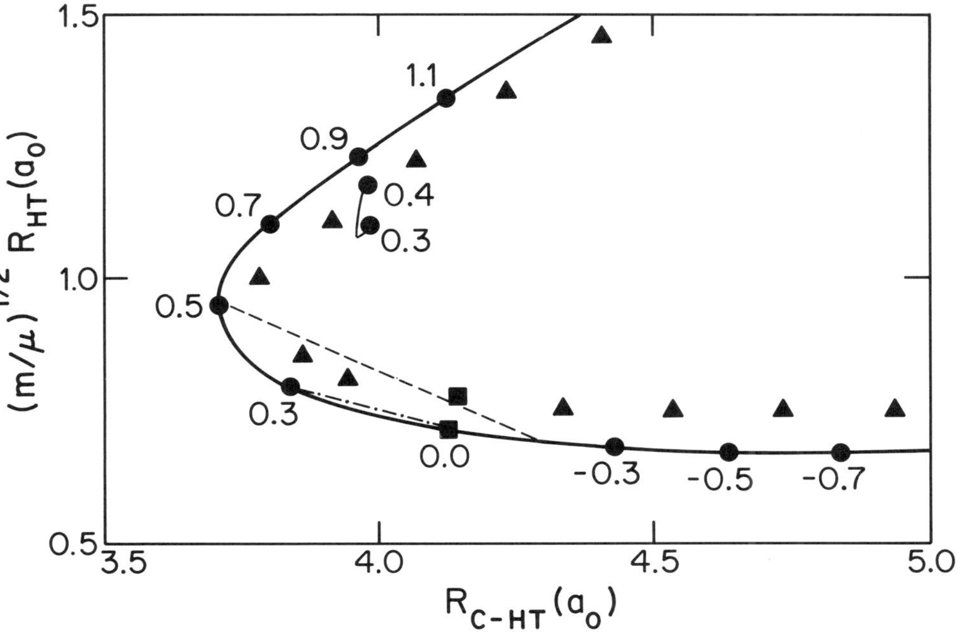

Fig. 4. Same as Fig. 3 except for CH_3 + HT → CH_4 + T, $m = m_H m_T / m_{HT}$, $\mu = m_{CH_3} m_{HT} / m_{CH_4T}$, the most probable tunneling energies are 32.07 and 34.40 kcal/mol, and the light curve is shown for s = 0.3-0.4 a_0.

of the MEP) of the generalized normal mode corresponding to the bound C-H-H' motion. We calculated the ordinate and abscissa of Fig. 3 for these geometries and plotted these points as triangles. A curve (not shown explicitly) passing through these triangles is an analog of the Marcus-Coltrin tunneling path originally found[5,171] variationally for collinear H + H_2. The loci where the locally multivalued region of the reaction-path coordinate system, as estimated by the curvature of just the C-H-H' vibrational mode, gets closest to the MEP are shown in Fig. 3 as a short light curve. Two LC3 tunneling paths are also shown. In this case neither the analog of the Marcus-Coltrin path nor the LC3 tunneling paths reach into the region of locally multivalued coordinates, i.e., they are between the MEP and the light curve. Thus it appears sufficient to know the potential in regions where the RPH is valid. In Fig. 4 we give an analogous set of curves for CH_3 + HT → CH_4 + T, again using the Raff surface. The location where the boundary for the multivalued region gets closest to the MEP is shown between s = 0.3 and 0.4 a_0; again neither the analog of the Marcus-Coltrin path nor the LC3 tunneling paths enter the multivalued region. We conclude that an RPH formulation is adequate for CH_3 + H_2 → CH_4 + H and isotopic analogs, at least according to the Raff surface. A surface with a broader barrier would have more widely spread turning points, however, and for such a surface the LC3 path might enter the multivalued region. Furthermore, for other systems with much smaller skew angles, it becomes very likely that the tunneling paths will enter the multivalued region where the reaction-path coordinates break down. (Examples have been observed for atom-diatom collisions.[19-21]) In such cases, no RPH can be valid and we must consider a whole reaction swath, not just a local expansion about the MEP.

5. ACKNOWLEDGMENTS

The authors are grateful to several coworkers for work on various VTST projects. We would like to acknowledge Kenneth Dykema, Bruce C. Garrett, Mark S. Gordon, Nancy J. Kilpatrick, Alan W. Magnuson, John Overend, Sachchida N. Rai, Susan C. Tucker, and Trina Valencich for help with work discussed here or especially closely related to it. This work was supported in part by the U.S. Department of Energy, Office of Basic Energy Sciences, under contract no. DOE-AC02-79ER10425.

APPENDIX

In this appendix we give some examples of the coordinate transformation coefficients that appear in eqs. (40)-(43). For convenience we define $\bar{X}_{ij} = X_{ij}/r^{(0)}_{A_iA_j}$.

For the bond stretch $\Delta r_{AB} = r_{AB} - r^{(0)}_{AB}$, the coefficients can be obtained by differentiation of eq. (38) with respect to the difference cartesians X_{ij}. For $A_i = A_k = A_m = A$, $A_j = A_\ell = A_n = B$, and $\gamma_i \neq \gamma_k \neq \gamma_m$, this yields:[67]

$$\partial r_{AB}/\partial X_{ij} = \bar{X}_{ij} ,$$

$$\partial^2 r_{AB}/\partial X_{ij}^2 = (1 - \bar{X}_{ij}^2)/r_{AB}^{(0)} ,$$

$$\partial^2 r_{AB}/\partial X_{ij}\partial X_{k\ell} = -\bar{X}_{ij}\bar{X}_{k\ell}/r_{AB}^{(0)} ,$$

$$\partial^3 r_{AB}/\partial X_{ij}^3 = 3\bar{X}_{ij}(\bar{X}_{ij}^2 - 1)/(r_{AB}^{(0)})^2 ,$$

$$\partial^3 r_{AB}/\partial X_{ij}^2\partial X_{k\ell} = \bar{X}_{k\ell}(3\bar{X}_{ij}^2 - 1)/(r_{AB}^{(0)})^2 ,$$

$$\partial^3 r_{AB}/\partial X_{ij}\partial X_{k\ell}\partial X_{mn} = 3\bar{X}_{ij}\bar{X}_{k\ell}\bar{X}_{mn}/(r_{AB}^{(0)})^2 ,$$

$$\partial^4 r_{AB}/\partial X_{ij}^4 = -3(5\bar{X}_{ij}^4 - 6\bar{X}_{ij}^2 + 1)/(r_{AB}^{(0)})^3 ,$$

$$\partial^4 r_{AB}/\partial X_{ij}^3\partial X_{k\ell} = -3\bar{X}_{ij}\bar{X}_{k\ell}(5\bar{X}_{ij}^2 - 3)/(r_{AB}^{(0)})^3 ,$$

$$\partial^4 r_{AB}/\partial X_{ij}^2\partial X_{k\ell}^2 = (-15\bar{X}_{ij}^2\bar{X}_{k\ell}^2 + 3\bar{X}_{ij}^2 + 3\bar{X}_{k\ell}^2 - 1)/(r_{AB}^{(0)})^3 ,$$

and

$$\partial^4 r_{AB}/\partial X_{ij}^2\partial X_{k\ell}\partial X_{mn} = -3\bar{X}_{k\ell}\bar{X}_{mn}(5\bar{X}_{ij}^2 - 1)/(r_{AB}^{(0)})^3 .$$

For the angle deformation $\Delta\Phi_{ABC} = \Phi_{ABC} - \Phi_{ABC}^{(0)}$, we carry out a similar differentiation process with respect to eq. (39). For $A_i = A_k = A_m = A$, $A_j = A_\ell = A_n = A_{i'} = A_{k'} = B$, $A_{j'} = A_{\ell'} = C$, $\mathcal{C} = \cos\Phi_{ABC}$, $\mathcal{S} = \sin\Phi_{ABC}$, and $\gamma_i = \gamma_{i'} \neq \gamma_k = \gamma_{k'} \neq \gamma_m = \gamma_{m'}$, this yields:[67]

$$\partial\Phi_{ABC}/\partial X_{ij} = (\mathcal{C}\bar{X}_{ij} - \bar{X}_{i'j'})/\mathcal{S}r_{AB}^{(0)} ,$$

$$\partial\Phi_{ABC}/\partial X_{i'j'} = (\mathcal{C}\bar{X}_{i'j'} - \bar{X}_{ij})/\mathcal{S}r_{BC}^{(0)} ,$$

$$\partial^2\Phi_{ABC}/\partial X_{ij}^2 = [2\bar{X}_{ij}\bar{X}_{i'j'}/(r_{AB}^{(0)})^2 + \mathcal{C}(1 - 3\bar{X}_{ij}^2) - \\ - \mathcal{C}(\partial\Phi_{ABC}/\partial X_{ij})^2]/\mathcal{S} ,$$

$$\partial^2\Phi_{ABC}/\partial X_{ij}\partial X_{k\ell} = [(\bar{X}_{ij}\bar{X}_{k'\ell'} + i'j'k\ell - 3\bar{X}_{ij}\bar{X}_{k\ell}\mathcal{C})/(r_{AB}^{(0)})^2 - \\ - \mathcal{C}(\partial\Phi_{ABC}/\partial X_{ij})(k\ell)]/\mathcal{S} ,$$

$$\partial^2\Phi_{ABC}/\partial X_{ij}\partial X_{i'j'} = [(\bar{X}^2_{ij} + i'j' - \bar{X}_{ij}\bar{X}_{i'j'}\mathcal{C} - 1)/r^{(0)}_{AB}r^{(0)}_{BC} - \mathcal{C}(\partial\Phi_{ABC}/\partial X_{ij})(i'j')]/\mathcal{S} \quad ,$$

$$\partial^2\Phi_{ABC}/\partial X_{ij}\partial X_{k'\ell'} = [(\bar{X}_{ij}\bar{X}_{k\ell} + i'j'k'\ell' - \bar{X}_{ij}\bar{X}_{k'\ell'}\mathcal{C})/r^{(0)}_{AB}r^{(0)}_{BC} - \mathcal{C}(\partial\Phi_{ABC}/\partial X_{ij})(k'\ell')]/\mathcal{S} \quad ,$$

$$\partial^3\Phi_{ABC}/\partial X^3_{ij} = 3[\bar{X}_{i'j'} - 3\bar{X}_{ij}(\mathcal{C} + \bar{X}_{ij}\bar{X}_{i'j'}) + 5\mathcal{C}\bar{X}^3_{ij}]/\mathcal{S}(r^{(0)}_{AB})^3 + (\partial\Phi_{ABC}/\partial X_{ij})^3 - 3\mathcal{C}(\partial\Phi_{ABC}/\partial X_{ij})(\partial^2\Phi_{ABC}/\partial X^2_{ij})/\mathcal{S} \;,$$

$$\partial^3\Phi_{ABC}/\partial X^2_{ij}\partial X_{k\ell} = \{\bar{X}_{k'\ell'} - 3[\mathcal{C}\bar{X}_{k\ell} + 2\bar{X}_{ij}\bar{X}_{i'j'}\bar{X}_{k\ell} + \bar{X}^2_{ij} \times (\bar{X}_{k'\ell'} - 5\mathcal{C}\bar{X}_{k\ell})]\}/\mathcal{S}(r^{(0)}_{AB})^3 - \mathcal{C}[2(\partial\Phi_{ABC}/\partial X_{ij}) \times (\partial^2\Phi_{ABC}/\partial X_{ij}\partial X_{k\ell}) + (\partial\Phi_{ABC}/\partial X_{k\ell})(\partial^2\Phi_{ABC}/\partial X^2_{ij})]/\mathcal{S} + (\partial\Phi_{ABC}/\partial X_{ij})^2(\partial\Phi_{ABC}/\partial X_{k\ell}) \;,$$

$$\partial^3\Phi_{ABC}/\partial X_{ij}\partial X_{k\ell}\partial X_{mn} = 3[5\mathcal{C}\bar{X}_{ij}\bar{X}_{k\ell}\bar{X}_{mn} - (\bar{X}_{i'j'}\bar{X}_{k\ell}\bar{X}_{mn} + ijk'\ell'mn + ijk\ell m'n')]/\mathcal{S}(r^{(0)}_{AB})^3 - \mathcal{C}[(\partial\Phi_{ABC}/\partial X_{ij})(\partial^2\Phi_{ABC}/\partial X_{k\ell}\partial X_{mn}) + k\ell ijmn + mnijk\ell]/\mathcal{S} + (\partial\Phi_{ABC}/\partial X_{ij})(k\ell)(mn) \;,$$

$$\partial^3\Phi_{ABC}/\partial X^2_{ij}\partial X_{i'j'} = [3\bar{X}_{ij}(1 - \bar{X}^2_{ij} + \mathcal{C}\bar{X}_{ij}\bar{X}_{i'j'}) - 2\bar{X}_{ij}\bar{X}^2_{i'j'} - \mathcal{C}\bar{X}_{i'j'}]/\mathcal{S}(r^{(0)}_{AB})^2 r^{(0)}_{BC} - 2\mathcal{C}[(\partial\Phi_{ABC}/\partial X_{ij})(\partial^2\Phi_{ABC}/\partial X_{ij}\partial X_{i'j'}) + i'j'ijij]/\mathcal{S} + (\partial\Phi_{ABC}/\partial X_{ij})^2(\partial\Phi_{ABC}/\partial X_{i'j'}) \;,$$

$$\partial^3\Phi_{ABC}/\partial X^2_{ij}\partial X_{k'\ell'} = [\bar{X}_{k\ell} - 2\bar{X}_{ij}\bar{X}_{i'j'}\bar{X}_{k'\ell'} - \mathcal{C}\bar{X}_{k'\ell'} + 3\bar{X}^2_{ij}(\mathcal{C}\bar{X}_{k'\ell'} - \bar{X}_{k\ell})]/\mathcal{S}(r^{(0)}_{AB})^2 r^{(0)}_{BC} - 2\mathcal{C}[(\partial\Phi_{ABC}/\partial X_{ij}) \times (\partial^2\Phi_{ABC}/\partial X_{ij}\partial X_{k'\ell'}) + k'\ell'ijij]/\mathcal{S} +$$

$$+ (\partial\Phi_{ABC}/\partial X_{ij})^2(\partial\Phi_{ABC}/\partial X_{k'\ell'}) \ ,$$

$$\partial^3\Phi_{ABC}/\partial X_{ij}\partial X_{k\ell}\partial X_{i'j'} = [\bar{X}_{k\ell}(1 - 3\bar{X}_{ij}^2 + 3C\bar{X}_{ij}\bar{X}_{i'j'}) - \bar{X}_{i'j'} \times$$
$$\times (\bar{X}_{i'j'}\bar{X}_{k\ell} + ijk'\ell')]/S(r_{AB}^{(0)})^2 r_{BC}^{(0)} - C[(\partial\Phi_{ABC}/\partial X_{ij}) \times$$
$$\times (\partial^2\Phi_{ABC}/\partial X_{k\ell}\partial X_{i'j'}) + k\ell iji'j' + i'j'ijk\ell]/S +$$
$$+ (\partial\Phi_{ABC}/\partial X_{ij})(k\ell)(i'j') \ ,$$

$$\partial^3\Phi_{ABC}/\partial X_{ij}\partial X_{k\ell}\partial X_{m'n'} = -[\bar{X}_{m'n'}(\bar{X}_{i'j'}\bar{X}_{k\ell} + k'\ell' ij) + 3\bar{X}_{ij}\bar{X}_{k\ell} \times$$
$$\times (\bar{X}_{mn} - C\bar{X}_{m'n'})]/S(r_{AB}^{(0)})^2 r_{BC}^{(0)} - C[(\partial\Phi_{ABC}/\partial X_{ij}) \times$$
$$\times (\partial^2\Phi_{ABC}/\partial X_{k\ell}\partial X_{m'n'}) + k\ell ijm'n' + m'n'ijk\ell]/S +$$
$$+ (\partial\Phi_{ABC}/\partial X_{ij})(k\ell)(m'n') \ ,$$

and so on for the higher derivatives. In writing these expressions we used the convention that if a term or factor is identical to the previous one except for the lower case subscripts, we only repeat those. Because of the S^{-1} factors in these equations they cannot be used for a linear angle. One possible set of modifications that is useful in such a case is described in Ref. 67. Similar equations to those given above but for out-of-plane bending and torsional angle internal coordinates are given elsewhere.[74,78]

REFERENCES

1. Potential Energy Surfaces and Dynamics Calculations, edited by D.G. Truhlar (Plenum, New York, 1981).
2. J.N. Murrell, S. Carter, S.C. Farantos, P. Huxley, and A.J.C. Varandas, Molecular Potential Energy Functions (John Wiley and Sons, Chichester, 1984).
3. Dynamics of Molecular Collisions, Part B, edited by W.H. Miller (Plenum, New York, 1976).
4. D.G. Truhlar, F.B. Brown, D.W. Schwenke, R. Steckler, and B.C. Garrett, in Comparison of Ab Initio Quantum Chemistry with Experiment, edited by R.J. Bartlett (D. Reidel, Dordrecht, Holland), in press.
5. R.A. Marcus, J. Phys. Chem. 83, 204 (1979).
6. K. Fukui, Acc. Chem. Res. 14, 363 (1981).
7. K. Morokuma and S. Kato, in Potential Energy Surfaces and Dynamics Calculations, edited by D.G. Truhlar (Plenum, New York, 1981), p. 243.
8. W.H. Miller, J. Phys. Chem. 87, 3811 (1983).
9. D.G. Truhlar and B.C. Garrett, Acc. Chem. Res. 13, 440 (1980).

10. D.G. Truhlar, A.D. Isaacson, R.T. Skodje, and B.C. Garrett, J. Phys. Chem. 86, 2252 (1982), 87, 4554E (1983).
11. D.G. Truhlar and B.C. Garrett, Annu. Rev. Phys. Chem. 35, 159 (1984).
12. D.G. Truhlar and A.D. Isaacson, J. Chem. Phys. 77, 3516 (1982).
13. D.C. Clary, B.C. Garrett, and D.G. Truhlar, J. Chem. Phys. 78, 777 (1983).
14. D.G. Truhlar, W.L. Hase, and J.T. Hynes, J. Phys. Chem. 87, 2664, 5523E (1983).
15. B.C. Garrett and D.G. Truhlar, J. Chem. Phys. 81, 309 (1984).
16. D.G. Truhlar, B.C. Garrett, P.G. Hipes, and A. Kuppermann, J. Chem. Phys. 81, 3542 (1984).
17. A.D. Isaacson, M.T. Sund, S.N. Rai, and D.G. Truhlar, J. Chem. Phys. 82, 1338 (1985).
18. B.C. Garrett and D.G. Truhlar, J. Phys. Chem. 89, 2204 (1985).
19. B.C. Garrett, N. Abusalbi, D.J. Kouri, and D.G. Truhlar, J. Chem. Phys. 83, in press.
20. B.C. Garrett, D.G. Truhlar, A.F. Wagner, and T.H. Dunning, Jr., J. Chem. Phys. 78, 4400 (1983).
21. D.K. Bondi, J.N.L. Connor, B.C. Garrett, and D.G. Truhlar, J. Chem. Phys. 78, 5981 (1983).
22. B.C. Garrett and D.G. Truhlar, J. Chem. Phys. 79, 4931 (1983).
23. R. Steckler, D.G. Truhlar, B.C. Garrett, N.C. Blais, and R.B. Walker, J. Chem. Phys. 81, 5700 (1984).
24. F.B. Brown, R. Steckler, D.W. Schwenke, D.G. Truhlar, and B.C. Garrett, J. Chem. Phys. 82, 188 (1985).
25. B.C. Garrett and D.G. Truhlar, J. Phys. Chem. 86, 1136 (1982), 87, 4554E (1983).
26. B.C. Garrett, D.G. Truhlar, R.S. Grev, G.C. Schatz, and R.B. Walker, J. Phys. Chem. 85, 3806 (1981).
27. R.T. Skodje, D.W. Schwenke, D.G. Truhlar, and B.C. Garrett, J. Phys. Chem. 88, 628 (1984).
28. R.T. Skodje, D.W. Schwenke, D.G. Truhlar, and B.C. Garrett, J. Chem. Phys. 80, 3569 (1984).
29. B.C. Garrett, D.W. Schwenke, R.T. Skodje, D. Thirumalai, T.C. Thompson, and D.G. Truhlar, in Resonances, edited by D.G. Truhlar (American Chemical Society, Washington, 1984), p. 375.
30. F.B. Brown, S.C. Tucker, and D.G. Truhlar, J. Chem. Phys. 83, in press.
31. W.H. Miller and S.-H. Shi, J. Chem. Phys. 75, 2258 (1981).
32. C.J. Cerjan, S. Shi, and W.H. Miller, J. Phys. Chem. 86, 2244 (1982).
33. K. Morokuma, S. Kato, K. Kitaura, S. Obara, K. Ohta, and M. Hanamura, in New Horizons of Quantum Chemistry, edited by P.-O. Lowdin and B. Pullman (D. Reidel, Dordrecht, Holland, 1983), p. 221.
34. W.H. Miller, N.C. Handy, and J.E. Adams, J. Chem. Phys. 72, 99 (1980).
35. G. Natanson, Mol. Phys. 46, 481 (1982).
36. D.G. Truhlar, Int. J. Quantum Chem. Symp. 17, 77 (1983).
37. G.L. Hofacker, Z. Naturforsch. A 18, 607 (1963).
38. R.A. Marcus, J. Chem. Phys. 45, 4493, 4500 (1966).

39. D.G. Truhlar, J. Chem. Phys. 53, 2041 (1970).
40. K. Fukui, A. Tachibana, and K. Yamashita, Int. J. Quantum Chem. Symp. 15, 621 (1981).
41. A. Tweedale and K.J. Laidler, J. Chem. Phys. 53, 2045 (1970).
42. B.C. Garrett and D.G. Truhlar, J. Phys. Chem. 83, 1079 (1979), 84, 682E (1980), 87, 4553E (1983).
43. B.C. Garrett and D.G. Truhlar, J. Amer. Chem. Soc. 101, 4534 (1979).
44. B.C. Garrett, D.G. Truhlar, R.S. Grev, and A.W. Magnuson, J. Phys. Chem. 84, 1730 (1980), 87, 4554E (1983).
45. S. Glasstone, K.J. Laidler, and H. Eyring, Theory of Rate Processes (McGraw-Hill, New York, 1941).
46. J.C. Keck, Adv. Chem. Phys. 13, 85 (1967).
47. I. Shavitt, Theoretical Chemistry Laboratory Report WIS-AEC-23, University of Wisconsin, 11 August 1959 (unpublished).
48. D.G. Truhlar and A. Kuppermann, J. Amer. Chem. Soc. 93, 1840 (1971).
49. K. Fukui, S. Kato, and H. Fujimoto, J. Amer. Chem. Soc. 97, 1 (1975).
50. H.F. Schaefer III, Chem. Brit. 11, 227 (1975).
51. B.C. Garrett and D.G. Truhlar, J. Phys. Chem. 83, 1052, 3058E (1979), 87, 4553E (1983).
52. R.T. Skodje, D.G. Truhlar, and B.C. Garrett, J. Phys. Chem. 85, 3019 (1981).
53. R.T. Skodje, D.G. Truhlar, and B.C. Garrett, J. Chem. Phys. 77, 5955 (1982).
54. D.G. Truhlar, A.D. Isaacson, and B.C. Garrett, in The Theory of Chemical Reaction Dynamics, edited by M. Baer (CRC Press, Boca Raton, FL, 1985), Vol. 4, p. 1.
55. V.K. Babamov and R.A. Marcus, J. Chem. Phys. 74, 1790 (1978).
56. P. Pechukas and F.J. McLafferty, J. Chem. Phys. 58, 1622 (1973).
57. B.C. Garrett and D.G. Truhlar, J. Phys. Chem. 83, 1915 (1979).
58. B.C. Garrett and D.G. Truhlar, J. Amer. Chem. Soc. 101, 5207 (1979).
59. A.D. Isaacson and D.G. Truhlar, J. Chem. Phys. 76, 1380 (1982).
60. G.C. Schatz and H. Elgersma, Chem. Phys. Lett. 73, 21 (1980).
61. S.P. Walch and T.H. Dunning, J. Chem. Phys. 72, 1303 (1980).
62. W.B. Wilson, Jr., J.C. Decius, and P.C. Cross, Molecular Vibrations (McGraw-Hill, New York, 1955), p. 19, (a) p. 172.
63. D.G. Truhlar, J. Mol. Spectrosc. 38, 415 (1971).
64. A.D. Isaacson, D.G. Truhlar, K. Scanlon, and J. Overend, J. Chem. Phys. 75, 3017 (1981).
65. B.C. Garrett and D.G. Truhlar, J. Chem. Phys. 72, 3460 (1980).
66. H.H. Nielsen, Encycl. Phys. 37/part one, 173 (1959).
67. M.A. Pariseau, I. Suzuki, and J. Overend, J. Chem. Phys. 42, 2335 (1965).
68. D.G. Truhlar, R.W. Olson, A.C. Jeannotte, and J. Overend, J. Amer. Chem. Soc. 98, 2373 (1976).
69. D. Papousek and M.R. Aliev, Molecular Vibrational-Rotational Spectra (Elsevier, Amsterdam, 1982), p. 38.
70. B.C. Garrett, D.G. Truhlar, and A.W. Magnuson, J. Chem. Phys. 76, 2321 (1982).
71. H.S. Johnston and C.A. Parr, J. Amer. Chem. Soc. 85, 2544 (1963).

72. A.R. Hoy, I.M. Mills, and G. Strey, Mol. Phys. **24**, 1265 (1972).
73. M.A. Pariseau, I. Suzuki, and J. Overend, J. Chem. Phys. **44**, 3561 (1966).
74. S. Califano, Vibrational States (Wiley, London, 1976).
75. B. Crawford, Jr. and J. Overend, J. Mol. Spectrosc. **12**, 307 (1964).
76. W.B. Brown and E. Steiner, J. Mol. Spectrosc. **10**, 348 (1963).
77. S. Brodersen and J. Christoffersen, J. Mol. Spectrosc. **12**, 303 (1964).
78. K. Machida, J. Chem. Phys. **44**, 4186 (1966).
79. I. Suzuki and J. Overend. Spectrochim. Acta **3**, 1093 (1981).
80. N.W. Bazley and D.W. Fox, Phys. Rev. **124**, 483 (1961).
81. K.S. Pitzer and W.D. Gwinn, J. Chem. Phys. **10**, 428 (1942).
82. A.D. Isaacson and D.G. Truhlar, J. Chem. Phys. **75**, 4090 (1981).
83. D.A. McQuarrie, Statistical Mechanics (Harper and Row, New York, 1976).
84. D.L. Bunker, J. Chem. Phys. **37**, 393 (1962).
85. W.L. Hase and D.G. Buckowski, Chem. Phys. Lett. **74**, 284 (1980).
86. J.E. Adams and J.D. Doll, J. Chem. Phys. **74**, 5332 (1981).
87. A.E. Barton and B.J. Howard, Faraday Discuss. Chem. Soc. **73**, 45 (1982).
88. A.F. Voter, J. Chem. Phys. **82**, 1890 (1985).
89. D.L. Freeman and J.D. Doll, J. Chem. Phys. **80**, 5709 (1984).
90. P. Pulay, Mol. Phys. **17**, 197 (1969).
91. P. Pulay, in Applications of Electronic Structure Theory, edited by H.F. Schaefer (Plenum, New York, 1977), p. 153.
92. P. Pulay, in The Force Concept in Chemistry, edited by B.M. Deb (Van Nostrand Reinhold, New York, 1981), p. 449.
93. J.A. Pople, R. Krishnan, H.B. Schlegel, and J.B. Binkley, Int. J. Quantum Chem. Symp. **13**, 225 (1979).
94. J.D. Goddard, N.C. Handy, and H.F. Schaefer, J. Chem. Phys. **71**, 1525 (1979).
95. S. Kato and K. Morokuma, Chem. Phys. Lett. **65**, 19 (1979).
96. B.R. Brooks, W.D. Laidig, P. Saxe, J.D. Goddard, Y. Yamaguchi, and H.F. Schaefer, J. Chem. Phys. **72**, 4652 (1980).
97. R. Krishnan, H.B. Schlegel, and J.A. Pople, J. Chem. Phys. **72**, 4654 (1980).
98. Y. Osamura, Y. Yamaguchi, and H.F. Schaefer, J. Chem. Phys. **77**, 383 (1982).
99. M. Page, P. Saxe, G.F. Adams, and B.H. Lengsfield, J. Chem. Phys. **81**, 434 (1984).
100. G. Fitzgerald, R. Harrison, W.D. Laidig, and R.J. Bartlett, J. Chem. Phys. **82**, 4379 (1985).
101. J.W. McIver, Jr. and A. Komornicki, Chem. Phys. Lett. **10**, 303 (1971).
102. D. Poppinger, Chem. Phys. Lett. **34**, 332 (1975).
103. D. Poppinger, Chem. Phys. Lett. **35**, 550 (1975).
104. A. Komornicki, K. Ishida, K. Morokuma, R. Ditchfield, and M. Conrad, Chem. Phys. Lett. **45**, 595 (1977).
105. B. Schlegel, J. Comput. Chem. **3**, 214 (1982).
106. A. Banerjee, N. Adams, J. Simons, and R. Shepard, J. Phys. Chem. **89**, 52 (1985).

107. Y. Osamura, Y. Yamaguchi, P. Saxe, M.A. Vincent, J.F. Gaw, and H.F. Schaefer, Chem. Phys. 72, 131 (1982).
108. Y. Yamaguchi, Y. Osamura, G. Fitzgerald, and H.F. Schaefer, J. Chem. Phys. 78, 1607 (1983).
109. R.N. Camp, H.F. King, J.W. McIver, Jr., and D. Mullally, J. Chem. Phys. 79, 1088 (1983).
110. D.J. Fox, Y. Osamura, M.R. Hoffman, J.F. Gaw, G. Fitzgerald, Y. Yamaguchi, and H.F. Schaefer, Chem. Phys. Lett. 102, 17 (1983).
111. M.R. Hoffman, D.J. Fox, J.F. Gaw, Y. Osamura, Y. Yamaguchi, R.S. Grev, G. Fitzgerald, H.F. Schaefer, P.J. Knowles, and N.C. Handy, J. Chem. Phys. 80, 2660 (1984).
112. P. Pulay, J. Chem. Phys. 78, 5043 (1983).
113. P. Jørgensen and J. Simons, J. Chem. Phys. 79, 334 (1983).
114. J.F. Gaw, Y. Yamaguchi, and H.F. Schaefer, J. Chem. Phys. 81, 6395 (1984).
115. S. Kato, H. Kato, and H. Fukui, J. Amer. Chem. Soc. 99, 684 (1977).
116. S.K. Gray, W.H. Miller, Y. Yamaguchi, and H.F. Schaefer, J. Chem. Phys. 73, 2733 (1980).
117. K. Yamashita and Y. Yamabe, Int. J. Quantum Chem. Symp. 17, 177 (1983).
118. S.M. Colwell and N.C. Handy, J. Chem. Phys. 82, 1281 (1985).
119. A. Tachibana, T. Okazaki, M. Koizumi, K. Hori, and T. Yamabe, J. Amer. Chem. Soc. 107, 1190 (1985).
120. D.G. Truhlar, N.J. Kilpatrick, and B.C. Garrett, J. Chem. Phys. 78, 2438 (1983).
121. H.S. Johnston and J. Heicklen, J. Phys. Chem. 66, 532 (1962).
122. B.C. Garrett and D.G. Truhlar, J. Phys. Chem. 83, 2921 (1979).
123. M. Quack and J. Troe, Ber. Bunsenges. Phys. Chem. 78, 240 (1974).
124. M. Quack and J. Troe, Ber. Bunsenges. Phys. Chem. 79, 170, 469 (1975).
125. D.G. Truhlar and A. Kuppermann, J. Chem. Phys. 52, 2232 (1970).
126. M.J. Stern, A. Persky, and F.S. Klein, J. Chem. Phys. 58, 5697 (1973).
127. J.T. Muckerman, Theor. Chem. Adv. Perspect. A 6, 1 (1981).
128. J.W. Duff and D.G. Truhlar, J. Chem. Phys. 62, 2477 (1975).
129. T. Carrington, Jr., L.M. Hubbard, H.F. Schaefer III, and W.H. Miller, J. Chem. Phys. 80, 4347 (1984).
130. M.V. Basilevsky and V.M. Ryaboy, Chem. Phys. 41, 461 (1979).
131. B.C. Garrett, R.T. Skodje, and D.G. Truhlar, unpublished calculations.
132. L.M. Raff, J. Chem. Phys. 60, 2220 (1974).
133. T. Valencich and D.L. Bunker, J. Chem. Phys. 61, 21 (1974).
134. S. Chapman and D.L. Bunker, J. Chem. Phys. 62, 2890 (1975).
135. W.A. Lathan, W.J. Hehre, L.A. Curtiss, and J.A. Pople, J. Amer. Chem. Soc. 93, 6377 (1971).
136. S. Ehrenson and M.D. Newton, Chem. Phys. Lett. 13, 24 (1972).
137. K. Morokuma and R.E. Davis, J. Amer. Chem. Soc. 94, 1060 (1972).
138. K. Fukui, S. Kato, and H. Fujimoto, J. Amer. Chem. Soc. 97, 1 (1975).

139. K. Niblaeus, B.O. Roos, and P.E.M. Siegbahn, Chem. Phys. 26, 59 (1977).
140. P. Cársky and R. Zahradnik, J. Mol. Struct. 54, 247 (1979).
141. P. Cársky and R. Zahradnik, Int. J. Quantum Chem. 16, 243 (1979).
142. P. Cársky, Coll. Czech. Chem. Comm. 44, 3452 (1979).
143. S.P. Walch, J. Chem. Phys. 72, 4932 (1980).
144. G.C. Schatz, S.P. Walch, and A.F. Wagner, J. Chem. Phys. 73, 4536 (1980).
145. G.C. Schatz, A.F. Wagner, and T.H. Dunning, Jr., J. Phys. Chem. 88, 221 (1984).
146. M.S. Gordon, D.R. Gano, and J.A. Boatz, J. Amer. Chem. Soc. 105, 5771 (1983).
147. S. Sana, G. Leroy, and J.L. Villaveces, Theoret. Chim. Acta 65, 109 (1984).
148. B. Maessen, P. Bopp, D.R. McLaughlin, and M. Wolfsberg, Z. Naturfor. 39a, 1005 (1984).
149. M.W. Schmidt, M.S. Gordon, and M. Dupuis, J. Amer. Chem. Soc. 107, 2585 (1985).
150. K. Ishida, K. Morokuma, and A. Komornicki, J. Chem. Phys. 66, 2153 (1977).
151. S.M. Colwell, Mol. Phys. 51, 1217 (1984).
152. D.G. Truhlar and C.J. Horowitz, J. Chem. Phys. 68, 2466 (1978), 71, 1514E (1979).
153. M. Dupuis, D. Spangler, and J.J. Wendoloski, Program QG01, in NRCC Software Catalog (Lawrence Berkeley Laboratory technical report LBL-10811, Berkeley, CA, 1980), p. 60.
154. P.J. Kuntz, E.M. Nemeth, J.C. Polanyi, S.D. Rosner, and C.E. Young, J. Phys. Chem. 44, 1168 (1966).
155. C.A. Parr and D.G. Truhlar, J. Phys. Chem. 75, 1844 (1971).
156. N.C. Blais and D.G. Truhlar, J. Chem. Phys. 58, 4186 (1974), 65, 3803E (1976).
157. F.B. Brown, R. Steckler, D.W. Schwenke, D.G. Truhlar, and B.C. Garrett, J. Chem. Phys. 82, 188 (1985).
158. W.H. Miller, J. Phys. Chem. 87, 21 (1983).
159. D.L. Bunker and W.L. Hase, J. Chem. Phys. 59, 4621 (1973), 69, 4711E (1978).
160. W.L. Hase, G. Mrowka, R.J. Brudzynski, and C.S. Sloane, J. Chem. Phys. 69, 3548 (1978).
161. W.L. Hase, D.M. Ludlow, R.J. Wolf, and T. Schlick, J. Phys. Chem. 85, 958 (1981).
162. E.R. Grant and D.L. Bunker, J. Chem. Phys. 68, 628 (1978).
163. C.S. Sloane and W.L. Hase, Discuss. Faraday Soc. 62, 210 (1977).
164. P.J. Nagy and W.L. Hase, Chem. Phys. Lett. 54, 73 (1978).
165. W.L. Hase and K.C. Bhalla, J. Chem. Phys. 75, 2807 (1981).
166. R.J. Duchovic, W.L. Hase, and H.B. Schlegel, J. Phys. Chem. 88, 1339 (1984).
167. C.J. Cobos and J. Troe, Chem. Phys. Lett. 113, 41 (1985).
168. H.S. Johnston and P. Goldfinger, J. Chem. Phys. 37, 700 (1962).
169. G.W. Burton, L.B. Sims, J.C. Wilson, and A. Fry, J. Amer. Chem. Soc. 99, 3371 (1977).
170. L.B. Sims and D.E. Lewis, Isotopes in Organic Chemistry, Vol. 6,

"Isotopic Effects: Recent Developments in Theory and Experiment", edited by E. Buncel and C.C. Lee (Elsevier, Amsterdam, 1984), p. 161.
171. R.A. Marcus and M.E. Coltrin, J. Chem. Phys. 67, 2609 (1977).

LIGHT-HEAVY-LIGHT CHEMICAL REACTIONS

D. C. Clary and J. P. Henshaw
University Chemical Laboratory
Cambridge University
Lensfield Road
Cambridge CB2 1EW
United Kingdom

ABSTRACT. The approximations that have been proposed for light-heavy-light chemical reactions are reviewed. Their accuracy is examined by comparison with close-coupling results for the H+BrH→HBr+H reaction in three dimensions. Calculations are also presented of the photodissociation spectrum for $H_2O \rightarrow OH(^2\Pi)+H$ in which the $^1B_1$ dissociative potential energy surface is reactive. Reactive scattering resonances produce some structure in this calculated spectrum.

1. INTRODUCTION

Chemical reactions of the form

$$L' + HL \rightarrow L'H + L \qquad (I)$$

where L and L' are light atoms, and H is a heavy atom, have been the subject of several theoretical investigations. The main reason for this, is that these gas-phase reactions offer a mathematical simplification to theory which enables three-dimensional (3D) quantum mechanical calculations to be performed. The simplification is that the centre of mass of both the reactant and product diatomic molecules can be placed, to a good approximation, on the heavy atom H. This enables quantum-mechanical wavefunctions for the entrance and reaction channels to be matched with ease. It is this matching problem that has been such a stumbling block in the extension of 3D quantum theories to reactions more complicated than $H+H_2$. Thus light-heavy-light reactions offer a special opportunity for performing accurate 3D quantum-mechanical calculations and using the results to test the accuracy of approximate theories.

To date, our knowledge of the accuracy of those approximate 3D reaction theories which can be readily extendable to a wide variety of reactions, is almost completely based on comparisons with accurate results for the $H+H_2$ reaction. A light-heavy-light reaction such as

$$H + BrH \rightarrow HBr + H \qquad (II)$$

D. C. Clary (ed.), The Theory of Chemical Reaction Dynamics, 331–358.

has the important difference, when compared to the $H+H_2$ reaction, that the rotational states of the reactant and product molecule, HBr, are much more closely spaced than those for H_2; the rotational constant B for HBr is $8.46cm^{-1}$ while that for H_2 is $60.85cm^{-1}$ [1]. Furthermore, in the entrance or exit channel scattering calculations on $H+H_2$, the H_2 rotational states with even values of j are not coupled with odd j states, while they are strongly coupled in H+BrH. Thus, calculations on a reaction such as (II) should be a much sterner test of rotationally adiabatic theories than computations on $H+H_2$.

Light-heavy-light reactions are also of experimental significance. For example, molecular beam measurements of angular distributions [2], and kinetic experiments yielding rate constants [3], have been carried out on the D+BrH→DBr+H reaction. Furthermore, there has been much recent interest in the competition between non-reactive and reactive energy transfer in collisions such as H+ClH[4]. Vibrational quenching reactions like

$$H + FH(v=2) \rightarrow HF(v=1)+H$$

are also thought to be important in chemical laser systems [5].

The calculation of potential energy surfaces for light-heavy-light reactions has been a difficult task [6,7], and hence any good scattering calculations on proposed potential energy surfaces have value, when compared with reliable experimental results, in testing the accuracy of the potential energy surfaces. Indeed, several semiempirical potential energy surfaces that have been developed are known to have a reasonable form for the abstraction reaction (e.g. $H+HF \rightarrow H_2+F$) while having reaction barriers considerably in error for the exchange reaction [8] (e.g. H+FH → HF+H).

Quantum mechanical tunnelling effects are likely to be very important in light-heavy-light reactions. This is because the reduced mass of the colliding system is very close to the mass of the light atom, and the barriers to reaction are known to be large in many cases [7]. This would suggest that the most sophisticated of quantum theories are required in calculating reaction cross sections over the energy range required for rate constants. Thus, accurate quantum-dynamical results should present a valuable test of tunnelling methods incorporated into modern transition state theories (TST) [9].

Another rapidly developing area where light-heavy-light reaction theory will be important is in the direct photodissociation of symmetric triatomic molecules. If the triatomic molecule is in a given vibrational-rotational state n, and it is photoexcited to a new electronic state which is directly dissociative to an atom plus a diatomic molecule in vibrational-rotational state m, then the photodissociation cross section is

$$I = C < \psi_m^{(-)} | \mu_D | \phi_n >^2$$

Here, C is a constant, $\psi_m^{(-)}$ is the scattering wavefunction on the upper electronic potential energy surface which satisfies incoming boundary conditions, μ_D is the transition dipole operator and ϕ_n is the triatomic bound state n. For a symmetric triatomic molecule, such as H_2O or H_2S, the potential energy surface for the exchange channel on the excited potential surface must be reactive. Thus, for the photodissociation of H_2O, the process H+OH→HO+H must be treated using a theory appropriate for light-heavy-light reactions.

After reviewing, in Section 2, previous theories proposed for light-heavy-light reactions we describe, in Section 3, our own theoretical contribution in this area. A hierachy of sudden approximations have been developed and these are described and compared to the exact theory. The relative computational merits of the various theories are also discussed. In Section 4 we describe a range of computations of integral cross sections and rate constants we have recently carried out on the H+BrH → HBr+H reaction using a semi-empirical potential energy surface. In that Section we use accurate close-coupling results to test the accuracy of various sudden approximations, together with a variational transition state theory [9] and a bending-corrected-rotating linear model (BCRLM) [12,13]. The comparisons show some important differences to those obtained for the $H+H_2$ reaction. In Section 5 we describe how a version of the light-heavy-light theory can be applied to photodissociation, and present results for the photodissociation of the H_2O molecule into the $^1B_1$ state. Our conclusions are in Section 6.

2. PREVIOUS THEORIES FOR LIGHT-HEAVY-LIGHT REACTIONS

The mathematical simplifications in the quantum theory of light-heavy-light reactions were discussed in a series of papers by Baer and Kouri [14,15]. They used an integral equation formalism [16]. The potential energy surfaces they used were highly simplistic. However, they did discuss the types of reaction integral and differential cross sections [14], and rotational product distributions, [15] that are likely to arise in these reactions. They stressed the importance of a "selection rule" for light-heavy-light reactions, with isotropic potentials, namely that $\ell_\alpha \to j_\beta$ and $j_\alpha \to \ell_\beta$, where ℓ_α is an orbital angular momentum quantum number for the α (entrance) reaction channel, and j_β is a rotational angular momentum quantum number for the exit (β) reaction channel (and vice-versa). They also noted an interesting consequence of this selection rule: light-heavy-light reactions with an isotropic potential energy surface will give an isotropic differential cross section. Thus, the measurement of differential cross sections for light-heavy-light reactions gives direct information on the anisotropy in the potential energy surface.

These studies were extended in a different direction by Baer [17]. He performed close coupling calculations on the

$$H + FH(v=0) \to HF(v=0) + H$$

reaction in coplanar geometry. Unfortunately, the reaction barrier of 0.078 eV, in the potential energy surface he used was unrealistically low; ab initio calculations suggest that the reaction barrier for this exchange reaction is greater than 1.7 eV [6]. However, the study did highlight some interesting features, such as the drop in reaction cross section as j is increased.

The theory of the infinite-order-sudden approximation (IOSA) for 3D light-heavy-light reactions was considered in detail by Barg and Drolshagen [18] in the body-fixed frame. The IOSA, for light-heavy-light reactions, is described in more detail in the next Section, and involves the solution of the quantum scattering problem for fixed orientations Θ_γ of the atom-molecule centre-of-mass vector with respect to the diatomic molecule vector for channel γ. They noted that $\Theta_\alpha = \Theta_\beta$ for light-heavy-light reactions, and this leads to particular simplifications in applying the IOSA to these reactions. A preliminary application of this approach to the D + CℓH → DCℓ + H reaction was made by Drolshagen [19]. However, the matching condition $\overline{j_\alpha} = \overline{j_\beta}$ was used, which is not appropriate for light-heavy-light reactions.

In the next Section, we describe our contributions to the 3D theory of light-heavy-light reactions, and also refer to the previous calculations we have performed using these theories.

3. RECENT THEORY FOR LIGHT-HEAVY-LIGHT REACTIONS

3.1 Close-Coupling Theory

In Section 3 we first of all describe the approach we have used for performing coupled-channel (CC) calculations on 3D light-heavy-light reactions. Then we go on to describe some approximate theories for these reactions.

We denote the entrance and exit reaction channels by α and β, respectively. In space-fixed coordinates, denoted by (xyz), the internuclear distance vector of the reactant molecule is $\bar{\underset{\sim}{r}}_\gamma$ and the vector of the colliding atom with respect to the centre of mass of the diatomic molecule is $\bar{\underset{\sim}{R}}_\gamma$. The angle between $\bar{\underset{\sim}{r}}_\gamma$ and $\bar{\underset{\sim}{R}}_\gamma$ is Θ_γ. We define the mass-scaled coordinates

$$\underset{\sim}{R}_\alpha = \bar{\underset{\sim}{R}}_\alpha, \quad \underset{\sim}{r}_\alpha = (\mu'_\alpha/\mu_\alpha)^{1/2}\, \bar{\underset{\sim}{r}}_\alpha$$

$$\underset{\sim}{R}_\beta = (\mu_\beta/\mu_\alpha)^{1/2}\, \bar{\underset{\sim}{R}}_\beta, \quad \underset{\sim}{r}_\beta = (\mu'_\beta/\mu_\alpha)^{1/2}\, \bar{\underset{\sim}{r}}_\beta$$

where μ'_γ is the reduced mass of the diatomic molecule and μ_γ is the reduced mass of the collisional system. The Hamiltonian for arrangement channel γ in space-fixed coordinates is then

$$H_\gamma = \frac{-\hbar^2}{2\mu_\alpha}\frac{1}{R_\gamma}\frac{\partial^2}{\partial R_\gamma^2}R_\gamma - \frac{\hbar^2}{2\mu_\alpha}\frac{1}{r_\gamma}\frac{\partial^2}{\partial r_\gamma^2}r_\gamma$$

$$+ \frac{\underset{\sim}{\ell}_\gamma^2}{2\mu_\alpha R_\gamma^2} + \frac{\underset{\sim}{j}_\gamma^2}{2\mu_\alpha r_\gamma^2} + V(R_\gamma, r_\gamma, \Theta_\gamma) . \qquad (1)$$

Here $\underset{\sim}{\ell}_\gamma^2$ is the orbital angular momentum operator, $\underset{\sim}{j}_\gamma^2$ is the rotational angular momentum operator and V is the potential energy surface.

For a given total energy E_{Tot} we then wish to solve the Schrodinger equation

$$(H_\gamma - E_{Tot})\, \psi_\gamma^{JM} = 0 \tag{2}$$

where J is the total angular momentum quantum number and M is the projection of the total angular momentum along the z axis.

Calculations on reactions are usually easier to do using body-fixed coordinates [20], which we denote by XYZ for the γ channel. We obtain these by a rotation through the Euler angles ($\overline{\alpha}$, $\overline{\beta}$, 0) such that Z points along $\underset{\sim}{R}_\gamma$, and Y remains in the xy plane. It is now convenient to replace $\underset{\sim}{\ell}_\gamma^2$ by $|\underset{\sim}{J}-\underset{\sim}{j}_\gamma|^2$, where $\underset{\sim}{J}$ is the total angular momentum operator.

For light-heavy-light reactions, it is straightforward to use a curvilinear coordinate system [21, 22] for the reaction. In the strong interaction region, polar coordinates (u_γ, v_γ) are used with an origin at (R_γ^*, r_γ^*). We have

$$R_\gamma = R_\gamma^* - \eta_\gamma \sigma_\gamma \sin \Theta_\gamma$$

$$r_\gamma = r_\gamma^* - \eta_\gamma \sigma_\gamma \cos \Theta_\gamma$$

$$\Theta_\gamma = \pi/2 - \varepsilon_\gamma - u_\gamma/\sigma_\gamma$$

$$\eta_\gamma = 1 + v_\gamma/\sigma_\gamma . \tag{3}$$

Where σ_γ is the radius of the circular reference curve with centre at (R_γ^*, r_γ^*), and ε_γ is a parameter that must be chosen to ensure that, for $u_\gamma = 0$, $\varepsilon_\alpha + \varepsilon_\beta = \Theta_S$, where Θ_S is the skew angle for the collision. These polar coordinates are defined to match smoothly with the cartesian coordinates

$$u_\gamma = \sigma_\gamma (\pi/2 - \varepsilon_\gamma) + R_\gamma - R_\gamma^*$$

$$v_\gamma = r_\gamma^o - r_\gamma \tag{4}$$

where r_γ^o is the mass-scaled equilibrium distance of the diatomic molecule. Note that u_γ and v_γ are "natural" translational and vibrational coordinates respectively.

In the polar region, the wavefunction can be expanded in the coupled-channel form

$$\psi_{\gamma}^{JM} = (\eta_{\gamma})^{1/2} \sum_{nj\Omega} f_{nj\Omega}^{JM} (u_{\gamma}) \, g_{n} (v_{\gamma}) \, D_{M\Omega}^{J*} (\overline{\alpha},\overline{\beta},0)$$

$$\times \; Y_{j}^{\Omega} (\Theta_{\gamma},\phi_{\gamma})/(R_{\gamma} r_{\gamma}) \tag{5}$$

where f is a "translational" function. The vibrational function g_n is determined by setting u_{γ} to u_{γ}^{i} and then solving the equation

$$[- \frac{\hbar^2}{2\mu_{\alpha}} \frac{d^2}{dv_{\gamma}^2} + V^{o} (u_{\gamma}^{i}, v_{\gamma})] \; g_{n} (v_{\gamma}) = E_{n} \, g_{n} (v_{\gamma}), \tag{6}$$

where V^{o} is a suitably chosen potential that could obtained, for example, by setting $\Theta_{\gamma} = 180^{o}$. Furthermore, $D_{M\Omega}^{J*}$ is a Wigner rotation function, and Y_{j}^{Ω} is a spherical harmonic, where $(\Theta_{\gamma}, \phi_{\gamma})$ are the spherical polar angles describing the orientation of $\underset{\sim}{r}_{\gamma}$ in the body-fixed coordinate system. In the cartesian region, a similar expansion to equation (5) is used, with the omission of the $(\eta_{\gamma})^{1/2}$ term.

Substituting equation (5) into equation (2), multiplication by the internal basis functions and integration over $(v_{\gamma},\overline{\alpha},\overline{\beta},\Theta_{\gamma},\phi_{\gamma})$ gives a set of close-coupled equations in the translational f functions [23]. These can be solved by a variety of methods. We prefer to use the R-matrix method of Light and Walker [22].

Once the close-coupled equations have been solved for each arrangement channel, it is necessary to ensure that the wavefunction and its derivative are continuous at a suitably chosen matching surface. For light-heavy-light reactions, an appropriate surface to use is $u_{\alpha} = u_{\beta} = 0$ as, in the limit of an infinitely heavy central atom,

$$\Theta_{\alpha} = \Theta_{\beta}, \qquad r_{\alpha} = R_{\beta}, \quad r_{\beta} = R_{\alpha} .$$

The matching matrices $\underset{\sim}{C}_1^{J}$ and $\underset{\sim}{C}_2^{J}$ are required in

$$\overline{\underset{\sim}{f}}^{\alpha J} (u_{\alpha} = 0) = \underset{\sim}{C}_1^{J} \, \overline{\underset{\sim}{f}}^{\beta J} (u_{\beta} = 0)$$

and

$$\frac{d}{du_{\alpha}} \overline{\underset{\sim}{f}}^{\alpha J} (u_{\alpha} = 0) = -\underset{\sim}{C}_2^{J} \, \frac{d}{du_{\beta}} \overline{\underset{\sim}{f}}^{\beta J} (u_{\beta} = 0), \tag{7}$$

where $\overline{\underset{\sim}{f}}$ is a vector of solutions that refer to adiabatic states that are obtained by diagonalising the close-coupling matrix at the matching surface. The appropriate matching matrices in primitive basis functions can be obtained from Schatz and Kuppermann [24] and are also discussed, for light-heavy-light reactions, by Barg and Drolshagen [18]. For a symmetric, light-heavy-light reaction the matching matrices $\underset{\sim}{C}_1^J$ and $\underset{\sim}{C}_2^J$ are simply equal to the identity matrix.

Once the global wavefunction has been obtained, the boundary conditions for the reactive solution

$$f^{JM}_{n'j'\Omega'}(R_\beta) \xrightarrow[R_\beta \to \infty]{} -(k_{n'j'}|k_{nj})^{1/2}$$

$$\times \exp(i(k_{n'j'} R_\beta - (J+j')\pi/2)\; S^J_{nj\Omega n'j'\Omega'} \qquad (8)$$

are applied, where $(nj\Omega)$ is the initial state, and k_{nj} is the wave-number for state (nj). This yields the S matrix element $S^J_{nj\Omega n'j'\Omega'}$. Note that we can set [24]

$$\Omega = -m_j,\; \Omega' = m_j'$$

where the helicity representation is used in which the axis of quantization of the incoming and outgoing rotational states, m_j and m_j', coincide with the direction of the incident and final wave-vetors respectively.

The differential and integral cross sections for reactive scattering are then obtained from the S matrix elements [24]. For example, the degeneracy averaged integral cross section is

$$\sigma(n, j \to n', j') = \frac{\pi}{k^2_{nj}(2j+1)} \sum_{J\Omega\Omega'} (2J+1)\, |S^J_{nj\Omega n'j'\Omega'}|^2 \qquad (9)$$

In the actual calculations, the $D^{J*}_{M\Omega}$ functions with the same absolute value of Ω can be combined to give basis functions of definite parity. This enables the coupled-channel equations to be separated into two independent sets of definite parity.

The above approach has been described in some detail to illustrate clearly how the approximations described in the next sections are related to the accurate close-coupling (CC) theory. The theory has been described for light-heavy-light reactions with complete neglect of the abstraction channel

$$L + LH \to L_2 + H.$$

This neglect of coupling between the two channels should be reasonable for reactions with potentials that are collinearly dominated, and have significant repulsive bending potentionals away from collinearity.

3.2 The Centrifugal Sudden Approximation

In the centrifugal sudden approximation (CSA), which is also known as the coupled states approximation, the operator $|\tilde{J}-\tilde{j}|^2$ is replaced by a value such as $[J^2+j^2 - 2\,J_z\,j_z]$ such that there is no coupling between different Ω states [20,24,25]. The S matrix elements are then of the form

$$S^{J\Omega}_{nj\,n'j'}$$

and the integral reaction cross sections are obtained from

$$\sigma(n,j \rightarrow n',j') = \frac{\pi}{k^2_{nj}\,(2j+1)} \sum_{J\Omega} (2J+1)\,|S^{J\Omega}_{nj\,n'j'}|^2 \qquad (10)$$

The matching can be done using the procedure described after equation (7).

The CSA calculations are cheaper than the CC because the basis set is not coupled in Ω. However, the calculations do have to be repeated for all values of J and Ω contributing to equation (10).

3.3 The ESA-CSA Method

In the ESA-CSA method [26] the energy sudden approximation (ESA) is applied to the entrance reaction channel and the CSA to the exit channel. This technique gives considerable computational simplications compared to the CC and CSA methods, although retaining a good accuracy as the results of Section 4 show.

The coordinates for the ESA-CSA method are defined in Figure 1.

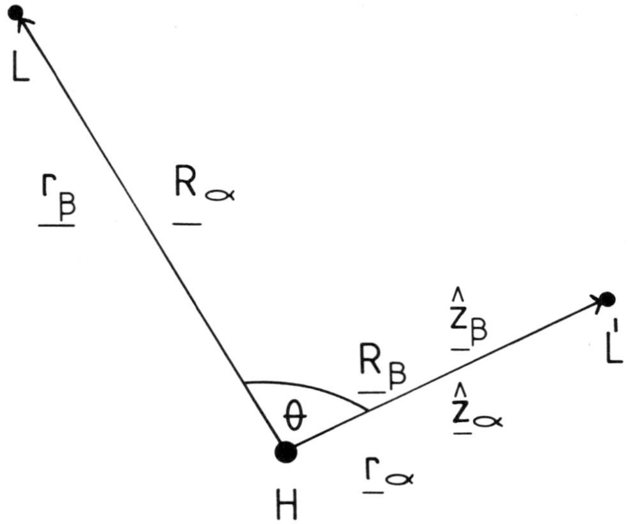

Figure 1. Coordinates for L-H-L' system

For the entrance channel α, the coordinates

$$\hat{z}_{\alpha} = \frac{\mathbf{r}_{\alpha}}{r_{\alpha}} \qquad \hat{y}_{\alpha} = \frac{\mathbf{r}_{\alpha} \times \mathbf{R}_{\alpha}}{|\mathbf{r}_{\alpha} \times \mathbf{R}_{\alpha}|} \qquad \text{and } \hat{x}_{\alpha} = \hat{y}_{\alpha} \times \hat{z}_{\alpha}$$

are used with the $\hat{z}_{\alpha}$ axis unconventionally placed along $\mathbf{r}_{\alpha}$ and not $\mathbf{R}_{\alpha}$. For the exit channel the normal body fixed coordinates

$$\hat{z}_{\beta} = \frac{\mathbf{R}_{\beta}}{R_{\beta}}, \quad \hat{y}_{\beta} = \frac{\mathbf{R}_{\beta} \times \mathbf{r}_{\beta}}{|\mathbf{R}_{\beta} \times \mathbf{r}_{\beta}|} \quad \text{and} \quad \hat{x}_{\beta} = \hat{y}_{\beta} \times \hat{z}_{\beta}$$

are used. In the case of an infinitely heavy central atom, $\mathbf{r}_{\alpha} = \mathbf{R}_{\beta}$ and $\mathbf{r}_{\beta} = \mathbf{R}_{\alpha}$. Thus the two sets of coordinate axes are entirely equivalent in that limit.

In the application of the ESA to the entrance channel α, the operator j_{α}^2 in the Hamiltonian of equation (1) is replaced by the average value $\hbar^2\ \bar{j}\ (\bar{j}+1)$ [27,28]. If $\bar{M}$ is the projection of $\boldsymbol{\ell}_{\alpha}$ along z_{α}, then the problem reduces to obtaining solutions for each separate $\bar{M}$ value. We have the coupled channel expansion

$$\Psi^{\alpha}_{\bar{M}\,\bar{j}}\,(u_{\alpha},v_{\alpha},\theta_{\alpha}) = \sum_{\ell=|\bar{M}|} \sum_{n} f^{\alpha}_{\bar{M}\,\bar{j}\,\ell\,n}\,(u_{\alpha})\ \ g_n(v_{\alpha})\ \ Y^{\bar{M}}_{\ell}\,(\theta_{\alpha},0) \tag{11}$$

This gives a set of close-coupling equations coupled in the orbital quantum number ℓ and n. As before, we solve these equations using the R-matrix propagator method [22,26].

For the exit reaction channel β, the CSA is applied in which the operator $\boldsymbol{\ell}_{\beta}^2$ is replaced by $\hbar^2\ \bar{\ell}\ (\bar{\ell}+1)$ [20,25]. If $\bar{N}$ is the projection of $\mathbf{j}_{\beta}$ along $\mathbf{z}_{\beta}$, then we now have solutions decoupled in $\bar{N}$. The appropriate coupled-channel expansion is

$$\Psi^{\beta}_{\bar{N}\bar{\ell}}\,(u_{\beta},v_{\beta},\theta_{\beta}) = \sum_{j=|\bar{N}|} \sum_{n} f^{\beta}_{\bar{N}\,\bar{\ell}\,j\,n}\,(u_{\beta})\ g_n(v_{\beta})\ \ Y^{\bar{N}}_{j}\,(\theta_{\beta},0) \tag{12}$$

and the equations are coupled in the rotational quantum number j and n.

In the limit of an infinitely heavy central atom, we have

$$z_{\alpha} = z_{\beta},\ \boldsymbol{\ell}_{\alpha} = \mathbf{j}_{\beta} \quad \text{and} \quad \boldsymbol{\ell}_{\beta} = \mathbf{j}_{\alpha}$$

It is thus appropriate to set

$$\bar{\ell} = \bar{j} \text{ and } \bar{M} = \bar{N}$$

when matching the primitive basis functions of equations (11) and (12) so that the matching matrices equivalent to those of equation (7) can be determined.

In the ESA-CSA method, the reactive translation functions for the transition $n\ \ell \rightarrow n'j'$ are

$$f^{\bar{j}\,\bar{M}}_{n\ell n'j'}(R_\beta) \xrightarrow[R_\beta\to\infty]{} -\left(\frac{k_{n'j'}}{k_{n\bar{j}}}\right)^{1/2} \exp\left[i(k_{n'j'}R_\beta - \bar{j}\pi/2)\right]$$

$$\times\ S^{\bar{j}\,\bar{M}}_{n\ell n'j'} \qquad (13)$$

Comparison with equation (8) shows that a transformation must be performed to rotate the entrance channel z_α axis so that its direction is along the normal body-fixed coordinate $\tilde{R}_\alpha$ instead of $\tilde{r}_\alpha$. This transformation is presented in [26] and the final result is

$$S^{J}_{nj\Omega\, n'j'\Omega'} = \left[\frac{(2j+1)}{(2J+1)}\right]^{1/2} (-1)^{J+\Omega}\ i^{j-J-j'}$$

$$\times \sum_{\ell}\ C(J\ j\ \ell\ -\Omega\ \Omega\ 0)\ \ C(\ell\ j\ J\ \Omega'\ 0\ \Omega')\ S^{j\Omega'}_{n\ell n'j'} \qquad (14)$$

where we have set $\bar{j}=j$ and $\bar{M}=\Omega'$. The C coefficients are Clebsch-Gordan coefficients.

Reaction integral and differential cross sections can then be obtained from these S matrix elements. For example, substitution of equation (14) into equation (9), and using the sum rule for the Clebsch-Gordan coefficients to sum over Ω, and the series rule to sum over J, we get the ESA-CSA reaction cross section

$$\sigma(n,j \rightarrow n',j') = \frac{\pi}{k^2_{nj}} \sum_{\Omega'} \sum_{\ell} |S^{j\Omega'}_{n\ell n'j'}|^2 \qquad (15)$$

This ESA-CSA formula has a distinct advantage over that obtained using the CSA, (see equation (10)). In the CSA, the calculations have to be repeated for many values of J. This is not necessary in the ESA-CSA as the coupled-channel expansion in the entrance channel contains the orbital spherical harmonics, the sum over which in equation (15) achieves the partial-wave expansion. For light-heavy-light reactions at room temperature energies, the maximum value of J required is about 20 and the ESA-CSA computations will then be 20 times cheaper than the CSA

3.4 The Infinite-Order-Sudden Approximation

The theory of the IOSA for light-heavy-light reactions was considered by Barg and Drolshagen [18]. It was then extended and implemented by Clary and Drolshagen [29]. Here we present an alternative derivation, based on the ESA-CSA method.

In the IOSA we set both $\tilde{j}^2_\gamma$ and $\tilde{\ell}^2_\gamma$ in the entrance and exit channels to the average values $\hbar^2\ \bar{j}_\gamma(\bar{j}_\gamma + 1)$ and $\hbar^2\ \bar{\ell}_\gamma\ (\bar{\ell}_\gamma + 1)$. Furthermore, if we use the matching conditions $j = \bar{j}_\alpha = \bar{\ell}_\beta$, $j' = \bar{\ell}_\alpha = \bar{j}_\beta$, and $\theta = \theta_\alpha = \theta_\beta$, we have the IOSA S matrix element

$$S^{j\ j'}_{n\ n'}(\theta).$$

To obtain the IOSA equivalent of the ESA-CSA S matrix element we take the matrix elements

$$S^{j\Omega'}_{n\ell n'j'} = < Y^{\Omega'}_{\ell}(\theta,0)\ |\ S^{jj'}_{nn'}(\theta)\,|\ Y^{\Omega'}_{j'}(\theta,0)> \qquad (16)$$

Substitution of this formula into equation (15), and using the completeness of the spherical harmonics to sum over ℓ, and the sum-rule to sum over Ω', we have the IOSA reaction cross section

$$\sigma(n,j \rightarrow n',j') = \frac{\pi}{k^2_{nj}}\ \frac{(2j'+1)}{2} \int_0^{\pi} |S^{jj'}_{nn'}(\theta)|^2 \sin\theta d\theta \qquad (17)$$

This formula is much cheaper to compute than the ESA-CSA formula (15) because there is no coupled-channel expansion in angular momentum states, and calculations do not have to be repeated for different values of the projection quantum number Ω'. Another important advantage of the IOSA for light-heavy-light reactions, is that boundary conditions can be applied using the correct entrance and exit channel wavenumbers k_{nj} and $k_{n'j'}$ and hence closed rotational states cannot be populated. This arises from the $\bar{\ell}_\alpha = \bar{j}_\beta$ matching condition, which is not appropriate in the IOSA theory for more general reactions where closed rotational states can be populated (see, for example, [30,31]). For light-heavy-light reactions, the IOSA clearly offers special simplifications that do not arise in other types of reactions (see the Chapter by Baer and Kouri in this book).

3.5 The BCRLM method

The rotating-linear model with corrections for bending zero-point energy (the BCRLM) has been applied recently, to reactions such as $H+H_2$ and $F+H_2$ by Walker and Hayes [12,13]. In this approach, collinear close-coupling calculations are performed using a hamiltonian which has had a centrifugal term, labelled by the orbital quantum number ℓ, added to it.

The potential also contains the zero-point energy for bending of the triatomic molecule. In the Walker-Hayes approach, which is described by Walker in his Chapter in this book, the bending zero-point energy is obtained by constructing an approximate hamiltonian in the valence bending angle θ_b and not the atom-molecule centre of mass orientation angle θ. Diagonalisation of this hamiltonian using a basis set of harmonic oscillators at grid-points along the reaction coordinate gives the bending energy [12].

For light-heavy-light reactions, however, $\theta = \theta_b$ to an excellent approximation. Thus, in this special case, an appropriate hamiltonian to use in obtaining the bending zero-point energy is

$$< g_o(v_\gamma)\,|\ \frac{\tilde{j}_\gamma^2}{2\mu_\alpha r_\gamma^2} + V(u_\gamma,v_\gamma,\theta)\,|\ g_o(v_\gamma)>$$

This can be diagonalised using a basis set of spherical harmonics to give a set of adiabatic curves labelled by the projection quantum number Ω and i (the i'th eigenstate in the Ω manifold).

For a given (i,Ω) adiabatic potential, a coupled basis set of vibrational basis functions is used for each ℓ value and the S matrix elements

$$S^{\ell i\Omega}_{nn'}$$

are obtained. For the initial rotational state j=0, the BCRLM cross section is obtained from

$$\sigma(j=0,\ n\rightarrow n') = \frac{\pi}{k^2_n} \sum_\ell \sum_i (2\ell+1)\ |S^{\ell i\Omega=0}_{nn'}|^2 \qquad (18)$$

One can also define the "rotationally accumulated" cross section

$$\sigma(n\rightarrow n') = \frac{\pi}{k^2_n} \sum_\ell \sum_\Omega \sum_i (2\ell+1)\ |S^{\ell i\Omega}_{nn'}|^2 \qquad (19)$$

which involes a summation over all bending Ω states.

From $\sigma(j=0,n\rightarrow n')$, the BCRLM rate coefficient can be obtained by Maxwell-Boltzmann averaging over these cross sections. As is shown in the Chapters by Bowman, Walker and Pollak, a more realistic rotationally averaged BCRLM (RBCRLM) rate coefficient is obtained by Maxwell-Boltzmann averaging the rotationally accumulated cross section $\sigma(n\rightarrow n')$ over translational energy, and dividing by the rotational partition function

$$\sum_j (2j+1)\ \exp[-Bj(j+1)/(k_\beta T)]$$

where B is the rotor constant, k_β is Planck's constant, T is temperature, and a rigid rotor has been assumed. Since (19) represents a sum over all possible bending states, which is not true in equation (18), then it is consistent to divide by the rotational partition function in the RBCRLM approach. In practice, the $\Omega=0$ state dominates the cross section summation for collinearly dominated reactions at lower collisional energies.

Below we present a summary of the various methods we have described for light-heavy-light reactions. Differences between the methods can also be clarified by examining the cross section equations (9), (10), (15), (17) and (18). We note also that variational-transition-state theories can also be applied to light-heavy-light reactions [32] but as these require no special modification for these reactions, we do not discuss their details here.

Summary of Methods for Light-Heavy-Light Reactions

Method	Acronym	Approximation	Basis Set
Close-Coupling	CC	None	n x j x Ω
Centrifugal-Sudden Approximation	CSA	Neglect Ω coupling	n x j for each Ω
(Energy-Sudden (entrance channel), Centrigugal-Sudden (exit channel) Approximations)	ESA-CSA	$\underset{\sim}{j}_{\alpha}^2=\hbar^2 j(j+1)$ $\underset{\sim}{\ell}_{\beta}^2=\hbar^2 \ell(\ell+1)$	n x ℓ (entrance channel) n x j (exit channel) for each Ω
Infinite-Order Sudden Approximation	IOSA	$\underset{\sim}{j}_{\gamma}^2=\hbar^2 j(j+1)$ $\underset{\sim}{\ell}_{\gamma}^2=\hbar^2 \ell(\ell+1)$ for $\gamma=\alpha$ and β	n, for quadrature in θ
Bending Corrected Rotating Linear Model	BCRLM	Neglect Rotations Add bending (Ω=0) zero-point energy	n
Rotationally Averaged BCRLM	RBCRLM	As for BCRLM but include all Ω states and the rate constant is rotationally averaged	

4. CALCULATIONS ON THE H+BrH REACTION

4.1 Potential Energy Surface

In this Section we discuss and compare the results obtained using the various theories of Section 3, and a variational transition state theory with a tunnelling correction [9]. The reaction we concentrate on is H+BrH → HBr+H. We emphasize the rate constants, but also discuss reaction cross sections. The potential energy surface we used in the H+BrH computations is of a semiempirical diatomics in molecules type, the form of which (called DIM-3C) is due to Last and Baer [11]. The surface contains a three-centre integral term that has been parameterised [33] by comparing ESA-CSA calculations with an experimental [3] room temperature rate constant for the D+BrH → DBr+H reaction. The minimum potential energy path is collinear, and there is a strong bending potential away from collinearity. This potential is of a realistic form, but we should emphasize that it is not likely that it is of a quantitative accuracy, and any results obtained using it should be compared cautiously with experiment.

For a given value of $u_\gamma = u_\gamma^i$, the potential energy surface is fitted to the form [34]

$$A(v_\gamma) + B(v_\gamma)\ [1 + \cos\theta_\gamma] \tag{20}$$

This gives a bending zero-point energy contribution to the reaction barrier of 0.044eV, using the methods described in 3.5. The 3D adiabatic barrier for this potential is 0.232eV [32]. We note that the methods of Garrett and Truhlar also give a bending zero point energy of 0.044eV [32] for this potential.

4.2 Previous ESA-CSA and IOSA Computations

An extensive series of 3D ESA-CSA and IOSA computations have been performed previously using DIM-3C potential energy surfaces. The reactions studied have been [33,34,35]

$$D+HC\ell(v=0,1) \rightarrow DC\ell+H \tag{III}$$

$$D+HI(v=0) \rightarrow DI+H \tag{IV}$$

$$H+BrH(v=0,1) \rightarrow HBr+H \tag{V}$$

and

$$D+BrH(v=0,1) \rightarrow DBr+H \tag{VI}$$

The first ESA-CSA calculations [34] on the D+HCℓ reaction were also done using a LEPS potential for comparison with the DIM-3C results. Calculations of integral and differential cross sections and rate constants were reported. It was found that the ESA-CSA calculations of a room temperature rate constant for the LEPS surface agreed well with those computed using the quasiclassical trajectory method [36]. However, the contribution from energies below the reaction barrier height to the room temperature rate constant was small and thus tunnelling is not important for this LEPS surface. This was not true in the ESA-CSA computations using the DIM-3C potential energy surface. Here, the reaction barrier is much narrower and the main contribution to the room temperature rate constant came from tunnelling energies. The less anisotropic LEPS potential gave a differential cross section that was distributed in both the forward and backward regions. The DIM-3C differential cross section, however, showed only backward scattering - which was in good agreement with experiment [2]. Since the anisotropy in the differential cross section directly reflected aristropy in the molecular potential, thus comparison suggested that the H+CℓH potential should be very anisotropic.

These ESA-CSA calculations on the DIM-3C potential surface were extended to the D+BrH and D+IH exchange reactions [33], and the HBrH potential was obtained as discussed in 4.1. There remains considerable doubt [37] concerning the barrier height for the HCℓH reaction, and thus H+BrH is a more suitable test case for a realistic study. Once again, ESA-CSA computations using the DIM-3C potential give a good agreement with the experimental [2] differential cross section, with strong backward peaking in the distribution. This is illustrated in Figure 2.

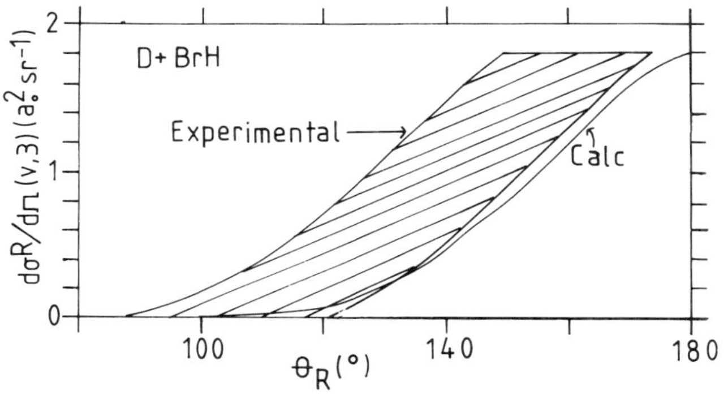

Figure 2. Differential Cross Sections for the D+BrH reaction, normalized to value at 180°.

The ESA-CSA calculations on the LEPS surface for D+CℓH were also extended [33] to the H+CℓH, H+CℓD and D+CℓD exchange reactions. Here it was found that the room temperature rate constants agreed to within a factor of two with those obtained using the classical trajectory method and a variational state theory with semiclassical adiabatic ground-state transmission coefficients [38].

The ESA-CSA computations on the DIM-3C surface were extended [35] to the vibrationally excited exchange reactions H+BrH(v=1) and D+BrH (v=1). IOSA calculations on these reactions were also reported [35] and were found to give cross sections and rate constants in good agreement with the ESA-CSA results. This is illustrated in Figures 3 and 4 where ESA-CSA and IOSA comparisons of integral cross sections, summed over product rotational states, for these vibrationally excited reactions are shown. The comparison is also good for product rotational distributions [35].

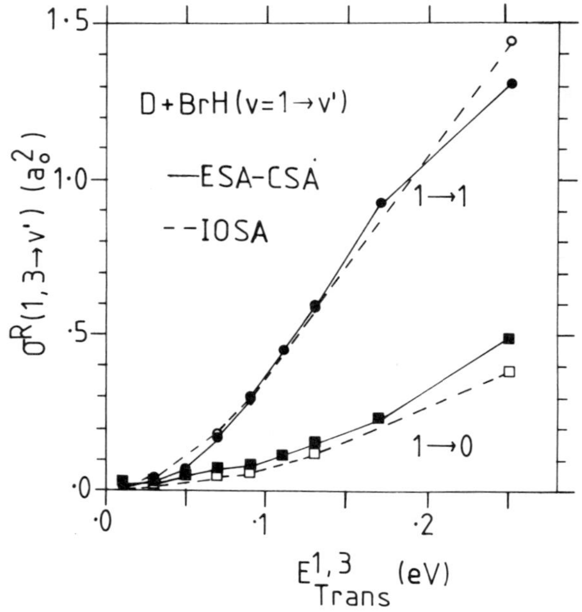

Figure 3. Cross Sections for the D+BrH(v=1) reaction

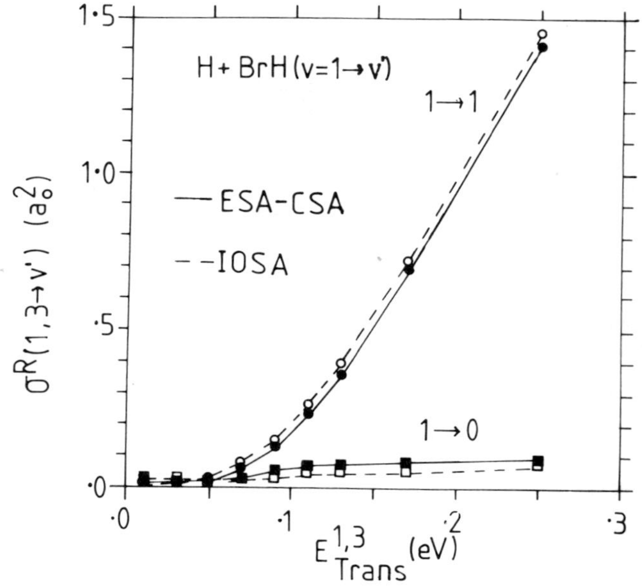

Figure 4. Cross Sections for the H+BrH(v=1) reaction

This study also produced results on isotope effects, which are illustrated in Table 2.

TABLE 2

Rate Constants for the H+BrH(v)→HBr(v')+H and D+BrH(v)→ DBr(v')+H reactions calculated using the ESA-CSA method [35]. The temperature is 300K. Units are $cm^3 molec^{-1} s^{-1}$.

System	v'=1	v'=0
H+BrH(v=1)	0.45(-12)[a)]	0.20(-12)
H+BrH(v=0)		0.16(-13)
D+BrH(v=1)	0.68(-12)	0.26(-12)
D+BrH(v=0)		0.13(-13)

a) Numbers in parentheses are powers of 10

These results show that, for the v=0 reactions, the H+BrH reactions has a slightly larger rate constant than that for D+BrH, and this is because tunnelling is so important. For the v=1 reactions, however, the effective barrier is not so large, and tunnelling is not so important. In this case the vibrationally adiabatic barrier is smaller for D+BrH(v=1) than for H+BrH(v=1), and consequently the D reaction has the larger rate constant. The results of Table 2 demonstrate that the vibrationally excited reactions have rate constants over an order of magnitude larger than those for v=0. They also show that the v=1 reactions are not totally vibrationally adiabatic, in the sense that the v=1→v'=0 rate constants are quite large.

4.3. A Comparison of Results for H+BrH(v=0)

In this Section we use results obtained using the close-coupling method of Section 3.1 to test the accuracy of those obtained using the CSA, ESA-CSA, IOSA, BCRLM and RBCRLM. We also compare with rate constants obtained using an improved canonical variational transition state theory combined with a small curvature tunnelling approximation and semi-classical adiabatic ground state transmission coefficient (ICVT-SCTSAG method) [39]. The reaction is H+BrH(v=0)→HBr(v=0)+H with the DIM-3C potential energy surface described in 4.1 and [33].

The numerical details of the CC calculations are published elsewhere [23], where some preliminary comparisons of the methods were made. The present comparison is more complete as we include CSA, BCRLM and RBCRLM results. Furthermore, all the IOSA and ESA-CSA computations reported in this Section used the initial rotational parameter $\bar{j}$=0. All of the results discussed in Section 4.2 were obtained from calaculations using $\bar{j}$=3 but, as was discovered recently [23], it is more appropriate to use $\bar{j}$=0 so that artificial lowering of the effective reaction barrier, due to bond stretching at the transition state, is avoided.

The CC calculations of the reaction cross sections

$$\sigma(j\rightarrow j') = \sigma(n=0,\ j\rightarrow n'=0,\ j') \tag{21}$$

and

$$\sigma(j) = \sum_{j'} \sigma(j\rightarrow j') \tag{22}$$

were done [23] over a range of energies sufficient to provide a room temperature rate constant. The CSA cross sections give very good agreement with the CC results. This is illustrated in Table 3.

TABLE 3

Comparison of CC and CSA cross sections $\sigma(j)$ for H+BrH. Numbers in parentheses are powers of ten.

j	$\sigma(j)/a_0^2$			
	E_{Tot} = 0.22eV		E_{Tot} = 0.34eV	
	CC	CSA	CC	CSA
0	0.753(-4)	0.757(-4)	.378(-1)	.363(-1)
1	0.594(-4)	0.597(-4)	.347(-1)	.330(-1)
2	0.355(-4)	0.356(-4)	.288(-1)	.270(-1)
3	0.161(-4)	0.161(-4)	.209(-1)	.188(-1)
4	0.526(-5)	0.528(-5)	.134(-1)	.115(-1)
5	0.111(-5)	0.111(-5)	.745(-2)	.600(-2)
6	0.120(-6)	0.120(-6)	.339(-2)	.250(-2)

One reason for the good accuracy of the CSA in this reaction is that the reaction is dominated by $\Omega=0\rightarrow\Omega'=0$ transitions [40] due to the strong bending potential. The Ω coupling, which is ignored in the CSA, will not thus be important. It is also observed that the CSA works well for the $H+H_2$ reaction [40,41] when compared with accurate CC results (see the Chapter by Schatz).

Figure 5 shows cross sections $\sigma(j=0)$ as a function of the initial translational energy E_j^{Trans}. Results obtained using the CC, IOSA and BCRLM methods are compared in this figure. At very low collision energies, the BCRLM results give a superior agreement with the CC cross sections than those obtained using the IOSA. However, at higher energies, which includes the energy range most relevant to rate constants for temperatures from 150-300K, the IOSA cross sections are much more accurate than the BCRLM.

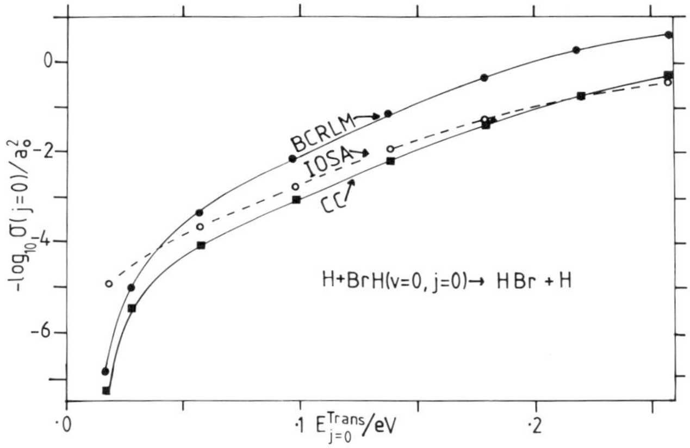

Figure 5. Cross Sections for the H+BrH(j=0) reaction

This comparison is extended in Table 4, where we report reaction rate constants k(T) obtained by Maxwell-Boltzmann averaging the cross sections.

TABLE 4 Rate Constants for the 3D H + BrH reaction. Units are $cm^3 s^{-1} molec^{-1}$

Temp/K	Method					
	ICVT-SCTSAG[32]	BCRLM	RBCRLM	IOSA	ESA-CSA	CC [23]
150	4.0(-16)[a)]	1.2(-15)	1.0(-16)	3.6(-16)	2.2(-16)	1.3(-16)
200	1.7(-15)	8.9(-15)	6.0(-16)	1.5(-15)	1.1(-15)	7.6(-16)
250	6.0(-15)	4.1(-14)	2.5(-15)	5.2(-15)	4.2(-15)	3.4(-15)
300	1.7(-14)	1.4(-13)	7.7(-15)	1.4(-14)	1.3(-14)	1.1(-14)

a) Numbers in parentheses are powers of ten.

Results obtained using the CC, IOSA, ESA-CSA, BCRLM, RBCRLM and ICVT-SCTSAG methods are compared in this Table. It can be seen that all the approximate methods overestimate the CC result apart from the RBCRLM. At 300K, the ESA-CSA, IOSA, RBCRLM and ICVT-SCTSAG rate constants all agree with the CC value to within a factor of two, which must be considered a very satisfactory agreement considering that the rate constant increases by a factor of 85 over the temperature range 150-300K.

The agreement between the CC and IOSA and ESA-CSA rate constants improves markedly as the temperature increases. This is to be expected from the cross-section results shown in Figure 5. Conversely, the BCRLM rate constants are too high by an order of magnitude. However, the RBCRLM rate constants show excellent agreement with the CC results at 150 K, although the agreement gets slightly worse as the temperature is increased. This is because the rotational partition function appears to weight the rate constant too heavily at higher temperatures in the RBCRLM approach.

The suitability of the BCRLM for a reaction such as H+BrH is further tested by determining the lowest bending wavefunction at the transition state, and determining the contribution from each j state (see 3.5). We find that the square of the spherical harmonic basis function coefficients have the values 0.075, 0.193, 0.234, 0.208, 0.145, 0.082, 0.039, 0.015 and 0.005 for j=0,1,...,7,8 respectively. Thus less than 10% of the transition-state bending wavefunction is j=0 in character. Hence, although j=0 is assigned to the BCRLM cross section (18), the rotationally averaged formalism is clearly more appropriate.

The main reason why the IOSA and ESA-CSA rate constants are slightly larger than the CC is that these approximate theories are not properly state-selective in j. The calculated CC rate constants $k_j(T)$, selected in the initial rotational state, showed a slight decrease as j is increased, while the ESA-CSA k(T) agreed very well with the CC $k_j(T)$ for j=0 [23]. The reason behind the decrease in $k_j(T)$ as j is increased is that the potential energy surface is collinearly dominated, and any initial rotational excitation will make it difficult for the attacking atom to "lock in" to the minimum potential energy path.

The agreement between the ICVT/SCTSAG and CC rate constants must be considered to be very satisfactory when it is realised that the SCTSAG transmission coefficient is 1050 at 150K and 5.9 at 300K [32]. Furthermore, alternative methods [42] for computing the transmission coefficients have recently been developed and these might give improved agreement with the CC results. Also, we should emphasise that there are error bars, concerned with numerical convergence, on the rate constants of Table 4. We estimate the error to be 12% for the CC results [23], although since the BCRLM, IOSA, ESA-CSA and CC calculations were done with very similar numerical procedures [23], the errors involved in comparing the results of these methods will be very slight indeed.

Since the ESA-CSA method gives such good rate constants, it is interesting to compare the rotational product distribution obtained using this method with those computed using the CC technique. Such a comparison is presented in Figure 6. It can be seen that the ESA-CSA calculations give too much rotational excitation in the HBr products. Thus, the sudden approximations used in the ESA-CSA, although giving a good accuracy for j' summed cross sections, are not quite so accurate for the rotational product distributions. This is also true for the IOSA, which gives rotational distributions that agree well with the ESA-CSA [35].

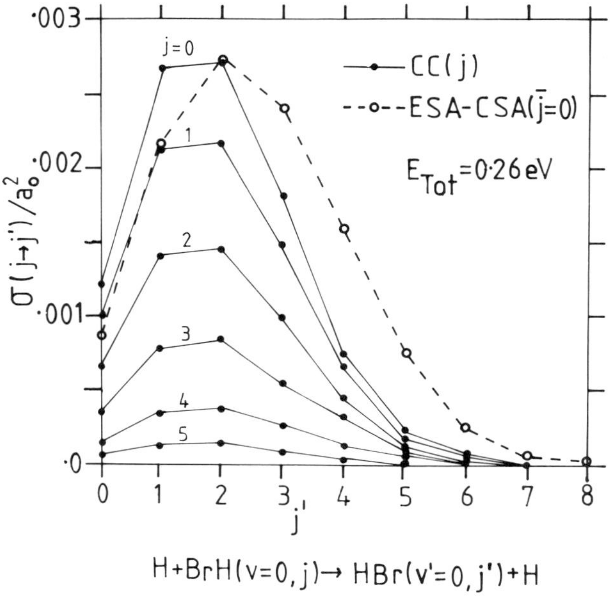

Figure 6: Rotational cross sections for H+BrH.

5. PHOTODISSOCIATION

As we emphasised in the Introduction, the theories for light-heavy-light reactions can be applied to the calculation of photodissociation spectra of symmetric triatomic molecules such as

$$H_2O \ (X\,^1A_1) \xrightarrow{h\upsilon} OH(X^2\pi) + H, \qquad \text{(VII)}$$

where the excited state of $^1B_1$ symmetry has a normal reactive scattering potential energy surface. Ab initio calculations suggest that the barrier height of the excited surface is 1.97eV [43]. Since the overlap between the ground and scattering state wavefunctions will only be large when the scattering energy is close to the energy of the barrier height, a large number of vibrational channels have to be included in the reactive scattering calculation. This makes accurate close-coupling or CSA computations very expensive, but IOSA and BCRLM calculations are tractable. Since the IOSA gave accurate cross sections at higher energies for the H+BrH reaction (see Figure 4), then it would seem appropriate to apply the IOSA to the photodissociation problem.

The photodissociation of H_2O into the $\tilde{A}\ ^1B_1$ state is of considerable experimental interest [44-46]. In particular Andresen et.al. have recently measured the rotation and vibration distributions of the product $OH(X^2\pi)$ diatomic radicals [45,46]. They used an excimer laser at 157nm to dissociate the H_2O molecule. Furthermore, ab initio calculations of points on the reactive potential energy surface for the $H(^2S) + OH(X^2\pi)$ system have been performed using the CEPA method [43]. We have used this ab initio data to parameterise a LEPS potential energy surface. Thus our predictions should be realistic.

There have been several high quality 3D calculations on the photodissociation of triatomic molecules using non-reactive potential energy surfaces [47-51]. Indeed, Segev and Shapiro calculated [47] the photoabsorption spectrum for the process

$$H_2O(X^1A_1) \overset{h\nu}{\rightarrow} OH(A^2\Sigma^+) + H \qquad \text{(VIII)}$$

using a non-reactive potential energy surface for the product channel which has B^1A_1 symmetry. As far as we are aware, however, there have been no previous 3D calculations of photodissociation cross sections for a reactive product potential energy surface. However, there have been some reactive photodissociation calculations for collinear CO_2 by Kulander and Light [10] and for fixed-angle H_2S by Kulander [52].

In 5.1, we describe how the IOSA is used to calculate photodissociation cross sections. Particular care has to be taken in averaging over the orientation-angle dependent scattering wavefunction using the bending wavefunction of the ground state. To do this we apply, to the reaction problem, a procedure developed by Segev and Shapiro for the vibrational predissociation of Van der Waals complexes [51]. Then in 5.2, we present our calculations of the photoabsorption spectrum for the H_2O photodissociation (VII).

5.1 IOSA for Reactive Photodissociation

Here we describe the essential features of the application of the IOSA to reactive photodissociation. More details of the approach will be published elsewhere [53]. We wish to calculate the photodissociation cross sections I discussed in Section 1. For most problems, the transition dipole moment surface μ_D will not be known. It is unlikely that μ_D will have a significant effect as it is not a highly oscillatory function like the scattering wavefunction, nor a highly localized function like the bound state. Thus, it should be a reasonable approximation to compute

$$I = C' < \psi_m^{(-)} | \phi_n >^2$$

Here we describe the theory for initial total angular momentum J=0, although the theory for J>0 is straightforward.

In computing the bound-state wavefunction we continue to use the body-fixed coordinates defined in 3.3. We choose a fixed orientation angle θ, neglect the j_γ^2 and ℓ_γ^2 terms of equation (1), and use a coupled basis set of harmonic oscillators to solve the resulting bound-state Schrodinger equation which depends on R_γ and r_γ only. The energy levels obtained in this way are called $\varepsilon_{n'}(\theta)$ and correspond approximately to those for the symmetric and asymmetric stretch normal coordinates. The wavefunction obtained from this calculation is $\phi^S_{n'}(R,r;\theta)$. Once a grid of $\varepsilon_{n'}(\theta)$ has been obtained for various values of θ, the Schrodinger equation

$$\left[\frac{\underset{\sim}{j}^2}{2\mu_\alpha}\left[\frac{1}{r_\gamma^2} + \frac{1}{R_\gamma^2}\right] + \varepsilon_{n'}(\theta) \right] \phi^b_{n''n'}(\theta) = \varepsilon_{n''n'}\ \phi^b_{n''n'}(\theta) \tag{24}$$

is solved using a basis set of Legendre polynomials with r_γ and R_γ set to the equilibrium values. The resulting wavefunction in this "IOS-bound state" approximation [51] is

$$\phi_n \equiv \phi^s_{n'}(R,r;\theta)\ \phi^b_{n''n'}(\theta) \tag{25}$$

Note that the bound state label $n \equiv (n'',n')$. For many problems, including the one of interest here, only the ground state n=0 will be of interest.

For the scattering state, the IOSA is applied in the way described in 3.4 with the $\bar{\ell}$ and $\bar{j}$ quantum numbers set to zero. We obtain the θ dependent scattering wavefunction

$$\psi^{(-)}_{m'}(R,r;\theta),$$

where m' defines the final diatomic state and the boundary conditions appropriate to photodissociation have been applied [10]. These computations are repeated for various values of θ, and the IOSA wave-function

$$\psi_m \equiv \psi^{(-)}_{m'}(R,r;\theta)(2\pi)^{1/2}\ Y^o_j(\theta,0) \tag{26}$$

is obtained where $m \equiv (m',j)$. The photodissociation overlap is then given by

$$\langle\psi^{(-)}_m|\phi_n\rangle = \int_0^\infty R^2\,dR \int_0^\infty r^2\,dr \int_0^\pi \sin\theta\, d\theta$$

$$\times\ \psi^{(-)*}_{m'}(R,r;\theta)\,(2\pi)^{1/2}\ Y^o_j(\theta,0)\,\phi^s_{n'}(R,r;\theta)\ \phi^b_n(\theta). \tag{27}$$

Substitution of this expression into (23) and using the completeness of the spherical harmonics to sum over j gives

$$I(\omega,m,n) = C' \int_0^\pi \sin\theta\, d\theta\ \ |\phi_n^b(\theta)|^2\ |F_{mn'}(\theta)|^2 \tag{28}$$

where

$$F_{mn'}(\theta) = \int_0^\infty R^2\,dR \int_0^\infty r^2\,dr\ \ \psi^{(-)*}_m(R,r;\theta)\ \phi^s_{n'}(R,r;\theta) \tag{29}$$

where m and n' now refer to vibrational states only. Hence the problem reduces to the calculation of the overlaps of equation (29) for each θ, and averaging these overlaps using the bending wavefunction $\phi_n^b(\theta)$ as shown in equation (28).

Kulander and Light [10] described how $F_{mn'}(\theta)$ can be calculated efficiently within the framework of the R-matrix propagator method [22]. This technique involves dividing up the scattering coordinate R into small sectors and propagating the R-matrix, the ratio between the scattering wavefunction and its derivative, from sector-to-sector. Kulander and Light showed that integrals similar to $F_{mn'}(\theta)$ can be propagated numberically at the same time as the R-matrix. Multiplication by an appropriate complex matrix at the end of the propagation ensures that $\psi_m^{(-)}(R,r;\theta)$ has the correct boundary conditions [10,48,54].

The IOSA theory for reactive photodissociation described above, together with the Kulander-Light technique for calculating photodissociation integrals, has been incorporated into a general computer program for symmetric triatomic molecules. The program can be used for various mass combinations. For example, calculations of the 3D photoabsorption spectrum for

$$CO_2 \rightarrow CO(^1\Sigma^+) + O(^1S)$$

have been performed [53]. It should be noted that the angular range of θ contributing to the integral (28) is highly localized about the equilibrium bond angle for the ground vibrational state and hence the IOS approximation of holding θ fixed should be very reasonable, even for non light-heavy-light reactions. However, the technique is still best applied to light-heavy-light systems for which θ is a natural orientation angle for the reactive scattering calculations. Clearly H_2O is such a system.

5.2 Calculations on H_2O

We have calculated [53], the full photodissociation spectrum for the dissociation of H_2O into $OH(X^2\pi)$+H on the first absorption band, the $\tilde{A}\ ^1B_1$ state. A LEPS potential was used for the dissociating state in which the Sato parameters were made θ dependent and were varied to obtain a good fit to the ab initio data of Staemmler and Palma [43]. This potential had a barrier height of 1.97eV and, at the transition state, the O-H bond length was $R_{OH} = 1.08$Å and the bond angle was $\theta(H_2O) = 105^\circ$. The Sorbie-Murrell potential [55] was used for the H_2O ground state and here $R_{OH} = 0.957$Å and $\theta(H_2O) = 104.5^\circ$.

Figure 7 shows our calculated photodissociation spectrum for H_2O, summed over all final OH vibrational states. The energy on the axis corresponds to the OH(v=0) reference energy and the energies 2-3.7eV correspond to the laser wavelength 141-174nm.

The shape of the calculated spectrum is in good agreement with experiment [56], although the calculated peak occurs at 0.5eV higher than the experimental peak. This suggests that the barrier height in the potential used is probably too high. We also computed the OH(v) vibrational state population at an energy of 2.79eV which corresponds to a photon at 157nm, which is the wavelength at which Andresen et.al. [46] measured the OH(v) distributions. For the relative populations in the v=0, 1 and 2 states we obtained the ratio 1:1.07:0.65 which compares fairly well with the experimental ratio of 1:1:0.15.

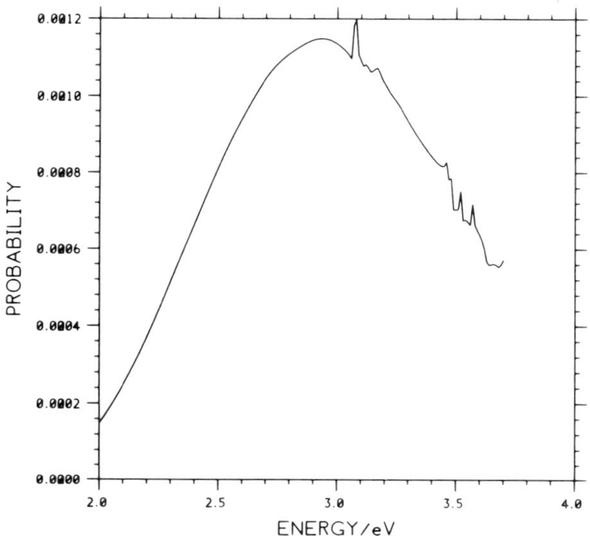

Figure 7. Photodissociation spectrum for H_2O

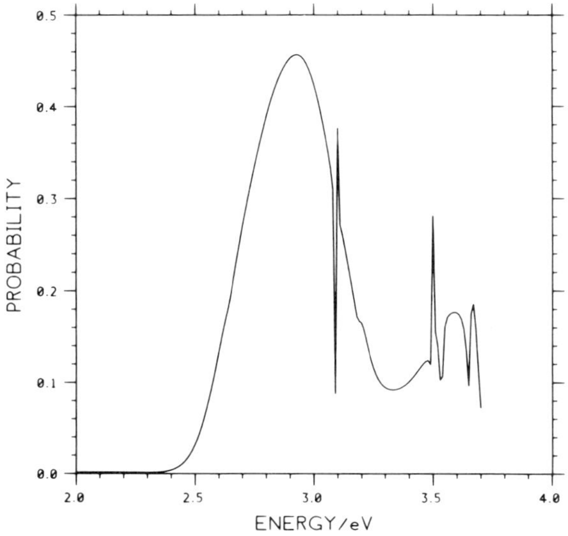

Figure 8. Reaction probabilities for $H+OH(v=4) \rightarrow HO(v'=4)+H$ at $\theta = 103^{o}$.

One interesting feature of our calculated spectrum, is the observation of structure at higher energies. This is due to reactive scattering resonances. Figure 8 shows a calculation of reaction probabilities for H+OH(v=4)→HO(v'=4)+H at $\theta = 103^{\circ}$. It can be seen that the sharp extremes in these reaction probability curves correspond exactly to the energies at which the structure is seen in the photodissociation spectrum. Furthermore, the IOS angle averaging of equation (28) does not completely wash out the resonance structure. The reactive resonance is caused by vibrationally adiabatic potential energy curves having local minima for higher vibrational states which cause flux to get temporarily trapped at certain collision energies [53]. These types of vibrationally adiabatic local minima are a common feature of chemical reaction potential energy surfaces [57] and are due to the vibrational force constant reducing in magnitude along the reaction path. The experimental search for reactive resonances has largely been confined to molecular beam experiments of the angular distributions for the products of chemical reactions [58]. Our calculations suggests that a more positive identification of reactive resonances might occur in high-resolution photodissociation experiments, particularly if the spectrum for each final vibrational state can be resolved. Photodissociation experiments explicitly probe the transition-state, and the photodissociation process does not suffer so heavily from the angular momentum summation that tends to average out resonances in 3D reaction cross sections [59]. Indeed, we predicted the importance of reactive resonances in photodissociation in a model calculation [60] and our realistic calculations discussed here suggest that they should be detectable. More details will be presented elsewhere [53].

6. CONCLUSION

Light-heavy-light chemical reactions are a rare class of reaction in that close-coupling calculations, and a variety of approximations, can be used readily to calculate quantities such as cross sections and rate constants. Such computations are possible because the centre-of-mass in both entrance and exit reaction channels can be placed on the heavy atom, and the matching between entrance and exit channel wavefunctions is much simplified. We have reported close-coupling calculations for the H+BrH reaction in three dimensions, and these results have been used to test the accuracy of a variety of approximate theories including an extended transition state theory, some sudden approximations and two rotating linear models. Light-heavy-light reactions are systems which offer the most comprehensive and tractable test of these approximations for reactions beyond $H+H_2$.

An application of light-heavy-light reaction theory to a problem of considerable current interest is to the computation of the photodissociation spectra for symmetric triatomic molecules having reactive dissociative potential energy surfaces. We have reported IOSA computations of the photodissociation spectrum for the first absorption band of H_2O. Reactive scattering resonances produce some structure in this calculated spectrum. The results suggest that photodissociation experiments on symmetric triatomics might detect reactive scattering resonances directly.

ACKNOWLEDGEMENTS: This research was supported by the U.K. Science and Engineering Research Council (SERC). JPH acknowledges a Research Studentship from the SERC. The computations were performed on the CRAY-1 Computer of the University of London Computer Centre.

REFERENCES

[1] K. P. Huber and G. Herzberg, Constants of Diatomic Molecules, (Van Nostrand, 1979).
[2] J. D. McDonald and D. R. Herschbach, J. Chem.Phys., **62**, 4740 (1975).
[3] H. Endo and G. P. Glass, J. Phys.Chem., **80**, 1519 (1976).
[4] C. A. Wight, F. Magnotta and S. R. Leone, J.Chem.Phys., **81**, 3951 (1984).
[5] R. W. F. Gross and J. F. Bott, Handbook of Chemical Lasers, (John Wiley, 1976).
[6] P. Botschwina and W. Meyer, Chem.Phys., **20**, 43(1977).
[7] T. H. Dunning Jr., J.Phys.Chem., **88**, 2469(1984).
[8] C. F. Bender, B. J. Garrison and H. F. Schaefer III, J.Chem.Phys., **62**, 1188(1975).
[9] D. G. Truhlar and B. C. Garrett, Acc.Chem.Res., **13**, 440(1980).
[10] K. C. Kulander and J. C. Light, J.Chem.Phys., **73**, 4337(1980).
[11] I. Last and M. Baer, J.Chem.Phys., **75**, 288(1981).
[12] R. B. Walker and E. F. Hayes, J.Phys.Chem., **87**, 1255(1983).
[13] E. F. Hayes and R. B. Walker, J.Phys.Chem., **88**, 3318(1984).
[14] M. Baer and D. J. Kouri, J.Chem.Phys., **56**, 1758(1972); J.Chem.Phys., **57**, 3441(1972).
[15] M. Baer, Molec.Phys., **26**, 369(1973).
[16] M. Baer and D. J. Kouri, Phys.Rev., **A4**, 1924(1971).
[17] M. Baer, J.Chem.Phys., **65**, 493(1976).
[18] G. D. Barg and G. Drolshagen, Chem.Phys., **47**, 209(1980).
[19] G. Drolshagen, Diplom. Thesis, Bericht, Max-Planck Institut fur Stromungsforschung, Gottingen, FRG (1979).
[20] R. T. Pack, J.Chem.Phys., **60**, 633 (1974).
[21] B. R. Johnson, Chem.Phys.Lett., **13**, 172 (1972).
[22] J. C. Light and R. B. Walker, J.Chem.Phys., **65**, 4272(1976).
[23] D. C. Clary, J.Chem.Phys., **83**, 1685(1985).
[24] G. C. Schatz and A. Kuppermann, J.Chem.Phys., **65**, 4642(1976).
[25] P. McGuire and D. J. Kouri, J.Chem.Phys., **60**, 2488(1974).
[26] D. C. Clary, Molec.Phys., **44**, 1067(1981).
[27] S. I. Chu and A. Dalgarno, Proc.R.Soc.London, A342, 191(1975).
[28] V. Khare, J.Chem.Phys., **68**, 4631(1978).
[29] D. C. Clary and G. Drolshagen, J.Chem.Phys., **76**, 5027(1982).
[30] J. M. Bowman and K. T. Lee, J.Chem.Phys., **72**, 5071(1980).
[31] D. J. Kouri, V. Khare and M. Baer, J.Chem.Phys., **75**, 1179(1981).
[32] D. C. Clary, B. C. Garrett and D. G. Truhlar, J.Chem.Phys., **78**, 777(1983).
[33] D. C. Clary, Chem.Phys., **71**, 117(1982).
[34] D. C. Clary, Molec.Phys., **44**, 1083(1981); Chem.Phys.Lett., **80**, 271(1981).

[35] D. C. Clary, Chem.Phys., **81**,379(1983).
[36] T. Valencich, J. Hsieh, J. Kwan, T. Stewart and T. Lenhardt, Ber. Bungsenges, Phys.Chem., **81**, 131(1977).
[37] R. E. Weston Jr., J.Phys.Chem., **83**, 61(1979).
[38] B. C. Garrett, D. G. Truhlar and A. W. Magnuson, J.Chem.Phys., **74**, 1029(1981)
[39] B. C. Garrett, D. G. Truhlar, R. S. Grev and A. W. Magnuson, J.Phys.Chem., **84**, 1730(1980); R. T. Skodje, D. G. Truhlar and B. C. Garrett, J.Phys.Chem., **85**, 3019(1980); J.Chem.Phys., **77**, 5955(1982).
[40] A. Kuppermann, G. C. Schatz and J. P. Dwyer , Chem.Phys.Lett., **45**, 71(1977).
[41] G. C. Schatz, Chem.Phys.Lett., **108**, 532(1984).
[42] B. C. Garrett and D. G. Truhlar, J.Chem.Phys., **79**, 4931(1983).
[43] V. Staemmler and A. Palma, Chem.Phys., **93**, 63(1985).
[44] K.-H. Welge and F. Stuhl, J.Chem.Phys., **46**, 2440(1967).
[45] P. Andresen, G. S. Ondrey and B. Titze, Phys.Rev.Lett., **50**, 486(1983)
[46] P. Andresen, G. S. Ondrey, B. Titze and E. W. Rothe, J.Chem.Phys., **80**, 2548(1984).
[47] E. Segev and M. Shapiro, J.Chem.Phys., **77**, 5601(1982).
[48] R. W. Heather and J. C. Light, J.Chem.Phys., **78**, 5513(1983); **79**, 147(1983).
[49] J. A. Beswick and M. Shapiro, Chem.Phys., **64**, 333(1982).
[50] I. F. Kidd and G. G. Balint-Kurti, J.Chem.Phys., **82**, 93(1985).
[51] E. Segev and M. Shapiro, J.Chem.Phys., **78**, 4969(1983).
[52] K. Kulander, Chem.Phys.Lett., **103**, 373(1984).
[53] J. P. Henshaw and D. C. Clary, to be published.
[54] K. Takatsuka and M. Gordon, J.Chem.Phys., **74**, 5718(1981).
[55] K. S. Sorbie and J. N. Murrell, Molec.Phys., **29**, 1387(1975).
[56] K. Watanabe and M. Zelikoff, J.Opt.Soc.Am., **43**, 753(1953).
[57] B. C. Garrett, D. G. Truhlar, R. S. Grev, G. C. Schatz and R. B. Walker, J.Phys.Chem., **85**, 3806(1981).
[58] D. M. Neumark, A. M. Wodtke, G. N. Robinson, C. C. Hayden and Y. T. Lee, J.Chem.Phys., **82**, 3045(1985).
[59] M. J. Redmon and R. E. Wyatt, Chem.Phys.Lett., **63**, 209(1979).
[60] D. C. Clary, J.Phys.Chem., **86**, 2569(1982).

ARRANGEMENT CHANNEL QUANTUM MECHANICAL APPROACH TO REACTIVE SCATTERING

D. J. KOURI
Department of Chemistry
University of Houston - University Park
Houston, Texas 77004
United States

M. BAER
Applied Mathematics
Soreq Nuclear Research Center
Yavne 70600
Israel

ABSTRACT. The arrangement channel quantum mechanical formalism is discussed as a general approach to reactive scattering. Features of the equations are examined and the positive and negative aspects are noted. Both differential and integral equation versions of the theory are presented (with primary emphasis on the solution of the integral equations), as well as time independent and time dependent approaches. Explicit algebraic equations suitable for numerical solution are derived for a general 3-dimensional reactive collision system. Previous applications of the approach are also reviewed.

1. INTRODUCTION

The central difficulty in reactive scattering which complicates it compared to nonreactive scattering is the fact that there exist multiple asymptotia, each of which corresponds to a distinct possible clustering of the particles comprising the system. Associated with each clustering, there are in general sets of variables or degrees of freedom which describe internal motions of the cluster, there are variables describing the motion of various clusters relative to one another and there are the variables describing the motion of the overall system center of mass. For each arrangement of the particles into a set of distinct clusters, the total wavefunction must satisfy appropriate scattering boundary conditions in the limit that the relative separations of all particle clusters in that arrangement become large. The simultaneous imposition of these different asymptotic boundary conditions is quite complicated when standard Jacobi coordinates for each arrangement and the usual Schrodinger equation are employed. A second, related difficulty is the fact that the natural basis states for describing various internal degrees of freedom are not orthogonal for different arrangements, except in the limit of infinite separations. There are

D. C. Clary (ed.), The Theory of Chemical Reaction Dynamics, 359–381.

several ways in which this difficulty can be addressed. The first was implemented many years ago in the early days of quantum mechanics and has subsequently been used in the only successful converged full three dimensional (3D) calculations for reactive scattering for a realistic atom-diatom reaction, $H+H_2$. (1-9) In this approach, configuration space is divided up into separate regions in which the Schrodinger equation is solved using appropriate coordinates for each region. Then the resulting wavefunctions and their directional derivatives are matched on surfaces which divide the various regions. Finally, the wavefunction is required to satisfy the proper boundary conditions in the various asymptotic limits. The matching of the wavefunction and its directional derivative is in general very complicated and time consuming.

A second approach to this problem of multiple asymptotia is to introduce specialized coordinates which change smoothly from one arrangement to the others in such a way as to enable one to analyze the wavefunction in the various arrangement limits. This leads to the definition of generalized "reaction coordinates". (10-16) The leading examples of these are the natural collision coordinates first introduced by Marcus (12) and the hyperspherical coordinates pioneered by Delves (10). These have led to important advances in reactive scattering theory but are discussed by others active in their study so we do not pursue them further here.

The third approach to this problem was initiated in work due to Eyges (18), whose idea was later utilized by Faddeev (19) in the first rigorous mathematical treatment of the three body problem. In this approach, the Schrodinger equation is replaced by a system of equations for arrangement channel components of the wavefunction. (17-43) As we shall see, each piece or component of the wavefunction describes one (and only one) asymptotic region of the multiparticle system, when the collision energy is below the breakup threshold. In this formalism, the single Hilbert space associated with the Schrodinger equation is replaced by a direct product of Hilbert spaces, one for each arrangement or asymptotic limit possible. The Hamiltonian operator is also replaced by a matrix in arrangement channel space, which acts on a column vector the components of which are the pieces of the wavefunction associated with the various arrangements. There turns out to be considerable latitude possible in defining these wavefunction arrangement channel components and we shall focus on the one which, in our opinion, lends itself most simply to the treatment of reactive scattering. The Faddeev approach has, however, been employed in a number of studies also (44). These formalisms can be utilized in either differential or integral equation form as well as time independent or dependent form. We shall discuss all of these but our primary focus will be on the integral equation approach. This paper is organized as follows. In Section 2, we present the basic arrangement channel quantum mechanics formalism, both in differential and integral equation form. We discuss also some of the positive and negative features of the approach. Then in Section 3, we show in detail how the equations are cast into a form suitable for actual computations. In Section 4 we briefly review the calculations done to date using the integral equation form of the equations. In Section 5 we discuss the status of the implementation of these equations.

2. ARRANGEMENT CHANNEL QUANTUM MECHANICS

2.1 Time Independent Differential Equations

We begin by considering the time independent Schrodinger equation for a system having three possible arrangements:

$$A + BC \rightarrow A + BC \tag{1}$$
$$\rightarrow B + AC \tag{2}$$
$$\rightarrow C + AB \tag{3}$$

We shall employ Greek labels to denote an arbitrary arrangement (clustering). When all three particles are close to one another, the hamiltonian is taken to be

$$H = T + V, \tag{4}$$

where T is the total kinetic energy of the system and V is the total potential energy of the system. We associate with each arrangement λ a vector $\underset{\sim}{R}_\lambda$ between atom λ and the center of mass of the diatom in arrangement λ, and a vector $\underset{\sim}{r}_\lambda$ between the two atoms of the diatom in arrangement λ. Then we define H_λ by

$$\lim_{R_\lambda \to \infty} H \equiv H_\lambda; \tag{5}$$

H_λ is the unperturbed hamiltonian describing the atom and diatom in arrangement λ. From equation (4),

$$H_\lambda = T + V_\lambda^{(\lambda)}, \tag{6}$$

where $V^{(\lambda)}$ is the binding potential for the diatom in arrangement λ. Then the perturbation responsible for scattering in arrangement λ is

$$V_\lambda = H - H_\lambda. \tag{7}$$

We notice that by equation (5),

$$\lim_{R_\lambda \to \infty} V_\lambda \equiv 0. \tag{8}$$

The time independent Schrodinger equation is

$$(E - H)\psi = 0, \tag{9}$$

and we note that by equation (8),

$$H = H_\lambda + V_\lambda, \text{ all } \lambda. \tag{10}$$

Let us now divide the total wavefunction ψ into 3 pieces,

$$\psi = \sum_{\lambda=1}^{3} \psi_\lambda . \tag{11}$$

Then Schrodinger's equation can be written as

$$(E - H_1 - V_1)\psi_1 + (E - H_2 - V_2)\psi_2 + (E - H_3 - V_3)\psi_3 = 0 \tag{12}$$

Up to this point, the ψ_λ are arbitrary. We remove this arbitrariness by requiring that

$$(E - H_1)\psi_1 - V_2\psi_2 = 0, \tag{13}$$

$$(E - H_2)\psi_2 - V_3\psi_3 = 0, \tag{14}$$

and

$$(E - H_3)\psi_3 - V_1\psi_1 = 0. \tag{15}$$

It is clear that we are simply requiring that equation (12) be satisfied by a particular cancellation of terms. It is the specification of how the different pieces in equation (12) cancel that determines the particular version of arrangement channel quantum mechanics (ACQM) that one obtains (i.e., the above choice is the "channel permuting" choice of Baer and Kouri (21-23), Kouri and Levin (25) and Tobocman (26), the BKLT-ACQM formulism). A different specification of the cancellation (i.e., a different specification of the components ψ_λ of ψ) leads to a variant of the Faddeev equations. Rewriting equations (13)-(15),

$$(E - H_1)\psi_1 = V_2\psi_2, \tag{16}$$

$$(E - H_2)\psi_2 = V_3\psi_3, \tag{17}$$

$$(E - H_3)\psi_3 = V_1\psi_1, \tag{18}$$

which are the BLKT-ACQM equations for a 3-arrangement system. It is of interest to examine how the ψ_λ behave in the various relevant asymptotia. We begin by considering equation (18) in the limit that R_1 becomes large. We note that H_3 contains the full kinetic energy for the system plus the binding potential for atoms 1 and 2. The limit under consideration is one in which the distance of atom 1 from the center of mass of atoms 2 and 3 becomes large. When this occurs, first V_1 tends to zero so the equation becomes

$$\lim_{R_1\to\infty} (E - H_3)\psi_3 = 0. \tag{19}$$

However, V_{12} also will tend to zero in this limit since particle 1 is receding away from both particles 3 and 2. Thus, the limiting form of the equation is

$$\lim_{R_1\to\infty} (E - T)\psi_3 = 0. \tag{20}$$

so that ψ_3 must have the character of a totally free wave at large separations of particle 1 from particles 2 and 3. However, at energies <u>below</u> the 3 body breakup, the wavefunction cannot display totally free character at large separation (this is a closed channel and therefore not observable at large separation). Thus, our conclusion is that

$$\lim_{R_1 [illegible]} \psi [illegible] \; 0. \tag{21}$$

Technically, [illegible] y that (E - T) cannot support any bound subclusters as $R_1 \to \infty$ so tha[illegible] low the breakup threshold, ψ_3 must tend to zero in this limit. Ar[illegible] with this result, we now examine equation (17). The right hand side [illegible]anishes as $R_1 \to \infty$ due to the appearance of the factor ψ_3. Then we have

$$\lim_{R_1 \to \infty} (E - H_2)\psi_2 = 0; \tag{22}$$

but H_2 involves the total kinetic energy and the binding potential V_{13}. This binding potential also tends to zero since particle 1 is receding from both particles 2 and 3. We are left with the equation

$$\lim_{R_1 \to \infty} (E - T)\, \psi_2 = 0, \tag{23}$$

so that ψ_2 can only involve free wave motion in this limit. If the energy is below the breakup threshold, then the totally free channel is closed and cannot be observed in any asymptotic limit. Therefore, we conclude that

$$\lim_{R_1 \to \infty} \psi_2 = 0 \tag{24}$$

also. Finally, we use this result in equation (16) to find

$$\lim_{R_1 \to \infty} (E - H_1)\psi_1 = 0. \tag{25}$$

Now, however, the binding potential is that for particles 2 and 3, V_{23}, and this remains nonzero as particle 1 recedes from 2 and 3. Thus, in the limit $R_1 \to \infty$, $(E-H_1)$ can support bound clusters of atoms 2 and 3, with particle 1 free. This is just the type of state associated with arrangement 1 so that ψ_1 asymptotically contains <u>all</u> the scattering information in which atoms 2 and 3 are bound and atom 1 is free. ψ_2 and ψ_3 contain <u>none</u> of this information and indeed tend to zero in the region of configuration space associated with this arrangement. One may do an analogous analysis of equations (16)-(18) in the limit that either R_2 or R_3 tends to infinity. The result is that

$$\lim_{R_\lambda \to \infty} \psi_{\lambda'} = 0 \tag{26}$$

unless $\lambda = \lambda'$ and ψ_λ contains all the scattering information associated with arrangement λ, in which atom λ is free and the other two atoms are bound.

In the computational studies to be described in this paper, we actually deal with the so called amplitude densities, which are related to the arrangement channel wavefunction components by

$$\zeta_\lambda = V_{\lambda+1} \psi_{\lambda+1}, \tag{27}$$

where the index ranges from 1 to 3 on the left and modulus 3 on the right. It is noteworthy that while the ψ_λ tend to zero in the limit $R_{\lambda'} \to \infty$, $\lambda' \neq \lambda$, the ζ_λ tend to zero in the limit that <u>any</u> of the scattering coordinates $R_{\lambda'} \to \infty$. This is ensured by the fact that $V_\lambda \psi_\lambda$ tends to zero as $R_{\lambda'} \to \infty$, $\lambda' \neq \lambda$ due to the factor ψ_λ and as $R_\lambda \to \infty$ due to the factor V_λ (see equation (8)). Thus, provided the V_λ is well behaved as $R_\lambda \to 0$ (i.e., not too strongly divergent), the ζ_λ can be expanded in terms of quadratically integrable basis functions.

Therefore, bound state-like solution methods can be employed in calculating the ζ_λ. In place of the differential equations for ψ_λ, equations (16)-(18), it is therefore convenient to solve instead integral equations for the ζ_λ. To derive these most easily, we write the formal solution for the ψ_λ's in matrix form as

$$\underline{\psi}(\lambda i) = \underline{\phi}(\lambda i) + \underline{\underline{G}}_0^+ \underline{\underline{V}} \underline{\psi}(\lambda i) \tag{28}$$

where

$$[\underline{\psi}(\lambda i)]_{\lambda'} = \psi_{\lambda'}(\lambda i) \tag{29}$$

$$[\underline{\phi}(\lambda i)]_{\lambda'} = \delta_{\lambda\lambda'} \phi(\lambda i) \tag{30}$$

$$[\underline{\underline{G}}_0^+]_{\lambda\lambda'} = \delta_{\lambda\lambda'}/(E-H_\lambda+i\varepsilon) \tag{31}$$

$$\underline{\underline{V}} = \begin{pmatrix} 0 & V_2 & 0 \\ 0 & 0 & V_3 \\ V_1 & 0 & 0 \end{pmatrix} \tag{32}$$

Here (λi) denote the initial arrangement and quantum numbers. We then form

$$\underline{\zeta}(\lambda i) = \underline{\underline{V}} \underline{\psi}(\lambda i) \tag{33}$$

so that

$$\underline{\zeta}(\lambda i) = \underline{\underline{V}} \underline{\phi}(\lambda i) + \underline{\underline{V}} \underline{\underline{G}}_0^+ \underline{\zeta}(\lambda i) \tag{34}$$

or explicitly

$$\zeta_1(\lambda i) = \delta_{\lambda 2} V_2 \phi(2i) + V_2 G_2^+ \zeta_2(\lambda i) \tag{35}$$

$$\zeta_2(\lambda i) = \delta_{\lambda 3} V_3 \phi(3i) + V_3 G_3^+ \zeta_3(\lambda i) \tag{36}$$

$$\zeta_3(\lambda i) = \delta_{\lambda 1} V_1 \phi(1i) + V_1 G_1^+ \zeta_1(\lambda i). \tag{37}$$

It is clear that the inhomogeniety is zero only for two of these equations, but this is sufficient to ensure a nontrivial solution to the set of equations due to the cyclical coupling. The sequential nature of the coupling can be viewed in terms of the channel coupling array introduced by Baer and Kouri (23) and corresponds to the channel permuting array choice (21-23, 25-26). This choice ensures a connected kernel in the equations, such that after two iterations, the kernel contains no terms in which there exists a spectator particle, i.e., all particles interact. Equations (28) or (35)-(37) are particular examples of the matrix generalizations of the ordinary Lippmann-Schwinger equations for systems permitting rearrangement processes (25, 45-46). The positive features of such equations include 1) all asymptotic boundary conditions are displayed explicitly due to the appearance of all possible partially interacting Green's functions, G_i^+, i=1,2,3; 2) iteration of the equations is well behaved due to the absence of so called disconnected diagrams after a finite number of iterations; 3) the equations can be explicitly decoupled so far as the arrangement channel labels are concerned by a simple back substitution; 4) due to explicit arrangement channel coupling, there occurs no need to match the wavefunction in different arrangements at some boundary surface; 5) Basis set expansions may be made in terms of bound state (L^2) type functions due to the nature of the $\zeta_{\lambda'}(\lambda i)$; 6) algebraic solution methods are appropriate and these should be highly suited to implementation on supercomputers (essentially the natural solution method involves the solution of simultaneous algebraic equations); 7) much of the computational algebra can be set up as energy independent; 8) it is straight forward to cast the equations into a purely real form so that only in the last stage is complex arithmetic required; 9) a single formalism is able to describe all types of reactive systems. Negative features of the formalism include 1) The arrangement channel matrix generalization of the Hamiltonian is not Hermitian (this is true of all such formalisms including the celebrated Faddeev equations (27,36,42)); 2) the optimum choice of expansion basis functions for the dependence on the scattering variables is not necessarily obvious, nor is the optimum choice of a distortion potential; 3) because of the great generality of the equations, there may exist specific methods which are more easily applied to certain special systems; 4) the coupled equations for the arrangement channel wavefunction components ψ_λ can have solutions which are not permissable solutions of the original Schrodinger equations; 5) the permutative coupling of arrangement channels is highly asymmetric for systems having more than 2 arrangements. In the remainder of this paper, we shall explicate the above points.

2.2 Detailed Properties of the Equations

We now wish to consider the equations in some detail in order to expose the features alluded to in the preceeding Section. The asymptotics of the equations are most easily seen using the arrangement channel component ψ_λ equations in integral form. From equation (28), we note that

$$\psi_{\lambda'}(\lambda i) = \delta_{\lambda\lambda'}\,\phi(\lambda i) + G^{+}_{\lambda'}\,V_{\lambda'+1}\,\psi_{\lambda'+1}\,(\lambda i) \tag{38}$$

For the sake of discussion, but without loss of generality let us take $\lambda=1$. Then we have

$$\psi_1(1i) = \phi(1i) + G^{+}_{1}\,V_2\,\psi_2\,(1i) \tag{39}$$

$$\psi_2(1i) = G^{+}_{2}\,V_3\,\psi_3\,(1i) \tag{40}$$

$$\psi_3(1i) = G^{+}_{3}\,V_1\,\psi_1\,(1i) \tag{41}$$

It is clear that only ψ_1 (1i) has an incoming wave, ϕ(1i). Furthermore, the outgoing scattered waves in $\psi_{\lambda'}$ (1i) are generated by the partially interacting Green's function $G^{+}_{\lambda'}$. However, an unusual feature of these equations is that the scattering amplitude in arrangement 1 is generated by the amplitude density $V_2\psi_2$(1i)! This has been demonstrated computationally for the $H+H_2$ collinear reaction (43). Because the asymptotic behavior for a given arrangement λ' is concentrated in the single component $\psi_{\lambda'}(\lambda i)$, one may use either standard Jacobi coordinates in computational studies or one may employ any of a variety of natural or reaction coordinates.

In discussing connectivity of the kernel, it is simplest to write the potential in terms of pairwise and three body interactions:

$$V = V_{12} + V_{13} + V_{23} + V_{123} \tag{42}$$

where V_{123} tends to zero if the distance between any pair of particles becomes large. If we label arrangements by the particle index of the (asymptotically) free particle, then

$$V_\lambda = V_{\lambda\lambda'} + V_{\lambda\lambda''} + V_{\lambda\lambda'\lambda''} \tag{43}$$

$$H_\lambda = T + V_{\lambda'\lambda''} \tag{44}$$

where $\lambda=1,2,3$ and $\lambda\neq\lambda'\neq\lambda''\neq\lambda$. Iteration of equation (28) twice yields the equation

$$\underline{\psi}(\lambda i) = \phi(\lambda i) + \underline{\underline{G}}^{+}_{o}\,\underline{\underline{V}}\,\underline{\underline{G}}^{+}_{o}\,\underline{\underline{V}}\,\underline{\underline{G}}^{+}_{o}\,\underline{\underline{V}}\,\underline{\psi}\,(\lambda i) \tag{45}$$

and the kernel $\underline{\underline{\kappa}}$ of this equation is

$$\underline{\underline{\kappa}} = \underline{\underline{G}}^{+}_{o}\,\underline{\underline{V}}\,\underline{\underline{G}}^{+}_{o}\,\underline{\underline{V}}\,\underline{\underline{G}}^{+}_{o}\,\underline{\underline{V}}. \tag{46}$$

We wish to examine each element of this 3X3 matrix in order to see if it contains any terms in which one of the particles does not interact with any others. $\underline{\underline{G}}^{+}_{o}$ involves the Green's functions G^{+}_{λ}, given by

$$G_\lambda^+ = 1/(E - H \quad + i\varepsilon) \tag{47}$$

$$= 1/(E - T - V_{\lambda'\lambda''} + i\varepsilon), \tag{48}$$

with $\lambda' \neq \lambda'' \neq \lambda \neq \lambda'$. Now

$$G_\lambda^+ = G_o^+ + G_o^+ V_{\lambda'\lambda''} G_\lambda^+ \tag{49}$$

where

$$G_o^+ = 1/(E - T + i\varepsilon) \tag{50}$$

is the totally free Green's function. Since it corresponds to all three particles propagating freely (without interaction), it is clear that the factors G_o^+ cannot in any way influence the connectivity properties of $\underline{\underline{\kappa}}$. Thus, for the purpose of seeing how the particles interact with one another in $\underline{\underline{\kappa}}$, it suffices to examine the product $\underline{\underline{V}}^3$. By equation (32), this is

$$\underline{\underline{V}}^3 = \begin{pmatrix} [V_2V_3V_1] & 0 & 0 \\ 0 & [V_3V_1V_2] & 0 \\ 0 & 0 & [V_1V_2V_3] \end{pmatrix} \tag{51}$$

Because of the purely multiplicative nature of the V_λ, the order of the products is irrelevant and we see that $\underline{\underline{V}}^3$ involves only $V_1V_2V_3$. By equation (42), this is

$$V_1V_2V_3 = (V_{12}+V_{13}+V_{123})(V_{12}+V_{23}+V_{123})(V_{23}+V_{13}+V_{123}). \tag{52}$$

All terms involving V_{123} are connected due to this being a three body interaction. Therefore,

$$V_1V_2V_3 = (V_{12}^2 + \text{manifestly connected terms})(V_{23}+V_{13}+V_{123}) \tag{53}$$

$$= V_{12}^2V_{23} + V_{12}^2\,V_{13} + \text{manifestly connected terms}. \tag{54}$$

But $V_{12}^2V_{23}$ involves 1 interacting with 2 twice followed by 2 interacting with three. Graphically, this is

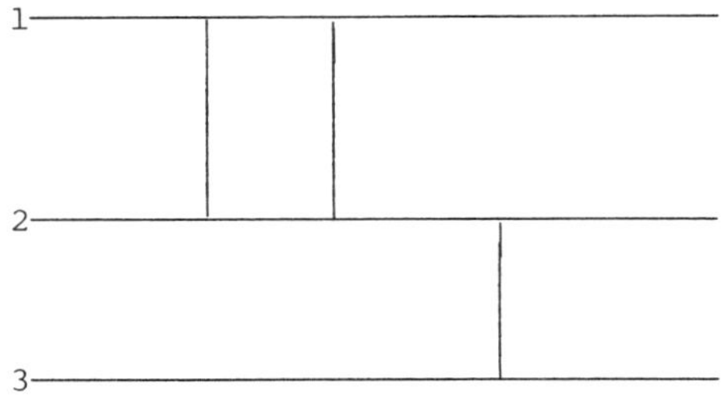

which is clearly connected. The term $V_{12}^2 V_{13}$ is

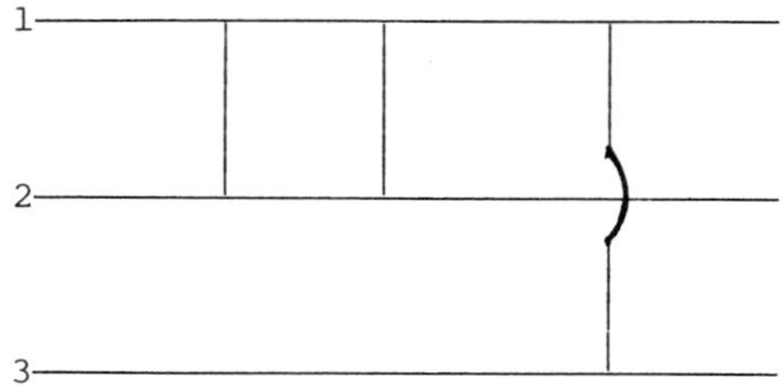

which is also connected. Thus, the kernel of the twice iterated equation is completely connected. The iteration of the equations also accounts for their becoming uncoupled. This is the result of the property of the matrix $\underline{\underline{V}}$ becoming a diagonal matrix when cubed (equation (51)). Equivalently, we simply back substitute equation (41) into (40), and this result into equation (39) to obtain

$$\psi_1(1i) = \phi(1i) + G_1^+ V_2 G_2^+ V_3 G_3^+ V_1 \psi_1(1i). \tag{55}$$

Similarly, one can obtain slightly more complicated expressions for $\psi_2(1i)$ and $\psi_3(1i)$ (which by the way are manifestly inhomogeneous equations due to the inhomogeniety in the $\psi_1(1i)$ equation). This decoupling of the equations upon iteration has potentially important significance when numerical calculations are performed since it reduces the size of the equations to be solved and converts the calculation of the other scattering amplitudes into simple matrix multiplications.

The avoidance of the need to match the wavefunction from the various arrangement has been shown explicitly for the BKLT type equations by Baer and Kouri (21) in the first calculations done for a simple waveguide model. Subsequently, extensive calculations for a variety of collinear systems have amply verified this fact. (21-23, 38-40,43) Basically, the boundary matching procedure explicitly couples the wavefunction in the various arrangement channels so that information of what is happening in one arrangement is communicated to the other arrangements. In general, this must be done for all arrangements simultaneously. The interarrangement coupling of the BKLT equations achieves this without having to define a hypersurface on which matching is carried out. Since the geometry of a matching hypersurface is strongly dependent on the potential for the system, one expects that different systems might require radically different matching surfaces. This is avoided in the present formalism.

The choice of basis functions for expanding the amplitude density functions is still a subject of intensive research. Rather than state the optimum choice, we can only indicate the types of functions used to date. In the initial studies of Baer and Kouri (21-22) for simple wave-guide models in which the atom common to two arrangements was taken to be infinitely massive, it could be shown analytically that the natural expansion functions, both for vibrational and translational (scattering coordinate) variables, were the asymptotic vibrational states in the two arrangements. When collinear reactions with smooth realistic potentials were studied recently, it was found again that vibrations were readily expanded in terms of the asymptotic vibrational states of the diatoms. For the translational or scattering variables, it was found that particle in a box states worked at least as well as did vibrational states. (38) The use of either type function is made possible by the fact that the $\zeta_{\lambda'}(\lambda i)$ tend to zero in all asymptotia.

Because one employs L^2 expansion bases in all degrees of freedom, the relevant integral equations are converted into simulations inhomogeneous algebraic equations which may then be solved by a variety of methods. However, it is also possible to convert the equations into integral equations in two dimensions and it may be possible to solve these equations noniteratively by a generalization of the Volterra integral equation algorithm developed by Sams and Kouri (47-51). Alternatively, one may utilize a reaction coordinate approach and convert the equations into the form of standard close coupling equations whose structure is analogous to that of purely inelastic scattering. (41,17) Finally, it may be possible to employ a generalization of a wave packet formalism and solve the time dependent version of the $\psi_{\lambda'}(\lambda i)$ equations. These equations are

$$i\hbar \frac{\partial}{\partial t} \psi_1 = H_1\psi_1 + V_2\psi_2. \tag{56}$$

$$i\hbar \frac{\partial}{\partial t} \psi_2 = H_2\psi_2 + V_3\psi_3 \tag{57}$$

$$i\hbar \frac{\partial}{\partial t} \psi_3 = H_3\psi_3 + V_1\psi_1, \tag{58}$$

which sum up to the time dependent Schrodinger equation

$$i\hbar \frac{\partial}{\partial t} (\psi_1 + \psi_2 + \psi_3) = H(\psi_1 + \psi_2 + \psi_3). \tag{59}$$

The question of the energy independence of a large portion of the computation will be discussed in detail in the next Section when we examine the explicit equations in the straightforward algebraic equation approach. However, it is essentially a consequence of the fact that only the translational portion of the Green's function involves any energy dependence. By means of appropriate changes of variables, we are able to isolate this dependence so that the same amount of energy dependence occurs in the three dimensional problem as arises in the collinear one.

In order to work only with real arithmetic, it is standard to introduce the reactance operator, related to the transition operator by the Heitler damping equation. In inelastic scattering, the relation hinges on the equation

$$G_\lambda^+ = G_\lambda^p - i\pi\delta(E - H_\lambda) \tag{60}$$

where G_λ^p is the principal value or standing wave boundary condition Green's function which is purely real. Then

$$T_\lambda = V_\lambda + V_\lambda G_\lambda^+ T_\lambda \tag{61}$$

where by T_λ we refer to the ordinary transition operator for nonreactive scattering in arrangement λ:

$$T_\lambda = V_\lambda + V_\lambda G^+ V_\lambda \tag{62}$$

$$G^+ = \frac{1}{E - (H_\lambda + V_\lambda) + i\varepsilon} \, . \tag{63}$$

Then the reactance operator is <u>defined</u> by

$$K_\lambda = V_\lambda + V_\lambda G_\lambda^p K_\lambda . \tag{64}$$

We form the difference of equations (62) and (64), and use equation (60) to write

$$T_\lambda - K_\lambda = V_\lambda G_\lambda^p (T_\lambda - K_\lambda) - i\pi V_\lambda \delta(E-H_\lambda) T_\lambda . \tag{65}$$

The formal solution of equation (64) for K_λ is

$$K_\lambda = (1 - V_\lambda G_\lambda^p)^{-1} V_\lambda , \tag{66}$$

and solving equation (65) for $T_\lambda - K_\lambda$, we obtain

$$T_\lambda - K_\lambda = -i\pi (1 - V_\lambda G_\lambda^p)^{-1} V_\lambda \, \delta(E-H_\lambda) T_\lambda \tag{67}$$

or

$$T_\lambda = K_\lambda - i\pi K_\lambda (E-H_\lambda) T_\lambda . \tag{68}$$

Now because $\underline{\underline{G}}_o^+$ is <u>diagonal</u>, one can readily verify that

$$\underline{\underline{G}}_o^+ = \underline{\underline{G}}_o^p - i\pi \, \delta(E-\underline{\underline{H}}_o) \tag{69}$$

where by $\delta(E-\underline{\underline{H}}_o)$, we mean

$$\delta(E-\underline{\underline{H}}_o) = \begin{pmatrix} \delta(E-H_1) & 0 & 0 \\ 0 & \delta(E-H_2) & 0 \\ 0 & 0 & \delta(E-H_3) \end{pmatrix} \tag{70}$$

and

$$\underline{\underline{G}}_o^p = \begin{pmatrix} G_1^p & 0 & 0 \\ 0 & G_2^p & 0 \\ 0 & 0 & G_3^p \end{pmatrix} \tag{71}$$

Now we can, e.g., consider equation (34) and write it as

$$\underline{\zeta}(\lambda i) = \underline{\underline{V}}\ \underline{\phi}\ (\lambda i) + \underline{\underline{V}}\ \underline{\underline{G}}_o^+\ \underline{\zeta}\ (\lambda i) \tag{72}$$

$$= \underline{\underline{T}}\ \underline{\phi}\ (\lambda i). \tag{73}$$

Then

$$\underline{\underline{T}} = \underline{\underline{V}} + \underline{\underline{V}}\underline{\underline{G}}_o^+\underline{\underline{T}} \tag{74}$$

and we define

$$\underline{\underline{K}} = \underline{\underline{V}} + \underline{\underline{V}}\underline{\underline{G}}_o^p\ \underline{\underline{K}}. \tag{75}$$

As before, we form the difference of $\underline{\underline{T}}$ and $\underline{\underline{K}}$, employ equation (69) and formally solve for $\underline{\underline{T}}-\underline{\underline{K}}$ to obtain

$$\underline{\underline{T}} = \underline{\underline{K}} - i\pi\ \underline{\underline{K}}\ (E-\underline{\underline{H}}_o)\underline{\underline{T}}. \tag{76}$$

Associated with $\underline{\underline{K}}$ there is a real amplitude density $\underline{\tilde{\zeta}}(\lambda i)$ defined as

$$\underline{\underline{K}}\ \underline{\phi}(\lambda i) = \underline{\underline{V}}\ \underline{\tilde{\zeta}}(\lambda i) \tag{77}$$

such that

$$\underline{\tilde{\zeta}}(\lambda i) = \underline{\underline{V}}\ \underline{\phi}\ (\lambda i) + \underline{\underline{V}}\ \underline{\underline{G}}_o^p\ \underline{\tilde{\zeta}}(\lambda i). \tag{78}$$

Thus, the formal structure is analogous to that of nonreactive (inelastic) scattering. All the calculations done recently for collinear systems with smooth, realistic potentials are based on equation (78) and the physical transition amplitudes are computed using the arrangement channel generalized damping relation equation (76).

Finally, with regard to the desirable features, the above formalism has made no use of any special features of the reaction process. It is completely general and in principle can be applied to any reaction.

Some of the negative features of these equations are as follows. First, and perhaps foremost, is the fact that the arrangement channel matrix generalized Hamiltonian is no longer Hermitian. This is a general feature of such formalisms, including that due to Faddeev (27, 36,42). The consequence of this is seen if one considers the time dependent equations, (56)-(58), in matrix form. One has

$$i\hbar \frac{\partial}{\partial t}\ \underline{\psi}\ (\lambda i) = \underline{\underline{H}}\ \underline{\psi}\ (\lambda i) \tag{79}$$

with

$$\underline{\underline{H}} = \underline{\underline{H}}_o + \underline{\underline{V}}, \tag{80}$$

$$\underline{\underline{H}}_o = \begin{pmatrix} H_1 & 0 & 0 \\ 0 & H_2 & 0 \\ 0 & 0 & H_3 \end{pmatrix} \tag{81}$$

The formal solution of this equation is

$$\underline{\psi}(\lambda i) = \exp(-i\underline{\underline{H}}t/\hbar)\ \underline{\psi}(\lambda i|0). \tag{82}$$

However, due to the lack of Hermiticity, the spectrum of $\underline{\underline{H}}$ includes, in addition to the normal spectrum of the ordinary Hamiltonian H, the possibility of complex eigenvalues. The occurence of such then causes the general operator $\exp(-i\underline{\underline{H}}t/\hbar)$ to be undefined unless one considers it in a projected subspace (42). Within such a subspace comprised of all physical solutions to the time dependent Schrodinger equation, equation (82) is perfectly well behaved. Let us consider again the time independent equation

$$\underline{\underline{H}}\underline{\psi} = E\underline{\psi}. \tag{83}$$

Let us suppose that the components of $\underline{\psi}$ are such that

$$\sum_\lambda \psi_\lambda \neq 0 \tag{84}$$

except perhaps for a set of points of zero measure. Then it follows by equation (83) that ψ defined by

$$\psi = \sum_\lambda \psi_\lambda \tag{85}$$

must be a solution of the Schrodinger equation with energy E. The only solutions of equation (93) which are not also solutions of the Schrodinger equation are those for which

$$\sum_\lambda \psi_\lambda = 0 \tag{86}$$

everywhere. These are called spurious solutions (36,42) and they can have complex eigenvalues. However, they are readily identified by the property (86). The degree to which such spurious solutions create problems for the BKLT formalism is not clear as yet. Certainly no problems have been encountered in the time independent solutions carried out to date. These have all been for systems having 2 rather than 3 arrangements, although both collinear and 3-dimensional problems have been solved successfully. Wave packet solutions of equation (79) are now under study. Since this method requires the evaluation of the action of $\exp(i\underline{\underline{H}}t/\hbar)$ on the packet, perhaps more understanding of the consequences, if any, of the spurious eigenstates of $\underline{\underline{H}}$ will be forthcoming.

A second area of question for this approach revolves around the choice of expansion basis functions. All calculations done to date have employed asymptotic vibrational states rather than allowing them to change as the reaction path is traversed. However, recently some work has been done on trying to generalize the treatment to include an adiabatic vibrational basis (17). At this point, no system has been attempted for which solutions have not been obtained but the convergence rate might be much faster with a more sophisticated basis. Similarly, the translational basis functions used most have been the simple particle in a box functions. It may be that more rapid convergence can be gotten with a different basis. This is still a matter of study. Related to this issue, most of the recent studies have made use of distortion potentials for the translational motion. This has the effect of reducing the size of the perturbation, as well as driving the translational part of the Green's functions to zero properly in non-classical regions. The result was a significant acceleration in convergence (in fact until recently, the only successful BKLT calculations for three finite mass atoms interacting collinearly via a smooth potential were done with a distortion potential; now successful calculations have been done without a distortion potential and the convergence is found to be slower). The problem is that no systematic procedure for choosing the distortion potential for the 3-dimensional reactive scattering problem exists, although there is hope for finding one. The inherent assymmetry of the BKLT equations has not yet been studied since this requires that the system involve more than 2 arrangements. It may create problems with the rate at which solutions converge with basis set.

Finally, whenever one develops a completely general procedure for solving problems, it often occurs that there are specific special problems for which hand tailored methods may be more efficient. In spite of this, it is still important to have both approaches available in order to be able to address any type of system which might be encountered.

3. SOLUTION OF THE BKLT EQUATIONS

In the following we shall focus on the 3-dimensional case since the simpler collinear problem can be understood readily in terms of it. An attractive feature of the BKLT formalism is the fact that the basic approach to solving the equations is the same for both collinear and 3-dimensional problems. In addition, we shall concentrate on equations (35)-(37) assuming that the initial arrangement λ is labeled as arrangement 1. Since we haven't specified which of the three arrangements of a general ABC system,

A + BC

B + AC

C + AB,

is 1, this entails no loss of generality. Our initial quantum index i represents the set of quantum numbers $n_o j_o m_o J_o M_o$, respectively the initial vibration, diatom angular momentum and total angular momentum quantum numbers. The initial state (1i) is given (in the coordinate representation) by

$$\phi(|i|\underset{\sim}{R}_1\underset{\sim}{r}_1) = \frac{2\mu}{\hbar^2}\sum_m \chi_{n_o j_o}(r_1)\, D^J_{mM}(\phi_1\theta_1\psi_1) Y_{j_o m}(\gamma_1,o)$$

$$\times\ \psi^{reg}_{Jj_o m m_o n_o}(R_1), \tag{87}$$

where $\underset{\sim}{R}_{\lambda'}$ is the vector from the center of mass of the diatom to the atom in arrangement λ', $\underset{\sim}{r}_{\lambda'}$, is the internuclear vector of the diatom in arrangement λ',

$$\psi^{reg}_{Jj_o m m_o n_o}(R_1)$$

is a solution of the unperturbed radial equation in the body frame (if no distortion potential is employed, then

$$\psi^{reg}_{Jj_o m m_o n_o}(R_1)$$

is just the body frame regular Bessel function (52)), D^J_{mM} is the usual representation coefficient for the 3-dimensional rotation group, $Y_{j_o m}$ is a normalized spherical harmonic associated with the internal angle γ_1 between $\underset{\sim}{R}_1$ and $\underset{\sim}{r}_1$; ϕ_1, θ_1 and ψ_1 are the Euler angles associated with orienting the 3 atom triangle in space (with the angles θ_1, ψ_1 normally being chosen as the angles of $\underset{\sim}{R}_1$ relative to a space oriented coordinate system) and

$$\chi_{n_o j_o}(r_1)$$

is the initial diatom vibrational state (including effects of centrifugal distortion when the diatom is initially in rotor state j_o) satisfying the equation

$$\left\{-\frac{\hbar^2}{2\mu}\left[\frac{1}{r_1^{\,2}}\frac{\partial}{\partial r_1}\left(r_1^2\frac{\partial}{\partial r_1}\right) - \frac{j(j+1)}{r_1^{\,2}}\right] + V(R_{1,0}, r_1)\right\}\chi_{nj}(r_1)$$

$$= \varepsilon_{nj}\,\chi_{nj}(r_1). \tag{88}$$

Here $R_{1,0}$ is a sufficiently large value of R_1 that the atom and diatom in arrangement 1 do not interact and μ is the system reduced mass common to all arrangements (2-3); i.e., $\underset{\sim}{R}_{\lambda'}$, $\underset{\sim}{r}_{\lambda'}$ are mass scaled coordinates. We form the cordinate representatives of the principal value version of equations (35)-(37) by forming the scalar products

$$\langle \underset{\sim}{R}_{\lambda'}\underset{\sim}{r}_{\lambda'}|\zeta_{\lambda'}(1i)\rangle$$

and making use of the locality of the potentials V_λ. This implies that

$$\langle R_{\lambda'} r_{\lambda'} | V_{\lambda'+1} \chi \rangle = V_{\lambda'+1}(R_{\lambda'+1}(R_{\lambda'} r_{\lambda'})\ r_{\lambda'+1}(R_{\lambda'} r_{\lambda'})$$
$$\chi(R_{\lambda'+1}(R_{\lambda'} r_{\lambda'})\ r_{\lambda'+1}(R_{\lambda'} r_{\lambda'})) \qquad (89)$$

for any state vector χ, where $R_{\lambda'+1}(R_{\lambda'}, r_{\lambda'})$ means the vector $R_{\lambda'+1}$ corresponding to the configuration $R_{\lambda'}$, $r_{\lambda'}$ and similarly for $r_{\lambda'+1}(R_{\lambda'}, r_{\lambda'})$. We employ the short hand notation $f_{\lambda'+1}(R_{\lambda'}, r_{\lambda'})$ to denote a function associated with the $\lambda'+1$ arrangement but expressed in terms of the $R_{\lambda'}$, $r_{\lambda'}$ variables. Then our equations are given by

$$\zeta(\lambda'|1n_oj_om_oJ_oM_o|R_{\lambda'} r_{\lambda'}) = \delta_{\lambda'3} V_1(R_3 r_3)\phi(1n_oj_om_oJ_oM_o|R_3 r_3)$$
$$+ \frac{2\mu}{\hbar^2}\int dR'_{\lambda'+1}\int dr'_{\lambda'+1}\, V_{\lambda'+1}(R_{\lambda'} r_{\lambda'}) G^p_{\lambda'+1}(R_{\lambda'} r_{\lambda'}|R'_{\lambda'+1} r'_{\lambda'+1})$$
$$\times\ \zeta(\lambda'+1|1n_oj_om_oJ_oM_o|R'_{\lambda'+1} r'_{\lambda'+1}), \qquad (90)$$

where again we recall that if $\lambda'=3$, then $\lambda'+1$ is 1. We now expand the amplitude densities in basis sets according to

$$\zeta(\lambda'|1n_oj_om_oJ_oM_o|R_{\lambda'} r_{\lambda'})$$
$$= \sum_{\substack{njm\\JM}} \chi_{nj}(r_{\lambda'})\ D^J_{mM}(\phi_{\lambda'}, \theta_{\lambda'}, \psi_{\lambda'})\ Y_{jm}(\gamma_{\lambda'}, 0)$$
$$\times\ \zeta(\lambda' njmJM|1n_oj_om_oJ_oM_o|R_{\lambda'}). \qquad (91)$$

In addition, the Green's functions are expanded as

$$G^p_{\lambda'+1}(R_{\lambda'+1} r_{\lambda'+1}|R'_{\lambda'+1} r'_{\lambda'+1}) = \sum_{\substack{J'M'\\n'j'm'\\m''}} D^{J'}_{m'M'}(\phi_{\lambda'+1}\theta_{\lambda'+1}\psi_{\lambda'+1})$$
$$\times\ D^{J'*}_{m''M'}(\phi'_{\lambda'+1}\theta'_{\lambda'+1}\psi'_{\lambda'+1})\ Y_{j'm'}(\gamma_{\lambda'+1}, 0)\ Y^*_{j'm''}(\gamma'_{\lambda'+1}, 0)$$
$$\times\ g^{J'n'j'm'm''}_{\lambda'+1}(R_{\lambda'+1}|R'_{\lambda'+1})\ \chi_{n'j'}(r_{\lambda'+1})\ \chi^*_{n'j'}(r'_{\lambda'+1}), \qquad (92)$$

where the radial Green's function is given by

$$g^{J'n'j'm'm''}_{\lambda'+1}(R_{\lambda'+1}|R'_{\lambda'+1}) = \sum_{m'''} \psi^{reg}_{J'n'j'm'''m'}(R^<_{\lambda'+1})$$
$$\times\ \psi^{irreg}_{J'n'j'm'''m''}(R^>_{\lambda'+1}), \qquad (93)$$

and

$$\psi^{irreg}_{J'n'j'm'''m''}$$

is the irregular solution of the unperturbed body frame radial equation. We substitute these expansions into equation (90) and employ the relation

$$D^{J'}_{m'M'}(\phi_{\lambda'+1}\,\theta_{\lambda'+1}\,\psi_{\lambda'+1}) = \sum_{\bar{m}} d^{J'}_{m'\bar{m}}(\Delta_{\lambda',\lambda'+1})\,D^{J'}_{\bar{m}M'}(\phi_{\lambda'},\theta_{\lambda'},\psi_{\lambda'}) \tag{94}$$

which expresses the $\lambda'+1$ rotation matrices in terms of the λ' ones with the angle $\Delta_{\lambda',\lambda'+1}$ defined by (2-3)

$$\cos\Delta_{\lambda',\lambda'+1} = \hat{R}_{\lambda'}\cdot\hat{R}_{\lambda'+1}. \tag{95}$$

This can be expressed in terms of λ' arrangement coordinates only. The resulting expression is multiplied on both sides by the relevant basis functions in arrangement λ' and integrated over $r_{\lambda'},\phi_{\lambda'},\theta_{\lambda'},\psi_{\lambda'},\gamma_{\lambda'}$, use being made of the orthogonality of the basis functions to obtain

$$\begin{aligned}
&\zeta(\lambda'njmJM|1n_oj_om_oJ_oM_o|R_{\lambda'}) = \frac{2\mu}{\hbar^2}\,\delta_{\lambda'3}\,\delta_{JJ_o}\,\delta_{MM_o}\sum_{\bar{m}}\int dr_3 r_3^2 \\
&\times \int d(\cos\gamma_3)\,\chi^*_{nj}(r_3)\,Y^*_{jm}(\gamma_3,0)\,V_1(R_3r_3\gamma_3)\,\chi_{n_oj_o}(r_1)\,Y_{j_o\bar{m}}(\gamma_1,0)\,d^J_{\bar{m}m}(\Delta_{31}) \\
&\times \psi^{reg}_{Jj_o\bar{m}m_on_o}(R_1) + \frac{2\mu}{\hbar^2}\sum_{\substack{n''j''\\ m''\bar{m}}} \int dr_{\lambda'}r_{\lambda'}^2 \int d(\cos\gamma_{\lambda'}) \int dR'_{\lambda'+1} \\
&\quad R'^2_{\lambda'+1}\;\chi^*_{nj}(r_{\lambda'}) \\
&\times Y^*_{jm}(\gamma_{\lambda'},0)\,d^J_{\bar{m}m}(\Delta_{\lambda',\lambda'+1})\;V_{\lambda'+1}(R_{\lambda'},r_{\lambda'},\gamma_{\lambda'})\,Y_{j''\bar{m}}(\gamma_{\lambda'+1},0) \\
&\times \chi_{n''j''}(r_{\lambda'+1})\;g^{Jn''j''\bar{m}m''}_{\lambda'+1}(R_{\lambda'+1}|R'_{\lambda'+1}) \\
&\times \zeta(\lambda'+1n''j''m''JM|1n_oj_om_oJ_oM_o|R'_{\lambda'+1})
\end{aligned} \tag{96}$$

In this equation, we point out that the argument $R_{\lambda'+1}$ occuring in

$$g^{Jn''j''\bar{m}m''}_{\lambda'+1}$$

must be interpreted as a function of $R_{\lambda'},r_{\lambda'},\gamma_{\lambda'}$. Thus, it is under the integrals over $r_{\lambda'}$ and $\gamma_{\lambda'}$. Similarly, $V_{\lambda'+1}(R_{\lambda'},r_{\lambda'},\gamma_{\lambda'})$ means $V_{\lambda'+1}$ expressed at the point defined by $R_{\lambda'},r_{\lambda'},\gamma_{\lambda'}$, so it also is under the integrals over $r_{\lambda'},\gamma_{\lambda'}$. Furthermore, we note that no coupling in J,M occurs in the equation and the equations are homogeneous when $J\neq J_o$, $M\neq M_o$. It follows for scattering energies that the

$$\zeta(\lambda'njmJM|1n_oj_om_oJ_oM_o|R_{\lambda'}) = \delta_{JJ_o}\,\delta_{MM_o}\,\xi^J(\lambda'njm|1n_oj_om_o|R_{\lambda'}), \tag{97}$$

where use has also been made of the fact that the equation for ξ^J,

$$\xi^J(\lambda'njm|1n_oj_om_o|R_{\lambda'}) = \frac{2\mu}{\hbar^2}\delta_{\lambda'3}\sum_{\bar{m}}\int dr_3r_3^2\int d(\cos\gamma_3)\chi^*_{nj}(r_3)$$

$$\times Y^*_{jm}(\gamma_3,0)V_1(R_3r_3\gamma_3)\chi_{n_oj_o}(r_1)\ Y_{j_o\bar{m}}(\gamma_1,0)d^J_{\bar{m}m}(\Delta_{31})$$

$$\times \psi^{reg}_{Jn_oj_o\bar{m}m_o}(R_1) + \frac{2\mu}{\hbar^2}\sum_{\substack{n"j"\\m"\bar{m}}}\int d\ r_{\lambda'}\ r_{\lambda'}^2\ \int d(\cos\gamma_{\lambda'})\int dR'_{\lambda'+1}R'^2_{\lambda'+1}$$

$$\times \chi^*_{nj}(r_{\lambda'})Y^*_{jm}(\gamma_{\lambda'},0)d^J_{\bar{m}m}\ (\Delta_{\lambda',\lambda'+1})\ V_{\lambda'+1}\ (R_{\lambda'}r_{\lambda'}\gamma_{\lambda'})$$

$$\times Y_{j"\bar{m}}\ (\gamma_{\lambda'+1},0)$$

$$\times \chi_{n"j"}(r_{\lambda'+1})\ g^{Jn"j"\bar{m}m"}_{\lambda'+1}\ (R_{\lambda'+1}\ |\ R'_{\lambda'+1})$$

$$\times \xi^J\ (\lambda'+1\ n"j"m"|1n_oj_om_o|R'_{\lambda'+1}) \qquad (98)$$

is independent of M (the result of the Wigner-Eckart theorem). From this point on, the analysis of the equations is completely parallel to that of the collinear case. The only effect of the fact we deal here with 3-dimensional reactive scattering is the occurrance of the extra integral over $\cos\gamma_{\lambda'}$ and the extra factors of $d^J_{\bar{m}m}(\Delta_{\lambda'\lambda'+1})$ and $Y_{jm}(\gamma_{\lambda'},0)$ or $Y_{j"\bar{m}}(\gamma_{\lambda'+1},0)$.

We now expand the radial amplitude densities in terms of an appropriate basis of L^2-type functions

$$\xi^J(\lambda'njm|1n_oj_om_o|R_{\lambda'}) = \sum_t a^J(\lambda'njmt|1n_oj_om_o)\chi(\lambda't|R_{\lambda'}). \qquad (99)$$

We substitute appropriately, multiply by a particular basis function and integrate over $R_{\lambda'}$ to obtain

$$a^J(\lambda'njmt|1n_oj_om_o) = \frac{2\mu}{\hbar^2}\delta_{\lambda'3}\sum_{\bar{m}}\ \int dR_3R_3^2\int dr_3r_3^2\int d(\cos\gamma_3)$$

$$\times \chi^*(3t|R_3)\ \chi^*_{nj}(r_3)\ Y^*_{jm}(\gamma_3,0)\ V_1(R_3r_3\gamma_3)\chi_{n_oj_o}(r_1)Y_{j_o\bar{m}}(\gamma_1,o)$$

$$\times d^J_{\bar{m}m}\ (\Delta_{31})\psi^{reg}_{Jj_on_o\bar{m}m_o}(R_1) + \frac{2\mu}{\hbar^2}\sum_{\substack{n"j"\\m"\bar{m}t'}}\int dR_{\lambda'}R^2_{\lambda'}\int dr_{\lambda'}r^2_{\lambda'}$$

$$\times \int d(\cos\gamma_{\lambda'})\ \int dR'_{\lambda'+1}R'^2_{\lambda'+1}\chi^*(\lambda't|R_{\lambda'})\ \chi^*_{nj}(r_{\lambda'})Y^*_{jm}(\gamma_{\lambda'},0)$$

$$\times d^J_{\bar{m}m}(\Delta_{\lambda',\lambda'+1})V_{\lambda'+1}(R_{\lambda'}r_{\lambda'}\gamma_{\lambda'})Y_{j"\bar{m}}(\gamma_{\lambda'+1},0)\chi_{n"j"}(r_{\lambda'+1})$$

$$\times\, g^{Jn''j''\bar{m}m''}_{\lambda'+1}(R_{\lambda'+1}|R'_{\lambda'+1})\chi(\lambda'+1t|R'_{\lambda'+1})a^J(\lambda'+1n''j''m''t'|1n_oj_om_o). \tag{100}$$

The next step is to change the variables from $R_{\lambda'}$, $r_{\lambda'}$, $\cos\gamma_{\lambda'}$ to $R_{\lambda'+1}$, $r_{\lambda'+1}$, $\cos\gamma_{\lambda'+1}$ and define the following energy independent quantities:

$$E^J(\lambda'njmt|\lambda'+1n''j''\bar{m}|R_{\lambda'+1}) = \frac{2\mu}{\hbar^2}\int dr_{\lambda'+1}r^2_{\lambda'+1}\int d(\cos\gamma_{\lambda'+1})$$

$$\times\, \chi^*(\lambda't|R_{\lambda'})\chi^*_{nj}(r_{\lambda'})Y^*_{jm}(\gamma_{\lambda'},0)d^J_{\bar{m}m}(\Delta_{\lambda',\lambda'+1})Y_{j''\bar{m}}(\gamma_{\lambda'+1},0)$$

$$\times\, \chi_{n''j''}(r_{\lambda'+1})\, V_{\lambda'+1}(R_{\lambda'+1}\, r_{\lambda'+1}\, \gamma_{\lambda'+1}). \tag{101}$$

In terms of this definition, we obtain the final equation

$$a^J(\lambda'njmt|1n_oj_om_o) = \delta_{\lambda'3}\sum_{\bar{m}}\int dR_1R_1^2\; E^J(3njmt|1n_oj_o\bar{m}|R_1)$$

$$\times\, \psi^{reg}_{Jj_o\bar{m}m_on_o}(R_1) + \sum_{\substack{n''j''\\ m''\bar{m}t'}}\int dR_{\lambda'+1}R^2_{\lambda'+1}\int dR'_{\lambda'+1}R'^2_{\lambda'+1}$$

$$\times\, E^J(\lambda'njmt|\lambda'+1n''j''\bar{m}|R_{\lambda'+1})\chi(\lambda'+1\;t'|R'_{\lambda'+1})$$

$$\times\, g^{Jn''j''\bar{m}m''}_{\lambda'+1}(R_{\lambda'+1}|R'_{\lambda'+1})a^J(\lambda'+1n''j''m''t'|1n_oj_om_o). \tag{102}$$

This has <u>exactly</u> the same structure as occurs in the collinear scattering problem. The only difference is the functions $E^J(\lambda'n'j'm't'|\lambda'+1\, njmt|R_{\lambda'+1})$ here involve the extra integral over $\cos\gamma_{\lambda'+1}$, and of course there are additional quantum numbers associated with the rotational degrees of freedom. However, it is important that the additional integral occurs <u>only</u> in the energy independent part of the problem rather than in the energy dependent part. Further, the structure of the equations which lead to the decoupling discussed in the preceeding section also applies to the solution of the algebraic equations represented by equation (102). Thus, simultaneous algebraic equations in only <u>one</u> arrangement must be solved with a^J-coefficients for other arrangements being obtained by an additional matrix multiplication. Of course, the size of the algebraic equations can be very large since the dimension is determined by the product of the number of vibrational basis states, times the number of rotational basis states, times the number of translational basis states. In the collinear case, converged results were obtained typically with 8 to 10 vibrational basis functions and a similar number of translational basis states, for a total of around 100 algebraic equations. The 3-dimensional problem will be much larger. However, we believe casting the problem as either the solution of a large set of algebraic equations by matrix inversion (if one wants results for all possible initial conditions) or by other methods (if one desires only a subset of initial conditions) is well suited to the latest generation of supercomputers and we are optimistic that solutions will be forthcoming for the 3-dimensional problem.

4. CALCULATIONS

In this section, we shall briefly review the systems for which calculations have been successfully made by the above method. The initial studies done by Baer and Kouri (21-22) focussed on piecewise constant potentials for either collinear or 3-dimensional reactive scattering models. Excellent agreement was obtained for the collinear waveguide model (21) and the first 3-dimensional model of an atom-heteronuclear diatom reactive system was studied (22). Finally, preliminary studies of the 3-dimensional e^-+H system were carried out (23). Subsequently Eccles and Secrest (33-34) have done further studies of the e+H system obtaining converged results with a differential equation version of the formalism. Top and Shapiro (41) carried out a study of the collinear H+H_2 exchange reaction also using a differential equation ACQM approach along with a particular reaction coordinate and obtained excellent results. Baer and Kouri (32) suggested the use of distortion potentials as a means of improving the basis set convergence in the integral equation approach and the first calculations were done by Shima and Baer (38) for the collinear H+FH→HF+H displacement reaction. In this system, the central F atom is sufficiently massive that the skewing angle is close to 90°, as it had been in the earlier waveguide models studied by Baer and Kouri (21-22). Next, Shima, Baer and Kouri (39-40,43) in a series of studies have generalized the distortion potential to treat a general collinear three-finite-mass-atom system and applied it to the H+H_2, F+H_2 and D+H_2 collinear exchange reactions. In all cases, excellent agreement with more standard approaches was obtained. In addition, several alternate forms of the equations, including that utilized in the discussion of Section II, were shown to yield the correct results. Most recently, Shima, Baer and Kouri (53) have shown that collinear 3-finite-mass-atom systems with realistic smooth potential surfaces can be solved without use of a distortion potential, albeit this does require a larger basis set as indicated earlier by Baer and Kouri (32). The fact that successful applications have been made both to 3-D reactions, the e^-+H system and general collinear reactions with smooth surfaces and all atoms having finite mass increases confidence that this formalism can provide a general approach to real 3-dimensional reactive collisions.

5. STATUS OF 3-D REACTIVE STUDIES

The initial steps in applying this formalism to realistic 3-dimensional reactive systems have already been taken (17). These include the adaptation of programs to generate the vibrational states $\chi_{nj_\lambda}(r_\lambda)$, rotation matrices $d^J_{mm'}$, and spherical harmonics $Y_{jm'}$, translational basis states and the regular and irregular radial functions comprising the

Green's functions. Current efforts are being devoted to assembling these programs to construct the coefficient and inhomogeniety matrices and to invert the coefficient matrix. Initial calculations will be done for the $H+H_2$ system for which close coupling results are available (6-8). However, the major effort will focus on studies of some asymmetric reaction systems.

6. ACKNOWLEDGEMENTS

D.J.K., gratefully acknowleges support under NSF Grant CHE82-15317 during the time this paper was being prepared.

7. REFERENCES

1. H. M. Hulbert and J. O. Hirschfelder, J.Chem.Phys., **11**, 276(1943). For later developments, see the following reviews, articles and references contained therein.
2. M. Baer, 'A Review of Quantum Mechanical Approximate Treatments of Three Body Reactive Systems', Advances in Chemical Physics, Vol. 49, eds., I Prigogine and S. A. Rice (Wiley, N.Y., 1982)191.
3. M. Baer, 'General Theory of Reactive Scattering. The Differential Equation Approach', in Theory of Chemical Reaction Dynamics, ed. M. Baer (C.R.C. Press, Boca Raton, Fl., 1985).
4. R. E. Wyatt, 'Direct-Mode Chemical Reactions I. Methodology for Accurate Quantal Calculations', in Atom-Molecule Collision Theory. A Guide for the Experimentalist, Ed. R. B. Bernstein (Plenum, N.Y., 1979)567.
5. A. Kuppermann, G. C. Schatz and M. Baer, J.Chem.Phys., **65**, 4596(1976).
6. G. C. Schatz and A. Kuppermann, J.Chem.Phys., **65**, 4668(1976).
7. A. B. Elkowitz and R. E. Wyatt, J.Chem.Phys., **62**, 2504(1975).
8. R. B. Walker, E. B. Stechel and J. C. Light, J.Chem.Phys., **69**, 2922 (1978).
9. R. B. Walker, J. C. Light and A. Altenberger-Siczek, J.Chem.Phys., **64** 1166(1976).
10. L. M. Delves, Nucl.Phys., **9**, 391(1959).
11. G. L. Hofacker, Naturforschung, **189**, 607(1963).
12. R. A. Marcus, J.Chem.Phys., **45**, 4493(1966).
13. A. Kuppermann, J. A. Kaye and J. P. Dwyer, Chem.Phys.Lett., **74**, 257 (1980).
14. J. Romelt, Chem.Phys.Lett., **74**, 263(1980).
15. J. Manz and J. Romelt, J.Chem.Phys., **73**, 5040(1980); Chem.Phys.Lett., **76**, 337(1980) and **77**, 172(1981).
16. J. A. Kaye and A. Kuppermann, Chem.Phys.Lett., **77**, 573(1981) and **78**, 546(1981).
17. See the review and references in D. J. Kouri, 'General Theory of Reactive Scattering. The Integral Equation Approach' in Theory of Chemical Reaction Dynamics, ed., M. Bear (CRC Press, Boca Raton, Fl., 1985).
18. L. Eyges, Phys.Rev., **115**, 1643(1959).
19. L. D. Faddeev, Mathematical Aspects of the Three Body Problem in Quantum Scattering Theory, (Davey, N.Y., 1965).

20. Y. Hahn, Phys.Rev., **169**, 794(1968).
21. M. Baer and D. J. Kouri, Phys.Rev., **A4**, 1924(1971).
22. M. Baer and D. J. Kouri, J.Chem.Phys., **56**, 1758 and 4840(1972).
23. M. Baer and D. J. Kouri, J.Math.Phys., **14**, 1637(1973).
24. D. J. Kouri, M. Craigie and D. Secrest, J.Chem.Phys., **60**, 1851(1974).
25. D. J. Kouri and F. S. Levin, Phys.Rev., **A10**, 1616(1974) and Nucl.Phys., **A250**, 127(1975); Phys.Lett., **B50**, 421(1974).
26. W. Tobocman, Phys.Rev., **C9**, 2466(1974) and **C11**, 43(1975).
27. R. Goldflam and D. J. Kouri, Chem.Phys.Lett., **34**, 594(1975).
28. D. J. Kouri, H. Kruger and F. S. Levin, Phys.Rev., **D15**, 1156(1977).
29. S. Rabitz and H. Rabitz, J.Chem.Phys., **67**, 2965(1977).
30. R. Goldflam and W. Tobocman, Phys.Rev., **C17**, 1914(1978) and **C18**, 1857(1978).
31. D. K. Hoffman, D. J. Kouri and Z. H. Top, J.Chem.Phys., **70**, 4640 (1979).
32. M. Baer and D. J. Kouri, proposal to U.S.-Israel Binational Science Foundation, 1979, unpublished.
33. J. Eccles and D. Secrest, <u>private communication</u>, 1980.
34. J. Eccles, 'Reactive Scattering of an Atom by a Diatom: The Coupled Arrangement Wavefunction Method', Ph.D. Thesis, University of Illinois-Urbana, 1980.
35. F. S. Levin, Phys.Rev., **C21**, 2199(1980).
36. J. W. Evans, J.Math.Phys., **22**, 1672(1981).
37. J. W. Evans and D. K. Hoffman, J.Math.Phys., **22**, 2858(1981).
38. Y. Shima and M Baer, Chem.Phys.Lett., **91**, 43(1982) and J.Phys., **B16**, 2169(1983).
39. Y. Shima, M. Bear and D. J. Kouri, Chem.Phys.Lett., **94**, 321(1983).
40. Y. Shima, D. J. Kouri and M. Baer, J.Chem.Phys., **78**, 6666(1983).
41. Z. H. Top and M. Shapiro, J.Chem.Phys., **77**, 5009(1983).
42. J. W. Evans, D. K. Hoffman and D. J. Kouri, J.Math.Phys., **24**, 576 (1983).
43. N. AbuSalbi, D. J. Kouri, Y. Shima and M. Baer, J.Chem.Phys., **81**, 1813(1984).
44. See the review by D. A. Micha in <u>Theory of Chemical Reaction Dynamics</u>, ed. M. Baer (CRC Press, Boca Raton, Fl., 1985).
45. B. A. Lippmann and J. Schwinger, Phys.Rev., **79**, 469(1950).
46. M. Gell-Mann and M. L. Goldberger, Phys.Rev., **91**, 398(1953).
47. W. N. Sams and D. J. Kouri, J.Chem.Phys., **51**, 4809 and 4815(1969).
48. R. A. White and E. F. Hayes, J.Chem.Phys., **57**, 2985(1972) and Chem.Phys.Lett., **14**, 98(1972).
49. D. Secrest in <u>Methods of Computational Physics</u>, Vol. 10, eds. B. Alder, S. Fernbach and M. Rotenberg (Academic, N.Y., 1971).
50. E. R. Smith and R. J. W. Henry, Phys.Rev., **A7**, 1585(1973).
51. M. A. Morrison in <u>Electron-Molecule and Photon-Molecule Collisions</u>, eds. T. Rescigno, V. McKoy and B. Schneider (Plenum, N.Y., 1978).
52. D. J. Kouri, T. G. Heil and Y. Shimoni, <u>J.Chem.Phys.</u>, **65**, 226(1976).
53. Y. Shima, M. Baer and D. J. Kouri, to be published.

RESONANCES IN REACTIONS: A SEMICLASSICAL VIEW

Vincenzo Aquilanti
Dipartimento di Chimica dell'Universita'
06100 Perugia, Italy

ABSTRACT. Recent work on quantum coupled oscillators and the collinear dynamics of three bodies, as models for unimolecular and bimolecular reactions, is reviewed with special reference to the role of resonances. The approach, semiclassical in spirit, exploits the approximate separability of the radius of hyperspherical formulations and allows to localize the breakdown of adiabaticity at "ridges in the potential", where transitions between modes occur.

1. INTRODUCTION

The understanding of the elementary processes of chemical kinetics involves the development of advanced quantum mechanical techniques to deal with complicated scattering problems, including rearrangement. However, as often in chemistry, one can exploit with success the relatively large mass of nuclei with respect to that of electrons and use a classical picture for nuclear motion. A proper blend of classical and quantum mechanics is thus appropriate, both for a qualitative discussion of the main features associated with the reactive events and for the development of efficient computer codes for the reliable assessments of quantities such as rate constants or cross sections for state to state processes. These techniques, loosely referred to as semi-classical, can all be reduced within a common denominator by observing that they all depend on the possibility of treating Planck's constant as a small parameter, and it is this character of asymptotic behaviour with respect to this parameter which makes the semiclassical techniques a useful reference frame for the discussion of some important features which have recently emerged in the theoretical investigations of unimolecular and bimolecular reactions.

Among these features, we will focus our attention in Section 2 on the role which resonance phenomena [1] play on the rate of flow of the energy within a molecule, thus affecting the characteristics of unimolecular decay. Mode specificity (i.e. strong dependence of the resonance lifetimes not only on energy) may lead to nonstatistical intramolecular vibrational relaxation and selective unimolecular

D. C. Clary (ed.), The Theory of Chemical Reaction Dynamics, 383–413.

decomposition, a topic of great theoretical and practical interest. Along the same line, a semiclassical view of resonances and interference effects in collinear reactions will be present in Section 3.

Due to the exploratory nature of these investigations at this stage, rather than pretending to study realistic systems, we will limit ourselves to the consideration of simple models, which we believe contain an indication of methods to be further pursued and a useful phenomenology in nuce. Accordingly, in this Introductory Section, we will show that the semiclassical approach naturally leads to a search for a quasiseparable variable, and to adiabatic and diabatic representations (see the Appendix). A general analysis of transition between modes will then illustrate the role of local breakdown of adiabaticity and a semiclassical study of the pendulum motion will provide an introduction to the mathematical techniques involved.

1.1 Semiclassical analysis of asymptotic separability and of its local breakdown

Because of the masses and the interactions involved, molecular behaviour is typically a problem in semiclassical mechanics: quantum effects are too important to be neglected altogether, but Planck's constant is definitely a parameter so small that appropriate asymptotic techniques can be effectively exploited. The paradigmatic example is the WKB approach: it is useful both for bound states and for scattering [2] whenever the problem is essentially one-dimensional, and we will base on it many of our considerations in the following.

The extension of the asymptotic approach to multidimensional nonseparable systems is not so straightforward: EBK quantization acts only on classical quasiperiodic trajectories, thus yielding only part of the spectrum, and cannot be generalized to scattering states. Among the techniques developed for dealing explicitly with inelastic scattering and reactions, a generalization of the Born-Oppenheimer separation of nuclear and electronic motion has recently been proved very successful [3]. It involves the search for a nearly separable variable, in terms of which the time independent Schroedinger equation reduces to an infinite set of coupled second order ordinary differential equations. Besides offering an effective computational scheme, the procedure allows to be implemented semiclassically, and each step is amenable to qualitative interpretation, such as is needed for deepening our insight of complicated quantum systems.

Regular, quasiperiodic behaviour of a quantum system is definitely associated with some at least approximate separability of the equations of motion. (How the converse, i.e. 'chaos' whatever may be its definition, can be associated with nonseparability, is a matter of current research [4]). Separability, on the other hand, is always a manifestation of some symmetry: this, in quantum mechanics, corresponds to the existence of operators commuting with the Hamiltonian, and leads to the possibility of defining good quantum numbers. Although for intrinsically nonseparable problems separation cannot be carried out exactly (globally), it is nonetheless possible to find important examples where quasiseparability (approximate commuting operators, nearly good quantum numbers) can be obtained. In the description of diatomic molecules, for example, electronic, vibrational and rotational modes are progressively considered separately following a well established hierarchy, which allows to arrange modes according to the characteristic frequencies. The underlying idea is that, if a mode is much slower than others, it can be considered as freezed, while studying the fast ones. Thus, in the familiar Born-Oppenheimer separation, internuclear distances are slow coordinates with respect to electron-nucleus and electron-electron ones. The whole of quantum chemistry capitalizes on this idea.

A key observation for fruitful generalization is that, to achieve approximate separation, one employs, more or less rigorously, asymptotic expansions with respect to some parameters (mass ratios, frequency ratios). In the present investigations, we start from the consideration that it is often possible, for problems of definite chemical and physical interest, to find some representation which allows to obtain an approximate separation (at least locally), by expansions which are asymptotic in Planck's constant, treated as a small parameter: it is a natural choice, since this corresponds to what is commonly understood as the semiclassical regime. Although these approximately separated representations will fail somewhere, we found [5] that the localization of failure may lead to a source of quantum irregular behaviour and that the search for special asymptotic techniques for dealing with local nonseparability is particularly promising.

In the Appendix, an outline of the formalism leading to adiabatic and diabatic representations is given: these manipulations of the Schroedinger equation are particularly useful for a discussion of properties of systems from the point of view of asympototic methods. It is immediate to see from equation (A2) that whenever elements of P are small, since $\hbar^2/2m$ is a small parameter, equations adiabatically decouple into one-dimensional problems for the effective potentials $\varepsilon_\nu(\rho)$. In turn, these problems can be analyzed by the Liouville-Green WKB technique, which requires special care whenever $\varepsilon_\nu = E$ (turning points): but this problem is to be considered as effectively solved by the method of comparison equations. It is important to realize that proper coordinate choices may lead to wide regions of ρ space where this decoupling is very effective: in such a case, approximate quantum numbers can be assigned, and it is possible to compute semiclassically bound or resonance states and scattering properties.

As will be illustrated in the following, the success of the procedure critically depends on how appropriate the definition of the ρ coordinate is, to exhibit as more localized as possible any breakdown of approximate separability. This breakdown is measured by the P matrix [equation (A4)], and therefore a study of its analytical structure is an important step of the present program. In fact, equation (A4) shows that around any poles of P matrix elements sufficiently close to the real ρ axis their neglect is not warranted, however small $\hbar/2m$ is considered. In the following, therefore, we will find examples where such features of the P matrix have been characterized. Around these features, adiabatic conditions fail, and nonadiabatic corrections must be considered.

1.2 Transition between modes: the semiclassical pendulum

Strongly nonadiabatic behaviour is often localized where the actual character of systems changes drastically. So, when the interaction between two atoms is considered as a function of the internuclear distance R, it is found that the transition between the typical behaviour of separated atoms and that of a diatomic molecule is often localized around sharp maxima in the elements of a P(R) matrix [6]. These maxima correspond to poles near the real R axis in a proper analytic continuation of P(R), and mark the breakdown of the Born-Oppenheimer separation.

Several examples can be put forward in order to show that transition between modes, i.e. a local breakdown of adiabaticity, typically takes place at well defined characteristic features of the potential. For problems involving more than two bodies, several investigations have identified as a good candidate for near separability the hyperradial variable ρ [3,7]. Low values of ρ correspond to closeness of all particles, and the various possible rearrangement channels correspond to large ρ. It will be seen in Section 3 that, at least for the simplified situation that the three particles are constrained to be on a line, a rearrangement process, such as a chemical reaction, can be described in a time independent picture as the transition between two types of modes, one corresponding to an intermediate complex (transition state) which may dissociate into channels corresponding to reactants and products: the transition can be described adiabatically, nonadiabaticity being important only along a line in the potential energy surface (the ridge) which separates the valleys of reactants and products. Implementing semiclassically these ideas, it has been possible to obtain not only qualitative descriptions, but also quantitative results for resonance positions and widths, and for interference effects in the probability for reactive collisions (Section 3). A classical study of these problems points at a connection between chaotic behaviour and temporary trapping in the transition state: again, a connection between local non-separability and irregular modes is emerging.

The basic physics and the related mathematics associated with mode transitions is illustrated by the pendulum, whose classical mechanics are described in many textbooks and reviews [9]. The two modes are designated vibrating or librating, and rotating or precessing at energies respectively lower and higher than the maximum in the potential. In the context of recent investigations of highly excited molecules, the two modes would correspond for example to normal and local vibrations, repectively. The two modes are sharply separated by a trajectory (the separatrix) corresponding to the maximum in the potential (the ridge, in our applications). As is often the case, in quantum mechanics the transition between modes is smoother [10].

Mathematically, the classical problem is completely soluble in terms of Jacobian elliptic functions, the quantum problem in terms of Mathieu, or elliptical cylinder, functions. The limiting behaviour of quantum solutions for the two modes are extremely well known, and the literature contains extensive discussions of these phenomena in the limiting cases: both can be handled by perturbation techniques [11], which however fail around the ridge where a connection problem arises. Therefore, a discussion of the transition regime is of specific interest, and can be carried out by a simple uniform asymptotic technique [12].

The Schroedinger equation for the physical pendulum [10]

$$\left[-\frac{\hbar^2}{2m\ell^2}\frac{d^2}{d\theta^2} + V_o\cos\theta\right]\psi_\nu(\theta) = E_\nu\psi_\nu(\theta) \tag{1}$$

is transformed into the standard Mathieu equation

$$\frac{d^2}{d\alpha^2} Y_\nu(\alpha) + (\Lambda_\nu - 2q\cos 2\alpha)\, Y_\nu(\alpha) = 0 \tag{2}$$

defining a new angle $\alpha = \theta/2$, the parameter $4m\ell^2 V_o/\hbar^2$ and the eigenvalue $8m\ell^2 E_\nu/\hbar^2$.

Eigenvalues and eigenfunctions for this equation can be generated by using expansions in Fourier series (Hill's method) [13]

$$Y_\nu(\alpha) = \sum_{\eta=-\infty}^{\infty} t_{2\eta,\nu} \exp[i(\beta+2\eta)] \tag{3}$$

(where β depends on boundary conditions, see below). This expansion inserted in (2) leads to a secular equation, giving Λ_ν as eigenvalues and the coefficients $t_{2\eta,\nu}$ as elements of eigenvectors.

For the pendulum, the boundary conditions require the solutions in α to have π as a period, and one obtains, as a function of q, even and odd eigenvalues usually denoted a_{2n} and b_{2n}, respectively. Equation (2) appears in many problems where potentials with different symmetries, such as three-fold periodic Henon-Heiles [5] and those describing restricted molecular rotations [14], are involved. In Section 2, we will use it for studying unimolecular decay. Also, equation (2), as the simplest case of a Hill equation, may represent a zero-order approach to problems where a potential is expanded in a Fourier series. In general, therefore, other periodic boundary conditions are of interest. A Floquet type [13] of analysis shows that for N-fold symmetric potentials, when N is even, solutions with period 2π are also acceptable: corresponding eigenvalues are denoted a_{2n1} and b_{2n1} for the even and odd cases, respectively. In the language of group theory, only states with the same periodicity and parity will belong to the same irreducible representation of the symmetry group of the potential.

For general N, doubly degenerate solutions also appear, and their eigenvalues are labelled by $2n+\beta$, where β is a rational fraction less than 2, $\beta=2k/_N$, and k=1,...,-1. They induce irreducible representations of type E. It is convenient to extend the definition of k (and β) to the nondegenerate cases, corresponding to k=0 (and β=0) for 2n states (π periodicity) and k=N/2 (and β=1) for 2n+1 states (2π periodicity).

The behaviour of the eigenvalues as a function of q, well documented in the literature [14,15], is very clearly exhibited by a semiclassical analysis. Following reference [16], it is possible to obtain by an extended WKB procedure, a quantization rule, which holds asympotically in $\hbar$:

$$\cos(\sigma-\phi)=[1+\exp(-2\eta)]^{-1/2}\cos(\pi\beta) \tag{4}$$

where σ is the phase integral for the well, and η is the tunnel integral. The factor

$$\phi(\eta) = \arg\,\Gamma\,(\tfrac{1}{2}+i\eta/\pi)-(\eta/\pi)(\ell n|\eta|/\pi-1) \tag{5}$$

is introduced to improve on the primitive WKB approximation, and it is based on a mapping of the potential maximum onto a parabola. This mapping makes this asymptotic approach uniform, because it holds both below and above the potential maximum (V_o for the physical pendulum).

The phase integrals in [4], easily defined for $\Lambda_\nu<2q$, must be analytically continued for $\Lambda_\nu>2q$. For the Mathieu equation (general sinusoidal potential) all the involved integrals can be evaluated in closed form, in terms of elliptic integrals [15]. Defining $m=\Lambda_\nu/4q+1/2$, we find for $0<|\Lambda|<2q$

$$\sigma = 4q^{1/2}\,[E(m) - (1-m)K(m)],$$

$$\eta =-4q^{1/2}\,[E'(m)-mK'(m)]$$

and for $\Lambda>2q>0$

$$\sigma = 2(\Lambda+2q)^{1/2}\,E(1/m),$$

$$\eta = 2(\Lambda+2q)^{1/2}\,[K'(1/m)-E'(1/m)].$$

The quantization rule (4) has been studied in detail recently [21]: it has been found that this approach is useful not only for the approximate computation of eigenvalues but also for describing most qualitative features of the transition between modes as a function of q, in particular the behaviour of the allowed and forbidden regions for the eigenvalues. The characterization of these regions (Figure 1) is of interest for an enormous variety of phenomena, ranging from bands in solids [17] and unimolecular reaction theory [18], to quadrupole mass spectrometry [19] and particle generation by electric fields in a vacuum [20].

1.3 The P matrix and ridge effect

As shown in Figure 1 and in many figures in books which describe Mathieu functions [13,14], the dependence of eigenvalues as a function of q shows an abrupt change in character as they go through a line corresponding to 2q, which is classically the locus of separatrix trajectories. This is perhaps the simplest manifestation of the ridge effect, and, according to the nomenclature now well established in atomic physics [3] and in chemical reaction theory [21-24], 2q is identified as the ridge line (and -2q is the valley bottom line).

In typical problems, q is the slow varying variable in an adiabatic treatment where the fast variables fail to be so at ridge. A specific example, the familiar Henon-Heiles potential, has been worked out: there, q was related to a slow varying radial variable, and nonadiabaticity, i.e. the possibility of transition between states as q varies, was measured by the matrix P (see Appendix) which is conveniently generalized and defined in the present context as

$$P_{\nu\nu'}(q) = \int_0^{2\pi} Y_\nu (\alpha\ q) \frac{d}{dq} Y_{\nu'} (\alpha,q)\, d\alpha \tag{6}$$

This matrix was introduced by F. T. Smith [25] for the treatment of nonadiabatic (diabatic) couplings in atomic collisions. It is now familiar also in molecular structure problems, to indicate local breakdowns of the Born-Oppenheimer approximation. Within the hyperspherical formalism, it has been introduced in the three-body Coulomb problem [20] and in chemical reactions [21-24], see also Section 3 . Also, from equation (A4)

$$P_{\nu\nu'}(q) = (\Lambda_\nu - \Lambda_{\nu'})^{-1} \int_0^{2\pi} 2Y_\nu (\alpha,q)\cos 2\alpha\, Y_{\nu'}(\alpha,q)\, d\alpha \tag{7}$$

It is then easy to show that, for q = 0,

$$P_{\nu\nu'}(0) = (\nu^2 - \nu'^2)^{-1}$$

The actual computation of the P matrix was performed using the matrices T of the coefficients $t_{2\eta,\nu}$ introduced in (3). From (6), which in matrix notation becomes

$$\underline{P}(q) = \underline{\tilde{T}}(q) \frac{d}{dq} \underline{T}(q)$$

we easily obtained P by computing T at two close q values:

$$\underline{P}(q) = \delta^{-1} [\underline{\tilde{T}}(q)\, \underline{T}(q+\delta) - \underline{1}] \tag{8}$$

where δ is a small number. From (7), we have, after some manipulations, an alternative formula

$$P_{\nu\nu'}(q) = [\Lambda_\nu(q) - \Lambda_{\nu'}(q)]^{-1} [\underline{\tilde{T}}(q) \frac{d\underline{V}}{dq} \underline{T}(q)]_{\nu\nu'} \tag{9}$$

where $dV_{\nu\nu'}/dq$ is simply $\delta_{|\nu-\nu'|,1}$. This latter formula has the advantage of requiring a single diagonalization at each q, and therefore becomes more convenient than (8) as the size of the secular problem increases. The left hand side of Figure 1 shows some computed P-matrix elements: maxima at ridge, to be expected from the corresponding minima in the eigenvalue differences (equation (9)) and the general properties are clearly displayed in Figure 1.

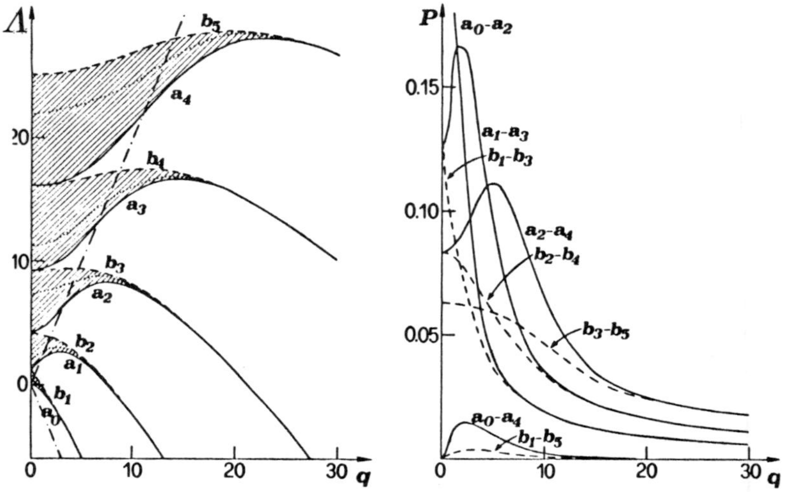

Figure 1. Some eigenvalues of the Mathieu equation, as a function of the parameter q are reported on the left, together with the ridge line 2q and the valley bottom line -2q (dash-dotted). Eigenvalues corresponding to β = 2/3 and 4/3 are shown by dotted lines. Allowed regions for eigenvalues are hatched. The right-hand side shows elements of the P matrix as a function of q.

We conclude by commenting briefly on the role of sequences of avoided crossings along the ridge, and on the related question whether analytic continuation would reveal true crossings for complex values of q. Recent results [26] on the analytic continuation of the eigenvalues of the Mathieu equation are motivated by the fact that their crossings in the complex q-plane are related to convergence radii of perturbation expansions. Therefore, it is not surprising to find that there is a correspondence between real parts of complex crossings, as listed in [26], and positions of maxima in elements of the matrix P as defined in this work.

Actually, the semiclassical formulas introduced in the previous section, although valid only asymptotically, are in the form which appears to be suitable for extensions in the complex q-plane. It would be interesting to further investigate this aspect, since analytic continuation plays a role in theories of nonadiabatic transition [27], a role which has not been firmly assessed until now because in actual problems the analytic structure of numerically generated eigenvalues is poorly understood [28].

2. MODE SPECIFIC RESONANCES AND UNIMOLECULAR DECAY

Recent advances in experimental techniques (in particular, laser and molecular beams) are promoting interesting developments in the study of intramolecular vibrational relaxation in molecules selectively excited by collisions or photon absorption [29]. These experiments can be carried out under conditions much more controlled than in the past, thus making it possible to test more accurately the statistical assumptions which underlie the current theories of unimolecular decompositions. An important goal also for practical purposes is to succeed in obtaining substantial deviations from statistical behaviour, thus opening the experimental possibility for great selectivity for the elementary processes of chemical kinetics and photochemistry.

An analysis of the specificity of unimolecular decompositions and of intramolecular vibrational relaxation can be developed by starting with some simple models for coupled oscillators, for which different modes and (transitions among them) have been well characterized, both in classical and in quantum mechanics.

2.1 Regular and irregular modes for coupled oscillators

The analysis of the pendulum motion in Section 1.3 can be immediately applied to provide a semiclassical discussion of a two dimensional (Henon-Heiles) model for coupled oscillators [5], conveniently written in polar coordinates

$$V(\rho,\theta) = \frac{1}{2}\rho^2 + \frac{\lambda\rho^3}{3}\cos 3\theta \qquad (10)$$

We refer to a previous paper [5] for a description of the model and of its relevance for a semiclassical discussion of regular and irregular modes in classical and quantum mechanics [30,31]. By relating the polar variable ρ and the parameter q

$$q = \frac{4}{27}\lambda\rho^5 \qquad (11)$$

the adiabatic potential energy curves $\varepsilon_m(\rho)$ are obtained by Mathieu eigenvalues $\Lambda_n(q)$ according to the formula

$$\varepsilon_n(\rho) = \frac{9}{8\rho^2}\Lambda_n(q) + \frac{1}{2}\rho^2 \qquad (12)$$

The elements of the nonadiabatic coupling matrix are likewise related:

$$P_{nm}(\rho) = \frac{dq}{d\rho}P_{nm}(q) \qquad (13)$$

Figures 2, 3, 4 are obtained by these formulas.

A useful aspect of this approach is to provide a classification scheme for levels. When $\lambda=0$ (the simple isotropic oscillator) a good quantum number exists, and it is designated by $\pm\ell$ in references [30] and [31]. For finite λ, the potential belongs to the C_{3v} symmetry groups, the wavefunctions are classified according to its irreducible representations A_1, A_2 and E. Mathieu functions Ce_{2n} and Se_{2n}, under the C_{3v}

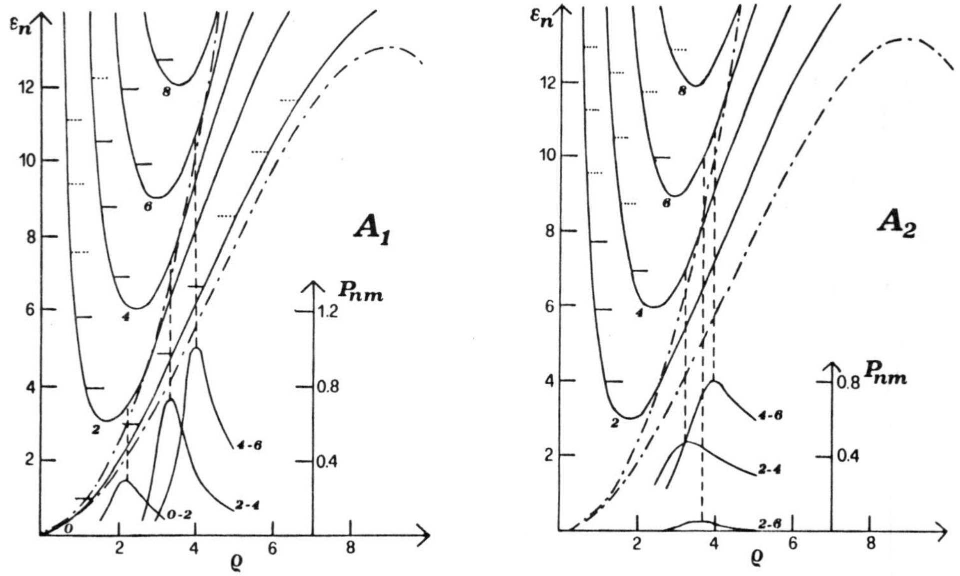

Figure 2. For the Henon-Heiles potential with $\lambda=80^{-\frac{1}{2}}$ (equation (2); upper broken curve, ridge profile ($\theta=0$, $2\pi/3$; $4\pi/3$); lower broken curve, valley bottom profile ($\theta=\pi/3$, π, $5\pi/3$)), adiabatic potential curves $\varepsilon_n(\rho)$ (equation 12) and corresponding nonadiabatic coupling matrix elements $P_{nm}(\rho)$ (equation (13)) as a function of radial coordinate ρ for A_1 and A_2 symmetry. Positions of levels indicated by continuous segments for those identified as quasiperiodic [31] and by dotted segments for those not identified as quasiperiodic. For further details, see [5].

symmetry operations behave respectively as A_1 and A_2. Their eigenvalues are labelled as A_{2n} and B_{2n+2}, where $n=0,1,2,...$ (Figure 2). They are periodic by π and correspond to $\beta=0$. The E respresentation is induced by Mathieu functions of fractionary order $Ce_{2n+\beta}$ and $Se_{2n+\beta}$, and the corresponding doubly degenerate eigenvalues will be designated as $\Lambda_{2n+\beta}$. From the quantization formula (4), we have that for this symmetry β can assume only the values 2/3 and 4/3, and therefore, in order that the proper boundary conditions are satisfied, the functions will have periodicities 3π and $3\pi/2$. Therefore, the levels supported by each adiabatic curve will conveniently be labelled, both by the proper index of corresponding Mathieu functions $2n+\beta$, and by a progressive number $v=0,1,2...$. The Mathieu index is related to ℓ by $|\ell|=3n+\beta/2$ and in the $\lambda=0$ limit the energy levels are given by

$$\varepsilon(v,2n+\beta) = 2v+1+|\ell|.$$

As illustrated in Figure 2, failures of the adiabatic picture, as measured by the elements of the matrix of nonadiabatic coupling P, occur at the ridge. The correlation between regular modes of classical investigation and the quantum mechanical states which are localized above the ridge has already been pointed out [5]. In our picture, quantum mechanical delocalization of the wavefunction is a process which is favoured by coupling between adiabatic eigenvalues in the proximity of the ridge, where a sequence of avoided crossings can be discussed within the familiar apparatus of curve crossing theory, and a striking similarity is apparent between these aspects and the theory of level perturbation for diatomic molecules [32]. Actually our current experience suggests that the semiclassical techniques introduced in such a context, are also extremely fruitful here.

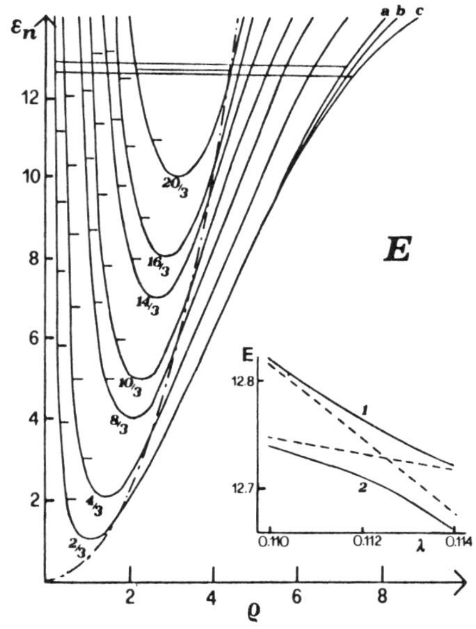

Figure 3. Adiabatic curves $\varepsilon_n(\rho)$, equation (12), for the E symmetry of Henon-Heiles potential (equation (10)) for λ close to $80^{-\frac{1}{2}}$=0.1118. Slight changes in λ affect mainly the large ρ region: for example, the curves labelled as a, b and c show how the 2/3 state varies for λ=0.110, 0.112, 0.144. The corresponding v=7 level varies as in inset, and thus would cross the v=2 level of the 20/3 state, practically unaffected by a change in λ [33] (dashed curves): actually, the crossing is avoided and the levels behave as the continuous curves 1 and 2.

Figure 3, which reports similar results for the E symmetry, focuses the attention onto a particularly interesting type of avoided crossing due to the interaction between almost degenerate levels, which the adiabatic curves support. This phenomenon leads to much more pronounced

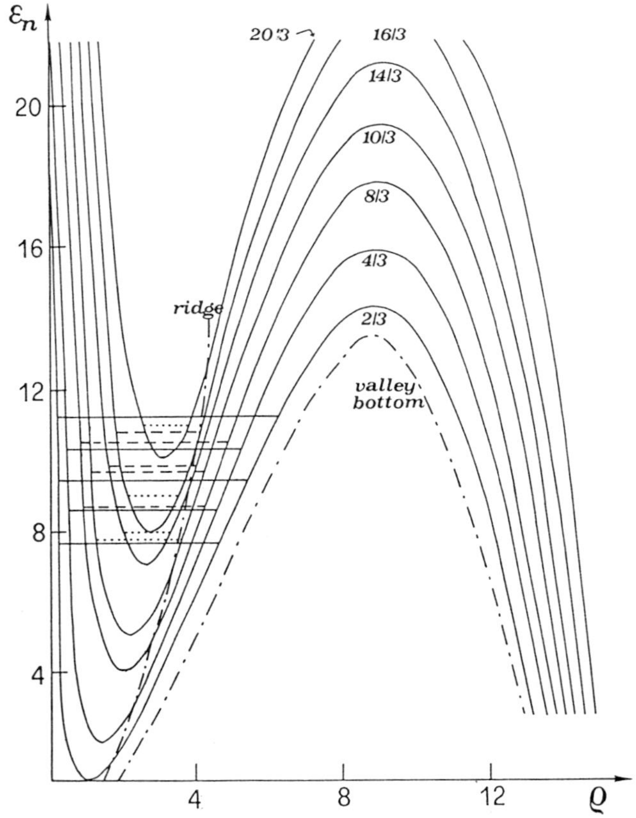

Figure 4. Behaviour, at larger ρ, of some curves as in Figure 3. Dotted, continuous and dashed lines indicate states designated as Q_I, Q_{II} and N, respectively, in [36].

delocalization of the wavefunctions, because of the strong mutual perturbation of the levels, and is strongly dependent on the parameter λ, which measures the strength of the coupling between the oscillators. Therefore, it has relevance with extended discussions [33] of the role of avoided crossings as a function of the parameter λ. In the present approach, such avoided crossings, formerly individuated as a road to quantum chaos by Percival [34], are seen to arise when, because of the increasing importance of anharmonicity for levels with high v quantum numbers, high v levels of lower curves enter into accidental resonance with low v levels of upper curves. The phenomenon is familiar in spectroscopy, leading to strong level repulsion [32]. For the model considered here, this phenomenon happens once in the neighbourhood of $\lambda=(80)^{-\frac{1}{2}}=0.118$ (Figure 3).

2.2 A resonance theory of unimolecular reactions

The previous analysis is of interest not only for general mode transition problems, but also for providing a useful model for unimolecular reaction theory.

Consider again the potential given by equation (10). If its behaviour is examined at large ρ values, it is seen that it has three symmetric saddles at height $(6\lambda^2)^{-1}$ at $\rho=\lambda^{-1}$, and eventually goes to minus infinity. Therefore all the states which it supports are actually metastable, and they will eventually decay by quantum mechanical tunnelling: in other words, they are typical quantum mechanical resonances, to which we may associate a width Γ and a lifetime $\tau = \hbar/\Gamma$. This model has already been considered [35,36] for unimolecular reaction theory, where the resonance lifetime is most naturally related to the inverse of the unimolecular rate constants $k=\tau^{-1}$.

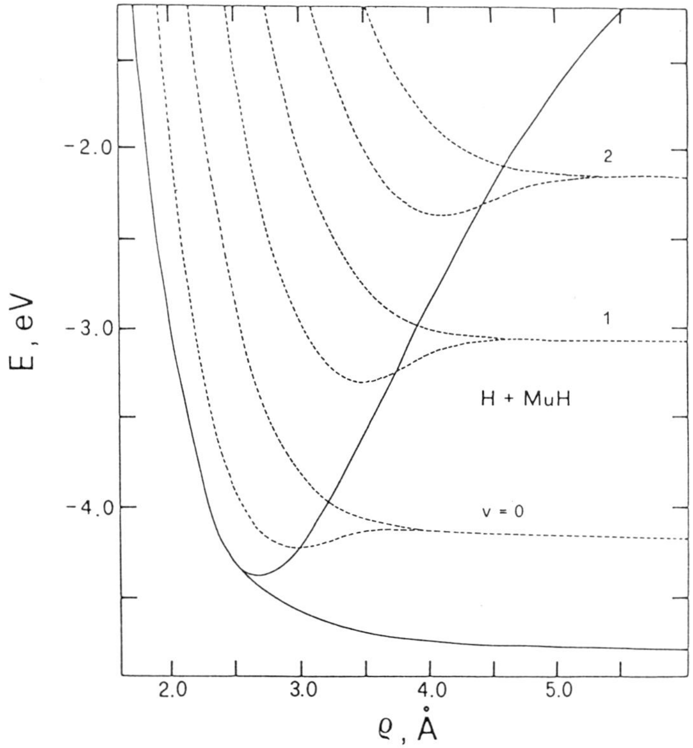

Figure 5. Adiabatic potential energy curves for H+MuH (dotted) [51, unpublished], see [21] for $H+H_2$. Valley bottom and ridge profile are shown as continuous lines.

In a search for mode specificity in resonances, we extend the previous analysis to larger ρ values, obtaining for example for the E symmetry the curves in Figure 5. Also shown in Figure 5 are some levels considered in reference [36] and their classification in terms of Q_I, Q_{II} and N states is also indicated. Bai et al. [36] find strong mode specificity for this system: the Q_{II} states show the largest unimolecular rate constants, the Q_I states the smallest, and the N states show an intermediate behaviour. (Here, as often [35-37], non-specificity means dependence on energy only: it is assumed to be an indication of full energy randomization in the molecules and, therefore, a good measure of the appropriateness of statistical arguments).

Figure 5 shows that the results of reference [36] can be qualitatively understood by considering again the ridge effect: in particular, it is apparent that Q_{II} states, being characterized by higher vibrational numbers of lower curves, have their outer turning points well beyond the ridge and, therefore, may undergo extensive nonadiabatic transitions to the lowest curve, for which tunnelling to dissociation is clearly easier. Conversly, Q_I states are confined inside the ridge and thus present the lowest decomposition rates. For the N states, for which the outer turning point is close to the ridge, there is clearly strong coupling between adiabatic states and the associated lifetimes are intermediate between the extremes.

A quantitative semiclassical analysis of these effects for this model and similar ones, in particular for a recent model for the reaction $CH_2O \rightarrow CO + H_2$ [35], is being carried out [38]. This analysis is based on a well known semiclassical formalism for resonance positions and widths [39]. It leads to quantitative agreement with the RRKM theory in the complete randomization hypothesis and it points out the role of ridge effects for such a randomization. The conditions for mode specificity are also analyzed: an important aspect of this approach is that mode specificity due to symmetries in the transition state [40] arises in a natural way through the group theoretical labelling of Mathieu functions outlined in the previous section.

3. RESONANCES AND INTERFERENCE EFFECTS IN COLLINEAR REACTIONS

Very extensive computations have been carried out on the dynamics of reactions of the type A+BC. They are either founded on classical or on quantum mechanics, are either to be considered 'exact' or involving more or less drastic approximations and have been based either in the real three dimensional world or in somewhat artificial spaces of lower dimensionality. These computations are thus attempts to solve the three-body problem more or less accurately. Other Chapters in this book extensively review this subject [41]. The papers, presented at a meeting celebrating "Fifty Years of Chemical Dynamics", held in Berlin in 1982 and published as an issue of the Berichte der Bunsen Gesellschaft in 1982, should be consulted, also for providing a historical perspective [42].

A main difficulty which arises in understanding the interplay of the various factors which influence a reaction is that both the description of reactants and that of the products is bound to fail somewhere in the course of a reaction, and it is necessary to perform a transformation whose nature and characteristics are hardly understood, although some formal progress has been made recently.

Representations symmetrical for at least some of the particles are being actively investigated [43]: none of them is, however, 'full range' in the reaction, because they do not correlate smoothly with reactants and products, and some kind of transformation is anyway to be performed in order to describe the transition. Roughly speaking, there are essentially two different ways of achieving this: the first one is based on the idea that a sudden switch during the collision from the reactant configuration to the product configuration could serve as a good starting point for following the evolution of the reaction: the second route, which is the one we pursue here, exploits the opposite view that the starting point could be the individuation of a smooth (adiabatic) path. In [41] an account is given of both approaches.

3.1 The hyperspherical adiabatic approach: curvature effects

As in the cases considered previously, the adiabatic idea attempts a simplification of the many-body problem by individuating a variable which can be approximately separated from the others: this is possible when the overall motion can be considered as taking place slowly with respect to this variable, so that the faster motion associated with the other ones can be effectively averaged. Nonadiabatic effects are to be introduced anyway for an exact description, but many significant features are likely to be displayed when the adiabatic coordinate is wisely chosen.

The traditional view for the treatment of chemical reaction involves the concept of the reaction path, a coordinate smoothly joining reactants and products. Most important progress [44] has been done recently along this line, although our approach will be an alternative. A central concept will be that of the reaction skewing angle (a measure of the curvature of the reaction path) introduced in the early thirties [42] and incorporated by F. T. Smith [45] into the general concept of kinematic rotations. Very recent work along these lines have been based on the fact, actually already present in Smith's earlier investigation [46], that the kinetic energy operator for many particles can be interpreted as the kinetic energy operator of a single particle in a space of a higher dimensionality than the physical one. The effect of the potential energy surface is then that of distorting the straight line trajectories of such a particle in the hyperspace and constraints on vibrational exchange can be deduced [45].

When the curvature along the reaction path is small, adiabatic evolution along it is a good approximation and a starting point for carrying out this approach for chemical reactions is to set up a hyperspherical coordinate system [47], for which there are several possibilities [48]. Most of the investigations carried out so far have dealt with the somewhat artificial constraint of particles moving on a line. Progress on the extension of these promising techniques to the full three-dimensional case has been limited to the development of analytical approaches and to the study of simple test cases. In the following, we will discuss in some detail what we have learned from the one-dimensional case and which we believe to be of interest also for the real three-dimensional world.

A configuration such as A+BC will be best described by mass scaling the two corresponding orthogonal Jacobi vectors: when the same is done also for the configuration of products AB+C, resulted from an exchange of the atom B, the properties of the kinetic energy operator are such that the new mass scaled vectors are still orthogonal, but rotated with respect to the previous ones. The potential energy surface for a collinear reaction is actually confined in a sector, defined by the reaction skewing angle, a function of the atomic masses

$$\gamma = \arctan[m_B(m_A+m_B+m_C)/(m_A+m_C)]^{1/2} \tag{14}$$

The confinement in such a sector sets boundaries to the dynamics, which physically correspond to the prohibition, for masses on a line, to overcome each other. It also shows that very different kinematic effects are likely to be associated with different mass combinations.

Considering then formula (14) we observe that the fully symmetrical exchange of an atom A in the process A+AA=AA+A involves an angle of 60 degrees. For the exchange of the atom B in the nearly symmetrical process A+BA=AB+A, the relative masses of A and B determine the full range of variation for the skewing angle from near zero when B is much lighter than A to near 90 when B is much heavier than A. The two limiting cases provide, as we shall see, very different kinematic effects, so that the full dynamics, obtained by introducing explicitly the potential energy surface, will be dramatically affected.

Consider first the case of very large skewing angle. The path from reactants to products involves a bent of nearly 90 degrees: any coordinate system which aims at describing the reaction path from reactants to products will introduce a centrifugal distortion at the bend, but may maintain some reasonable descriptive power in the qualitative treatment of the reactive process. From a computational point of view, setting up a coordinate system more or less based on the idea of following the evolution of the system along the reaction path entails the introduction of strong coupling between channels in a quantum mechanical framework, or strong centrifugal distortions requiring fine grid integration of trajectories in a classical mechanical framework; this effect is especially likely to be important in the region where the bend is sharpest, and this most often occurs when the system overcomes the saddle which separates the valley of reactants from that of products. Introducing a non-orthogonal system which follows the evolution of the system from reactants to products becomes increasingly difficult as the skewing angle decreases, because the distortion required to straighten the path into a Cartesian coordinate introduces terms in the Jacobian transformation matrix which physically correspond to centrifugal forces. Therefore, evolution along the reaction path provides a good description of what trajectory is really followed only in the adiabatic limit,i.e. for an infinitely slow process. It is apparent that the procedure becomes impractical even for computational purposes since extensive channel coupling has to be introduced explicitly. Actually, the practical computation of quantum mechanical one-dimensional chemical reaction rates for small skewing angle is the motivation which has lead various authors to use hyperspherical coordinates [49].

In the hyperspherical view, a radial coordinate ρ is defined by setting up a polar system in the properly mass scaled space of Jacobi vector components (see the Appendix). In terms of this coordinate, which is independent of the rearrangement channel, it is possible to follow the reaction as evolving from the intermediate state, where the particles are closer together, to reactant and product valleys. The numerical implementation of this approach has allowed the full characterization of reactions with small to moderate skewing angles. When matched with the more conventional method involving the reaction path coordinates which follow the reaction from reactants to products, now we have a complementary view for the full characterization of most of the features which collinear rearrangement processes may display (subthreshold behaviour, imputable to tunnel effects; resonance behaviour, which can be attributed to partial trapping in metastable states; oscillatory behaviour of cross sections for state to state transitions, attributable to interference between channels). Our semi-classical view of some of these features is described in the following.

3.2 Adiabatic energy levels

Following the formalism outlined in the Appendix, the first step, and perhaps the hardest one, is to map the interaction as adiabatic potential energy curves $\varepsilon(\rho)$ as a function of the hyperradius. This step is performed by solving a problem of lower dimensionality than the full one, and generates at the same time wavefunctions which vary parametrically with the hyperradius. These may then be used to obtain the elements of the matrix P which measures the nonadiabatic inter-channel coupling. Fully quantum and semiclassical recipes are being developed and tested for carrying out this program, which is preliminary to the dynamical calaculations, consisting in solving more or less exactly equation (A2), or (A7), or similar ones.

The investigations of these maps for model problems is of extreme interest, because it allows the full understanding within a unifying framework of most features which are observed in elaborated computational studies of atomic exchange processes. The paradigmatic situation considered here is the one of three particles on the line, involving the partially symmetric system A+BA=AB+A. For other similar systems, reactant and product valleys may be unsymmetrical and additional complications are introduced by the channel coupling. However, the general features outlined for the symmetric cases are confirmed [50].

The results for such maps, which are now available for many reactions in the original references (see the example of H+MuH in Figure 5 [51]), lead to the following view of the reactive process: instead of an evolution from reactants towards products, consider a reactive process as a decay of the intermediate state, located in the region where the interaction is strongest. This corresponds for simple situations to the transition state, characterized by a symmetric stretch vibration of a bound character, and an antisymmetric stretch vibration leading to dissociation. The system reaches this potential energy saddle configuration climbing from the reaction valley: sequentially in time it may come back again to give an elastic (vibrational conserving) collision or inelastic transition between vibrational states. Alternatively, it may descend to the product valley, either leading to the corresponding state of the products connected adiabatically to the original one in the reactants (adiabatic reaction), or may lead to vibrational states corresponding to nonadiabatic events. Again, the role of the ridge, which acts as the watershed between the valleys of reactants and products, has been found to be central.

The first step of calculating adiabatic energy levels at fixed values of the radial hyperspherical coordinate ρ is usually performed numerically (by us in [21-24] employing a harmonic expansion technique, seethe Appendix). Typically, as ρ varies from values lower than the saddle point to larger ones, one has to solve a problem of quantization in a single or in a double well. We have also examined [23] the accuracy of semiclassical quantization prescriptions, which (as reviewed in [32] for spectroscopic applications) require phase integrals σ for wells and η for barriers, and again as in equation (4) corrections ϕ (see equation (5)) for the coalescing of turning points.

3.3 Interferences and resonances for large curvature

As shown in [22] (similar results were independently obtained by others [52]), the simplest approach to the description of reaction probability is to assume full adiabatic decoupling and to treat the dynamics as scattering from the potentials generated by the adiabatic levels as ρ varies. Neglecting coupling in an adiabatic representation leads to the simplest description of the probabaility p_{vv} for a symmetric exchange reaction, in terms of phaseshifts $\delta_v^{\pm}$ from even (+) and odd (-) potentials [53]

$$p_{vv} = \sin^2 (\delta_v^+ - \delta_v^-) \tag{15}$$

Extremely accurate results for reaction probabilities, e.g. were obtained in a few cases (e.g. I+HI and isotopic variants, [22,52]; for H+MuH [22], see Figure 6), using purely semiclassical techniques for scattering phaseshifts. Formula (15) allows to predict oscillations due to interference between even and odd propagation in the energy dependence of probabilities. These oscillations have been found in numerical work [7]. As shown in Table I (from [22]) excellent agreement with exact calculation was obtained for resonances in H+MuH (see also Figure 5).

Finally, these adiabatic approaches have shown to be able to predict the formation of stable molecules trapped on a repulsive potential energy surface [54] (the stability of these molecules has been confirmed also by three dimensional calculations). This is a purely quantum mechanical effect, since classically such systems would dissociate: this prediction is a big success of the adiabatic approach to the three body problems, which therefore is being shown useful for a unified view of bound states and collisions.

As the skewing angle increases, however, the simple adiabatic hypothesis becomes less and less satisfactory, as the $H+H_2$ case shows, Figure 7 [51]. The clear inadequacy of this oversimplified approach to deal with the general case has pointed out the need for the semi-classical treatment of nonadiabatic coupling.

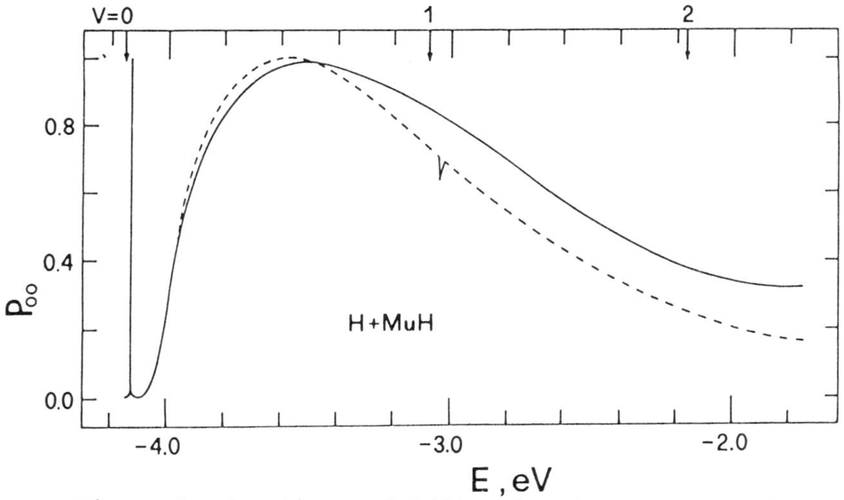

Figure 6. Reaction probability for the H+MuH reaction as computed from equation (15) (continuous line), is compared with exact results (J. Manz and J. Romelt, Chem. Phys. Letters **76**, 337 (1980)) by showing the latter as a dashed line. Energies are measured from the dissociation limit of MuH, and thresholds of vibrational levels are indicated by arrows [22].

TABLE 1. Resonances for the H+MuH reaction

Vibrational level	Position [a] eV	Width hwhm, eV	Method
0	- 4.1238	0.00023	Accurate quantum [b]
	- 4.1258	0.00027	Hyperspherical adiabatic [c]
	- 4.142	-	C.I. stabilization [d]
	- 4.097	-	SCF stabilization [d]
	- 4.061	-	Period orbits [e]
1	- 3.041	0.012	Accurate quantum [b]
	- 3.054	0.004	Hyperspherical adiabatic [c]
	- 3.105	-	Periodic orbits [e]
2	- 2.1501	0.0005	Accurate quantum [b]
	- 2.1673	0.0000	Hyperspherical adiabatic [c]
	- 2.205	-	Periodic orbits [e]

[a] Energies measured from the dissociation limit of MuH;
[b] J. Manz and J. Romelt, Chem.Phys.Lett., **76**, 337 (1980);
[c] Present work; [d] T. C. Thompson and D. G. Truhlar, J.Chem.Phys., **76**, 1790 (1982); [e] J. Manz, E. Pollak and J. Romelt, Chem.Phys.Lett., **86**, 26 (1982).

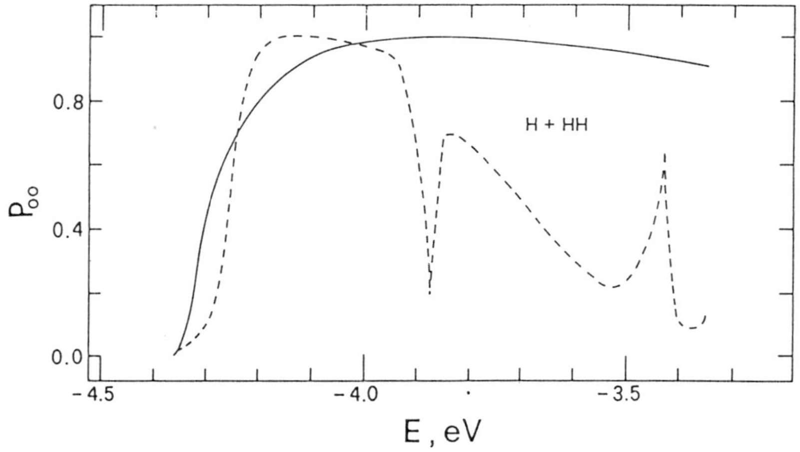

Figure 7. As in Figure 6, for $H+H_2$, [51, unpublished].

3.4 A nondiabatic model for resonances: the ridge effect

As a representative extreme example of failure of the adiabatic approach, we consider next a model for the H+FH reaction [23], as extensively studied by various authors [55]. Figure 8 shows the adiabatic curves for this reaction: for most of ρ values, they have a smooth, uncomplicated behaviour, except near the ridge which separates the valleys of reactants and products: wells and barriers, evidently due to mututal interactions, appear there. Arguments borrowed from spectroscopy of diatomics and atom-atom collision theory can be used to anticipate that couplings must be localized there.

Figure 9 shows the adiabatic levels for the isotopic variant D+FD [24]. A comparison with the H+FH case indicates the same overall qualitative behaviour, the difference being only due to the effect of the masses: according to our definition of the hyperradius (see the Appendix), the abscissa is slightly expanded here, resulting in a slight decrease of the steepness of the potential ridge. The density of adiabatic states is however higher, because of the larger reduced mass of DF as compared to HF. (Mass factors for all the reactions considered in this account are listed in Table 2, including the skewing angle, equation (14)).

In the process of computing the adiabatic eigenvalues, we generate the orthogonal matrix T of eigenvectors, and according to the Appendix and Section 1.3, we have obtained P by computing T matrices at very close values of ρ. Elements of the nonadiabatic matrix P which couple some of the adiabatic levels in Figure 9 are shown in Figure 10 [24].

These results are a striking demonstration of the effects associated to the potential ridge: couplings between states are localized there, and adiabatic behaviour holds on both sides far from it.

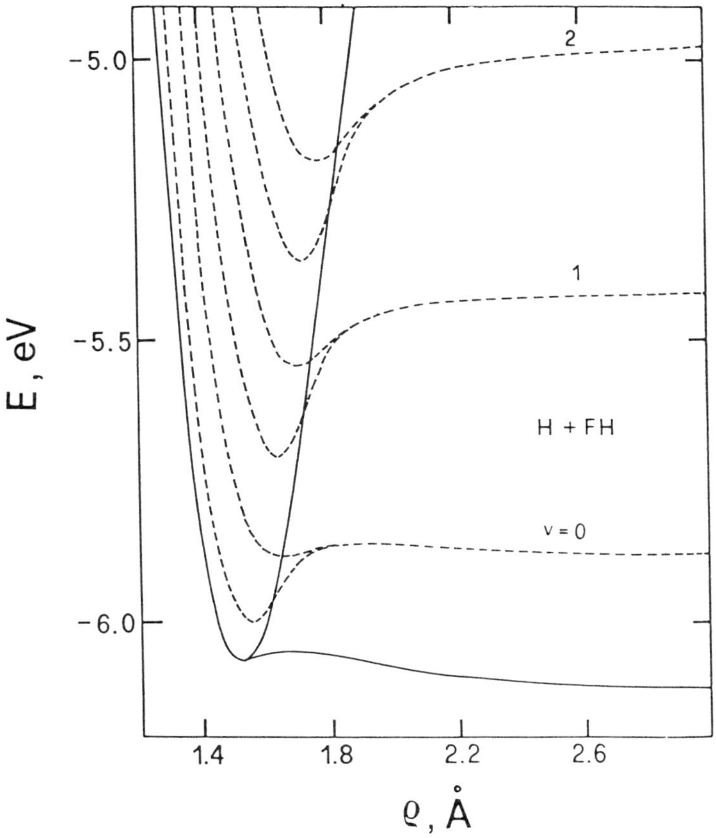

Figure 8. As in Figure 5, for H+FH [23].

TABLE 2. Mass factors

Reaction A + BA	m_{AB} (a.m.u.)	m_{ABA} (a.m.u.)	γ (degrees)
I + MuI	0.114	1.689	2.43
I + HI	1.000	7.982	7.20
H + MuH	0.101	0.231	25.95
H + HH	0.504	0.582	60
D + FD	1.821	1.829	84.50
H + FH	0.957	0.958	87.11

This can be readily understood by recalling (see equation (A4) that P is essentially a nondiagonal antisymmetric momentum matrix: its elements are those of a differential operator, and close scrutiny shows that eigenvectors change drastically at the ridge from those of the transition state (an intermediate complex) to those of the separated collision partners. It is easy to show that asymptotically P matrix elements decay as ρ^{-1}, and tend to overlaps of vibrational wave-functions of diatoms.

Alternatively, P can be viewed (equation (A4)) as a matrix element of the force along ρ divided by the difference of energy eigenvalues: according, as noted previously (Section 1.3), adiabatic levels are closest at the ridge. Actually we found that maxima in calculated P occur, to within 0.1Å, at ρ values where the difference of the corresponding levels has a minimum for all the cases studied (Table III lists as ρ_x and $\Delta\varepsilon_x$ these values for the ε_0^+ and ε_1^+ states).

TABLE III. Properties of the nonadiabatic interaction between ε_0^+ and ε_1^+.

Reaction	ρ_x (Å)	F_x (eV Å^{-1})	$\Delta\varepsilon_x$ (eV)	$P_x m^{-\frac{1}{2}}$ (Å^{-1} (a.m.u.)$^{-\frac{1}{2}}$) Approx.	Exact
I + MuI	20.8	0.25	0.54	0.28	<0.28
I + HI	11.0	1.0	0.16	2.2	0.45
H + MuH	3.50	1.3	0.82	3.3	4.33
H + HH	1.91	2.0	0.33	7.9	8.67
D + FD	1.69	2.8	0.15	13.8	12.72
H + FH	1.68	3.4	0.22	15.8	14.87

This avoided crossing structure around the ridge suggests we compare $F_x/\Delta\varepsilon_x$ and P_x (where x indicates "in ρ_x"), F_x being the local derivative ("steepness") of the ridge. Table III shows that these approximate values satisfactorily correlate with exact ones in the interesting cases that they are larger. (A scaling by the three-body reduced mass (Table II) has been performed, because P appears in Schroedinger's equation with a mass factor, see equation (A4)).

Finally, the role of the skewing angle as a measure of the purely kinematic effect is also apparent from Table II and III: classically, its increase favours temporary trapping in the transition zone, and obstacles direct dissociation. The increase in P as the skewing angle gets larger is then a manifestation that trapping is the classical analog for diabaticity, and direct dissociation for adiabaticity in the ρ variable.

It has been amply demonstrated that knowledge of P and its properties opens the possibility of explicitly introducing nonadiabatic effects within a semiclassical scheme. Actually, the results of this

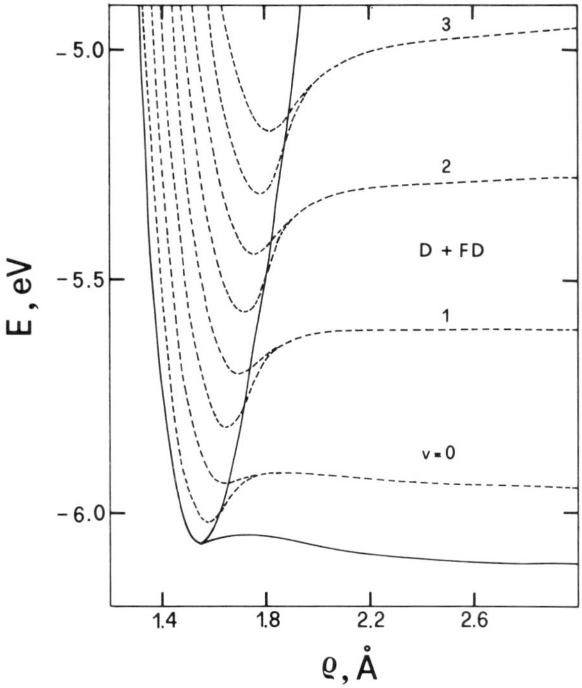

Figure 9. As in Figure 5, for D+FD [24].

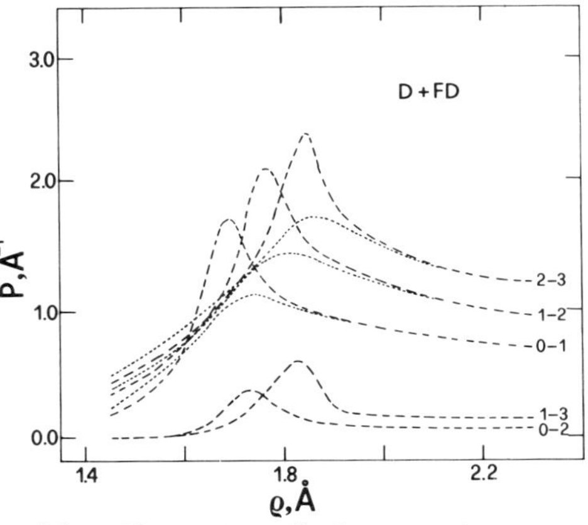

Figure 10. Elements of the matrix P, coupling some of the states for D+FD in Figure 9. ((+) states, dashed, (-) states, dotted) [24].

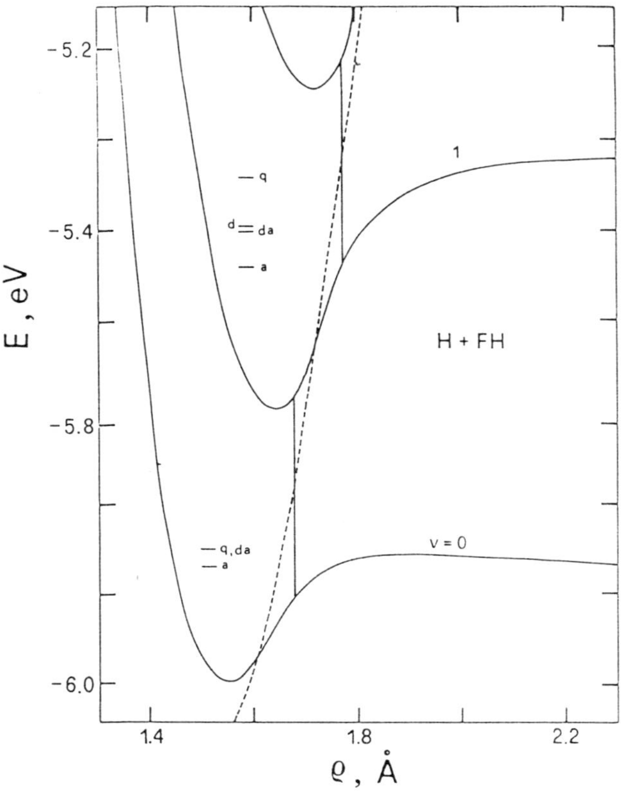

Figure 11. Resonance positions for H+FH (See Table IV). (q are exact, a adiabatic, d assuming diabatic behaviour at verical lines joining states, da assuming diabatic behaviour between 0 and 1 only). Continuous line is the potential ridge [23]

paper provide the foundation for a nonadiabatic model for resonances, outlined in reference [23], see Figures 11 and 12 and Table IV. The demonstrated locality of nonadiabaticity and its avoided crossing structure justifies the introduction of Landau-Zener type of treatments for diabatic-adiabatic behaviour at the ridge. (An estimate of the non-adiabatic transition probability by the Landau-Zener formula using the calculated P matrix confirms the diabatic behaviour postulated for the resonances in Figures 11, 12 and Table IV).

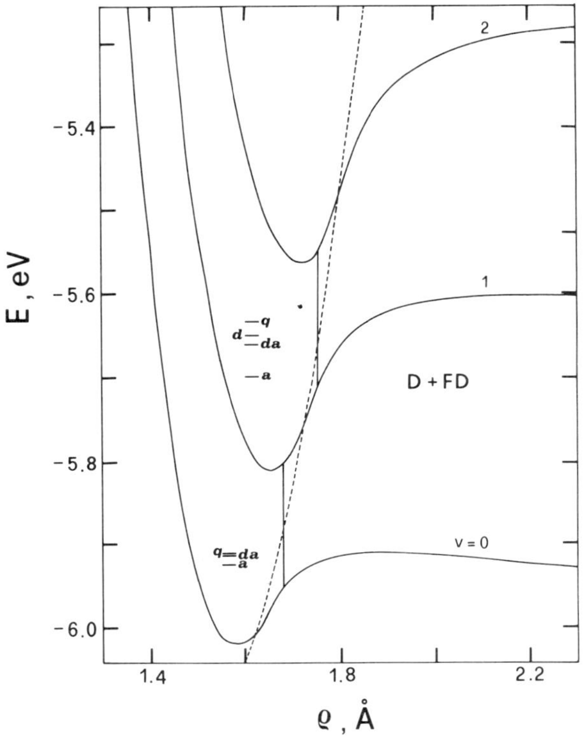

Figure 12. As in Figure 11, for D+FD [24].

TABLE IV. Resonances for the reactions H+FH and D+FD.

Position (eV)		Method
H + FH	D + FD	
-5.850	-5.912	Exact[a]
-5.848	-5.911	MEPVAG[a]
-5.841	-5.907	MEPSA[a]
-5.845	-	SCF[b]
-5.861	-	CI[b]
-5.869	-5.924	Hyperspherical adiabatic
-5.850	-5.910	Hyperspherical diabatic

[a]B.C. Garrett, D.G. Truhlar, R.S. Grev, G.C. Schatz, and R.B. Walker, J.Chem.Phys., **85** 3806 (1981); [b]T.C. Thompson and D.G. Truhlar, J.Chem.Phys., **76** 1790 (1982).

4. CONCLUDING REMARKS

The promising theoretical advances outlined above, especially the use of the hyperradius as a useful adiabatic coordinate both for unimolecular reaction models and collinear bimolecular reactions, are likely to be important tools for further progress. In particular, the actual computed P matrix elements for the systems considered here confirm indications from previous studies on the limits of the validity for the adiabatic approximation, and offer interesting clues on how to go beyond. This is particularly important if we are going towards applying these methods to the real 3-D world [56], where resonances in reactions can now be seen experimentally [57].

5. ACKNOWLEDGEMENTS

The author wishes to thank his collaborators on the work reported here, S. Cavalli, A. Lagana' and especially G. Grossi, who has also contributed to the writing of this paper.

APPENDIX: ADIABATIC AND DIABATIC REPRESENTATIONS

For definiteness, we consider the case that for an (n+1)-dimensional problem, the variable which we separate from the others is a distance ρ from a point in an (n+1)-dimensional space [21]. So, ρ is the hyper-radius of an n-dimensional sphere, and Ω_n will denote the remaining n coordinates (typically, hyperangles). Such a representation for the many body problem is being actively explored because the kinetic energy operator becomes essentially a Laplacian of a hyperspace. Also, ρ has proven to be a good choice as a nearly separable variable for many problems (see Section 3): other choices are of course possible, but may lead to the appearance of additional coupling terms in the following equations.

When the total wavefunction is expanded in an adiabatic basis set

$$\vec{\psi}(\rho,\Omega_n) = \rho^{-n/2}\, \underline{\phi}^{a}(\rho,\Omega_n)\, \vec{F}^{a}(\rho) \tag{A1}$$

the hyperradial adiabatic functions $\vec{F}^a$ are to be found as the solutions with proper boundary conditions of the infinite set of coupled linear differential equations [58]

$$\{-\frac{\hbar^2}{2m}[\underline{1}\frac{d}{d\rho} + \underline{P}(\rho)]^2 + \underline{\varepsilon}(\rho) - E\underline{1}\}\, \vec{F}^{a}(\rho) = 0 \tag{A2}$$

Here, the matrix of adiabatic eigenfunctions $\underline{\phi}^a(\rho,\Omega_n)$ and the diagonal matrix of adiabatic potential energy curves $\underline{\varepsilon}(\rho)$ are solutions of the eigenvalue problem

$$\{-\frac{\hbar^2}{2m\rho^2}[\underline{\Lambda}^2 + \frac{n^2}{4} - \frac{n^2}{2}] + V(\rho,\Omega_n) - \underline{\varepsilon}(\rho)\}\underline{\phi}^a(\rho,\Omega_n) = 0 \tag{A3}$$

where the operator $\underline{\Lambda}^2$ is the angular part of the Laplacian of the (n+1)-dimensional space, and $V(\rho,\Omega_n)$ is the interaction potential. The infinite sets are meant to be solved after proper truncation. The mass parameter m, which appears in equation (A2), depends on the definition of the hyperradius ρ: in Section 3, ρ has been defined so that m is the three-body reduced mass, $[m_A m_B m_C/(m_A+m_B+m_C)]^{\frac{1}{2}}$.

The adiabatic approximation consists of neglecting all coupling in equation (A2), i.e. in neglecting the elements of the antisymmetric matrix $\underline{P}(\rho)$

$$P_{vv'} = \langle\phi_v^a|\frac{\partial}{\partial\rho}\phi_{v'}^a\rangle = (\varepsilon_v - \varepsilon_{v'})^{-1}\langle\phi_v^a|\frac{\partial V}{\partial\rho}\phi_{v'}^a\rangle = -P_{v'v} \tag{A4}$$

the brackets denoting integration over hyperangles Ω_n, and use has been made of the Hellmann-Feynman theorem. Improved versions add to $-\varepsilon_v$ the diagonal terms of a matrix $\underline{Q}(\rho)$ given by $\underline{Q} = -\underline{P}^2 + \frac{d}{d\rho}\underline{P}$. A scheme for further corrections to the adiabatic approximation has also been developed [58].

Diabatic representations [25] correspond to an alternative expansion

$$\vec{\psi}(\rho,\Omega_n) = \rho^{-n/2}\underline{\phi}^d(\Omega_n)\,\vec{F}^d(\rho) \tag{A5}$$

By comparison with (A1), equation (A5) implies the definition of an orthogonal transformation matrix $\underline{T}(\rho)$

$$\underline{\phi}^a(\rho,\Omega_n) = \underline{\phi}^d(\Omega_n)\underline{T}(\rho),$$

and

$$\vec{F}^d(\rho) = \underline{T}(\rho)\vec{F}^a(\rho)$$

which can be obtained once equation (A3) has been solved by requiring that the orthogonal matrix $\underline{T}(\rho)$ satisfies the system

$$\underline{P}(\rho) = \underline{\tilde{T}}(\rho)\frac{d}{d\rho}\underline{T}(\rho) \tag{A6}$$

As a result, first derivatives disappear from equation (A2), which becomes

$$\{-\frac{\hbar^2}{2m}[\frac{d^2}{d\rho^2} + E]\underline{1} + \underline{V}(\rho)\}\vec{F}^d(\rho)=0 \tag{A7}$$

the coupling being transferred from the kinetic term in equation (A2) to the potential, which now is a non-diagonal diabatic matrix

$$\underline{V}(\rho) = \underline{T}(\rho)\,\underline{\varepsilon}(\rho)\,\underline{\tilde{T}}(\rho) \tag{A8}$$

The prescription is not unique, since any ρ-independent rotation of $\underline{T}$ is also a solution: boundary conditions are imposed on (A7) by requiring $\underline{V}(\rho)$ to coincide with $\underline{\varepsilon}(\rho)$ at some ρ (local diabaticity [5]).

It is often convenient to follow the opposite route, and obtain first a diabatic representation through an eigenfunction expansion: actually, by exploiting the algebraic advantages of the hyperspherical formulation, expansions in the harmonics of the n-dimensional hypersphere often provide a natural way for constructing the diabatic basis [21,48]. They can be constructed in several ways, corresponding to different parametrizations of hyperangles (see e.g. [48,49]; discrete representations can also be devised [60]) and should prove useful for obtaining alternative diabatic representations corresponding to alternative coupling schemes. These possibilities in the choice of coordinate frames are of interest both for the formation of exact treatments and for approximate approaches based on efficient decoupling schemes.

REFERENCES

[1] As an introduction to a very extensive literature, see the recent books Resonances-Models and Phenomena, Eds. S. Albeverio, L. S. Ferreira and L. Streit (Lecture Notes in Physics 211, Springer-Verlag, Berlin 1984); Resonances in Electron-Molecule Scattering, van der Waals Complexes and Reactive Chemical Dynamics, Ed. D. G. Truhlar, ACS Symposium Series, 263 (1984).
[2] See the books: M. S. Child, Molecular Collision Theory (Academic, New York, 1974) and Semiclassical Methods in Molecular Scattering and Spectroscopy, Ed. M. S. Child (Reidel, Dordrecht, 1980).
[3] U. Fano, Phys.Rev., **A22**, 2660 (1980); Rep.Progr.Phys., **46**, 97 (1983); in Atomic Physics, Vol. 8, Eds. I. Lindgren, A. Rosen and S. Svendberg (Plenum Press, New York, 1983); in Electronic and Atomic Collisions, Eds. J. Eichler, I. W. Hertel and N. Stolterfoht (North Holland, Amsterdam, 1984) p. 629.
[4] See the proceedings of the 1983 Como Conference, Chaotic Behavior in Quantum Systems, Ed. G. Casati (Plenum Press, New York, 1985), especially papers by W. P. Reinhardt, by E. J. Heller and R. L. Sundberg, and by R. A. Marcus.
[5] V. Aquilanti, S. Cavalli and G. Grossi, in Chaotic Behaviour in Quantum Systems, Ed. G. Casati (Plenum Press, New York, 1985) p. 299.
[6] V. Aquilanti, G. Grossi and F. Pirani, in Electronic and Atomic Collisions, Invited Papers, Eds. J. Eichler, I. W. Hertel and N. Stolterfoht (North Holland, Amsterdam, 1984) p. 441.
[7] See the contribution by J. Romelt in this book, and Section 3.
[8] Ch. Schlier, in Energy Storage and Redistribution in Molecules, Ed. J. Hinze (Plenum Press, New York, 1983).
[9] H. Goldstein, Classical Mechanics (Addison-Wesley, Reading, 1950); V. I. Arnold, Mathematical Methods of Classical Mechanics (Springer, Berlin, 1978); B. V. Chirikov, Phys.Rept., 52, 263 (1979).
[10] E. U. Condon, Phys.Rev., **31**, 891 (1928); S. Fluegge, Practical Quantum Mechanics (Springer, Berlin, 1984).
[11] A. H. Nayfeh, Perturbation Methods (Wiley, New York, 1973); W. Barrett, Phil.Trans.Roy.Soc.London, **A301**, 75,81,99,115,137 (1981).
[12] V. Aquilanti, S. Cavalli and G. Grossi, Chem.Phys.Lett., **110**, 43 (1984); also in Fundamental Processes in Atomic Collision Physics,

Ed. H. Kleinpoppen (Plenum Press, New York, in press).
[13] N. W. McLachlan, Theory and Application of Mathieu Functions (Clarendon, Oxford, 1947); J. Meixner and F. W. Schaefke, Mathieusche Funktionen und Sphaeroidfunktionen (Springer, Berlin, 1954); F. M. Arscott, Periodic Differential Equations (Pergamon Press, Oxford, 1964).
[14] C. C. Lin and J. D. Swalen, Rev.Mod.Phys., **31**, 841 (1959).
[15] M. Abramowitz and I. A. Stegun, Handbook of Mathematical Functions (Dover, New York, 1965).
[16] W. H. Miller, J.Chem.Phys., **48**, 1651 (1968); M. S. Child, Discussion Faraday Soc., **44**, 68 (1967); J.Mol.Spectry., 53, 280 (1974); J. N. L. Connor, T. Uzer, R. A. Marcus and A. D. Smith, J.Chem.Phys., **80**, 5095 (1984).
[17] J. C. Slater, Quantum Theory of Molecules and Solids, Vol. 2 (McGraw-Hill, New York, 1965).
[18] E. Thiele and D. J. Wilson, J.Chem.Phys., 35, 1256 (1961); see also Section 2.2.
[19] W. Paul and H. Steinwedel, Z. Naturforsch, **8a**, 448 (1953); W. Paul and M. Raether, Z.Physik, **140**, 262 (1955); W. Paul, H. P. Reinhard and U. von Zahn, Z.Physik, **152**, 143 (1958).
[20] V. M. Frolov, Differential Equations, **18**, 959 (1983).
[21] V. Aquilanti, G. Grossi and A. Lagana', Chem.Phys.Lett., **93**, 174 (1982).
[22] V. Aquilanti, S. Cavalli and A. Lagana', Chem.Phys.Lett., **93**, 179 (1982).
[23] V. Aquilanti, S. Cavalli, G. Grossi and A. Lagana', J.Mol.Struct., **93**, 319 (1983).
[24] V. Aquilanti, S. Cavalli, G. Grossi and A. Lagana', J.Mol.Struct., **107**, 95 (1984).
[25] F. T. Smith, Phys.Rev., **179**, 111 (1969).
[26] J. Meixner, F. W. Schaefke and G. Wolf, Mathieu Functions and Spheroidal Functions and Their Mathematical Foundations, Lecture Notes in Mathematics, Vol. 837 (Springer, Berlin, 1980).
[27] K. S. Lam and T. F. George, in Semiclassical Methods in Molecular Scattering and Spectroscopy, Ed. M. S. Child (Reidel, Dordrecht, 1980) Ch. 6.
[28] Analytical continuation of eigenvalues is central also for recent developments in the numerical computation of scattering resonances. See 1 , and also C. W. McCurdy and J. F. McNutt, Chem.Phys.Lett., **94**, 306 (1983); A. D. Isaacson and D. G. Truhlar, Chem.Phys.Lett., **110**, 130 (1984).
[29] For the role of reactive scattering resonances in photodissociation, see R. T Pack, J.Chem.Phys., **65**, 4765 (1976); D. C. Clary, J.Phys. Chem. **86**, 2569 (1982).
[30] D. W. Noid, M. L. Koszykowski and R. A. Marcus, Ann.Rev.Phys.Chem., **32**, 267 (1981).
[31] G. Hose and H. S. Taylor, J.Chem.Phys., **76**, 5356 (1982); Chem.Phys., **84**, 375 (1984); G. Hose, H. S. Taylor and Y. Y. Bai, J.Chem.Phys., **80**, 4363 (1984).
[32] M. S. Child, J.Mol.Spectry., **53**, 280 (1974), and in Semiclassical Methods in Molecular Scattering and Spectroscopy, Ed. M. S. Child (Reidel, Dordrecht, 1980).

[33] D. W. Noid, M. L. Koszykowski, M. Tabor and R. A. Marcus, J.Chem. Phys., **72**, 6169 (1980); see also T. Uzer, D. W. Noid and R. A. Marcus, J.Chem.Phys., **79**, 4412 (1983), and references therein.
[34] I. C. Percival, Advan.Chem.Phys., **36**, 1 (1977), and references therein.
[35] B. A. Waite and W. H. Miller, J.Chem.Phys., **74**, 3910 (1981); see also B. A. Waite, S. K. Gray and W. H. Miller, J.Chem.Phys., **78**, 259 (1983), and references therein.
[36] Y. Y. Bai, G. Hose, C. W. McCurdy and H. S. Taylor, Chem.Phys.Lett., **99**, 342 (1983).
[37] See papers by Marcus and co-workers, e.g. references [4] and [30], and references therein.
[38] V. Aquilanti, S. Cavalii and G. Grossi, to be published.
[39] J. N. L. Connor, in Semiclassical Methods in Molecular Scattering and Spectroscopy, Ed. M. S. Child (Reidel, Dordrecht, 1980) pp. 45-107; J. N. L. Connor and A. D. Smith, J.Chem.Phys., **78**, 6161 (1983).
[40] W. H. Miller, J.Am.Chem.Soc., **105**, 216 (1983).
[41] See also V. Aquilanti and A. Lagana', in Nonequilibrium Vibrational Kinetics, Ed. M. Capitelli (Springer, Berlin, in press).
[42] See also J. O. Hirschfelder, Ann.Rev.Phys.Chem., **34**, 1 (1983).
[43] F. T. Smith, Phys.Rev.Lett., **45**, 1157 (1980).
[44] See the contribution by W. H. Miller in this book.
[45] F. T. Smith, J.Chem.Phys., **31**, 1352 (1959).
[46] F. T. Smith, Phys.Rev., **120**, 1058 (1960).
[47] F. T. Smith, J.Math.Phys., **3**, 735 (1962); B. R. Johnson, J.Chem. Phys., **73**, 5051 (1980); **79**, 1906 (1983); **79**, 1916 (1983); R. T Pack, Chem.Phys.Lett., **108**, 333 (1984).
[48] See e.g. V. Aquilanti, G. Grossi and A. Lagana', J.Chem.Phys., **76**, 1587 (1982); G. Grossi, J.Chem.Phys., **81**, 3355 (1984).
[49] See the contributions by A. Kuppermann and J. Romelt in this book.
[50] J. M. Launay and M. LeDourneuf, J.Phys., **B15**, L455 (1982) and in Electronic and Atomic Collisions, Invited Papers, Eds. J. Eichler, I. W. Hertel and N. Stolterfoht (North Holland, Amsterdam, 1984) p. 635, introduced the hyperspherical adiabatic location of resonances and the role of the ridge for $F+H_2$; for hyperspherical semiclassical models, see: V. K. Babamov, V. Lopez and R. A. Marcus, J.Chem.Phys., **78**, 5621 (1983); **80**, 1812 (1984); Chem.Phys.Lett., **101**, 507 (1983); N. Abusalbi, D. J. Kouri, V. Lopez, V. K. Babamov and R. A. Marcus, Chem.Phys.Lett., **103**, 458 (1984); H. Nakamura, J.Phys.Chem., **88**, 4812 (1984).
[51] S. Cavalli, Tesi di Laurea, Universita' di Perugia, 1982.
[52] C. H. Hiller, J. Manz, W. H. Miller and J. Romelt, J.Chem.Phys., **78**, 3850 (1983).
[53] V. K. Babamov and R. A. Marcus, J.Chem.Phys., **74**, 1790 (1981).
[54] See e.g. J. Manz, R. Meyer, E. Pollak and J. Romelt, Chem.Phys.Lett. **93**, 184 (1982); D. C. Clary and J. N. L. Connor, J.Phys.Chem., **88**, 2758 (1984); V. Aquilanti, S. Cavalii, G. Grossi and A. Lagana', Hyperfine Interactions, **17**, 739 (1984).
[55] B. C. Garrett, D. G. Truhlar, R. S. Grev, G. C. Schatz and R. B. Walker, J.Chem.Phys., **85**, 3806 (1981).

[56] For sample calculations for 3-D at zero angular momentum, see M. Mishra and J. Linderberg, Mol.Phys., **50**, 91 (1983); Chem.Phys.Lett., **111**, 439 (1984).
[57] D. M. Neumark, A. M. Wodtke, G. N. Robinson, C. C. Hayden and Y. T. Lee, J.Chem.Phys., **82**, 3045 (1985); D. M. Neumark, A. M. Wodtke, G. N. Robinson, C. C. Hayden, K. Shobatake, R. K. Sparks, T. P. Schafer and Y. T. Lee, J.Chem.Phys., **82**, 3067 (1985).
[58] H. Klar, Phys.Rev., **A15**, 1542 (1977).
[59] V. Aquilanti, G. Grossi, A. Lagana', E. Pelikan and H. Klar, Lettere Nuovo Cim., **41**, 541 (1984).
[60] V. Aquilanti and G. Grossi, Letter Nuovo Cim., **42**, 157 (1985).

INDEX